Lecture Notes in Engineering

Edited by C. A. Brebbia and S. A. Orszag

53

A. S. Jovanović, K. F. Kussmaul, A. C. Lucia, P. P. Bonissone (Eds.)

Expert Systems in Structural Safety Assessment

Proceedings of an International Course
October 2–4, 1989, Stuttgart, FRG

Springer-Verlag
Berlin Heidelberg New York
London Paris Tokyo Hong Kong

Series Editors
C. A. Brebbia · S. A. Orszag

Consulting Editors
J. Argyris · K.-J. Bathe · A. S. Cakmak · J. Connor · R. McCrory
C. S. Desai · K.-P. Holz · F. A. Leckie · G. Pinder · A. R. S. Pont
J. H. Seinfeld · P. Silvester · P. Spanos · W. Wunderlich · S. Yip

Editors
Aleksandar S. Jovanović
Karl F. Kussmaul

Staatliche Materialprüfungsanstalt (MPA)
Universität Stuttgart
Pfaffenwaldring 32
7000 Stuttgart 80
Germany

Alfredo C. Lucia
Institute for Systems Engineering
JRC Ispra
I-21020 Ispra (Va)
Italy

Piero P. Bonissone
General Electric Company
CRD-K-1, 5C32A
Schenectady, NY 12301
USA

ISBN 3-540-51823-1 Springer-Verlag Berlin Heidelberg New York
ISBN 0-387-51823-1 Springer-Verlag New York Berlin Heidelberg

Printing: Mercedes-Druck, Berlin
Binding: B. Helm, Berlin
2161/3020-543210   Printed on acid-free paper.

Preface

Structural safety of industrial systems and components raises a steadily growing public, scientific and engineering interest, and causes permanent development of methods and techniques used for its assessment. In addition to the well established engineering methods, applied in the field, several new methods and tools have emerged recently. Among them, the most novel ones are probably those related to expert system applications, appearing as an important possible improvement of the current engineering practice. The issue has been addressed by the international course *EXPERT SYSTEMS IN STRUCTURAL SAFETY ASSESSMENT* organized by *MPA* Stuttgart and *JRC* Ispra (Stuttgart, October 2–4, 1989), and the proceedings of the course are contained in this volume of the *Lecture Notes in Engineering*.

The contributions (invited lectures) tackle the issues usually confronting developers and users of expert systems applied in structural engineering, i.e. in structural safety and integrity assessment. Both the book and the course are a combination of a tutorial and of presentation of the current achievements in the field. Starting from the basic elements of expert systems (knowledge based systems), the book should "guide" the reader up to the applications in various particular sub–domains.

The contributions in introductory part deal with the two essential aspects of structural safety. The first contribution (Kussmaul and Jovanovic) highlights the analytical and experimental work necessary for a comprehensive engineering assessment of structural safety of critical components, using as a representative example the research performed at *MPA* Stuttgart. The second contribution (Volta) shows that the assessment is a process strongly involving human judgment and experience, leading nowadays to the "revival of experts". Within the assessment, expert systems must be thus seen as a complement to the usual engineering analysis and as a tool leading to a complete, unique (non–partial) and "reference" decision base, shared by all the experts involved.

The tutorial part of the book starts with the fundamentals of expert systems. The contributions (Bonissone) are focused on the issues relevant for the main topic of the course. A reader interested to make a step backwards, is advised to consult other available literature: e.g., the D.T. Pham's book *Expert Systems in Engineering* (Springer 1988) for a survey of applications in process, civil, electrical, electronic, mechanical and manufacturing engineering, D. Nebendahl's book *Expert Systems* (Wiley, 1988), giving insight in the practical functionality of expert systems, or the book of Harmon, Maus and Morrissey (Wiley, 1988) *Expert Systems – Tools and Applications*, giving a very comprehensive survey of generic situation in the field.

The specific, knowledge engineering issues relevant for applications of expert systems in structural engineering (treatment of uncertainty, belief and evidence, and the decision making) are also presented in the tutorial manner (Bonissone, Zimmermann, Werners). They are illustrated with examples which can be easily "translated" to the application domain of the course, while the choice of the issues corresponds to the applications presented in the second part of the book. So, for instance, fuzzy set theory has been presented more in detail, as a possible tool for treatment of uncertainties and for coupling numerical and symbolic analysis in expert systems.

The contributions of Garribba and Bogaerts provide transition from the theory to practice and show that practical building of expert systems for structural safety and integrity assessment is a highly complex task. Expert systems of the second generation are tackled in the first of the two contributions.

The applications presented in the volume regard several main areas: pressure vessels, stress analysis, non–destructive examination, corrosion and knowledge engineering issues. The results and experiences from the application cases, presented in this part, are still not so numerous, but they clearly indicate main trends. So for instance, the applications of expert systems in the corrosion related structural problems (Bogaerts, Basden, Arents) show that high level of deployment has already been attained. Promising results, and above all, proven usefulness of expert systems, have been reported in areas of pressurized components (remaining life assessment, leak–before–break analysis), stress analysis and non–destructive examination (Lucia, Servida, Jovanovic, Zarka, Deuster). The contributions regarding the practical knowledge engineering issues deal mainly with the problem of acquisition and use of experts' knowledge (Servida, Khong, Jovanovic), while the contribution of Bersini tackles the connectionism and neural nets as one of the possible alternatives to *GOFAI* (Good Old–Fashioned Artificial Intelligence), i.e. to expert systems. Due to the quick development of science and technology in the domain, developers and users of expert systems should be aware of such alternatives. On the other hand, practical application of expert systems in this field is often connected with questions like what are the problems where the application pays best, what tools to develop/use, how to couple expert system with other elements of the assessment (e.g. with the engineering numerical calculi, standards), etc. We hope that this book will provide help in this sense too.

Realization of the book and of the course have involved a high degree of collaboration and interaction between the course coordinators and the invited lecturers. The lecturers have shown a great willingness to cooperate and their help can be only praised. This is gratefully acknowledged here. Help of the Commission of the European Communities and of Springer–Verlag, in organization of the course and in publishing of the proceedings, respectively, is highly appreciated.

It is our hope that this work will stimulate further communication among engineers involved in structural assessment and expert system applications, and help them both in their research efforts and in their usual practice.

A. S. Jovanovic
Stuttgart, August 1989

Contents

Contributors:
(alphabetical order)

Hans C. Arents
Faculty of Engineering - Dept. MTM, Katholieke Universiteit Leuven
W. de Croylaan 2, 3030 Leuven, Belgium

Andrew Basden
Informaton Technology Institute, University of Salford
Salford M5 4WT, United Kingdom

H. Bersini
IRIDIA - Université Libre de Bruxelles
50 av. F. Roosevelt CP 194/6, 1050 Bruxelles, Belgium

Walter F. L. Bogaerts
Faculty of Engineering - Dept. MTM, Katholieke Universiteit Leuven
W. de Croylaan 2, 3030 Leuven, Belgium

Piero P. Bonissone
General Electric Corporate Research and Development
Schenectady, N.Y. 12301, USA

G. Deuster
Fraunhofer-Institut für zerstörungsfreie Prüfverfahren
Universität, Gebaude 37, 6600 Saarbrücken 11, FR Germany

Sergio Garribba
CESNEF - Politecnico di Milano, v. Ponzio 34/3, 20133 Milano, Italy

J. M. Hablot
Laboratoire de Méchanique des solides, École polytechnique
91128 Palaiseau cédex, France

Marcel Hassler
Staatliche Matrialprüfungsanstalt (MPA)
Universität Stuttgart, Pfaffenwaldring 32, 7000 Stuttgart 80, FR Germany

Aleksandar S. Jovanović
Staatliche Matrialprüfungsanstalt (MPA)
Universität Stuttgart, Pfaffenwaldring 32, 7000 Stuttgart 80, FR Germany

Karl F. Kussmaul
Staatliche Matrialprüfungsanstalt (MPA)
Universität Stuttgart, Pfaffenwaldring 32, 7000 Stuttgart 80, FR Germany

Vee H. Khong
Institute for Systems Engineering, JRC Ispra
21020 Ispra (Va), Italy

Alfredo C. Lucia
Institute for Systems Engineering, JRC Ispra
21020 Ispra (Va), Italy

F. Perdieus
ESSO Refinery Antwerp, Polderfdijkweg
B-2030 Antwerp, Belgium

Andrea Servida
Institute for Systems Engineering, JRC Ispra
21020 Ispra (Va), Italy

Dietmar Sturm
Staatliche Matrialprüfungsanstalt (MPA)
Universität Stuttgart, Pfaffenwaldring 32, 7000 Stuttgart 80, FR Germany

Marc J.S. Vancoille
Faculty of Engineering - Dept. MTM, Katholieke Universiteit Leuven
W. de Croylaan 2, 3030 Leuven, Belgium

Giuseppe Volta
Institute for Systems Engineering, JRC Ispra
21020 Ispra (Va), Italy

F. Walte
Fraunhofer-Institut für zerstörungsfreie Prüfverfahren
Universität, Gebaude 37, 6600 Saarbrücken 11, FR Germany

Brigitte Werners
RWTH Aachen, Templergraben 55, 5100 Aachen, FR Germany

Joseph Zarka
Laboratoire de Méchanique des solides, École polytechnique
91128 Palaiseau cédex, France

Hans-Jürgen Zimmermann
RWTH Aachen, Templergraben 55, 5100 Aachen, FR Germany

Introductory Part

Structural Safety and Integrity Assessment and Use of Expert Systems at MPA Stuttgart

K. Kussmaul

Professor Dr.-Ing., Dr.techn.E.h. and Director of the
Staatliche Materialprüfungsanstalt (*MPA*)
Universität Stuttgart
Federal Republic of Germany

A. Jovanovic

Staatliche Materialprüfungsanstalt (*MPA*)
Universität Stuttgart
Federal Republic of Germany

Abstract

The paper gives a condensed review of current activities and research programs of *MPA* Stuttgart in the domain of structural safety and integrity assessment, and of the efforts directed towards the development and application of the related expert systems. The above mentioned activities are mostly related to pressurized nuclear and non-nuclear components. A large part of the research work has been performed within the scope of the Reactor Safety Research and Development Program of the German Federal Ministry for Research and Technology (BMFT). This program started in the early seventies, involving large scale experiments and analytical research. Issues like leak-before-break or pressurized thermal shock can be mentioned as the typical ones. The problems of fracture, crack initiation, crack propagation and crack arrest have been investigated, under simulated operational conditions, corresponding to the real ones in terms of internal pressure, temperature, corrosion, etc., which has been of decisive importance for the transferability of the test results to real components. In addition, a series of complementary issues has been investigated (material science aspects, non-destructive examination, etc.). Application of artificial intelligence, i.e. of expert systems, is one of these complementary activities, aimed at improving the actual possibilities to assess safety and integrity of structural components. The development and application of expert systems required first the definition of a general framework, i.e. of the environment for development of expert systems, as well as the identification of the specific practical application sub-domains. The modular structure of the framework and the first application sub-domains (leak-before-break analysis and high temperature pressurized components) are briefly outlined.

1. Introduction

The Staatliche Materialprüfungsanstalt *(MPA)* Universität Stuttgart *(founded in 1884)*, is an institute at the University of Stuttgart. Its staff of about 350 people comprises about 130 scientists (over 40 Ph.D.'s). About 90% of the budget of the institute is provided by third–party funding, by industry and/or public research. In the last two decades the activities of *MPA* have been concentrated mainly on safety research issues related to materials, components, structures and systems (failure and damage analysis). In this area, experience and know–how of the institute include both the specimen and the real component (full–scale) testing, as well as use and validation of sophisticated engineering methods and advanced software and hardware tools (e.g. CRAY2). Beside the usual material testing equipment, *MPA* disposes of some testing facilities which are unique in the world, such as the new underground building (1989) with structural features test facility for the testing of very large–scale complex structures (figure 1), the facility based on full–scale LWR power plant pressure vessels, or the 10,000 t (100 MN) universal tensile testing machine mainly used for the validation of fracture mechanics analyses, the high rate tensile testing machine with a capacity of 12 MN, the piping test facility for computer code validation (external hazards), the large set of autoclave systems for testing of steels used for the high temperature power plant components (including the simulation of water chemistry, static and cyclic loading and other relevant factors), the full–scale testing facility for the testing of pipe elbows at high temperature, etc. Some of these facilities allow a completely new kind of structural testing. Some *MPA* tests are performed at other test sites, such as Grosskraftwerk Mannheim, the army test field Meppen for high energy burst tests, at HDR Grosswelzheim (a decommissioned power plant, used now for tests on the pressure circuit and in the pressure vessel). *MPA* has also the long–term cooperation agreements with a number of laboratories and research organizations throughout the world (e.g. Battelle, CEA, EPRI, JAERI, JRC Ispra, JRC Petten, SWRI, UKAEA–Laboratories, VTT, etc).

A series of research activities in the domain of structural safety and integrity assessment, including the development and application of the related expert systems, is currently going on at *MPA*. These activities focus mostly on pressurized nuclear and non–nuclear components. A large part of the research work has been performed within the scope of the Reactor Safety Research and Development Program of the German Federal Ministry for Research and Technology (BMFT), which has been started in the early seventies and has pursued the following principal aims:

- further investigation of potential causes and sequences of accidents and analysis and evaluation of safety margins;

- further development of the methods used for a realistic safety assessment

- further development and optimization of safety technology.

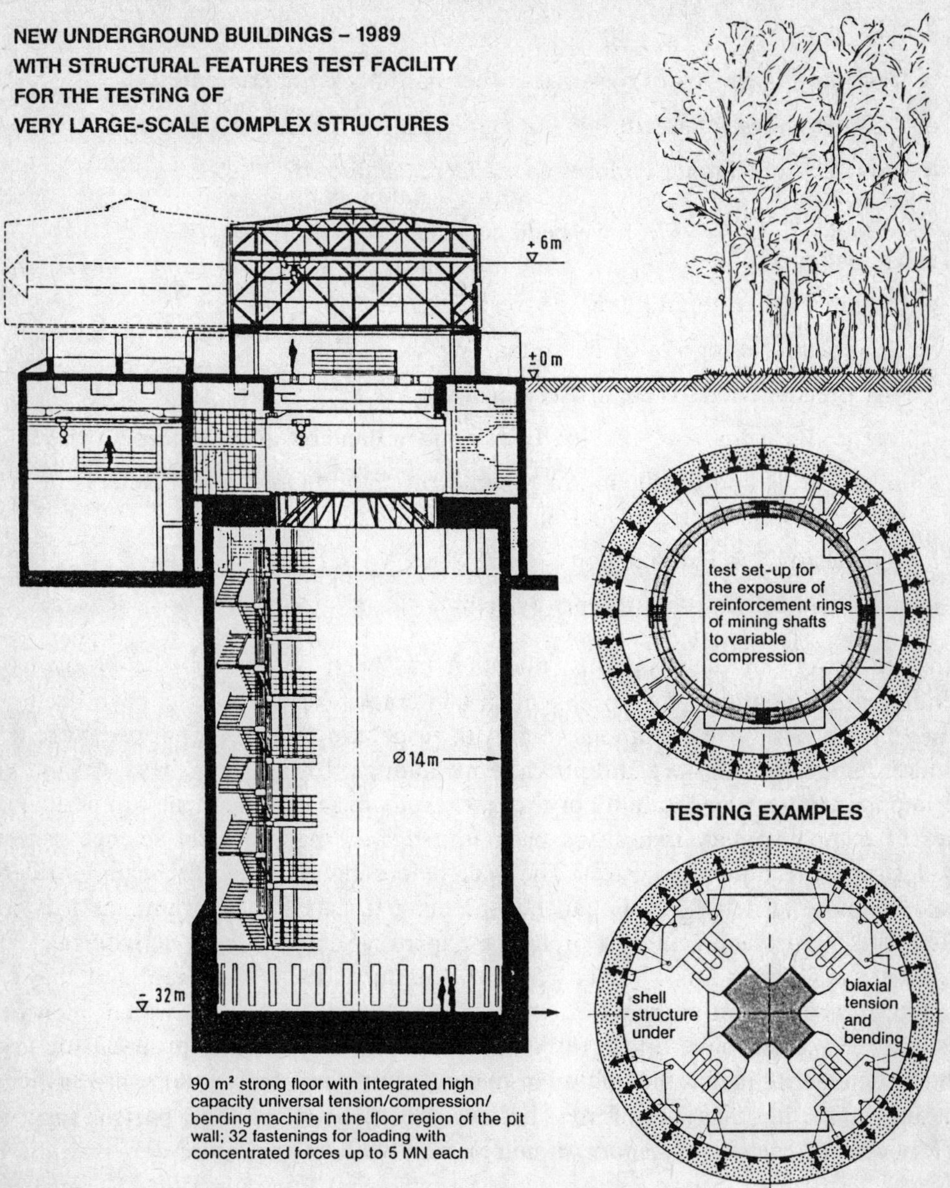

Figure 1 – MPA Stuttgart: New underground test facility

An overview of the specific research programs, related or performed within this program at *MPA*, is given further on. These specific programs, financed by the German government, industry and utilities, and highlighted in this paper are:

a. Pipe Tests

b. Thermal Fatigue and Pressurized Thermal Shock Experiments

c. HDR–Program, phase III

d. Main Coolant Piping Failure during Core Meltdown

e. Corrosion Assisted Crack Growth Tests

f. Irradiation Program

g. Dynamic Fracture and Crack Arrest Tests

h. Qualification Program of Dissimilar Welds

i. Fast Breeder Model Test of Reactor Tank Wall

j. Internal Pressure Tests at Pipe Bends under Bending in the Creep Regime

k. Investigation and Evaluation of Manufacturing and Operational Defects of Pressurized Components in Nuclear Power Plants

l. Non–destructive Examinations – Full Scale Vessel Tests

m. Artificial Intelligence (Expert Systems)

In the framework of the programs, attention has been focused on the problems of fracture, crack initiation, crack propagation and crack arrest. The tests have been performed under in–service conditions (e.g. with respect to scale, internal pressure, temperature, temperature history and pressure medium, corrosion, etc.). This was of decisive importance for transferability of the test results to real components. In addition, a series of complementary issues has been investigated too (material science aspects, non–destructive examination, etc.). The leak–before–break issue has been considered somewhat more in detail in this paper (pipe burst tests, HDR program), as it is connected with the development of a prototype expert system. All these activities are embedded in the German Basis Safety Concept (Kussmaul 1984: Kussmaul and co–workers, 1983), which is an approach to the nuclear power plant safety which allows the possibility of catastrophic failure to be excluded, without invoking probabilistic arguments. The essence is that the quality of manufacturing and design alone are sufficient to assure safety, but that is reinforced by four redundancies: multiple party testing, the "worst case" principle, in –service monitoring and validation (figure 2).

The application of artificial intelligence, i.e. expert systems, is one of these complementary activities, aimed at improving the actual possibilities to assess safety and integrity of structural components.The development and application of expert systems required first the definition of a general framework, i.e. environment for development and application of the expert systems and related methods and techniques, in the domain of structural safety integrity analysis and assessment. In addition, the specific practical application sub–domains have been identified and the developments started. The modular structure of the framework and the first applications in the sub–domains

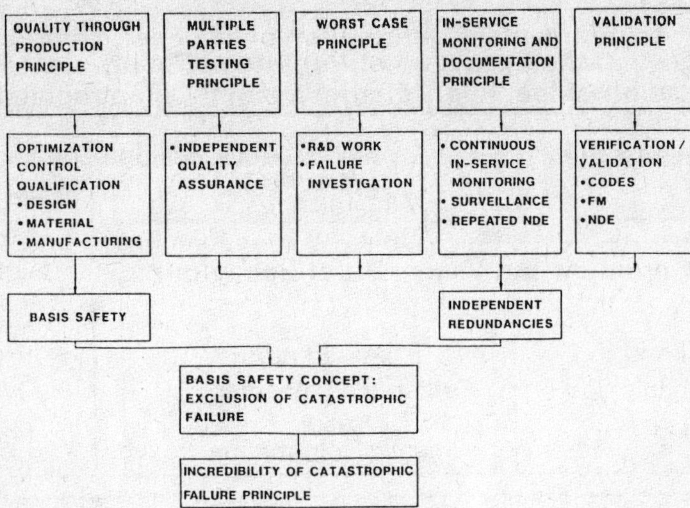

Figure 2 – Summary of the Basis Safety Concept

like leak–before–break analysis and high temperature pressurized components, are briefly outlined later on.

2. Pipe Burst Tests

2.1 Background

A research study, related to the nuclear reactor pressure vessel (RPV) issues, concluded by *MPA* Stuttgart in 1970, contributed considerably to the specification of the German Reactor Safety Research Program. One part of this research program deals with the pressure boundary of nuclear power plants, such as pressure vessels and piping systems in the primary circuit. The most important individual programs tackling the piping system are supported by BMFT and, partly, by German utilities, figure 3. The main aim of these individual projects is the experimental investigation of the phenomenology of fracture of pipes whose dimensions are in most cases equal to those of the primary piping of the pressurized water reactors. The test conditions, e.g. internal pressure and temperature, are chosen in accordance with the design and service loading conditions. A further goal of these research programs is the application, examination and validation of numerical and analytical computational models based on fracture mechanic concepts. The research project "Phenomenological Vessel Burst Tests", also called "Piping and Vessel Failure" or "BV Project" (Kussmaul, Sturm, Stoppler, 1985: Sturm, Stoppler, 1989), serves as experimental confirmation of the postulate of the incredibility of catastrophic failure of piping systems and thin–walled pressure vessels, as well as a means for the quantification of the safety margin against their disruptive (catastrophic) failure. The structure of the research project is shown in figure 4. The

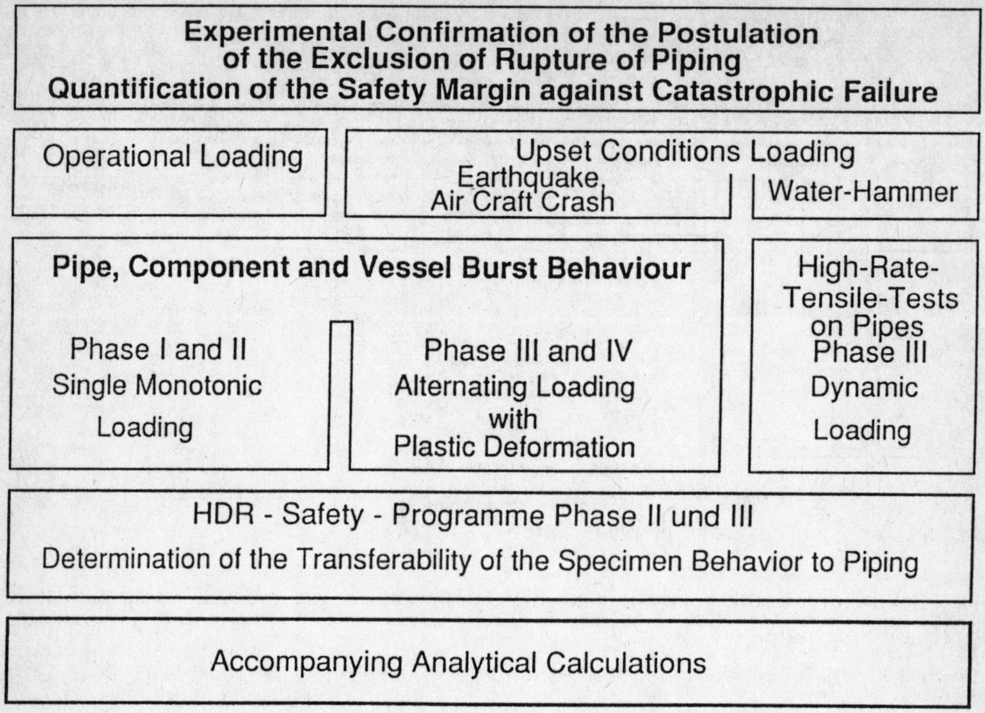

**Experimental Confirmation of the Postulation
of the Exclusion of Rupture of Piping
Quantification of the Safety Margin against Catastrophic Failure**

Operational Loading	Upset Conditions Loading Earthquake, Air Craft Crash	Water-Hammer

Pipe, Component and Vessel Burst Behaviour		High-Rate-Tensile-Tests on Pipes Phase III
Phase I and II Single Monotonic Loading	Phase III and IV Alternating Loading with Plastic Deformation	Dynamic Loading

HDR - Safety - Programme Phase II und III

Determination of the Transferability of the Specimen Behavior to Piping

Accompanying Analytical Calculations

Figure 3 – Research project for experimental confirmation of the posutlate
of incredibility of catastrophic failure in piping

crack growth, as well as the fracture behavior are to be examined in straight pipe sections, in pipe bends and in T–branches.

2.2 Scope

According to the specific tasks, the project has been divided into several phases. Phase I started in 1977. It was completed in 1984 and followed by phase II which was finished in 1987. Phase III has a duration of 4,5 years and will be completed in 1991. Phase IV started at the end of 1988 and should be finished by the end of 1992.

Phase I: The crack opening behavior had to be examined. The objectives of this phase were:

- Experimental determination of the crack behavior of pipes under pressurized and boiling water reactor conditions;

- Experimental determination of the leak–before–break curve;

- Experimental determination of the time history of internal pressure, gaping and crack opening.

- Examinations of the crack arrest behavior.

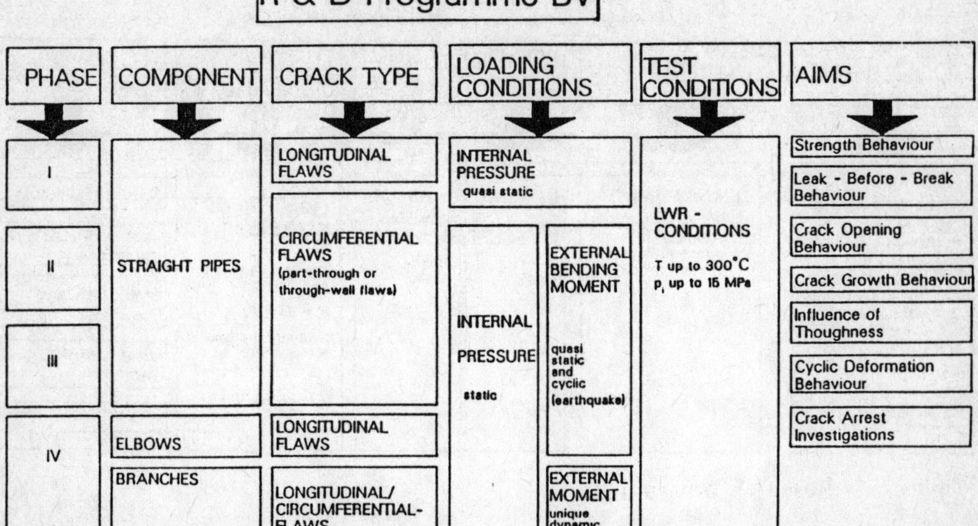

Figure 4 – Research project "BV"

- Comparison of the experimentally determined failure pressure with the one determined by the conventional calculation methods.
- Comparison of the fluid and structure dynamic behavior; experiment versus finite element analysis.

Phase II: Within phase II (figure 5) of the BV project it was to be shown whether the relationships, analogue to those which had been found in the case of pipes weakened by longitudinal flaws, exist also for pipes weakened by circumferential flaws. In particular, a proof was to be furnished, that up to the failure point of piping systems made of ductile materials (high upper shelf notch impact energy) and containing acceptable defects in circumferential direction, a bending moment can be applied, causing yielding over a large area of the pipe wall, before the failure due to leakage or rupture takes place. In the background of this research lies the requirement to absorb the kinetic energy which can affect the system during earthquake, water hammer or airplane crash.

The objectives of phase II of the BV project may be formulated as follows:

- Determination of the load bearing capacity of pipes weakened by (partial–) circumferential flaws under internal pressure and external bending moment, as a function of the toughness of the pipe material.
- Determination of the crack opening behavior and definition of the leak–before–break criterion.

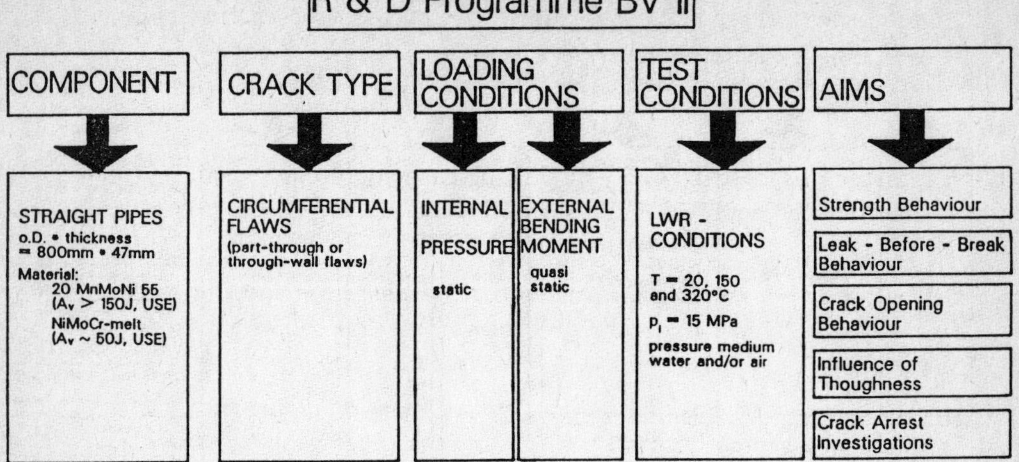

Figure 5 – Research program BV–II

> – Checking of the postulate of incredibility of catastrophic failure in the case of pipes weakened by circumferential flaws.

Phase III: The KTA–guidelines 3201.1 and.2 are applied for materials, design (both in the low and in the high cycle regimes), construction and the calculation of components of the primary circuit of light water reactors. The fatigue analysis which must be executed according to KTA 3201.2 chapter 7.8 is restricted to the determination of a maximum allowable number of stress cycles (the corresponding design curves are given). A component free of defects is taken for this kind of analysis. The objectives of phase III of the project are defined as follows:

- Determination of crack growth law for piping systems with circumferential defects under cyclic loading, both in the very low and in the high cycle regime by an external bending moment, in particular for the "over–realistic" strain amplitudes.

- Quantification of the influence of material toughness on the crack growth law.

- Quantification of the influence of a corrosive medium on the crack growth law.

- Determination of the leak–before–break behavior of piping systems and piping system components under cyclic bending load (both in the low and in high cycle regime).

2.3 Results

Pipes closed at both ends (inner diameter *700 mm*, wall thickness *47.5 mm* length *2,500* or *5,000 mm*) were used as test specimens. The chosen wall thickness of the test vessels enables reproduction of the nominal stress values in the pressurized water reactor piping (approximately *138 N/mm2*). As the energy stored in the pressurizing medium has an important influence on the crack development, the tests were executed using both the pressurized water and steam (hot air).

Phase I: The results obtained in this phase can be summarized as follows:

- Prediction concerning the load bearing capacity of pipes weakened by longitudinal flaws is possible, figure 6 (S_L = safety factor, testing temper. = 305 °C).

- The limit curve between the formation of leakage and catastrophic failure (leak–before–break curve) is dependent on the pipe geometry, the strength and the toughness of the material. Any influence of the crack criterion, static or dynamic, on the leak–before–break behavior cannot be noticed.

- The critical flaw length (slit) is at least 100% of the pipe diameter for the examined pipe geometry and a nominal stress of approximately *138 N/mm2* (service stress).

- For the flaws shorter than the critical slit length, only leakages or limited failures develop. In cases of crack arrest, the arrest takes place immediately after crack initiation.

- In the case of longitudinal flaws that reach or exceed the critical slit length it cannot be counted on crack arrest when using compressible pressure mediums

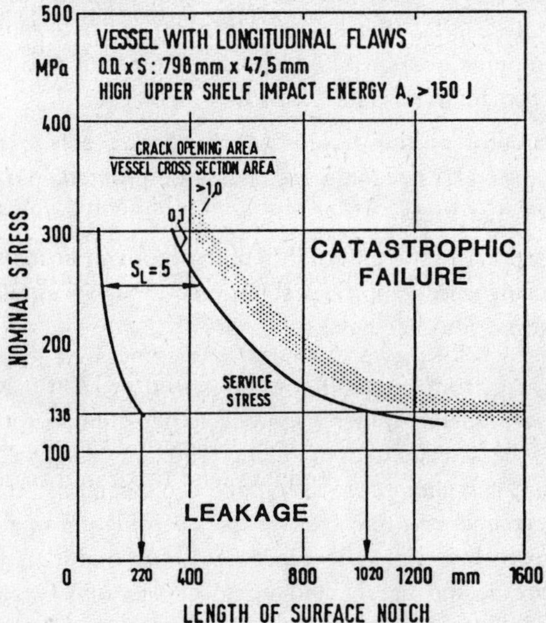

Figure 6 – Leak–before–break curves for pipes with longitudinal defects

(SWR or pressurized air conditions). In these cases the crack opening area will always be larger than the pipe diameter analogue.

– Flaws of a length of approx. *85%* of the critical slit length or less result in leakages with crack opening areas that are smaller than *10%* of the pipe cross-section area.

– The crack opening time for a 0,1F–crack is approx. *5 ms*.

– The internal pressure at crack initiation and the failure pressure (instability) of a pipe with longitudinal flaws can both be determined by using semi–empirical equations. The applied equations are based on failure theories that take linear–elastic and partial plastic material behavior into account.

Phase II: As the test specimens were to be stressed by internal pressure and by the external bending moment simultaneously, a test arrangement had to be designed so to approach most closely to the actual conditions in a piping system. So far, the results obtained in phase II can be summarized as follows:

– Prediction concerning the load bearing capacity of pipes weakened by circumferential defects is possible (figure 7).

– The leak–before–break curve depends on the pipe geometry, the strength and the toughness of the material.

– The critical defect length (slit) in circumferential direction, examined under internal pressure of *15 MPa* (the design pressure) and external bending moment of *11,7 MNm*, amounts to approx. *240 mm* for materials with high toughness and to approx. *100 mm* for materials with low toughness.

– In the case of circumferential defects which either reach or exceed the critical defect length, crack arrest cannot be taken for granted, particularly when the system is of a low stiffness.

– The internal pressure and the external bending moment at instability can both be determined by using semi–empirical equations, though without taking the toughness of the material into account.

Phase III: The bending rig from phase II was enhanced in such a way that tests with very low cycle bending (frequency under 1 load cycle per minute, with pipes of *700 mm* nominal diameter) can be executed both under load and displacement control. The pipes of *200 mm* nominal diameter stressed by high cycle bending (frequency approx. *8 Hz*), are loaded in the resonance area. During the examination of the corrosion influence on the crack growth behavior, the pipes (*700 mm* nominal diameter) are connected with a water circuit and are heated up to approximately *300 °C*. This water circuit provides purified water with an oxygen content of up to *8 ppm*. The component autoclave loop tests are accompanied by tests on small specimens. Cyclic load tests on smooth and notched specimens and tests with fracture mechanics specimens are per-

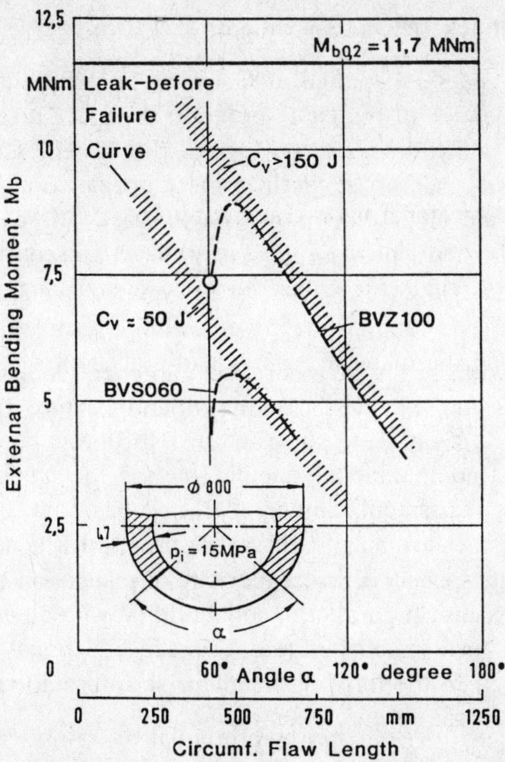

Figure 7 – Leak–before–break curves for pipes with circumferential defects

formed in order to determine the subcritical crack growth (under corrosive conditions, too).

2.4 Future Developments

In phase IV of the research project, the findings concerning critical flaw size, fracture opening behavior and pressure decay on straight pipes are to be extended to the loading case of the internally pressurized piping and T–branches, under static and cyclic loads in the very low cycle regime, as well as under the single dynamic external bending moment load. In such a way, a realistic component loading, as it might occur due to earthquake and water hammer, is taken into account. Furthermore, the postulate of incredibility of fracture in pipe bends and T–branches is to be validated experimentally, taking into consideration dissimilar welds in the case of connections of austenitic pipes/bends to branches made of ferritic steel. At the same time, the transferability of characteristics determined on specimens to the components, will be proved and the numerical procedures developed in such a way as to render possible an analytical description of the component behavior in the plastic range of cyclic loading too.

3. Pressurized Thermal Shock (PTS) Experiments

The PTS research in the Federal Republic of Germany has been initiated in the mid seventies. Within the framework of the HDR–program thermal fatigue experiments on plates and intermediate sized test vessels have been performed. These investigations led to the design of the real–scale cyclic thermal shock nozzle corner experiments in a feedwater line nozzle of the HDR reactor pressure vessel. These experiments have been extended later to the cylindrical vessel part as well. At present, nozzle and reactor pressure vessel (RPV) core region are subjected to so–called long–term cooling tests (PTS).

For all the experimental work, extensive theoretical/numerical analyses have been performed in order to develop and validate the analytical and fracture mechanics tools for solving the PTS issue. In 1982, a research program (NKS) was launched at *MPA* addressing experimental and computational ductile fracture mechanics applied to PTS loading of RPV structures (Kussmaul, Sauter, 1986). It has been completed in 1988 while the issue is being further investigated in a new program which studies the brittle dynamic crack events with special considerations to investigation of the crack phenomenology in brittle materials. In parallel, a collaboration with other institutions dealing with the PTS issue has been established (e.g. Oak Ridge National Laboratory, USA, Japan and JRC Ispra, Italy), in the form of an intensive information exchange and/or direct involvement based on the research contracts.

Within the *MPA* PTS research program, the loading of the cylindrical reactor pressure vessel part (core region) under emergency core cooling cold water injection has been simulated in large–scale specimens, figure 8. For this purpose a hollow cylinder has been designed as test specimen in which the behavior of circumferential, fatigue sharpened flaws could be examined. In the course of the program, five specimens, heated up to approximately *320 °C*, were subjected to tensile loading and internal pressure (pressurized water), while the inner surface was cooled down uniformly through the cold water injection.

With the specimen materials the whole range of RPV relevant heats, practically generated by special heat treatments, has been covered. Five different upper shelf energy (USE) levels with accordingly changing NDT–temperature were investigated, figure 9. The measured data like temperature profiles through the wall thickness, strains on inner and outer cylinder surface and crack mouth opening displacements were used in detailed post–test numerical computations. For the quantification of the crack extension the computational tools focused on two concepts. On the one hand, a best–estimate analysis method on the basis of a non–linear J–integral formulation was derived and implemented in a general purpose finite element code. By this means it was possible to simulate the stable crack growth of a 3–d surface flaw during its thermal–mechanical loading history. On the other hand, for a reduced–effort but conservative integrity analysis an R6–curve approach has been used with the Failure Assessment

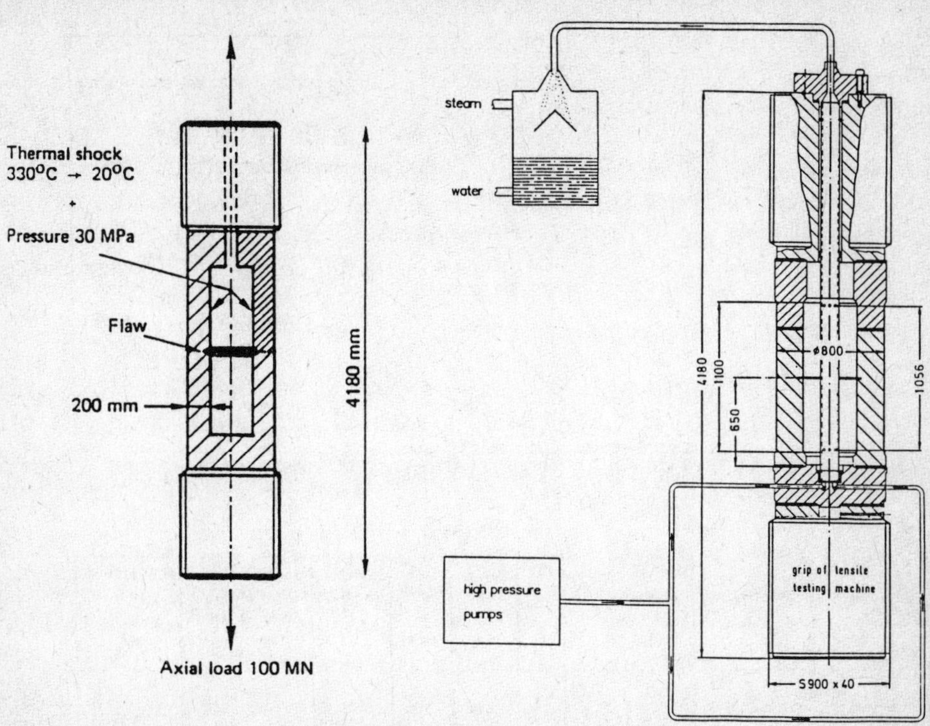

Figure 8 – Layout of the MPA PTS experiment

Diagram method. With this concept brittle as well as ductile crack extension can be assessed.

The crack extension was quantified by means of numerical analyses balancing out the crack tip loading, and the material–specific resistance to cracking for all time increments (corresponding to very small steps in loading). The two values are compared in figure 10. The J graphs plotted in the left half of figure 10 contain the crack driving forces. The J_R–curves determined on *20%* side–grooved *CT–25* specimen were intended to represent the material–specific cracking resistance under high deformation constraint, i.e. almost plane strain conditions. In comparison, a *CT–25* specimen tested at *160 °C*, not side–grooved and with a reduced thickness of *10 mm*, reflects the stress condition with less constraint.

Furthermore, the effectiveness of the *warm prestress effect (WPS)* could be demonstrated, when the crack tip temperature was lowered far below NDT–temperature under decreasing K_I–values but with K_I/K_k–ratios of the order of two to three. It has been also proved that the R6–curve approach in the Failure Assessment Diagram leads to conservative results, but it can be performed with essentially reduced computational

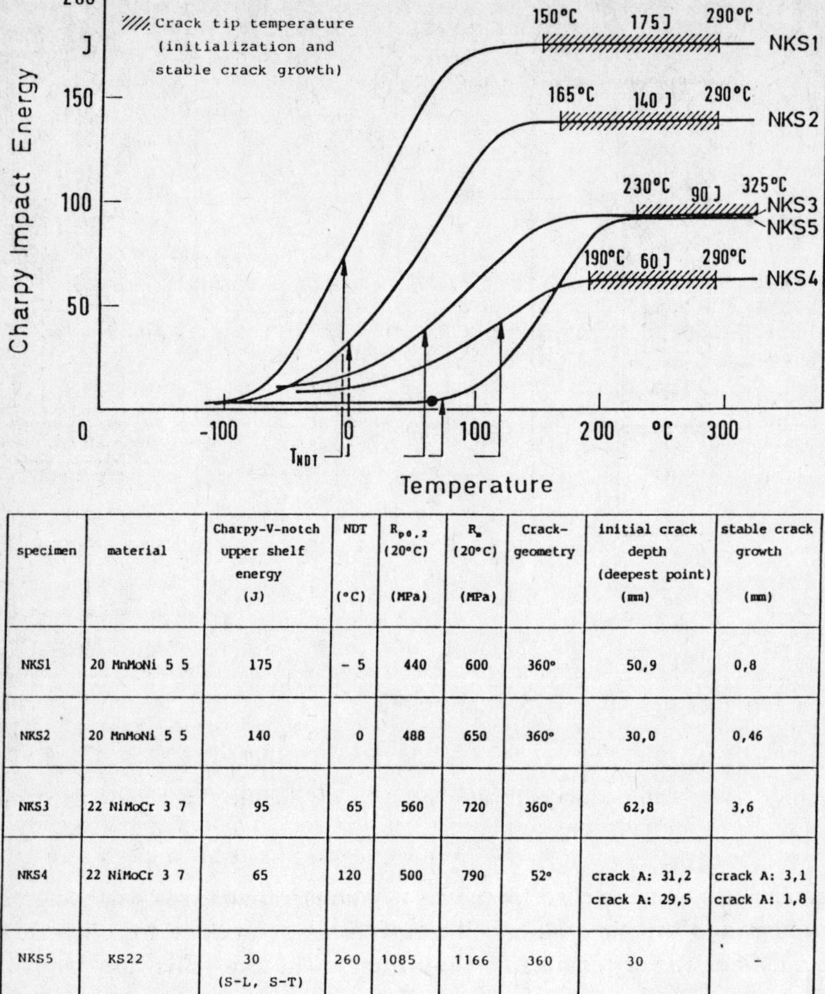

specimen	material	Charpy-V-notch upper shelf energy (J)	NDT (°C)	$R_{p0,2}$ (20°C) (MPa)	R_m (20°C) (MPa)	Crack-geometry	initial crack depth (deepest point) (mm)	stable crack growth (mm)
NKS1	20 MnMoNi 5 5	175	− 5	440	600	360°	50,9	0,8
NKS2	20 MnMoNi 5 5	140	0	488	650	360°	30,0	0,46
NKS3	22 NiMoCr 3 7	95	65	560	720	360°	62,8	3,6
NKS4	22 NiMoCr 3 7	65	120	500	790	52°	crack A: 31,2 crack A: 29,5	crack A: 3,1 crack A: 1,8
NKS5	KS22	30 (S-L, S-T)	260	1085	1166	360	30	−

Figure 9 – Properties of the materials used in the NKS experiments

effort. Furthermore, the treatment of brittle and ductile crack behavior can be dealt within one single analytical model and graphic representation.

For the JRC PTS experiment, a preprocessor for the ABAQUS finite element code has been developed (Jovanovic, Lucia, 1988), allowing an on-line analysis of the heat transfer coefficient and interactive preparation of an ABAQUS based analysis of the transient temperature, stress and strain fields, as well as of the J integral for the nozzle and safe-end cracks.

The ongoing research efforts in the PTS domain are concentrating on a highly brittle model material (specially heat treated *Mo-V* steel). Several types of specimens ranging

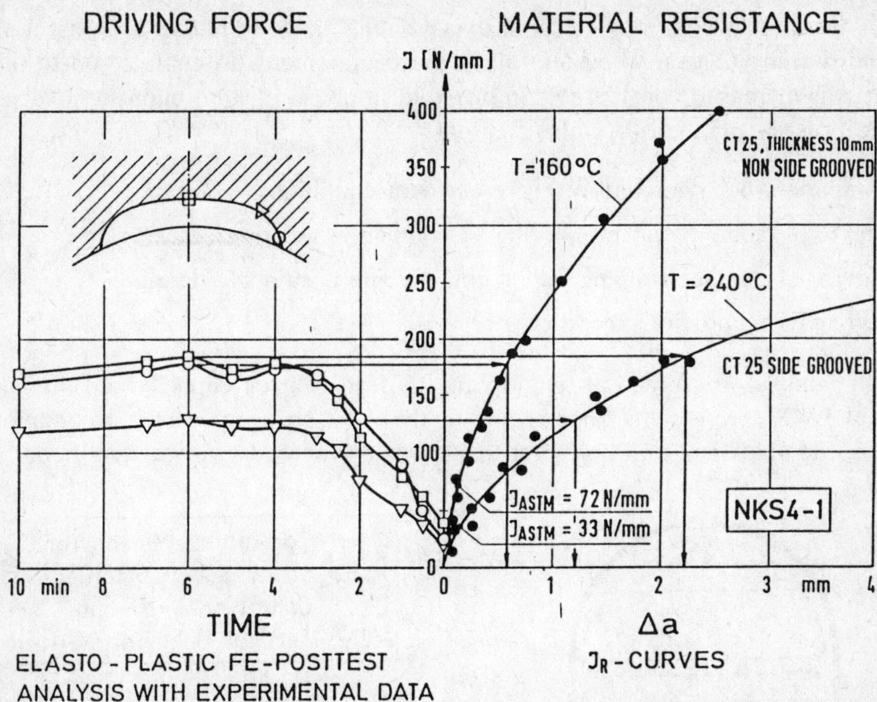

Figure 10 – J–based driving force for different crack positions in NKS4–1

from CT–like crack arrest specimens, to dynamically tested flat tensile specimens, wide plates with stationary temperature gradients, rotating disks and the above described hollow cylinder test piece, are going to be investigated experimentally as well as by means of sophisticated numerical (finite element) computations. Special emphasis will be put on the development of suitable material models and fracture characterizing parameters for the assessment of the non–linear dynamic and strain rate dependent crack behavior.

4. HDR Program

4.1 Background

The HDR Safety Program started in 1976, using as the test facility the decommissioned nuclear reactor HDR. Phase I of the research lasted until 1983, while Phase II ran from 1984 to 1988. Both phases dealt with the behavior of the safety–related systems and components of the Light Water Reactors (LWR): containment, reactor pressure vessel, piping, loss of coolant (blowdown) analysis, emergency core cooling (pressurized thermal shock), operational temperature transients (thermal fatigue, thermal stratification), slow strain rate cyclic varying bending loading of straight pipe sections and elbows in

high temperature water with elevated oxygen content, earthquakes and impulse loading, and burning of gas, wood and oil in the containment. In comparison to phase I,
loading was increased considerably in phase II. In phase III, the following 4 subjects of
investigation are planned (see also figure 11):

a) Containment behavior under severe accident conditions

b) Long-term damage and monitoring of components during service

c) Behavior of damaged piping under impulse and earthquake loading

d) Fire behavior and fire protection.

Items (a) and (c) are aimed at quantifying the load bearing capacity and the inherent
safety of LWR components under some of the most stringent accident conditions. In
addition, the analytical and non-destructive testing methods will be qualified.

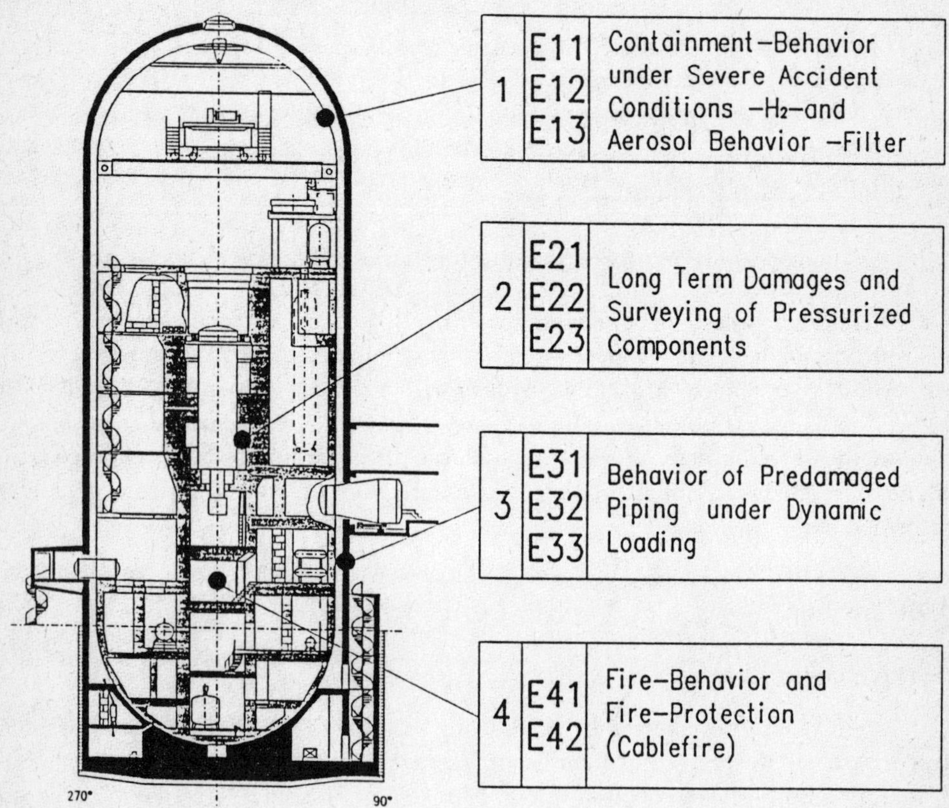

Figure 11 – Research areas in the HDR program – phase III

4.2 Scope

The topics of research in item (b) are:

- Predamaged piping under blowdown loading conditions,
- Predamaged piping under simulated earthquake loading,
- Piping under water hammer conditions induced by steam condensation.

In the course of the plant operation (startup, shutdown, load variation, etc.) piping is subjected to internal pressure and thermal loads including the loads caused by thermal stratification and mixing processes. These types of loading can lead to cracks in heavily stressed areas, e.g. pipe connections, nozzles and pipe bends. Unfavourable chemical composition of water, as e.g. increased oxygen concentration in the coolant may cause distinct reduction of the number of cycles for crack initiation as well as an accelerated crack growth.

As far as corrosion is concerned, temperature plays an important role. A relatively small amplitude of the load cycles (ΔT, ΔK) in the unfavourable temperature range (200 – 250 °C) and a high R–ratio (0,9) may possibly lead to a higher crack growth rate than more stringent transients (higher ΔT, with higher ΔK), as it has been shown by HDR thermal fatigue tests.

The thermal fatigue tests performed on nozzles of the reactor pressure vessel should provide more findings on the behavior of a "naturally" cracked component under complex loading conditions as described above. This enables to verify the analytical methods used for safety assessment of, e.g., T–branches and thermal sleeve dissimilar welds. Furthermore, the fatigue curve given by codes and standards cannot be directly used for the best–estimate analysis.

The investigation of the behavior of damaged pipes under dynamic accident loads, such as blowdown, earthquake and water hammer induced by steam condensation, will provide the basis for an improved evaluation of the remaining load bearing capacity. This leads to an improved leak–before–break criterion. In addition, it should be investigated how a single dynamic process affects the leak–before–break behavior determined in the static test. The earthquake tests have a similar objective. As opposed to the above mentioned tests, however, the dynamic impulses occur here repeatedly.

4.3 Results

Within Phase II of the HDR–Safety Program the pressure vessel and different piping systems were loaded intentionally by static and transient thermal and mechanical loads until damage and crack growth, in vessels, or catastrophic failure, in piping, took place. Laboratory tests on specimens, as well as extended calculations of the experiments were performed simultaneously to the component tests. The experimental and analytical results were evaluated with regard their transferability from specimen to component. The following statements summarize the conclusions drawn from the HDR–tests:

a)   An incipient crack in the austenitic cladding of the reactor pressure vessel (nozzle and cylinder wall) under thermal fatigue loading occurred at a number of cycles which corresponds approximately to the ASME–Design Curve, but it must be emphasized that the cladding of the nozzle is a hand–made one which is not typical for a modern production.

b)   In the case of piping components (incipient crack in an elbow) a good agreement between specimen and component was found.

c)   The crack growth under thermal fatigue loading in the reactor pressure vessel wall, under intensified corrosive conditions, is considerably over–estimated by the prediction. Main causes of the differences are the temperature transients not realized during the specimen tests and the flow velocity which cannot be exactly simulated. Simple crack growth laws are not sufficient to assure a reliable transferability of specimen results to components.

d)   The crack depths obtained by thermal fatigue were considerably smaller than expected and were sometimes over–estimated by factor 2.

e)   A relatively good transferability specimen/component could be stipulated for piping systems with constant temperature. The calculation on the basis of specimen values for crack extension was only slightly over–estimated.

f)   The occurrence of crack initiation and stable crack growth during long–time thermal shock (PTS) tests at the nozzle and the cylinder wall is dependent on the initial crack depth. As the crack depths assumed on the basis of non–destructive information at both the nozzle and the cylinder wall were smaller than projected, only an indication of beginning crack initiation could be determined. The calculations performed with larger crack depths did not furnish any comparable results either. The evaluation of the analytical procedures will only be possible after recalculation of the actual proportions. Influencing parameters stipulated up to now are as follows: material laws, modeling of the cracks, influence of multiple crack formations and crack branching.

g)   The leak–before–break behavior of piping systems made of a tough material could be confirmed in 2 tests with cyclic loading (DN 400) and 2 tests with earthquake loading (DN 100/250).

It has been demonstrated that the transferability specimen/component depends heavily on the type of component, on the kind of loading and on the environmental conditions. During some investigations, mainly at pipes, the pre–calculation agreed well with the test result. For the remaining cases such as e.g. thermal shock at the pressure vessel further investigations with regard to constitutive laws and calculation procedures for the improvement and validation of transferability are necessary and intended for Phase III.

The tests of Phase III will concentrate on determining the boundary conditions as exactly as possible and on deriving material laws which can be transferred from small specimens to components. This gives us way to validate and further develop analytical methods, especially the fracture mechanics concepts.

5. Main Coolant Piping Failure during Core Meltdown

The program was initiated to study the failure mechanisms and the short time creep behavior of the main coolant piping under the postulated conditions of the high pressure core meltdown sequences. The material behavior in the creep range (steel *20 MnMoNi 5 5*) is determined both by means of the small scale specimens and by means of the component test. In the small scale specimen tests, hot tensile tests, short–term creep tests and heating–up tests were carried out. The test temperature range extended up to *800 °C*. For the component test a pipe closed at both ends having a length of approx. 8 m and an inner diameter of 700 mm with a wall thickness of 47 mm was heated to approx. 700 °C under a constant internal pressure of 16,3 *MPa* at a heating rate of up to *7 K/min*. Once the desired test temperature had been achieved, the internal pressure and the temperature were kept constant until the failure of the vessel.

The test pipe pressurized with air failed catastrophically approx. *18 min* after reaching a temperature of *700 °C*, which was held constant. Initially a longitudinal crack formed which diverted into the circumferential direction when reaching the circumferential seam of the lower extension pipe, thus leading to its disruption. The maximum circumferential strain was approx. 30%; while at the location of the crack origin a decrease of the wall thickness of approx. 86% was observed. This component test proved that cracks which occur during short term creep rupture tests always result in catastrophic failure provided that the energy stored in the pressurizing medium is sufficiently high. The result of the pipe test falls within the scatter band of the corresponding small specimen tests, if the reference stress is calculated using the shear stress hypothesis.

In a subsequent program more tests with small scale specimens will be carried out. Stress–controlled short creep in a temperature range up to 1000 °C should supply the data necessary for the finite element computations.

6. Corrosion Assisted Crack Growth Tests

The corrosion behavior of pressurized ferritic components in high temperature water under static and/or transient mechanical loading can be a decisive factor considering the economy and safety of power generating plants (Kussmaul, Iskluth, 1988). Besides surface corrosion including erosion–corrosion, the stress corrosion cracking (SCC) has to be analyzed, too. A specific form of SCC, the so–called "Strain–Induced Corrosion Cracking (SICC)" was found to be responsible for a number of corrosion failures in piping of boiling water reactors which were investigated intensively at the end of the sixties.

The service time of components can be calculated with respect to fatigue from the knowledge of the cyclic crack initiation and growth behavior. In the meantime investigations have shown that crack growth under SICC conditions is also possible under static load, due to low temperature local creep effects at the crack tip. Starting from pre–existing cracks or notches crack growth can occur at stress intensities K_I which are under special conditions far beneath the critical stress intensities for crack initiation K_{Ic} and K_{IJ} in air. Many questions concerning crack initiation and crack growth in the case of static loading conditions considering a practical view on SICC remain open. The relationships between the leading parameters and the special conditions leading to SICC as well as the corrosion mechanisms from a mechanical point of view have to be investigated more deeply.

The main topic of the ongoing research work on SICC is the determination of characteristic values under special conditions, in particular the threshold values for crack initiation $K_{I\,SICC}$ under static loading conditions and the determination of the resulting crack growth velocities da/dt and the characteristic values for crack initiation and crack growth under very low cycle fatigue ($vLCF$) conditions.

Some investigations on crack growth in high pressure and high temperature water ($240°C$, $8\,ppm$ oxygen) were carried out on a low alloy ferritic $MnMoV$–steel. Advanced measuring techniques and data acquisition systems were used to measure the actual crack length during the tests carried out at statically loaded $CT50$–specimens.

The major result was crack growth developing with a certain growth rate under static loading conditions. The minimum K_I–value for crack growth in $8\,ppm$ oxygen high temperature water was determined to be about 25 MPa$\sqrt{}$m. In air the critical stress intensity K_{IJ} for crack initiation, resulting from a J–integral test on a CT 50–specimen, amounted to 100 MPa$\sqrt{}$m. The crack growth rates evaluated were found to be lying in the range of $4\cdot10^{-7}$ to $5\cdot10^{-5}$ mm/sec in water containing $8\,ppm$ oxygen. Detailed investigations on the fracture surfaces by means of the scanning electron microscope (SEM) revealed features which are characteristic for corrosion attack at slip lines as it had been observed on in–service cracks in plants.

Future research work has to be done to investigate the crack incubation and initiation period in detail.

7. Irradiation Program

The reduction of toughness of reactor pressure vessel materials due to neutron irradiation is still an issue of concern, especially for the older plants. Those plants contain base materials and welds which are more susceptible to neutron irradiation than the optimized materials.

The irradiation behavior of ferritic reactor pressure vessel steels is being investigated under two different aspects. The research program "Integrity of Components" (FKS) focuses mainly on general aspects of neutron irradiation of a broad spectrum of different heats and material states (Kussmaul, Foehl, Weissenberg, 1988). The results will serve as a basic matrix to assess the relevance of parameters with respect to the safety margin of the reactor pressure vessel as a function of operating time. The investigation addresses the issue of determination of the fracture mechanics properties on the basis of the results from Charpy impact testing from surveillance programs. The program includes also the investigation of the effect of combined service conditions (irradiation, temperature, corrosion and cyclic loading) on the fatigue crack growth rate.

The second aspect investigated is the transferability of surveillance results for the prediction of the reactor pressure vessel material state. Within a cooperative program with the US NRC trepans are being investigated taken from a decommissioned German boiling water reactor (figure 12). Data from those base metal and weld trepans are compared with surveillance data and results from short term test reactor irradiation experiments. This exercise realizes the most advanced techniques for neutron exposure calculation and focuses on fracture mechanics properties and the annealing behavior as well.

The results from the FKS program are only preliminary. The results obtained so far have proven that the K_{Ic} lower bound code curve and the procedure to determine the irradiation effects are conservative, for the materials investigated. The most recent development of US Regulatory Guide 1.99 Rev. 2 which does not consider phosphorus but copper and nickel as being responsible for irradiation sensitivity could not be supported by the FKS results.

The investigation of the trepan base material (forging) has brought up questions about different irradiation sensitivity depending on the main working direction of the material. The results obtained up to now have indicated that the present knowledge does not allow to explain all of the effects observed.

The irradiation experiments of the FKS program are planned to be completed in 1992. Main issues will be the testing of large scale *CT*–specimens, specimens from the low dose rate experiment and the evaluation of the results. The investigation of the trepan materials will be focused on fracture mechanics data, properties of the vessel cladding and microstructural investigations of both the base material and the weld.

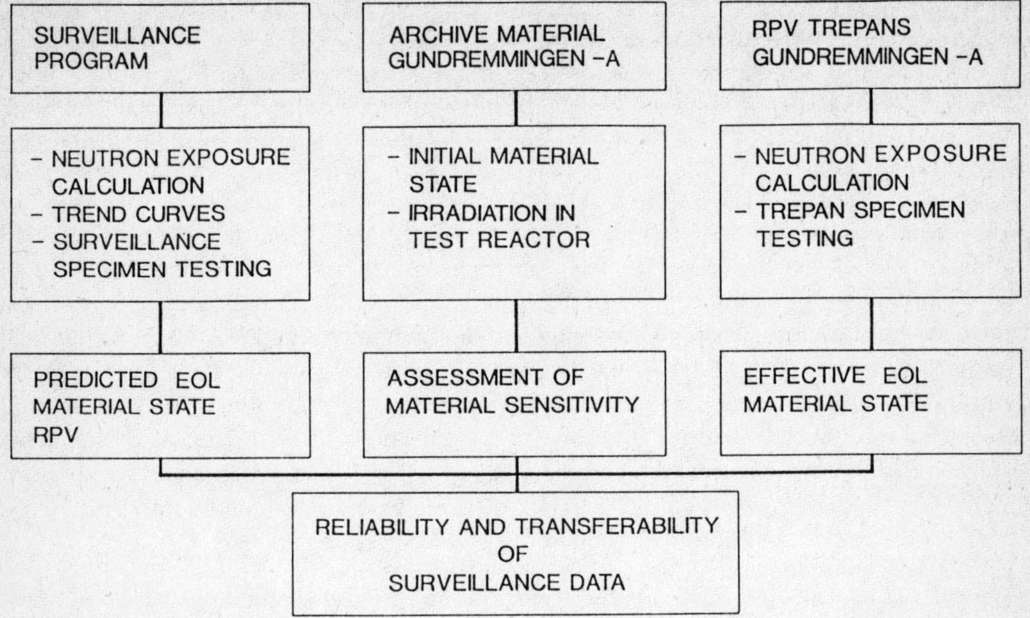

Figure 12 – The Gundremmingen power plant trepan exercise for verification of reliability and transferability of data from the irradiation surveillance programs

8. Dynamic Fracture and Crack Arrest Tests

A complete loading analysis of components containing cracks has also to include the modeling of crack initiation, crack propagation and crack arrest under the dynamic loads, which, in the power plants, could be caused, for instance, by possible water hammers, airplane crashes and similar events. For all the three stages of the crack behavior analysis the following subjects are considered within the research project performed at *MPA* and sponsored by BMFT:

– Determination of material parameters using small scale specimens

– Code development

– Verification of transferability

Emphasis is laid on the fact that the testing concepts developed in various laboratories for the determination of material parameters (LEFM, EPFM) must be evaluated and that the transferability of these parameters to components must be verified by means of different component–like specimen geometries with different states of stress. To describe the specimen and component behavior during dynamic crack initiation, crack propagation and crack arrest, various finite element codes (e.g. ADINA, ABAQUS, DYSMAS and VISCRACK) are available.

The tests are carried out on component–like specimens of different geometry and dimensions and pre–test as well as post–test calculations are performed. The comparison between experimental and numerical findings allows a further optimization of the testing techniques and codes. In the area of dynamic crack initiation it is intended to test the following specimen shapes

- Pipe strip specimens (servo–hydraulic high speed testing equipment)
- Wide plate specimens (*12 MN*–high speed testing machine)
- Straight pipes and pipe bends (*12 MN*–high speed testing machine, *10 MNm*–bending facility)

while crack arrest processes are investigated on

- Rotating discs (centrifugal test rig)
- Wide plate specimens (100 MN–testing facility) and
- Vessels (e.g. original feedwater vessel, thermal shock loaded thick–walled hollow cylinder).

As a lower bound investigation for model material with low upper shelf toughness modified melts of *17 MoV 8 4* and *15 NiCuMoNb 5* are employed. Large scale wide plate testing is done with the austenitic stainless steel *X6 CrNi 18 11*. For the ferritic steels the influence of the loading rate on the transition region of the impact fracture toughness–temperature curve has been demonstrated. The results obtained so far, for the dynamic loading under elastic–plastic conditions, show, for the low–toughness materials, an increase in the dynamic parameters in comparison to the static tests.

The results allow to conclude that there is a direct correlation between the Charpy toughness and K_{Ia} curves vs. temperature. This dependence was known for low temperatures and the transition region of the C_v–energy and is obviously transferable to the upper–shelf. Thus it can be expected that the K_{Ia}–values are continuously increasing in the transition region of the C_v–T–curve. On the other hand, however, a material–specific upper limit value which is analogous to the upper–shelf of the notch impact energy can also be anticipated for the crack arrest toughness, at least for the low USE material.

Wide plate tensile tests on X6 CrNi18 11 material with variable flaw geometry and flaw depth showed an increase in the load bearing capacity and in the case of surface flaws of small depth also an increase in the integral strain at fracture with the strain rate. It is intended to extend the testing of modified compact specimens up to temperatures of 250 °K above the NDT–temperature and to determine material characteristics at temperatures which correspond to the service temperature of a reactor pressure vessel. Additional arrest parameters shall be found using rotating discs and wide plate specimens with temperature or toughness gradients. Furthermore, the examination of transferability of properties determined on small and large scale specimens to thermal shock loaded hollow cylinders is intended.

9. Qualification Program of Dissimilar Welds

9.1 Background

The 12% Cr–steels are frequently used in German power plants for pipes, rotors and blades. Sufficient material toughness is required for the base metal to withstand sudden changes in load. This is of special interest for use in nuclear power plants. The R&D program MINERVA (Baumann et al., 1987) has been developed by the Brown Bovery Company (now ABB) in cooperation with *MPA*, to demonstrate safe operation of the safety–relevant components of the water–steam pressure circuit of the high temperature reactor. It is focused on dissimilar welds and is based on constitutive laws derived from basic studies of the materials under monotonic and cyclic loading as well as on a number of burst tests with pipes. It is expected that the findings of the program are generally applicable for advanced high temperature design and for code development. The program has been started in 1987 as a 20 000 h project under financial support of BMFT.

The specification of the individual tasks are shown in figures 13 and 9–2. The base material consists of 12% Cr steel, the weld deposit material of 32% Ni and 20% Cr with Al, stabilized with Ti. The purpose of this program is to qualify the dissimilar welds made of *X 20 CrMoV 12 1* and *X 10 NiCrAlTi 32 20* (Incoloy 800). It is aimed at investigating the degree of deformation and damage and to validate the predicted lifetime. The test sequence is terminated by a burst test with a representative circumferential crack. An important task within that validation trial is to produce reliable creep crack growth rate data and to evaluate the safety margin for leakage and catastrophic failure. In this was it possible to introduce an adequate in–service inspection procedure. Sophisticated test instrumentation including non–destructive equipment and mathematical modeling support the extensive numerical inelastic calculations and computer simulation.

The test program MINERVA is further supported by tests carried out on pipes made of *X 20 CrMoV 12 1* steel with similar metal welds. Toughness degradation in the base metal and deposited weld metal due to long–term service load in the high temperature regime are being quantified. Heat treatment procedures for recovery will be checked simultaneously.

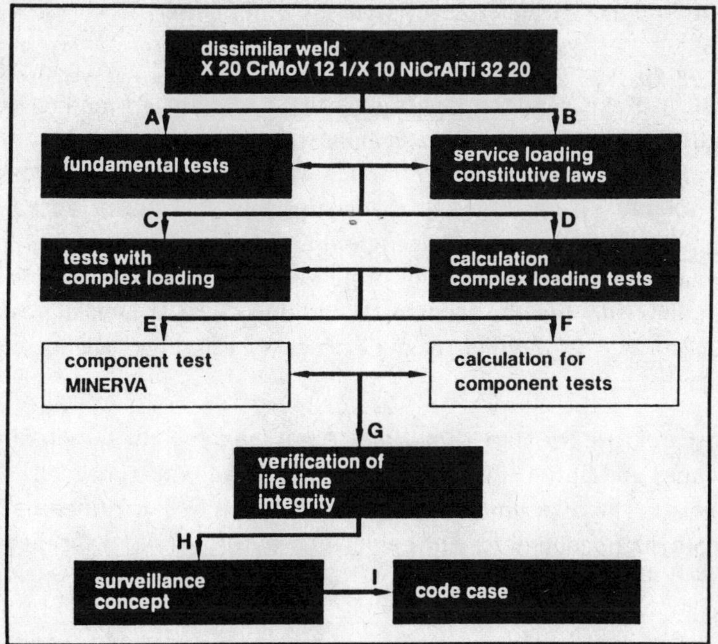

Figure 13 – Structure of the program MINERVA

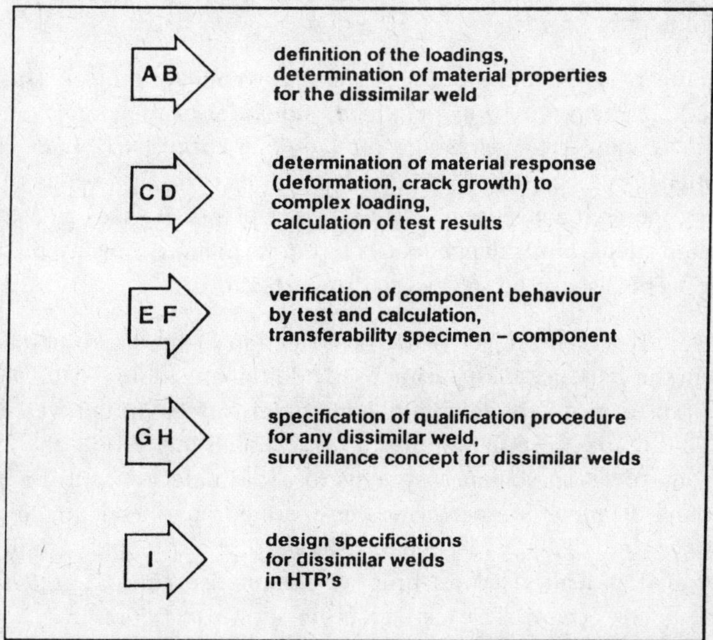

Figure 14 – Specification of the individual tasks within the program

10. Fast Breeder Model Test of Reactor Tank Wall

The research regards the reactor tank wall of the sodium–cooled fast breeder reactor SNR 300. An experimental investigation had to be performed to furnish the proof that the lifetime analysis of the reactor tank wall is reliable. Therefore, the long term tests are performed at large austenitic plates with longitudinal and cross seams. In order to obtain reliable assessment, the loads are only slightly increased compared with the actual ones. Increasing the test temperature up to 570 °C results in an acceleration of the damage. A tensile load with superimposed bending is applied. Measurement of deformations supplies the boundary conditions for the inelastic calculations which are to be performed in parallel. Non–destructive tests on crack formation are performed periodically.

This research project started in 1988. Analyses aimed at the optimization of the specimen shapes and at the stipulation of the test conditions preceded. The test rig has been constructed. The experimental investigations will be performed in the near future. The accompanying calculations will be verified by using the experimentally determined results.

11. Internal Pressure Tests at Pipe Bends made from Creep–Resistant Steel with Additionally Applied Bending Moments at Temperatures in the Creep Regime

In 1986 this program was started as a six year national project. This project is based on a cooperation between 10 associations, industrial institutes and universities. 50 % are financed by industrial means and 50 % are financed by BMFT. This research project was initiated by unexpected creep rupture defects at pipe bends of superheated steam piping in operated conventional steam power plants. Besides catastrophic failure due to exhaustion of the time–dependent creep deformation capacity different levels of pre–damage were discovered in operated pipe bends.

Aim of the research project is the investigation of the creep damage history, which is typical in the presence of superimposed additional bending load. The course of damage is to be reproduced experimentally and numerically and improved calculation bases for design and lifetime estimation are to be searched for. Within the experimental investigations improved surveillance systems to avoid defects are to be created. The experimental investigations are performed at operated pipe bends made of type *CrMoV*–steel with *1% Cr* and *12% Cr*, respectively. Internal pressure and a static moment at 500 °C are intended as load. The test time will amount to approx. *30,000 h* until fracture of the pipe bends occurs. The test set–up is shown in figure 15.

In order to define the test conditions extensive finite element calculations have been performed. Basis of these calculations is the uniaxially determined creep behavior until

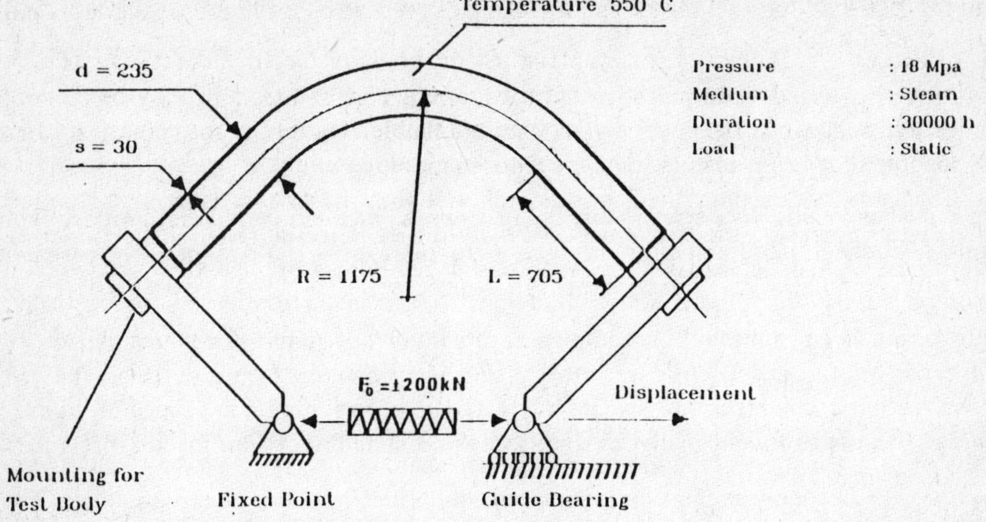

Figure 15 – Test set-up for pipe bend tests at elevated temperature

100,000 h. The analyses have shown that a static bending moment together with the internal pressure in connection with an oval pipe favors the damage observed in practice. Two large-scale test devices have been built up and the tests are running at *MPA*.

The numerical analyses are performed with the actual geometry of the test pipe bends. The calculations are to be verified by the experimentally determined data.

12. Investigation and Evaluation of Manufacturing and Operational Defects of Pressurized Components in Nuclear Power Plants

Failure experience in conventional and nuclear power plants has shown that during manufacturing and operation quality degrading influences, with the consequence of embrittlement and crack formation, may occur. The aim of this *MPA* project is the compilation and evaluation of manufacturing and operational defects of pressurized components in nuclear power plants. The analysis of the compiled failure phenomena indicates the frequencies of certain degrading mechanisms for typical components and the relevant influencing parameters. The improved understanding of the causes for crack formation and embrittlement leads to effective methods of failure prevention by optimized design, material and manufacturing as well as controlled operational conditions.

The program started with the evaluation of manufacturing defects in base metals and welded joints. The next steps comprised operational defects (cyclic mechanical or thermal loading, corrosion effects, etc. – figure 16).

The failure evaluation is performed in cooperation with governmental authorities, plant vendors, utilities, manufacturers and inspection authorities. The results of the research work are compiled in flaw and failure catalogues.

Future work is concentrated on complex combinations of failure mechanisms, for example corrosion assisted crack growth starting from manufacturing defects. The erosion/corrosion problems are of special interest. The central aim is the successful development and improvement of the design codes and standards regarding the life extension of power plants.

13. Nondestructive Examination – Full–Scale Vessel Tests (FSV)

Apart from the national R&D programs, under the auspices of the OECD Nuclear Energy Agency and of the Commission of the European Communities (CEC), the third international Program for the Inspection of Steel Components (PISC III) has been defined for the evaluation of non–destructive examination procedures. In PISC III, emphasis is laid on realistic geometries and conditions for the examination of the structures. The preceding PISC II program consisted of round robin tests on four plates

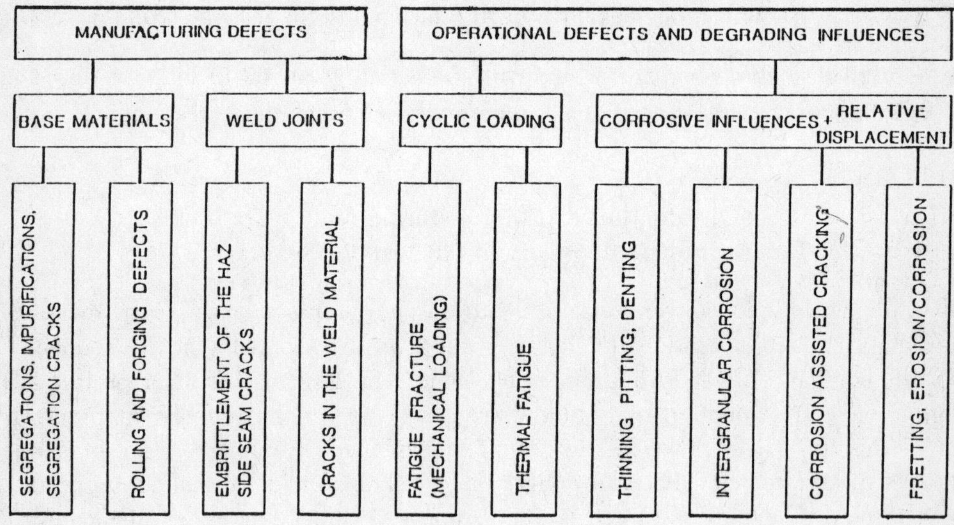

Figure 16 – Program for evaluation of defects

mainly under laboratory conditions and with associated laboratory parametric studies. It evaluated the performance of NDE techniques having corrected all human errors which were included during inspection. The aim of the new program is not only performance evaluation but also validation and reliability assessment. Thus, the Full Scale Vessel Tests (FSV) are the PISC III activities (Crutzen, Herkenrath, Kussmaul, Mletzko, 1988) most directly motivated by the PISC II results and recommendations. The FSV program will be composed of three phases, namely:

Phase 1: Advanced sizing technique – the application of all the techniques, devices and development work known for defect sizing.

Phase 2: ASME type procedure – the application by an international inspection team of the ASME type procedures parametrized by the recording level and supplementary techniques.

Phase 3: Full scale special procedures – the application of procedures characterized by combinations of standard and advanced techniques as routinely used for in-service inspection by the participating teams.

The structural components proposed for the validation activities and financed by BMFT and CEC are the 900 MW BWR Vessel (the Full Scale Vessel at *MPA* Stuttgart, figure 17) and the upper module, containing PWR nozzle and plate assemblies, added to the lower vessel (figure 18).

All assemblies contain artificially introduced defects, simulating service induced and fabrication defects. Some real fabrication defects, e.g. slag inclusions, also are present. Scanning devices are available for examinations from the outer surface. For the internal examinations, a central mast manipulator will be available for an agreed period of time. Local scanners will also be used.

The full scale vessel and the proposed full scale vessel components can be used by the participants during a period of three years. The full scale vessel is already available. Phase 1 is carried out in 1988 – 1989, Phases 2 and 3 are planned for 1989–1990. The Operating Agent will be responsible for the collection of inspection and test data. A standardized method will be used, adapted to the parameters of the specific tests. The PISC Referee Group established by the Operating Agent has the responsibility to ensure the confidentiality of the inspection results. The Reference Laboratory will be responsible for the certification of defects and for conducting or directing all destructive examinations. The analysis and evaluation of results is the responsibility of the Operating Agent and will be coordinated by a Data Analysis Coordinator following methodologies approved by the Evaluation Task Force reporting to the Managing Board.

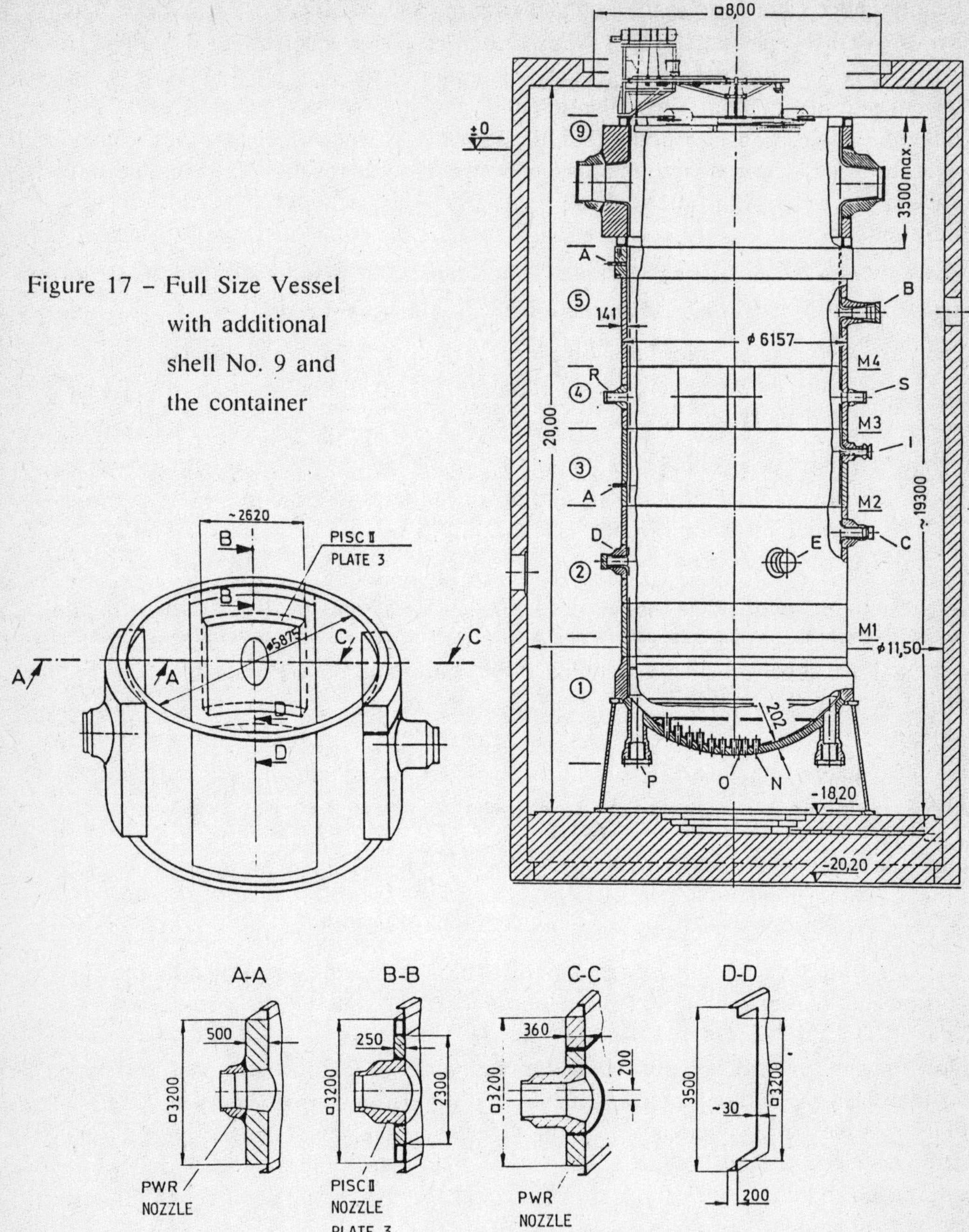

Figure 17 – Full Size Vessel with additional shell No. 9 and the container

Figure 18 – Details of the additional shell No. 9 of the FSV

14. Application of Artificial Intelligence: Expert Systems

14.1 The Basic Concept

There are several areas in the domain of structural integrity assessment, in which a possible application of artificial intelligence (AI), i.e. expert systems (ES), could provide substantial improvements of the current engineering practice. Coupling of heuristics, experience, expert opinions and qualitative data with the numerical analysis, and systematic and efficient engineering treatment of uncertain and imprecise information, are just two of the characteristic and important areas.

The *MPA* approach is based on the assumption that application of the AI/ES must be seen as a logical upgrade and improvement of the methodologies applied so far (numerical analysis, codes, experiments, failure analysis, etc.). In other words, the application should enable to perform better the engineer's tasks in the structural integrity assessment: for instance, to predict more precisely the remaining life of a pressurized component, or to make a more certain diagnosis of the actual structural state.

In order to obtain this goal, *MPA* has developed a modular concept which is both flexible when dealing with different application cases (e.g. leak–before–break analysis, corrosion problems, deep knowledge based structural diagnosis, high temperature pressurized components, etc.) and allows the application of sophisticated knowledge engineering (KE) methods and tools. The concept reflects the well known, empirical axiom of AI development, usually formulated as "Think big – start small" (figure 19).

Implementation of the concept (figure 20) resulted in the frame–like modular structure, in which the modules form three principal groups:

- The group of modules related to management and performance of the input/output operations and interfacing with the user, other expert systems, data bases, etc.
- The inference machine(s), represented on the current level of development and application by the commercially available KE shells.
- The group of application related modules, containing the specific knowledge bases and related methods and numerical routines for the supporting calculations.

Aside from being flexible, the structure puts only a few constraints regarding possible expansions/extensions of the system. Once the concept was established, *MPA* has continued to develop and test its elements/modules, and to gather experience in the use of various KE tools (AI languages, shells, etc.). The activities described below illustrate these developments.

14.2 The DSN Project

The DSN project (Deep Knowledge Expert System Using Coupled Symbolic–Numerical Analysis for Structural Diagnostics), was initiated as a proposal for the CEC ESPRIT

MPA Stuttgart — Expert System Applications		
Basic Axiom	Think Big —	Start Small
M P A Implementation	Have a clear GENERAL CONCEPT and precise GENERAL OBJECTIVES	Identify and choose cases where expert systems can bring QUICK and EFFICIENT improvements
Practical Application	1. Modular architecture 2. Expansible and versatile hardware and software environment (VAX, PS/2, KEE, GoldWorks, ...)	1. LBB Prototype 2. High Temperature Components — Life Assessment - Interest Club - ESR Project (BRITE) - German Utilities

Figure 19 – The MPA concept applied in development and application
of expert systems to be applied in the domain of structural analysis

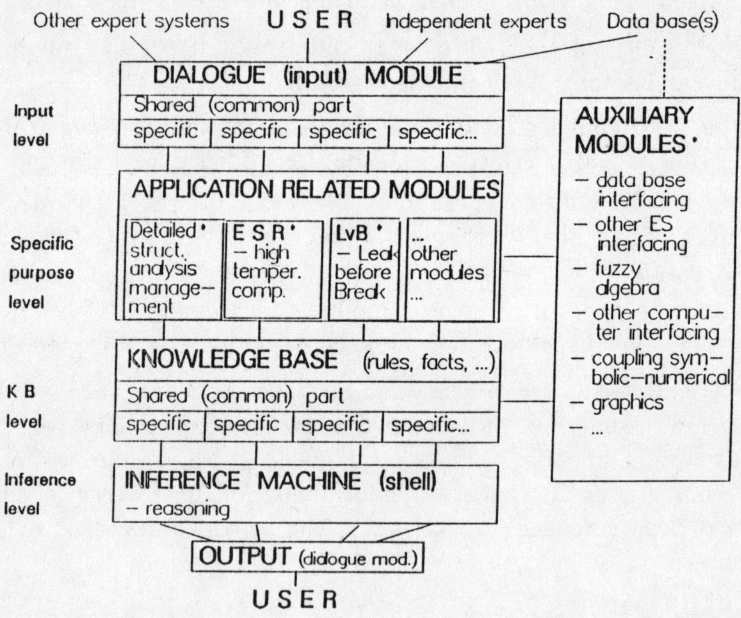

Figure 20 – Modular structure of the MPA frame for development and application
of expert systems to be applied in the domain of structural analysis

program. Its development has been backed by the fact that many power and process plants face nowadays the problem of unplanned and costly outages due to damage or structural failures of pressurized systems and components. The DSN expert system has been designed as a decision aid to plant engineers (Jovanovic et al., 1989). The principal task of the system, based on the structural diagnostics, is to indicate the optimum patterns of the future plant/component operation. To perform this task, all available information must take into account, i.e. both the standard, numerical, data (usually design data and resumed NDE data), and the relevant information contained in the humans' experience (e.g. experience with similar plants/components, experience from the material/component fabrication, independent expert opinion, etc.). This information is mainly qualitative and, very often, incomplete, uncertain and/or imprecise. The system must also explain its outcomes. Development of the system resulted mainly from the coordination of the research programs of *MPA* Stuttgart and JRC Ispra. Although the proposal as such has not been realized, many of its elements have been incorporated in the derived expert systems,e.g. in the leak–before–break expert system of *MPA*, further later on. The DSN objectives summarized in figure 21.

The DSN expert system is designed so to perform the structural safety and reliability assessment of the critical, heavy duty pressurized components of large power and proc-

APPLICATION DOMAIN	A I
<u>Area:</u> Flaw and damage assessment in pressurized components <u>Conditions:</u> mechanical and thermal loads, static, cyclic, under normal operation and transients <u>Geometries:</u> press. vessels, feedwater-tanks, large piping, nozzles <u>Result:</u> Safety and reliability diagnostics, analysis and life time prediction <u>Knowledge Base:</u> - General Facts - Facts being part of the current physical situation (e.g.: dimension, position and nature of those defects, environment, loads, material, etc.) - Rules: e.g. those regarding crack behaviour, NDE result interpretation, corrosion water/steel, material technology, Designer's/Operator's experience, etc.	- <u>Methods:</u> for treatment of imprecisions and uncertainties - <u>Non-standard logic</u> - <u>Qualitative modelling</u> (equations and constraints) - <u>Advanced logic,</u> e.g. for distinguishing between facts which are always true and facts which are true in the concrete situation only (modal logic), or making conclusions out of negative statements (negative inference), etc. - <u>New Rules Creation</u> by learning from application cases - <u>Deep Knowledge representation</u> methodologies

DSN Expert System Based Structural Safety Analysis

Figure 21 – Objectives of the DSN project

ess plants (e.g. large pressure vessels, piping, tanks). The assessment is dependent on structural diagnostics, which, in the DSN expert system, includes not only the numerical analysis, but also on the symbolic one. It means that both numerical and qualitative data are to be used and that uncertainties and imprecisions contained in both types of information will be treated. The system is in a dialogue with the user, basing its decisions and recommendations on both quantitative and qualitative elements of structural safety monitoring and diagnostics. The system should enable quicker assessment and use of non–numerical (qualitative), incomplete, imprecise and/or uncertain data, unconsidered in present methodologies which are mostly numerical. Incorporation of the deep knowledge (basic knowledge, used in the so called "reasoning from the first principles") should enable better handling of unexpected internal and external events in the component exploitation, such as material and structural degradation, or abnormal operation conditions. In the expert systems applied in diagnosis, the deep models offer many significant advantages, when compared to the models based on the shallow and/ or "compiled knowledge" only, in spite of some practical advantages of compiled knowledge based models. DSN relies on deep knowledge modelling mainly due to the following two reasons:

- Safety of a pressurized component is mostly affected by unexpected situations (transients, accidents) which, by definition, cannot be fully foreseen in advance. Shallow models are unable to deal with circumstances different from those explicitly anticipated.

- The processes forming the background of structural safety (e.g. crack behavior under stress gradients and/or complex loadings) are providing very poor, inconsistent and uncertain information for the formulation of premises. Reshaping this information is possible only in terms of "checking the compatibility with the first principles", which, again, leads to the use of deep knowledge.

The coupling between numerical and symbolic analyses in the area of structural diagnosis has to be done between the engineering numerical analysis and the symbolic analysis used in reasoning. Fuzzy algebra and concept of linguistic variable implemented in Lisp based software tools, are used in the symbolic analysis. The numerical, engineering calculations are mainly in Fortran. The two parts of the analysis interact either within the framework of the expert system shell software (numerical programs included as "methods" in the knowledge base), or the numerical part (e.g. Fortran subroutines) is invoked from Lisp–level of the program.

Principal elements of the DSN architecture are schematically represented in figure 22, showing that the system takes information from five sources: from the user, from the "independent expert", from the material data base, from the image knowledge base and from a case history data base. The user's requests provide guidelines for the whole session and they define its goals: structural failure risk assessment, lifetime prediction, etc. The user is supposed to provide the standard data regarding the component (loads,

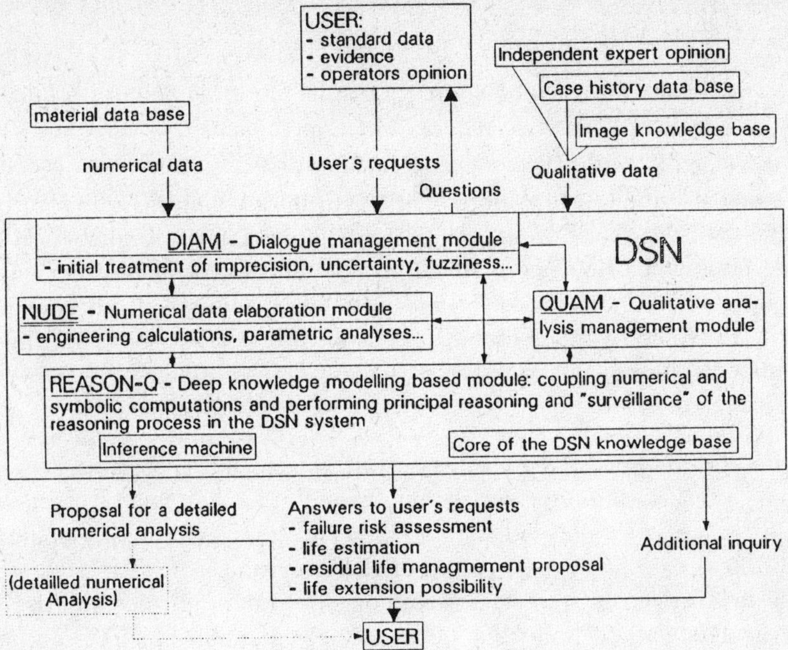

Figure 22 – Architecture of the DSN system

materials, known state of defects), as well as the evidence (description of the current situation) and the human opinion (e.g. the operator's one). DSN closes the loop towards the user providing three types of possible outputs:

- answer to user's questions,
- request for additional information and
- input for detailed analysis, consisting of pre-elaborated input data and the solution pattern proposal (e.g. sequence of codes to be used, type of analysis to be performed, etc.).

14.3 Leak–before–Break Expert System Prototype

Knowledge representation and reasoning issues are the issues particularly tackled in the work on the prototype expert system for the leak–before–break analysis developed and practically applied at *MPA*, Stuttgart (Jovanovic, Kobes, Sauter, Sturm, 1989). It has been necessary to develop an original knowledge representation scheme (Jovanovic, 1989), in order to enable the use of the *MPA* experimental evidence (Sturm, Stoppler, 1989) in an expert system for the leak–before–break analysis of pressurized pipes.

Special attention has been devoted to the development of an "intelligent recognition pattern" for identification and evaluation of the similarity between the analyzed case and the cases contained in the preceding experience, represented in the knowledge base of an object oriented system.

The solution is designed and first implemented as an open architecture, shell–free, intelligent module written in Lisp, interacting with the user in the search of the appropriate representation of various pieces of knowledge and in determination of its importance for the final conclusion (an identical version has also been developed in the KEE shell, later on). A weighting factor is attributed to each piece of knowledge, and its initial value is provided by the domain expert and reviewed in the dialog with the user, depending also on the newly acquired knowledge.

The reasoning has two basic levels: the one of the pre–established rules (the shallow knowledge and/or compiled deep knowledge rules), and the one of the self–generated and "implicit" rules, becoming effective due to the preceding and/or acquired experience. The inference is simple and robust, allowing the "reasoning by analogy" to be obtained. It allows to reach the diagnosis of the structural state (e.g. leak or break in the analyzed pressurized piping), taking into account both the rules and the results of the engineering calculation. Transition between the semantic (linguistic) and numerical probabilities, and vice versa, is obtained by means of the fuzzy algebra. The prototype expert system has been verified versus the *MPA* results of the preceding large scale leak–before–break tests on pressurized pipes (Table 1).

14.4 ESR Project – Life Assessment of High Temperature Components

The project has three complementary aspects:

- Development of a prototype, financed by a group of German utility companies.
- Collaboration in the framework of the Interest Club of the European Utilities (Partners: CEGB – Great Britain, ENEL – Italy, JRC Ispra – Italy, JRC Petten – The Netherlands, Laborelec – Belgium, *MPA* – Germany).
- BRITE/EURAM Proposal (a proposal for a CEC financed project).

Possibility of failure of high temperature pressurized components (pipework, headers, etc.) is one of the principal considerations in the safety and economy of power and other industrial plants. Catastrophic failure of such a component could both endanger the industrial plant and put at risk the population of the surrounding areas. The economic effects of improved/reduced availability are also considerable: the cost of one day of outage in a modern *500 MWe* power plant can be as much as *150,000 ECU*. In other industrial plants, the outage costs are usually much smaller (e. g. about *8,000 ECU/day*, for a *10 MW* plant), but the indirect costs due to disturbances in production can be much greater.

LBB expert system results, for unoptimized weighting factors and the analogy significance limits 0.3 and 0.7

Test Nr.	Test result	ES prediction	Probability descriptor
BVZ110	LEAK	BREAK	'meaningf.chance"
BVZ120	LEAK	LEAK	'meaningf.chance"
BVZ130	LEAK	LEAK	'meaningf.chance"
BVZ140	LEAK	BREAK	'meaningf.chance"
BVZ150	BREAK	BREAK	'meaningf.chance"
BVZ040	BREAK	BREAK	'meaningf.chance"
BVZ050	BREAK	LEAK	"it may be"
BVS060	BREAK	BREAK	'meaningf.chance"
BVS070	LEAK	LEAK	'meaningf.chance"
BVS080	BREAK	LEAK	'meaningf.chance"
BVS102	BREAK	LEAK	'meaningf.chance"
BVS050	BREAK	BREAK	'meaningf.chance"

Nr. of predictions vs. nr. of analyzed cases 12/12

Exact Predictions 58%

LBB expert system results, for optimized weighting factors and the analogy significance limits 0.1 and 0.9

Test Nr.	Test result	ES prediction	Probability descriptor
BVZ110	LEAK	not any	—
BVZ120	LEAK	LEAK	'meaningf.chance"
BVZ130	LEAK	not any	—
BVZ140	LEAK	not any	—
BVZ150	BREAK	not any	—
BVZ040	BREAK	BREAK	'meaningf.chance"
BVZ050	BREAK	not any	—
BVS060	BREAK	BREAK	'meaningf.chance"
BVS070	LEAK	not any	—
BVS080	BREAK	not any	—
BVS102	BREAK	not any	—
BVS050	BREAK	not any	—

Nr. of predictions vs. nr. of analyzed cases 3/12

Exact Predictions 100%

Table 1 – Sample results of the the LBB expert system prototype

Improving the methods and procedures used for assessment and management of remaining life of the high temperature pressurized components is therefore important. The development proposed in the ESR project should significantly improve the current practice by making it

- more comprehensive and reliable (a direct coupling of heuristics and specialists' knowledge with standard and/or advanced engineering calculi)

- quicker and more easily available (expertise resides in computer systems at the plants available round–the–clock)

- more uncertainty independent (systematic treatment of uncertainties involved).

To reach these goals, it has been proposed
  - to develop a "reference procedure", and
  - to base the assessment and the management on the expert system technology.

The development would thus improve safety (reduced risk of wrong assessment), reliability and availability (reduced outages caused by wrong management), and economy (see the figures) of the target components and plants. It is estimated that there are more than 25,000 components (each costing typically 500,000 ECU) operated in the countries which have participated in the preparation of the BRITE/EURAM project proposal.

The main tasks within the project are the following:

a. In-depth comparative analysis of current practice and preceding experience, with the goal to establish the "Reference Procedure" for assessment and management of remaining life of the target components. The procedure will include the regulatory requirements and the task will result in a summary of the "good practice" representing a contribution towards a "unified European approach" for the year 1992 (the draft of the flowchart is given in figure 23)

b. Identification and analysis, including interaction in the definition of the Reference Procedure, of 8-9 case studies, supplied by the utilities. The task will result in a "matrix" of cases with typical combinations of component types, materials, applications, etc.

c. Interviewing of domain experts (knowledge elicitation), with the particular aim to extract and formulate the heuristic rules, and to reconstruct the expert reasoning in sub-domains like non-destructive examinations, material testing, component disposition decision making, future operation regime definition, etc.

d. Development of a set of computer codes ("ESR/RP Library") covering numerical analysis in the "Reference Procedure" and chosen alternative algorithms for life assessment in terms of creep and fatigue. The tasks includes codes for elaboration of numerical data from service records.

e. Definition and preparation of the software and hardware environment. Detailed design (architecture) of the ESR expert system, including the design of modules and their interaction

f. Development of modules of the ESR system: interfacing with the user and use of stored numerical data (dialogue module), decision optimization (optimization module), module for the interactive treatment of uncertainties, the system data base, and the Hypertext-based "reference procedure" module.

g. Development of interfaces and final assembling of the ESR system.

h. Selection, on the basis of Task (b), and development of the "ESR – generic demonstrator", handling a generic case of a pressurized high temperature component.

i. Testing/verification/evaluation/modification loop including the establishment of the testing/validation procedure and testing of the generic demonstrator.

j. Development of "Specific Demonstrators" corresponding to the case studies from Task (b), and their testing and validation according to the procedure from Task 9.

k. Evaluation of the project and its results. Documentation.

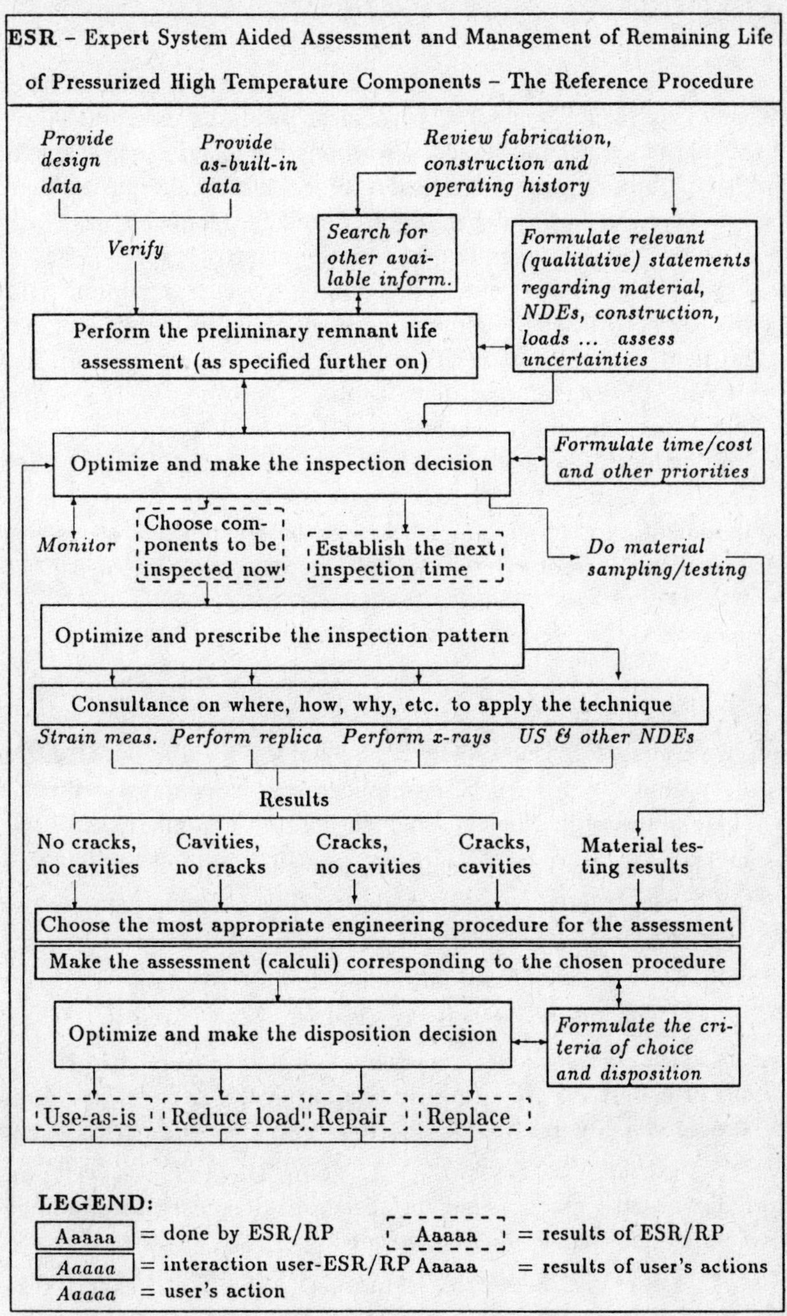

Figure 23 – Draft of the Reference Procedure (ESR expert system)

15. Conclusions

The overview presented in the paper gives an insight in a segment of the wide range of the *MPA* activities in the domain of structural safety assessment. As it can be seen, most of them are long-term oriented, encompassing a series of various research activities. The *MPA* tradition and the decades of experience in the field, as well as the considerable technical basis of the research and development work, are at the same time background and the incentive for development of expert systems. Background, because they represent an extremely rich source of necessary information: experimental, heuristic, numerical, etc. Incentive, because many of the existing and future problems in the field of structural safety and integrity engineering can be solved only if such experience is preserved and made usable – and the expert system technology is, for the time being, the most accessible and promising one, usable for engineering purposes on this level. However, in order to obtain all the expected results, it is necessary to reach also in the area of expert systems the same level of engineering proficiency (equipment, tools, personnel, etc.) as in the other projects described in the paper. Many of the current efforts at *MPA* are, therefore, oriented right in this direction.

References:

Baumann, B., Gnirss, G., Neureuther, K., Schneider, K. (1987): MINERVA – qualification of dissimilar welds fro HTR-application, Proceedings of the 13th *MPA* Seminar (also published in Nuclear Engineering and Design, 1989)

Crutzen, S., Herkenrath, H., Kussmaul, K., Mletzko, U. (1988): PISC Full scale reactor pressure vessel validation of non-destructive examination, 14th *MPA* Seminar, *MPA* Stuttgart

DSN – Deep knowledge expert system using coupled symbolic numerical analysis for structural diagnostic, ESPRIT-II Proposal Nr. 2224 (Area II.2.3), 1988

Jovanovic, A. (1989): Knowledge representation and reasoning in structural reliability assessment expert systems: Implementation in the Leak-before-Break expert system, Proceedings of the ICOSSAR Conference, San Francisco, August

Jovanovic, A., Lucia, A. C., Servida, A., Volta, G. (1989): Development of a deep knowledge based expert system for structural diagnosis, Proceedings of the ICOSSAR Conference, San Francisco, August

Jovanovic, A., Kobes, E., Sauter, A., Sturm, D. (1989): An expert system in the domain of fracture mechanics, Proc. of the SMiRT Conference, Los Angeles

KEE 3.0 Users' Manual, IntelliCorp, W. Mountain Veiw, CA.,1988

KTA 3201.2 – Komponenten des Primärkreises von Leichtwasserreaktoren. Teil 2: Auslegung, Konstruktion und Berechnung. Fassung 3/84

KTA 3201.3 – Komponenten des Primärkreises von Leichtwasserreaktoren. Teil 3: Herstellung. Fassung 12/87

Kussmaul, K. (1984): German Basis Safety Concept rules out possibility of catastrophic failure, Nuclear Engineering International, December

Kussmaul, K., Iskluth, B. (1988): Stress corrosion cracking of steels in high temperature water, Proceedings of the 14th *MPA* Seminar (to appear in Nuclear Engineering and Design)

Kussmaul, K., Foehl, J., Weissenberg, T. (1988): Investigation of material from decomissioned reactor pressure vessel, Proceedings of the 14th *MPA* Seminar (to appear in Nuclear Engineering and Design)

Kussmaul, K., Sauter, A. (1986): Ductile behaviour under pressurized thermal shock, Int. ANS/ENS Topical Meeting on Thermal Reactor Safety, San Diego, California

Kussmaul, K., Stoppler, W., Sturm, D., Julisch (1983): Exclusion of Fracture in Pipings of Pressure Boundary, International Symp. on Reliability of Reactor Pressure Components, IAEA–SM–269/7

Kussmaul, K., Sturm, D., Stoppler, W. (1985): Experimental and analytical investigations on the strength and fracture behaviour of longitudinal cracked piping, Proceeding of the PVPD Conference, New Orleans

Sturm, D., Stoppler, W. (1989): Strength behaviour of flawed pipes under internal pressure and external bending moment – Comparison between experiment and calculation, Specialist Meeting Leak–Before–Break in Water Reactor Piping and Vessels, Toronto, October 24–27

STRUCTURAL SAFETY ASSESSMENT AS COLLECTIVE COGNITIVE PROCESS

G.Volta

Commission of the European Communities, Joint Research Centre, Institute for Systems Engineering, 21020 Ispra, Varese, Italy

KEYWORDS

complexity, decision, expert, representation, safety assessment, uncertainty.

1. THE REVIVAL OF THE "EXPERTS"

For some thousands years structural engineering   has been the domain of "experts". These experts were people who were able to accumulate in their brains all the knowledge needed to build monuments or civil works for which we still feel with the greatest admiration.

Structural safety has been always considered and regulated in the sense that people had to respond of their works in case of negative consequences (one can see in the Hamurrabi's code that builders had to respond with their life of eventual collapse of buildings).

But at the extent that experts were at the same time keepers of the whole knowledge and of the full responsibility, the assessment of safety, as such, was an implicit aspect of the decision process of the single actor, the expert.

In the last hundred years, the practice of the division of labour and of the division of knowledge, and the need of public control of the negative side effects of technologies, modified the process of structural safety assessment: it has become the domain of many experts, of many interested parties, of many actors.

However, for a number of years, the conflicting potential of this multiplicity of actors did not rise particular problems.

In fact it was possible to eliminate the subjectivity of the "view" of each expert, founding the assessment process on a limited body of knowledge: precise data and algorithms codifiable in universally accepted procedures.

Possible conflicts were considered in general resolvable by more accurate and sophisticated computational efforts. Safety assessment has been for most people equated to search of precise data and computation.

The ill defined knowledge, the heuristics specific of each expert in approaching a problem, the "expertise" that distinguish a high level expert from a simple chartered engineer, disappeared from the scene of the explicit assessment process.

The expert, as such, has been overshadowed.

But the scientific and technical community under the pressure of some new problems raised by the industrial development, and with the help of new tools offered by this development, is now changing the perspective, and a renewed interest in experts, with their ill defined knowledge, judgements, opinions, is emerging.

The revival of the "experts", or, better, of the interest in pieces or forms of knowledge, that before were not considered acceptable in a scientific framework, has been caused by the recognition that safety assessment process consists of two parts.

A codified part based on algorithms and precise data, and a part, more qualitative, that includes the overall delineation of the problem, the choice of the starting hypotheses, the choice of what is relevant or not relevant, the choice of the appropriate rules in incompletely defined situations , the choice of the most suitable models of physical phenomena for which competing models exist.

The second part, which appeals to the intelligence, experience, deep culture, imagination of the assessors, is becoming the main source of uncertainty and therefore of conflicts.

This shift of importance from the "numerical and computational" part of the assessment process, to the " inferential and qualitative part" is the consequence of the dramatic increase of the complexity of the structural safety assessment as cognitive and decision process.

2. THE COMPLEXITY OF STRUCTURAL SAFETY ASSESSMENT

To illustrate the complexity of structural safety assessment as it is conceived today, I take the definition of "safe structure" given by a leading expert: "A structure is safe if it so withstands the loads that come upon it during its working life, that is it continues to serve the functions for which it was designed and does so to the satisfaction of its owners and users, without causing danger or undue disquiet to the general public" (Pugsley 1966, pag.2).

This apparently plain definition highlights the large spectrum of problems that the assessment brings forth, the large numbers of actors involved and the expectations today associated to the concept of safety.

Various bodies of knowledge concur to shape that concept.

First the knowledge about loads. The load history is dependent on the context of utilization. The prevision of a given load history is obtained by extrapolation of previous experience, by operational research on eventual loads on similar structures, by hypotheses about the context. But conditions and context are badly known in the extreme range of their stochastic variability. This is particularly true for a large class of external loads (earthquakes, external events) and for loads generated by accidents.

Then the working life. In general, for stationary structures, the working time is defined in terms of calendar time and for moving structures (for instance transport vehicles) in terms of service time. But the "service" history of stationary structures, is also important, and minor or major overhauls or replacements or maintenance can change drastically the life expectation.

The exploding interest in life-extension of structures, shows also the evolutionary character of the notion of working life.

The satisfaction of owners and users , in terms of "danger or disquiet of general public", depends on what is considered "public" and on the tolerability of risk at the moment of the use.

The knowledge that allows the determination of the boundary for the satisfaction, is partially embodied in regulations and codes of practices. But what is codified , in many cases is insufficient to take decisions.

Moreover codes and rules evolve during the lifetime of a structure.

Negotiations between the various actors of the safety assessment process: user, designer, manufacturer, safety authorities, are commonly required. In this negotiations technical and economical (cost of maintenance, cost of inspections) considerations are tightly linked.

In some cases the problem of safety and the problem of serviceability are positively correlated. In some other more intriguing cases they are independent or negatively correlated, and safety provisions, from the point of view of the specific function of the structure, are pure loss. The trade-off safety versus serviceability is an implicit constraint in any assessment.

The bodies of knowledge concurring to define the various aspects of the concept of structural safety are further extended by the increasing variety of materials and fabrication techniques available, by the evolving design criteria (for instance redundant or fail-safe criteria) by the availability of new inspection and diagnostic technologies.

The availability of new inspection technologies in some domains, for instance in transportation, has profoundly changed the basic approach to safety management: structures are conceived as systems with a feedback control, more than as "ballistic" systems.

These hints show the implacable trend toward a complexification of structural safety assessment.

3. MANAGEMENT OF COMPLEXITY BY SPECIALIZATION

The technical community, to manage this complexity and to simplify the decision process, has progressively segmented the overall domain of structural engineering   into different, more restricted fields.

Each field has its rules, its codes of practice, its underground knowledge that come to light as "expertise". In particular,   each field   has its areas of uncertainties and can develop its   more or less explicit ways for dealing with these uncertainties. For instance in the domain of metallic structures (see for instance the ASME Boiler and Pressure Vessel Code), one has embodied the   uncertainties in   safety factors applied to characteristic stress, while in the domain of concrete structures (see the CEB code) the uncertainties have been allocated to safety factors   applied   to   loads, materials properties and to their statistical   variability.

This   approach has drawbacks.

Decoupling various fields of expertise,   we loose the opportunity to   benefit from the knowledge accumulated in different domains. For example it has been argued that the famous Tacoma Bridge failure could be imputed to a lack of scientific and technical liaison between the structural engineering environment in which the project   has   been developed and the   aeronautic environment, well acquainted at the same time with the problems of aerodynamic oscillations (Pugsley 1972).

Moreover,   when dealing with   new structural aspects of new technologies, we are faced by the dilemma of putting them in preexisting unsuitable   frames or to set up new frames, to invent "new fields".   But inventing new separated   fields we risk to waste useful knowledge.

4. COMPLEXITY AS MULTIPLICITY OF NON COMPATIBLE
REPRESENTATIONS

A system-theoretic view of the nature   of complexity can show the deep reasons of specialization and can suggest how   a step forward, beyond the pragmatic specialization, is prepared by the emerging discipline called "knowledge engineering".

We state, in agreement with various modern system theorists (Casti 1984, Rosen 1985) that complexity is not an intrinsic property of a system, but a property   arising from the interaction of a system with its observer.

The complexity of the system S for the observer O is given by the number of   "non equivalent" representations that O   has of S. Conversely the complexity of the observer O for the system S, is given by the number of non equivalent representations that S has of O.

Talking of structural safety,   S can be the structure and its operational context , and O can be the designer or the safety authority or the public, or the collective entity composed by designer,   safety authority and public. For sake of simplicity we neglect the cognitive capacity of S and henceforth the interaction of the operational context with O that generate the so called "control complexity".

The reduction of complexity, a condition for decision, is the reduction of these non-equivalent, non-reducible representations.

When the observer is a collective entity, the problem is posed at two levels: the cognitive level of the single observer and the collective cognitive level.

The reduction of the number of non-equivalent representations can be obtained acting on the equivalence criteria or acting on the representations themselves.

The loosening of the equivalence criteria is an approach implicitly adopted in formal elicitation of probability judgements of experts about events and uncertain quantities (NRC 1989, von Winterfeldt 1989). This formal elicitation implies in fact the assumption that globally the judgements of various experts belong to the same "category", to the same statistical "population" of representations. The problem of the equivalence of representations is replaced by the problem of the equivalence of the experts.

Also the impressive development of methods and techniques, probabilistic and non probabilistic, for handling uncertain, imprecise, fuzzy information, answer to the same need of putting in the same basket, of accepting as exchangeable and equivalent different views of the same reality.

The representations themselves can be modified and reduced changing the boundaries of the system at hand, or the observables, or the measurement procedures, or the relation between observables (models). We point out that the reduction of non equivalent representations concerns steps of the cognitive process that are upstream of detailed analysis and computation. This reduction is in general the result of a cognitive negotiation, that produces successives and interactive adjustments, inside of the mental process of the single observer and between different observers. This reduction is unavoidably associated to a build-up of "matching" uncertainties, a class of uncertainties generated by the need of reducing complexity to take collective decisions.

5. WHAT WE EXPECT FROM EXPERT SYSTEMS

The system theoretic perspective offered in the preceding chapter shows safety assessment as a cognitive process, that through a progressive reduction of incompatible representations converge to a single representation and to a decision. The delimitation of empirical fields, the specialization, facilitates this reduction and ensure at some extent the repeatability and reproducibility of the process, in an implicit way. Knowledge engineering on the contrary, has the ambition of tackling the cognitive process in its structure and expert systems have the ambition of emulating the process itself.

All the Course will be dedicated to the theory and practice of these new tools. I would like to prospect, concluding this introductory lecture, some fundamental changes that the possibility emulating in an explicit way the various steps of structural safety assessment as a cognitive process, will introduce in the professional context of a structural engineer.

The consideration of to the qualitative, inferential, "human" aspects, will prevail over the computational aspects.

The identification and simulation of pattern of knowledge common to different specialized domains, will generate a reshuffling in the established order of engineering and humanistic disciplines, and will introduce many unexpected cross-links.

The resolution of conflicts raised by the "complexity" of the technological society will rely more and more on the cognitive negotiation allowed by the new instruments .

REFERENCES

Casti J., System similarities and natural laws. Laxenburg, Austria: IISA WP-84-1.

Nuclear Regulatory Commission, Severe Accident Risks, Summary Report Report of NUREG 1150. Washington, D.C.: Nuclear Regulatory Commission, 1989.

Pugsley A., The safety of structures. London: Edward Arnold LTD, 1966.

Pugsley A., "The engineering climatology of structural accidents", in Freudenthal A.M. (ed.), International Conference on structural safety and reliability. Oxford: Pergamon Press, 1972, p.335-340.

Rosen R., Anticipatory Systems. Oxford: Pergamon Press, 1985.

Von Winterfeldt D., "Eliciting and communicating expert judgement: methodology and application to nuclear safety", Annual Meeting of the American Statistical Association, Washington D. C., August 6-10, 1989.

Fundamentals of Expert Systems

KNOWLEDGE REPRESENTATION AND INFERENCE IN KNOWLEDGE BASED SYSTEMS (EXPERT SYSTEMS)

Piero P. Bonissone

Artificial Intelligence Program
General Electric Corporate Research and Development
Schenectady, New York 12301, USA

Abstract

Knowledge Based Systems (KBS) are computer programs in which knowledge and control are explicitly separated. In first generation KBS, the reasoning is usually monotonic and the control is procedural. Second generation KBS usually exhibit nonmonotonic reasoning, declarative control, and more sophisticated representations of uncertainty. We will focus on the first generation KBS and analyze their typical architecture, composed of a Knowledge Base (KB), a Working Memory (WM), and an Inference Engine (IE). The KB describes the domain knowledge; the WM describes a problem instance; the IE determines the applicability of different subsets of the KB to the current problem. The selection of a specific knowledge representation paradigm, used to build the KB, implicitly determines the selection of the inference mechanism to be used. We will briefly discuss Predicate Calculus, which uses unification and resolution, Frames, which use inheritance, and Production Rules, which use rule chaining.

1 Introduction

Knowledge Based Systems (KBS) or Expert Systems are problem solving systems whose (intelligent) behavior is mostly based on *declarative* rather than *procedural* knowledge. These systems derive their conclusions from explicit knowledge (i.e., knowledge-bases), rather than from hard-coded algorithms.

The key distinction between KBS and conventional programs is the explicit separation of knowledge and control. Unlike an algorithm in which the flow of control is interwoven with the information upon which the algorithm is based, a KBS has three distinct components: a Knowledge Base (KB), a Working Memory (WM) and an Inference Engine (IE). The KB is a set of rules (or other representation mechanisms such as frames) representing the domain knowledge. The Working Memory (WM) contains a specific problem instance, described by the original input data and augmented by all the inferences made by the system from the data. The Inference Engine accesses the elements in WM, scans the KB, determines the relevant rules and executes one or more of them to generate more conclusions associated with the problem instance. Usually, Knowledge Based Systems are classified according to their reasoning and control capabilities into first or second generation KBS.

1.1 First generation KBS

First generation KBS have usually handled tasks that are well scoped and bound, with a stronger emphasis on performance (based on shallow models) than understanding (based on deep models). The typical problems addressed by this early technology are described by the concept of *Classification Problem* introduced by Clancey in 1984 [1].
The *Classification Problem* (CP) consists of recognizing a situation from a collection of data and selecting the best action in accordance with some objectives. The classification problem has a recurrent solution structure:

1. A collection of data, generated from several sources, is interpreted as a predefined pattern.

2. The recognized pattern is mapped into a set of possible solutions.

3. One of these solutions is selected as the most appropriate for the given case.

This process is considered a *static* classification problem, since the input data are assumed to be invariant over time, or at least invariant over the time required to obtain the solution. This is illustrated in Figure 1.

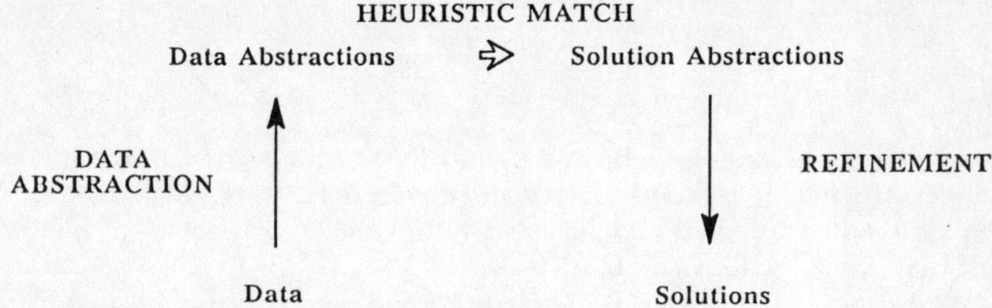

Figure 1: Static Classification Problem

First generation KBS have solved these problems by using monotonic reasoning, procedural meta-reasoning (control) embedded into the inference engine, and a crude representation of uncertainty (if any). Their reasoning has been based on behavioral rather than casual knowledge, and their knowledge representation has usually been flat rather than hierarchical. Examples of these first generation KBS are Mycin [2], [3], XCON [4], Prospector [5], and DELTA-CATS [6], [7].

1.2 Second Generation KBS

Second generation KBS handle tasks that are more open-ended than those described by the CP class. These tasks impose stronger requirements, such as non-monotonic reasoning, declarative control, a more complex representation of uncertainty, hypothetical reasoning using truth maintenance (choices), defeasible reasoning (defaults), model-based reasoning, case-based reasoning, automatic knowledge acquisition for limited perceptual tasks (neural nets), real-time performance, etc. The typical problems addressed by this new technology are described by the concept of *Dynamic Classification Problem* (DCP) introduced by Bonissone in 1988 [8].

The DCP is a class of classification problems where the environment from which data are collected changes at a rate comparable with the time required to obtain a refined solution, requiring real-time response. Examples of such *dynamic* classification problems include real-time situation assessment (e.g., air traffic control), real-time process diagnosis (e.g., airborne aircraft engine diagnosis), real-time planning, and real-time catalog selection (e.g., investment selection during market fluctuations). The characteristic structure of this class of dynamic classification problems is illustrated in Figure 2.

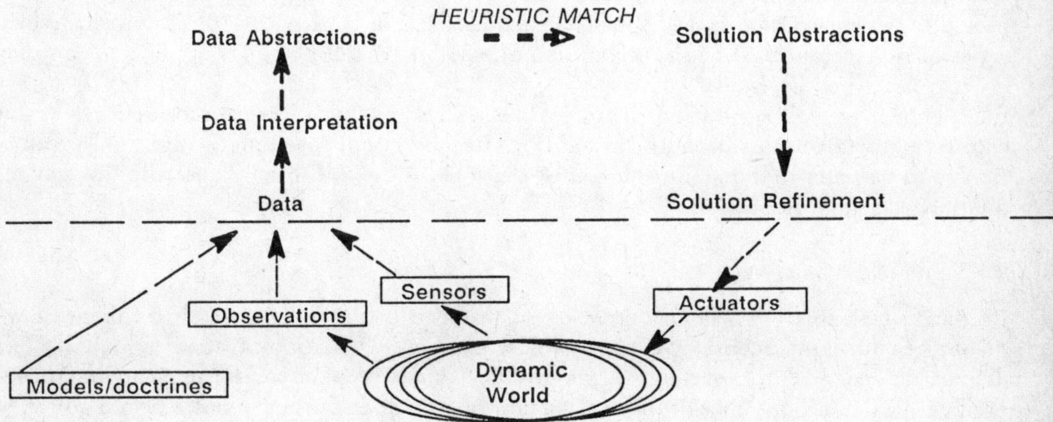

Figure 2: The Dynamic Classification Problem

A new set of KBS shells such as KEE [9], ART [10], RUM/RUMrunner [11], addresses subsets of the requirements induced by the DCP.

2   Knowledge Representation

We have observed that the most important characteristic of KBS is its explicit separation between knowledge and control. We can consider the development of a KBS application to be an exercise in programming using an embedded language. In the layered approach used in traditional Computer Science, a source program is written in a programming language; it is compiled by the corresponding compiler that transforms the source code into binary code (this step may be omitted for interpreted languages such as APL or LISP); finally the code is executed by an operating system (after the necessary linking and loading operations). Similarly, a KBS application is written in a representation language; the KB is then compiled by a knowledge compiler that transforms the declarative, unstructured information into a dependency graph of some sort (this is an optional step in most KBS); finally an inference engine executes the KB by interpreting it (if the compilation was omitted) or traversing the compiled graph (otherwise).

The most crucial aspect in a KBS application is the selection of a suitable knowledge representation language. A conflicting set of requirements is usually imposed upon it. The language must be very expressive and capable of representing high level of abstractions. This is required to avoid a major bottleneck during the Knowledge Acquisition step. The representation language must also be very flexible and modifiable. This is required to facilitate the many KB modifications during the Knowledge Design, Programming and Refinement [12].

According to the available structural information, three types of knowledge representation have been used to define the representation language: Predicate Calculus, Production Rules, and Frames. These representations are all equivalent in the Turing machine sense. However, the ease of their applicability changes drastically, based on the nature of the problem to be solved. Predicate calculus is suitable for very unstructured tasks, while production rules (with rule-classes) and frame systems increasingly capture the intrinsic structure of the problem domain. It should also be noted that the current trend, exemplified by recent commercial KBS shells such as KEE [9] and ART [10], is toward hybrid types of representation, the most common of which is the use of rules applied to complex objects described by frames.

Each knowledge representation paradigm has an associated inference mechanism. Predicate calculus uses unification and resolution, frame systems use inheritance, and production rules use rule chaining. In the rest of these notes, we will briefly describe these types of knowledge representation.

2.1   Predicate Calculus

Predicate Calculus (PC) is an extension of propositional logic, in which terms are augmented to be either constants or variables, existential ($\exists$) or universal ($\forall$) quantifiers define the scope of the variables, and predicates are applied to terms. In first order predicate calculus, the most common version of PC, quantifiers are not allowed to range over predicate variables. A more rigorous definition of the syntax of the language used in PC requires an alphabet, a recursive definition of *terms* and a recursive definition of *well-formed-formulae*. The reader is referred to [13] (chapter 6), [14] (chapter 2), or [15] for a more formal definition of the language.

The basic inference mechanisms used in conjunction with Predicate Calculus are unification and resolution. Simply stated, unification is the search for a unifier for a set of formulae, i.e., a substitution for the variables in the formulae that will make them identical strings.

Resolution is a general rule of inference that combines substitution, *modus ponens* and other logical syllogisms. Given the intersection of two clauses $c_1 \wedge c_2$, where $c_1 = \neg A \vee B$ and $c_2 = \neg B \vee C$, (i.e. $A \rightarrow B$ and $B \rightarrow C$), we can soundly deduce $\neg A \vee C$ (i.e., $A \rightarrow C$). This particular syllogism is extensively used in resolution to merge and simplify clauses in the *conjunctive normal form*.

2.2   Example of Unification and Resolution

To illustrate the use of unification and resolution principle, we will solve the following example, denoted as the Alpine Club Puzzle, which is proposed as an exercise in reference [13] (page 215):

> *Tony, Mike, and John belong to the Alpine Club. Every member of the Alpine Club is either a skier or a mountain climber or both. No mountain climber*

likes rain, and all skiers like snow. Mike dislikes whatever Tony likes and likes whatever Tony dislikes. Tony likes rain and snow. Is there a member of the Alpine Club who is a mountain climber but not a skier? Who?

The first step in the solution of this problem is to define a set of predicate and constants used to re-formulate the puzzle. The constants are *Tony, Mike, John, snow, rain*. The predicates are defined as:

L(x,w) x likes w
A(x)   x belongs to the Alpine Club
S(x)   x is a skier
C(x)   x is a mountain climber

The above puzzle can be reformulated as:

$F_1 : A(Tony) \wedge A(Mike) \wedge A(John)$	Tony, Mike, and John belong to the Alpine Club
$F_2 : \forall x[A(x) \rightarrow (S(x) \vee C(x))]$	Every member of the Alpine Club is either a skier or a mountain climber or both
$F_3 : \forall x[C(x) \rightarrow \neg L(x, rain)]$	No mountain climber likes rain
$F_4 : \forall x[S(x) \rightarrow L(x, snow)]$	All skiers like snow
$F_5 : \forall x[L(Tony, x) \rightarrow \neg L(Mike, x)]$	Mike dislikes whatever Tony likes
$F_6 : \forall x[\neg L(Tony, x) \rightarrow L(Mike, x)]$	Mike likes whatever Tony dislikes
$F_7 : L(Tony, rain) \wedge L(Tony, snow)$	Tony likes rain and snow
$Q : \exists w[A(w) \wedge C(w) \wedge \neg S(w)]$	Is there a member of the Alpine Club who is a mountain climber but not a skier?

The above facts must be converted to a standard clause form. The first step in this transformation is to replace the material implication ($A \rightarrow B$) by its logical equivalent ($\neg A \vee B$). At this point, the only connectives allowed in each clause are the disjunction $\vee$ and the negation $\neg$. Other steps may be necessary to complete this transformation. These steps are [13]:

- The scope of each negation $\neg$ is reduced so that it applies to at most one predicate letter. (This can be accomplished by using DeMorgan's Law, Involution, and the relationship between universal and existential quantifiers)[1]

- The variables that are in the scope of a quantifier are standardized, so that each quantifier has its own unique dummy variable.

- The existential quantifier $\exists$ is eliminated and replaced by a Skolem function. (See [13] (page 166) for further explanations of a Skolem function.)

---

[1]The substitution rules based on these identities are:

$$\neg(A \wedge B) \equiv \neg A \vee \neg B$$

$$\neg(A \vee B) \equiv \neg A \wedge \neg B$$

$$\neg\neg A \equiv A$$

$$\neg(\forall x)A \equiv (\exists x)\{\neg A\}$$

$$\neg(\exists x)A \equiv (\forall x)\{\neg A\}$$

- Each formula is converted to a *prenex* form, i.e., a *prefix* and a *matrix*. This conversion is simplified by the fact that in each formula there are no more existential quantifiers and the universal quantifiers are applied to their own unique dummy variables. Thus, from each formula we can extract the universal quantifiers, extend their scope to the entire formula, and write the universal quantifiers as the *prefix* of a quantifier-free formula (*matrix*).

- The matrix is converted to a *conjunctive normal form* (a conjunction of a finite set of disjunctions of predicates and/or their negations). This can be achieved by using the distributivity of the disjunction operator over the conjunction operator.[2]

- The universal quantifiers are removed, as all the variables appearing in the matrix are universally quantified.

- The conjunction operator $\wedge$ is removed and the clauses are simply separated by commas.

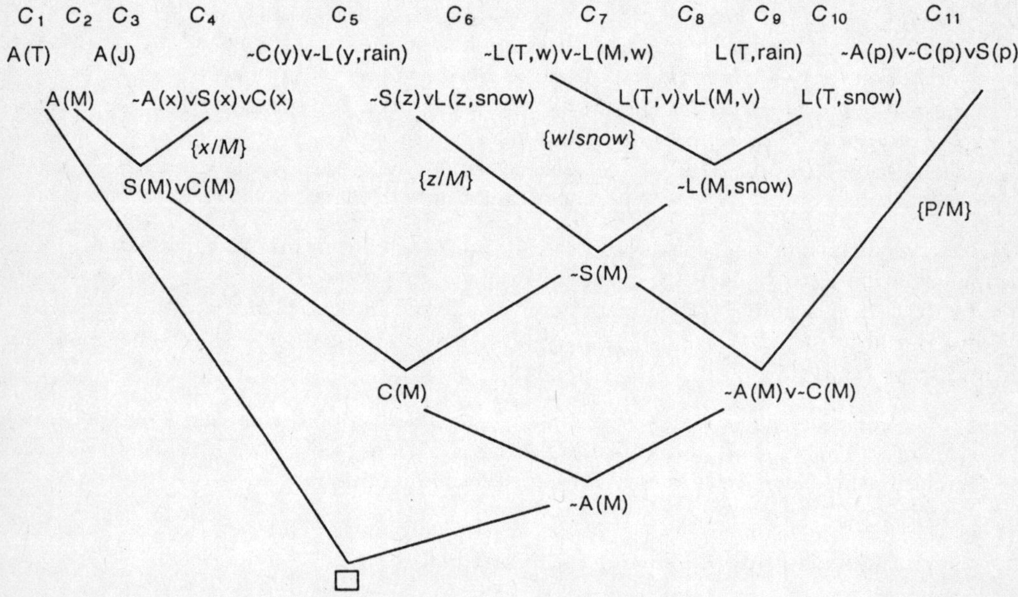

Figure 3: Refutation Tree

After these steps, all the formulae have been transformed into a normal form. At this point, before applying the resolution principle, we negate the question Q and add it as a clause to the set of normalized clauses. Our example becomes:

---

[2]The distributivity law is simply:

$$A \vee \{B \wedge C\} \equiv \{A \vee B\} \wedge \{A \vee C\}$$

$$F_1 : \quad A(Tony), A(Mike), A(John) \qquad C_1, C_2, C_3$$
$$F_2 : \quad \neg A(x) \lor S(x) \lor C(x) \qquad C_4$$
$$F_3 : \quad \neg C(y) \lor \neg L(y, rain) \qquad C_5$$
$$F_4 : \quad \neg S(z) \lor L(z, snow) \qquad C_6$$
$$F_5 : \quad \neg L(Tony, w) \lor \neg L(Mike, w) \qquad C_7$$
$$F_6 : \quad L(Tony, v) \lor L(Mike, v) \qquad C_8$$
$$F_7 : \quad L(Tony, rain), L(tony, snow) \qquad C_9, C_{10}$$
$$\neg Q : \quad \neg A(p) \lor \neg C(p) \lor S(p) \qquad C_{11}$$

We are now ready to apply unification and resolution to the transformed problem. This is done by generating the *refutation tree*. The *refutation tree* is the proof of the existence of a solution to the problem. It is obtained by negating the question Q, adding this negation (clause C_{11}) to the original set of clauses $S = \{C_1, \ldots, C_{10}\}$, and proving that the augmented set of clauses $\{C_1, \ldots, C_{11}\}$ is unsatisfiable. The augmented set of clauses is unsatisfiable if the result of applying unification and resolution to the augmented set yields the empty set. This is illustrated in Figure 3.

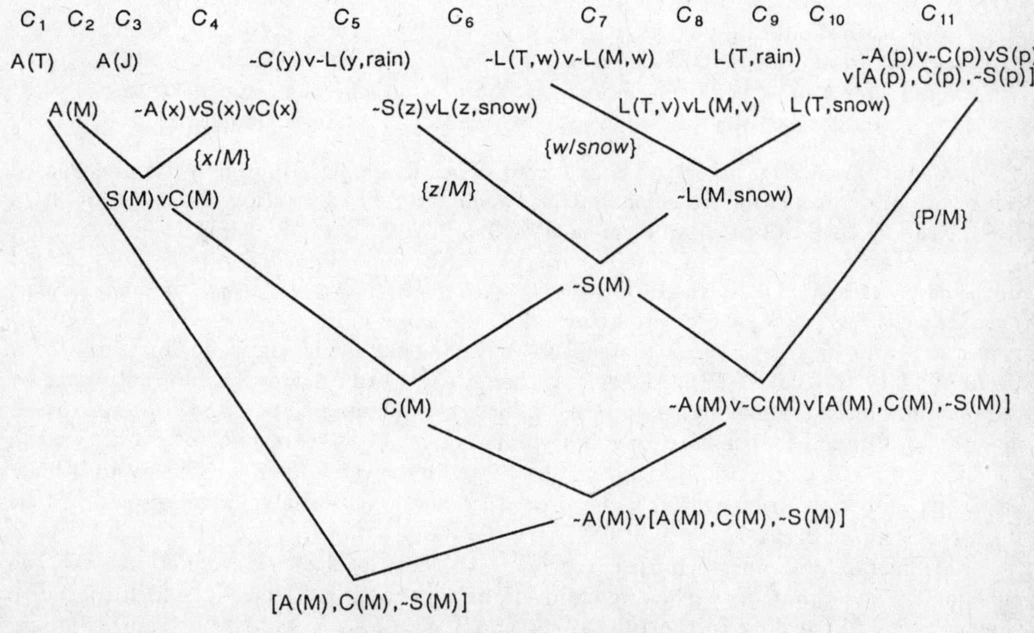

Figure 4: Modified Proof Tree

We have now proven that a solution exists. To obtain such a solution, a tautology is created by taking the disjunct of the original question Q with its negation $\neg Q$, i.e., $C_{11} = Q \lor \neg Q$. This tautology is added to the original set of clauses $S = \{C_1, \ldots, C_{10}\}$. The same sequence of unification and resolution steps used to generate the *refutation tree* will now produce the *modified proof tree* yielding the answer to the question Q. The is illustrated in Figure 4.

It should be noted that during the generation of the refutation tree (and subsequently during the generation of the modified proof tree) we did not use any strategy to guide the scarch for unifiers and resolvents. Had a link of the refutation tree been missed (due to the omission of a needed fact) it would have been rather difficult for a reasoning system based on resolution to identify the missing information and to elicit it from some other source, such as the user.

This type of problem solving mechanism is usually very suitable for proving the consistency of an existing set of constraints, as required, for instance, in program verification tasks. Resolution, on the other hand, is not well-suited for interactive tasks, such as diagnosis, in which the system tries to identify the most significant missing piece of evidence to complete the analysis.

2.3 Frames

The concept of frames was first introduced in by M. Minsky in a historical paper [16]. Frames are packets (higher level knowledge structures) that are part of a retrieval network. When one frame is invoked (accessed), other connected frames are considered potentially relevant and made available. In his original paper, Minsky defines a frame as:

> A *frame* is a data-structure for representing a stereotyped situation, like being in a certain kind of living room, or going to a child's birthday party. Attached to cach frame are several kinds of information. Some of this information is about how to use the frame. Some is about what one can expect to happen next. Some is about what to do if these expectations are not confirmed

The available knowledge structure is captured in the frame paradigm by grouping related information, by providing procedural attachments, and by connecting frames through an ISA hierarchy. Let us briefly analyze each of these three characteristics.

All the slots, i.e. pieces of information related to the same knowledge unit, are grouped in the same data structure. In the classic *restaurant frame*, concepts such as *menu, order, food, bill, waiter/waitress*, etc. are attached to the same unit.

A frame stores in the same data structure both the data and the procedures that can be applied to the data. Procedures attached to the same frame can be activated by requesting, adding, removing, modifying values in the frame slots. The procedures can generate additional information or modify existing one. In the *restaurant frame*, we can envision an *if-added* procedure that monitors the values in the *order* slot: for any additional value (i.e., any new ordered item), the procedure increments the value of the *bill* slot by its corresponding price.

The ISA hierarchy connects frames according to their level of generalization. We can envision the *restaurant frame* having children units in the ISA hierarchy, which specialize the restaurant according to the type of cuisine, e.g. *Chinese-restaurant frame, Italian-restaurant frame*, etc. The *restaurant frame* itself may have other sibling units such as *cafeteria frame, fast-food-parlor frame*, etc, all of which are specializations of a more abstract frame, the *eating-place frame*. The purpose of the ISA hierarchy is to share information across frames, so that partial results computed for a given frame need not be re-computed for specializations of the same frame. This sharing of information is called *inheritance*. For example, the procedure computing the value of the slot *bill* in the *restaurant frame* can be inherited by the *Chinese-restaurant frame* and the *Italian-restaurant frame* to be used when the computation of the bill must be performed.

Unlike logic theory, frame theory is not well defined. Minsky's paper has been a fruitful source of ideas, some of which have been captured, developed, or paraphrased in different implementations, such as KRL, KL-ONE, NICKL, etc. Many other researchers have defined similar concepts, such as scripts [17], schemas [18], and beta structures [19]. The concept of the ISA hierarchy has been originally introduced by Quillian [20] as part of the definition of semantic network, a knowledge representation paradigm that is very similar to the frame system.

Given the space and time constraints of our discussion, we will refer the reader to reference [21] for an excellent retrospective analysis of frame theory.

2.4 Production Rules

Production rules have been used as a notation in a variety of domains: in Post systems and Markov Algorithms they have described computational models; in the characterization of language types they have defined the generating grammars and their corresponding models of computation (from unrestricted grammar, i.e., Turing machines, to regular grammars, i.e., finite state machines); in Backus Naur Form (BNF) they have described the syntax of programming languages; in re-write rules they have been used to implement non-resolution theorem proving; in expert systems they have described the problem domain knowledge. Before introducing the use of production rules in expert systems, we will illustrate their use in Markov Algorithms [22]. Markov Algorithms (MA) represent a very simple and yet powerful application of production rules to compute functions. Furthermore, their role in MA is very similar to the role they play in expert systems.

2.4.1 Markov Algorithms

A Markov Algorithm is a simple programming language based on production rules and is computationally equivalent to a Turing machine. Each production rule causes the replacement of symbols in the input string. To avoid the indeterminacy caused by the applicability of multiple rules to the same input string, the execution of production rules is ordered according to the following meta-rules:

1. Execution test begins with rule 0

2. After the execution of an applicable rule, the next rule tested is rule 0

3. If for some i, $rule_i$ is found inapplicable, the next rule tested is $rule_{i+1}$

4. If no rule is applicable the algorithm terminates in a stalemate

5. If there is a period to the right of the right-hand side of an applicable rule, the algorithm terminates after the rule is applied

For a more formal description of Markov algorithms, see [23] (pages 263-264). In the sequel, we provide an illustrative example of MA.

2.4.2 Example of Markov Algorithm

Suppose we want to define algorithm $H(A)$. The purpose of $H(A)$ is to take any arbitrary substring of alphabet A as input and to make a copy of it, catenating it to the original string. For illustration purpose let us define alphabet $A = \{a, b\}$, the set of variables $V = \{\sigma, \eta\}$ and the set of markers $M = \{\alpha, \beta, \gamma\}$. The input string I is $\in A^*$, the variables in V can only take values $\in A$, the markers used in the production rules must be completely removed before $H(A)$ terminates, i.e., the output string O is $\in A^*$.

In our example $H(A)$ is defined by the following six production rules (where the symbol Λ denotes the empty string):

0: $\sigma\eta\beta \rightarrow \eta\beta\sigma$ (changes the order of the variables)
1: $\alpha\sigma \rightarrow \sigma\beta\sigma\alpha$ (makes a copy of the variable, marks it with β, shifts α to the right)
2: $\beta \rightarrow \gamma$ (changes β into γ)
3: $\gamma \rightarrow \Lambda$ (deletes γ)
4: $\alpha \rightarrow \Lambda.$ (deletes α and stops)
5: $\Lambda \rightarrow \alpha$ (begins by introducing marker α)

Given the input string I $= ab$, we can follow the following transformations:

$$ab \rightarrow^5 \alpha ab \rightarrow^1 a\beta a\alpha b \rightarrow^1 a\beta ab\beta b\alpha \rightarrow^0 a\beta b\beta ab\alpha \rightarrow^2 a\gamma b\beta ab\alpha \rightarrow^2 a\gamma b\gamma ab\alpha \rightarrow^3$$

$$ab\gamma ab\alpha \rightarrow^3 abab\alpha \rightarrow^4 abab$$

The analogy between MA and forward-chaining rule based expert system such as OPS5 [24], [25] is striking: the information is represented by symbols in the alphabet (facts in working memory); the input string (working memory) contains the original symbols and the results of all intermediate transformations (original facts and intermediate inferences); the inference mechanism is based on matching the left-hand side of a production rule with the input string (working memory) and on writing the symbols (facts) defined by the right-hand side of the applicable rule; the control of the inference to determine which rule to fire next is defined by the meta-rules (conflict resolution rules).

2.5 OPS5

OPS5 [24] is one of the earliest production rule based shells which has been used in a large number of applications (XCON,XSEl, etc) [25]. Its simple architecture is perhaps one of the most appealing factors is its widespread use.

OPS5 is composed of a global database (Working Memory), a set of rules (Knowledge Base) and a forward chaining inference engine with a simple control loop. We will briefly illustrate these components.

2.5.1 Knowledge Representation in OPS5

The language used to represent knowledge in OPS5 is succinctly described in [26] (pages 302-303) and is extensively analyzed in [25] (Chapter 2). The Working Memory is a set of vector of symbols or objects with associated attribute-value pairs.

The syntax of Working Memory is:

<Working Memory>   := <elements>*
<element>   := <vector-element>|<av-element>
<vector-element>   := $(\{$<value>$\}^+)$
<av-element>   := $($<object> $\{$ ˆ<attribute><value>$\}^+)$

Each rule in the rule set is composed of a Left-Hand Side (LHS), or antecedent, and a Right-Hand Side (RHS), or consequent. The rule LHS is compared against the elements in WM. If the match is successful and the rule is selected, the rule RHS is executed, changing the state of WM by writing new elements, or by modifying or deleting old ones. The syntax of a rule acting on objects of WM (*av-elements*) is:

```
<RULE SET>        := <rule>*
<rule>            := (P<rule-name> <antecedent>→<consequent>
<antecedent>      := <condition>*
<condition>       := (<object> { ^<attribute><value>}+)
<consequent>      := <action>+
<action>          := ( MAKE <object> { ^<attribute><value>}+) |
                     ( MODIFY <pattern-number> { ^<attribute><value>}+) |
                     ( REMOVE <pattern-number>) |
                     ( WRITE {<value>}+)
```

2.5.2 *OPS5 Inference Engine*

The control mechanism in OPS5 is a simple *recognize-select-act* loop which goes through the following steps:

1. Match the rules LHS with elements in WM, identify the rules with satisfied antecedents, and add them to as a set (referred to a the *conflict set*).

2. If the *conflict set* is empty, the inference engine stops. If the *conflict set* has only one rule, the rule is executed. If the *conflict set* is more than one rule, one rule is selected based on a set of meta-rules referred to as *conflict resolution* [27].

3. The actions of the selected rule are executed

4. Go back to the recognition phase described in step 1.

In this subsection we have tried to motivate the reader by appealing to the intrinsic simplicity of production rules. There are many other issues regarding the expressiveness of the rule language and the control of the inference engine that which have not been addressed in this discussion. In particular, we have only covered the data-driven use of the rules. An equally common rule firing strategy is backward chaining, extensively used in MYCIN [2] and other early versions of expert systems. A separate set of notes is devoted to a description of backward chaining.

3   Conclusions

We have briefly reviewed the characteristics of first and second generation KBS. We have described their commonality (separation of knowledge from control) and their differences (monotonic vs. nonmonotonic reasoning, static vs. dynamic problem domains, etc.) We have illustrated three different types of knowledge representation (predicate calculus, frames, and production rules) and we have described their corresponding inference mechanisms (resolution, inheritance, chaining).

References:

[1] William J. Clancey. Classification Problem Solving. In *Proceedings Fourth National Conference on Artificial Intelligence*, pages 49–55. AAAI, August 1984.

[2] E.H. Shortliffe and B. Buchanan. A Model of Inexact Reasoning in Medicine. *Mathematical Biosciences*, 23:351–379, 1975.

[3] B.G. Buchanan and E.H. Shortliffe. *Rule-Based Expert Systems*. Addison-Wesley, Reading, Massachusetts, 1984.

[4] Wendy B. Rauch-Hindin. *Artificial Intelligence in Business, Science, and Industry*. Prentice Hall, 1985.

[5] R.O. Duda, J. Gaschnig, P.E. Hart, K. Konolige, Reboh R., P. Barrett, and J. Slocum. Development of a computer-based consultant for mineral exploration. Sri project 5821 final report and sri project 6415 annual report, SRI International, Artificial Intelligence Center, Menlo Park, California, 1978.

[6] P. P. Bonissone. Delta: An expert system to troubleshoot diesel electric locomotives. In *Proceeding of ACM 83*, pages 44–45. ACM, New York, 1983.

[7] P. P. Bonissone and H. E. Johnson. Expert system for diesel electric locomotive repair. *Human Systems Management*, 4:255–262, 1984.

[8] Piero P. Bonissone and Nancy C Wood. Plausible Reasoning in Dynamic Classification Problems. In *Proceedings of the Validation and Testing of Knowledge-Based Systems Workshop*. AAAI, August 1988.

[9] Intellicorp. *KEE Software Development System User's Manual*, 1986. Version 3.0.

[10] Inference. *ART Reference Manual*, 1987. Version 3.0.

[11] Piero P. Bonissone, Stephen Gans, and Keith S. Decker. RUM: A Layered Architecture for Reasoning with Uncertainty. In *Proceedings 10th International Joint Conference on Artificial Intelligence*, pages 891–898. AAAI, August 1987.

[12] W.B. Gevarter. Expert Systems: limited but powerful. *IEEE Spectrum*, 20(8):39–45, 1983.

[13] N.J. Nilsson. *Problem-Solving Methods in Artificial Intelligence*. McGraw-Hill, 1971.

[14] Chin-Liang Chang and Richard Char-Tung Lee. *Symbolic Logic and Mechnaical Theorem Proving*. Acadermic Press, 1973.

[15] W.J. Rapaport. Predicate Logic. In Stuart Shapiro, editor, *Encyclopedia of Artificial Intelligence*, pages 538–544. John Wiley and Sons Co., New York, 1987.

[16] M. Minsky. A Framework for Representing Knowledge. In P. Winston, editor, *The Psychology of Computer Vision*, pages 211–277. McGraw-Hill, New York, 1975.

[17] R.Schank and R.P. Abelson. *Scripts, Plans, Goals, and Understanding*. Erlbaum, Hillsdale, N.J., 1977.

[18] F.C. Bartlett. *Remembering: A Study in Experimental And Social Psychology*. The University Press, Cambridge, 1932. Revised 1961.

[19] J. Moore and A. Newell. How can Merlin Understand? In L.W. Gregg, editor, *Knowledge and Cognition*, pages 201–252. Erlbaum, 1973.

[20] M.R. Quillian. Semantic Memory. In M. Minsky, editor, *Semantic Information Processing*. M.I.T. Press, Cambridge, MA., 1968.

[21] A.S. Maida. Frame Theory. In Stuart Shapiro, editor, *Encyclopedia of Artificial Intelligence*, pages 302–312. John Wiley and Sons Co., New York, 1987.

[22] A.A. Markov. The Theory of Algorithms (Russian). *Trudy Mathematicheskogo Instituta imeni V.A. Steklova*, 38:176–189, 1951. Translated from the Russian by J.J. Schorr-Kon, U.S. Department of Commerce, Office of Technical Services, number OTS 60-51085.

[23] H.R. Lewis and C. H. Papadimitriou. *Elements of the Theory of Computation*. Prentice Hall, 1981.

[24] C. Forgy and J. McDermott. OPS, a domain-independent production system language. In *Proc. Fifth Intern. Conf. on Artificial Intelligence*, pages 933–939, 1977.

[25] Lee Brownston, Robert, Elaine Kant, and Nancy Martin. *Programming Expert Systems in OPS5*. Addison-Wesley, Reading, Massachusetts, first edition, 1985.

[26] F. Hayes-Roth, D. Waterman, and D.Lenat (Eds.). *Builiding Expert Systems*. Addison-Wesley, 1983.

[27] J. McDermott and C. Forgy. Production System Conflict Resolution Strategies. In D. Waterman and F. Hayes-Roth, editors, *Pattern-Directed Inference Systems*, pages 177–199. Academic Press, New York, 1978.

A RULE BACKWARD CHAINER FOR
KNOWLEDGE BASED SYSTEMS (EXPERT SYSTEMS)

Piero P. Bonissone

Artificial Intelligence Program
General Electric Corporate Research and Development
Schenectady, New York 12301, USA

Abstract

Production rules are the most common knowledge representation paradigm used in Knowledge Based Systems (KBS). In these notes we will examine the structure of a rule Backward Chainer (BC), a particular rule firing strategy that is the basis for goal-directed search in the Knowledge Base. First, we will describe the recursive nature of a simple BC, without considering uncertainty. Then, we will extend the definition of the BC to a simple representation of uncertainty. Finally we will suggest enhancements for more complex versions of the BC.

1 Rule Representation Language

In another set of notes, we have briefly described the use of production rules in forward chaining mode, as used by OPS5 [1], [2], ART [3], and other systems. In these notes, we will focus on the recursive structure of a rule backward chainer (similar to the one used in MYCIN [4]) and we will illustrate its use with an example. The representation language used to write the rules will be simplified to reduce the example's complexity. As a result, we will only be concerned with *propositional* statements, i.e. statements with no variables or special symbols to constrain the pattern matcher.

1.1 Description of the Global Variables

Three global variables will be used in this example:

1. The *FACT_BASE*, representing the set of unconditional assertions that characterize the current problem, with its initial conditions (if any) and its partial or intermediate conclusions.

2. The *RULE_SET*, representing the set of conditional statements that define the knowledge in the problem domain.

3. The *HYPOTHESIS_SET*, representing the top-level goals to reach or prove.

The syntax of these global variables is defined as follows:

FACT BASE:

<*FACT_BASE*> := <fact>*
<fact> := <3_tuple> <cv>
<3_tuple> := <obj> <att> <val>
<obj> := *symbol*
<att> := *symbol*
<val> := *symbol*
<cv> := *real number* $\in$ [-1,1]

RULE BASE:

<*RULE_BASE*> := <rule>*
<rule> := <rule_id> <premise> <conclusion>
<premise> := <clause>*
<clause> := <3_tuple>
<conclusion> := <action>*(10368820)
<action> := <3_tuple> <cv>
<rule_id> := *symbol*
<3_tuple> := <obj> <att> <val>
<obj> := *symbol*
<att> := *symbol*
<val> := *symbol*
<cv> := *real number* $\in$ [-1,1]

HYPOTHESIS SET:

<*HYPOTHESIS_SET*> := <hypothesis>*
<hypothesis> := <3_tuple> <cv>
<3_tuple> := <obj> <att> <val>
<obj> := *symbol*
<att> := *symbol*
<val> := *symbol*
<cv> := *real number* $\in$ [-1,1]

1.2   Initialization of the Global Variables

To illustrate the syntax described in Subsection 1.1, we will initialize the global variables as follows:
FACT_BASE is initialized to *nil*, e.g. it is initially empty.

RULE_SET contains the following rules:

```
(R1:    ((animal has hair))                      (animal is carnivore)
        (((animal is mammal) 0.8)                (animal has dark_spots))
         ((animal is bird) -1.0)                (((animal is cheetah) 0.9)))
         ((animal is reptile) -1.0)))
                                         (R13:   ((animal is mammal)
(R2:    ((animal gives milk))                    (animal is carnivore)
        (((animal is mammal) 1.0)                (animal has black_stripes))
         ((animal is bird) -1.0)                (((animal is tiger) 0.9)))
         ((animal is reptile) -1.0)))
                                         (R14:   ((animal is mammal)
(R3:    ((animal lays eggs)                      (animal is carnivore)
         (animal has hard skin))                 (animal is pet ))
        (((animal is mammal) -1.0)              (((animal is dog) 0.9)))
         ((animal is bird) -1.0)
         ((animal is reptile) 1.0)))    (R15:   ((animal is reptile)
                                                 (animal is pet))
(R4:    ((animal lays eggs)                      (((animal is turtle) 0.7)))
         (animal can fly))
        (((animal is mammal) -1.0)      (R16:   ((animal is mammal)
         ((animal is bird) 1.0)                  (animal is ungulate)
         ((animal is reptile) -1.0)))            (animal has long_neck))
                                                 (((animal is giraffe) 1.0)))
(R5:    ((animal has feathers))
        (((animal is mammal) -1.0)      (R17:   ((animal is mammal)
         ((animal is bird) 1.0)                  (animal is ungulate)
         ((animal is reptile) -1.0)))            (animal has black_stripes))
                                                 (((animal is zebra) 0.9)))
(R6:    ((animal eats meat))
        (((animal is carnivore) 1.0)))(R18:    ((animal is mammal)
                                                 (animal can fly)
(R7:    ((animal has claws))                     (animal is ugly))
        (((animal is carnivore) 0.8)))          (((animal is bat) 0.9)))

(R8:    ((animal is mammal)             (R19:   ((animal is bird)
         (animal has hoofs))                     (animal flies well))
        (((animal is ungulate) 1.0)))           (((animal is albatross) 0.9)))

(R9:    ((animal is mammal)             (R20:   ((animal is bird)
         (animal chews cud))                     (animal runs fast))
        (((animal is ungulate) 0.7)))           (((animal is ostrich) 1.0)))

(R10:   ((animal lives with_people))
        (((animal is pet) 0.9)))

(R11:   ((animal lives at_zoo))
        (((animal is pet) -0.8)))

(R12:   ((animal is mammal)
```

HYPOTHESIS_SET contains the following hypotheses:

$$\begin{array}{ll}
\text{((animal is bat)} & 0.0) \\
\text{((animal is turtle)} & 0.0) \\
\text{((animal is cheetah)} & 0.0) \\
\text{((animal is tiger)} & 0.0) \\
\text{((animal is giraffe)} & 0.0) \\
\text{((animal is dog)} & 0.0) \\
\text{((animal is zebra)} & 0.0) \\
\text{((animal is ostrich)} & 0.0) \\
\text{((animal is albatross)} & 0.0)
\end{array}$$

2 Backward Chainer

2.1 Simple Backward Chainer (without uncertainty)

To better understand the structure of the Backward Chainer (BC), let us first assume that no uncertainty is present in either the *FACT_BASE* or the *RULE_BASE*. This can be achieved by changing the certainty values of the rules to 1.0 or -1.0 , according to their sign.

The purpose of the BC is to prove the hypotheses in the *HYPOTHESIS_BASE*. The BC will start from the first hypothesis and continue through the remaining ones, in their pre-determined order, until one hypothesis is proven true or all hypotheses have been examined.

The BC is a recursive procedure consisting of three basic steps:

1. *Ask the fact base*

2. *Try the rule base*

3. *Ask the user*

These steps are further described by the following cases:

Case 1: *CONDITION: The hypothesis (or sub-goal) is a member of the fact base*

If the hypothesis (or sub goal) R has previously been deduced, it will now be a fact in the fact base. The BC will search the fact base and, if successful, i.e., if $H \in$ *FACT_BASE*, the BC will return TRUE, unfolding the recursion; otherwise Case 1 fails.

The previous definition of success for Case 1, $H \in$ *FACT_BASE* could be rewritten as $F(H) \neq \phi$, where the subset $F(H)$ is defined as:

$$F(H) = \{\ fact \mid fact \in \ *fact\_base*,\ H = fact\ \}.$$

Case 2: *CONDITION: Case 1 failed, i.e., $F(H) = \phi$, but the hypothesis (or sub-goal) can logically be inferred (proven) by at least one rule in the rule base*

The rule set has already been sorted according to some pre-determined order (such as rule number or rule creation order). The BC will focus on the subset of rules whose *conclusion* contains the hypothesis (or sub-goal) H. Let $R(H)$ be such a subset, i.e.:

$$R(H) = \{\ rule \mid rule \in \text{*RULE\_SET*},\ H \in conclusion(rule)\ \}$$

Note that $R(H)$ does not change at run time. Therefore, at compile time, it is possible to determine the logical support for each hypothesis and compile this dependency information into a network to improve performance.

If $R(H) = \phi$, Case 2 fails. Otherwise, for each rule in $R(H)$, the BC will take the conjunction of clauses in its premise and, by calling itself recursively, will try to prove each and every clause in the order in which they appear in the premise, i.e., depth first. Because of the nature of the conjunction operator, if a clause were to fail, the remaining clauses in the same premise should be dropped. This procedure continues until hypothesis (or sub-goal) H is proven by one rule in $R(H)$, in which case the BC will write H in *FACT_BASE*, return TRUE and unfold the recursion, or until all the rules in $R(H)$ have been unsuccessfully tried, in which case the BC will return FAIL and unfold the recursion.

Case 3: *CONDITION: Case 1 and 2 failed:* $F(H) = \phi$, $R(H) = \phi$

(Note that Case 2 does not fail if there are rules which could prove hypothesis H, i.e. $R(H) \neq \phi$, but every rule in $R(H)$ happens to fail. Case 2 fails only if *there are no rules* which could potentially prove H, i.e., $R(H) = \phi$.

If H is a sub-goal, the BC will ask the user to determine whether or not H is TRUE. The answer is written in *FACT_BASE*, and its truth value is returned by the BC, unfolding the recursion.

If H is a top-level hypothesis, the BC will not ask the user and will display a message saying that such a hypothesis cannot be proven.

2.2 AND/OR Tree Analogy

The above procedure is analogous to generating an AND/OR tree. The root of the tree is the top level hypothesis H. The root is connected through an OR node to all the rules in $R(H)$, in their predefined order. Each rule is connected through an AND node to each clause in its premise. Each **clause** is now treated like the root: an OR node links the clause to the $R(clause)$, the subset of rules that could prove such a clause, an AND node links each of those rules to the clauses in their respective premises, etc.

The AND/OR tree is searched depth-first: if the clause under examination is not in *FACT_BASE*, its associated rule subset $R(clause)$, is examined. The frontier of the AND/OR tree represents the set of all possible questions that the user might be asked.

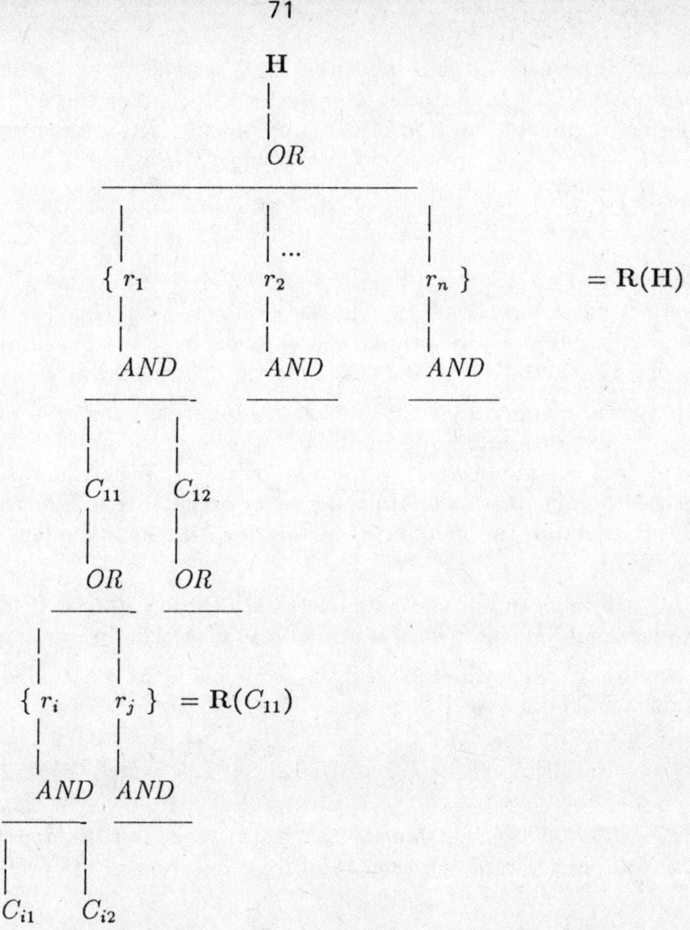

Figure 1: AND/OR Tree

2.3 Improved Backward Chainer (with Uncertainty)

Uncertainty is usually present in both knowledge base and facts. In this simple example, uncertainty is represented by *certainty values* (*cvs*), real numbers on the interval [-1,1]. The two extremes of the interval indicate CERTAINLY FALSE and CERTAINLY TRUE, respectively. 0, the mid-point of the interval, indicates UNKNOWN. The *cvs* are used to indicate the degrees of uncertainty attached to each statement. For unconditional statements (facts), the associated *cv* represents the degree to which that statement is believed to be true or false. For conditional statements (rules), the associated *cv* represents the degree to which the rule premise implies the rule conclusion. This certainty value is attached to the rule, assuming that the rule premise is completely satisfied by the environment. The functions required to aggregate and propagate certainty values are described in a separate section.

This improved version of the BC algorithm handles the certainty values of facts, rules, and hypotheses. For each hypothesis, the Backward Chainer will obtain a certainty value cv following an augmented version of the previously described recursive procedure. The augmented procedure is described by the following cases:

Case 1: *CONDITION: The hypothesis (or sub-goal) is a member of the fact base*

If the hypothesis (or sub-goal) **H** has previously been deduced, it will now be a fact in the fact base. Since each fact has an associated certainty value cv, it is necessary to define when a given hypothesis can be considered a *member* of the *fact base* (thus satisfying Case 1). To describe this condition the following definition is required:

Definition of Relevant Fact: a fact will be considered relevant to make an inference and therefore suitable to match a hypothesis (or sub-goal) only if the absolute value of its associated certainty value is greater or equal than β, i.e. $|cv(fact)| \geq \beta$. Threshold β indicates the *minimal degree of acceptance of a fact*. The absolute value is used to measure the *amount of certainty* with which the fact is considered to be true or false.

The BC will retrieve the associated certainty value $cv(fact)$. If $|cv| \geq \beta$, the BC will return it, unfolding the recursion; otherwise Case 1 fails.

The condition under which Case 1 succeeds could be written as $\mathbf{F(H)} \neq \phi$, where the subset $\mathbf{F(H)}$ is now defined as:

$$\mathbf{F(H)} = \{\ fact\ |\ fact \in \ *fact\_base*,\ H = fact,\ |cv(fact)| \geq \beta\ \}$$

Case 2: *CONDITION: Case 1 failed, i.e., $\mathbf{F(H)} = \phi$, but the hypothesis (or sub-goal) can logically be inferred (proven) by at least one rule in the rule base*

The rule set has already been sorted according to some pre-determined order (such as rule number or rule creation order). The BC will focus on the subset of rules whose *conclusion* contains the hypothesis (or sub-goal) **H**. However, rules can make conclusions about the given hypothesis (or sub-goal) with various degrees of certainty. Therefore, it is necessary to determine when the strength of the conclusion made by a rule is high enough to consider such a rule as part of the useful subset of rules. To describe this condition the following definition is required:

Definition of Useful Rule: If a rule is used to infer a hypothesis (or sub-goal), the computed certainty value associated with the inferred hypothesis (or sub-goal) will be a function of the degree of matching of the rule premise and the degree of implication of the rule. A detail explanation of this function will be given in next section. In the best of all cases, when the premise is completely satisfied, i.e., its aggregated certainty value is 1, the computed certainty value for the inference will only depend on the degree of implication associated with the rule. By establishing a threshold ϵ to indicate the *minimum acceptable level of implication* that a useful rule can provide, we can then define the subset of useful rules $\mathbf{R(H)}$ as:

$$\mathbf{R(H)} = \{\ rule\ |\ rule \in *\text{RULE\_SET}*,\ \mathbf{H} = action \in conclusion(rule),$$
$$|cv(action)| \geq \epsilon\}$$

If $R(H) = \phi$, Case 2 fails. Otherwise, for each rule in $R(H)$, the BC will take the conjunction of clauses in its premise and, by calling itself recursively, will try to prove each and every clause in the order in which they appear in the premise, i.e., depth first. Because of the nature of the conjunction operator, if a clause were to fail, the remaining clauses in the same premise should be dropped. A clause will fail if its computed certainty value is less than the predefined threshold β. This means that the clause is either false, unknown, or true to a very low degree. A more stringent constraint is to determine if the entire premise has not failed yet. This constraint is described under the heading of **Rule Firing** in the Control Issues section.

This procedure continues until hypothesis (or sub-goal) H is proven to a satisfactory degree by one or more rules in $R(H)$, or until all the rules in $R(H)$ have been tried. In the first case, the BC will write H (*and only H*)[1] in *FACT_BASE* with its computed certainty value,[2] will return such a certainty value and unfold the recursion.

In the second case, the BC will return the computed certainty value and unfold the recursion.

The determination of when a hypothesis (or sub-goal) has been sufficiently proven is described in the section on Control Issues.

Case 3: *CONDITION: Case 1 and 2 failed:* $F(H) = \phi$, $R(H) = \phi$

If H is a sub-goal, the BC will ask the user to provide its cv value. The user could just limit his/her answer to **Yes, Unknown, No**, which would then be translated to 1.0, 0.0, -1.0, respectively. Alternatively, the user could answer with a value on the scale [-1,1] that will be interpreted directly as the requested cv. The answer is written in *FACT_BASE* with its computed certainty value, then such a certainty value is returned by the BC, unfolding the recursion. Note that there is a potential problem with this approach. The user may be repetitively asked the same question to which he/she previously answered *Unknown*. This can be avoided by tagging the data structure with a label showing that the user already provided an answer.

If H is a top-level hypothesis, the BC will not ask the user and will display a message saying that such a hypothesis cannot be proven.

2.4   Aggregating And Propagating Certainty Values

Three types of functions must be defined to handle the certainty values of facts, sub-goals, rules, and hypotheses:

1. *CONJUNCTION* of cvs of clauses in the same premise

2. *PROPAGATION* of an aggregated cv through rule firing

3. *DISJUNCTION* of various cvs obtained from different rules

---

[1]If the conclusion of the rule to be fired contains other actions beside H, the BC should *only* write H and disregard all other actions.

[2]If two or more rule are asserting the same hypothesis or subgoal H, their certainty values will be combined according to the semantics of the operator corresponding to the OR node (see Section 2.4).

2.4.1 Conjunction

When a premise is formed by more than one clause, its certainty value (premise truth value) is the result of extending the intersection operation to the certainty values of its clauses. Based on the associativity of the conjunction, we have:

$$cv(premise_i) = \mathbf{T}(cv(C_{i1}), cv(C_{i2}), ..., cv(C_{in+1})) = \mathbf{T}(\mathbf{T}(cv(C_{i1}), ..., cv(C_{in})), cv(C_{in+1}))$$

For this example, we will use the *minimum* operator to aggregate the certainty values of the clauses in the same premise. The result is the degree to which the entire premise is satisfied by the information in *FACT_BASE*. This value is sometimes referred to as the *truth value* of the premise.

2.4.2 Propagation

This function is a special case of the conjunction. The function $\mathbf{W}(cv(premise_i), cv(rule_i))$ computes and propagates the certainty measure of the conclusion of $rule_i$. This weight represents the aggregation of the rule premise truth value $cv(premise_i)$, indicating the degree of fulfillment of the premise, with the strength of the rule $cv(rule_i)$, indicating the degree of weak implication of the rule. This function is only used with $cv(premise) \geq \delta(rule)$ and $cv(rule) \geq \epsilon$.

$$cv(y_i) = \mathbf{W}(cv(premise_i), cv(rule_i))$$

For this example, we will use the *product* operator to aggregate the certainty value of a premise with the certainty value of the rule itself. The result is the degree to which a given assertion can be inferred by that rule, given the current information in *FACT_BASE*.

2.4.3 Disjunction

When the same conclusion is made by more than one rule, its certainty value is the result of extending the union operation to the certainty values of the conclusion made by each rule. The function S(a,b) represents such an extension. Based on the associativity of the disjunction, we have:

$$cv(Y) = \mathbf{S}(cv(y_1), cv(y_2), ..., cv(y_{n+1})) = \mathbf{S}(\mathbf{S}(cv(y_1), ..., cv(y_n)), cv(y_{n+1}))$$

For this example, we can use the *maximum* operator to aggregate the certainty values of two or more rules regarding the same assertion $\mathbf{H}$. The result is the final degree to which a given conclusion can be asserted by the $\mathbf{R(H)}$, the subset of *RULE_SET* relevant to $\mathbf{H}$.[3]

---

[3]Note that the OR node is biased toward positive information, just as its dual (AND node) is biased toward negative information. This bias causes a stronger piece of negative evidence (e.g., -0.9) to be overridden by a weaker piece of information (e.g., -0.4). If we want to aggregate the certainty values on the basis of their strengths only, we could use the following operator:

$$\begin{array}{lll} S(a,b) & = a & \text{if } |a| > |b| \\ & = b & \text{if } |b| > |a| \\ & = \max\{a,b\} & \text{if } |b| = |a| \end{array}$$

S(a,b) takes the argument with the maximum absolute value, keeps its sign, and returns it with the maximum absolute value. The positive bias is used to distinguish between two equally strong pieces of evidence of opposite sign.

2.5 AND/OR Tree Analogy

The use of these aggregating and propagating functions to augment the BC algorithm, is analogous to generating a *parallel inference* structure, isomorphic with the AND/OR tree previously illustrated in Figure 1. This is described in Figure 2.

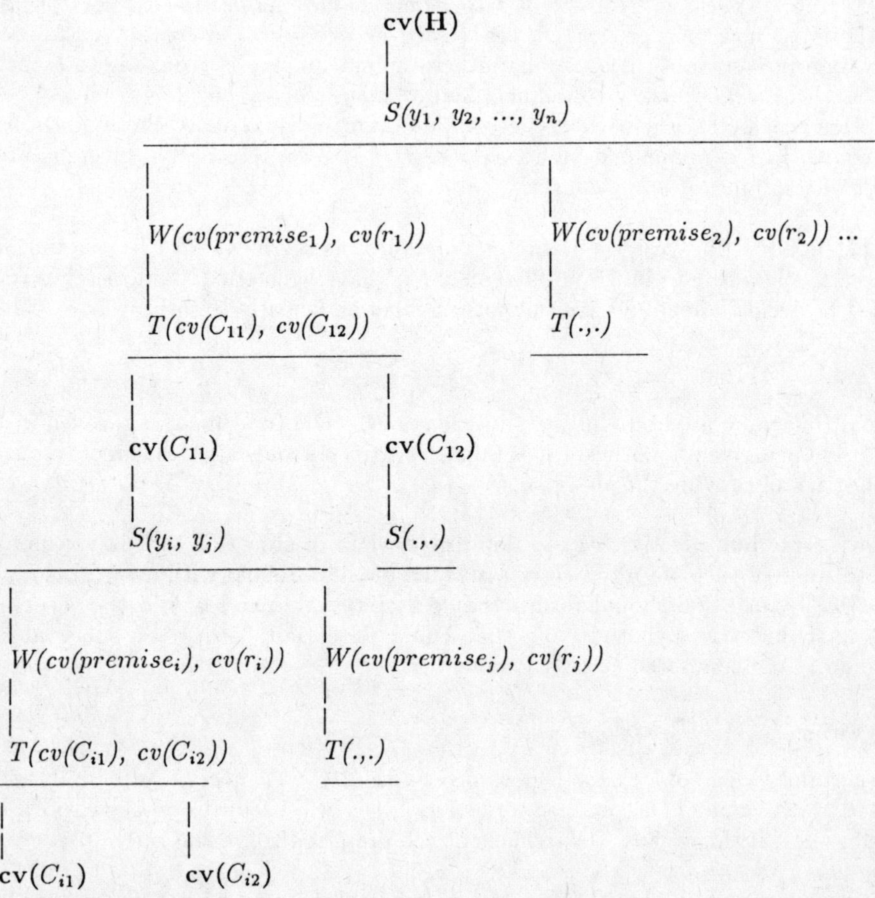

Figure 2: Parallel Inference Tree Isomorphic to AND/OR Tree

3 Control Issues

It is desirable to have an explicit, declarative representation of the control of inferences. In most first generation expert systems, however, the control of inferences is not in a declarative form but, for simplicity, is procedurally embedded in the inference engine. The following notes describe some relevant control issues, and suggest simple solutions to be implemented within the scope of this example.

3.1 Hypotheses

The purpose of BC algorithm is to prove one or more hypotheses from a given set, e.g., *HYPOTHESIS_SET*. When the hypotheses are *mutually exclusive*, as in the case of this example, it is useful to use this explicit knowledge to avoid needless inferences.

By setting a design parameter *me_hypo* to t one should enable the BC to stop after *one* hypothesis H_i has been proven true to a *satisfactory degree*. Because of the use of certainty values in facts and rules, any hypothesis might be proven true only to a partial degree. It is therefore necessary to define when a *satisfactory degree* has been achieved. This condition will be satisfied when the computed certainty value of the hypothesis is larger or equal than a predefined value, i.e., $cv(H_i) \geq \alpha$, where α is a given *minimal threshold for acceptance of a hypothesis*.

On the other hand, if the design parameter *me_hypo* is set to *nil*, the BC should only stop *after all* the hypotheses in *HYPOTHESIS_SET* have been tried. Parameter α could still be used to decide when any given hypothesis had sufficiently been proven.

3.2 Sub-goals

If a sub-goal (intermediate node in the equivalent AND/OR tree) must be proven in the process of proving a given hypothesis, it is important to control the allocation of resources spent in the task of proving the sub-goal.

A somewhat crude but effective way to determine when to stop this task is to compare the absolute value of the computed certainty value of the sub-goal with a threshold γ. If $|cv(sub\text{-}goal)| \geq \gamma$, the BC should return that certainty value and unfold the recursion; otherwise, it should try any other rule that could potentially prove the sub-goal and therefore modify the computed certainty value.

3.3 Rule Firing

When the certainty value of a clause is obtained by the BC, it is necessary to analyze the accumulated truth value of the premise to decide if it is still useful to proceed and prove the next clause or, if this was the last clause of the premise, to fire the rule.

In general, after the certainty value of each clause has been computed, e. g., $cv(clause_i)$, the truth value of the rule premise must be updated. This incremental update is based on the associativity of the aggregating function $T(.,.)$:

$$cv(premise) = T(cv(clause_i), cv(premise)).$$

The aggregated result must be compared with a threshold $\delta(rule)$ that indicates the minimum level of *meaningfulness of continuing the analysis or execution of a given rule*, i.e.:

$$T(cv(clause_i), cv(premise)) \geq \delta(rule).$$

If $T(.,.)$ is the *minimum* operator, it is sufficient to test that $cv(clause_i) \geq \delta(rule)$ to guarantee the success of the test.

Note that $\delta(rule)$ is not an independent parameter. Once β, the threshold to determine a *useful fact*, has been fixed, any meaningful addition of facts to *FACT_BASE* must have a certainty value whose absolute value is larger or equal to β. When a rule is fired, the computed certainty value of its conclusion is given by $W(cv(premise),\ cv(rule))$. If $W(.,.)$ is the *product* operator, we have that:

$$|cv(inferred\ statement)| = |cv(premise)\ cv(rule)| \geq \beta$$

Therefore:

$$|cv(premise)| \geq \delta(rule) = \beta\ \bigwedge |cv(rule)|$$

Any useful rule satisfies the condition $|cv(rule)| \geq \epsilon$. For consistency, we must also have $\epsilon \geq \beta$. Then $\delta(rule)$ is bounded by:

$$\beta \leq \delta(rule) = \beta/|cv(rule)| \leq \beta/\epsilon \leq 1$$

3.4   Summarizing the Thresholds

The thresholds used in the improved version of the BC algorithm are ordered as:

$$
\begin{array}{ccccccccc}
| & | & & | & | & | & | & | & \qquad | \quad | \\
-1 & -\gamma & & -\epsilon & -\beta & 0 & \beta, & \epsilon & \qquad \alpha,\gamma \quad 1
\end{array}
$$

where:

$\alpha =$ parameter to determine when a top-level hypothesis has been proven to a satisfactory degree such that no other hypothesis should be further analyzed (assuming the the hypotheses are mutually exclusive).

$\beta =$ parameter to determine when a fact (or is negation) has enough certainty to be relevant and useful in an inference and therefore suitable to be matched with a hypothesis or subgoal

$\gamma =$ parameter to determine when a fact (or its negation) has enough certainty that no further effort (i.e., trying more relevant rules) should be devoted in trying to improve its certainty

$\delta(rule) =$ parameter to determine if a premise has enough strength to trigger a rule. $\delta(rule) \in [\beta,\ \beta/\epsilon]$

$\epsilon =$ parameter to determine if a rule has enough strength to infer a given hypothesis (or sub-goal), assuming that the premise has been completely satisfied

For display's convenience, parameters α and γ are shown in the diagram with the same value. While this is a reasonable assignment, usually we have $\alpha \leq \gamma$.

4   An Exercise For The Reader

One of the best way to learn new concepts is to reduce them to practice. Thus, we would like to encourage the reader to implement the improved version of the BC following the specifications described in Section 2.3. A reasonable set of values for the parameters used to control the BC are:

$$
\begin{aligned}
me\_hypo &= \text{t} \\
\alpha &= 0.7 \\
\beta &= 0.2 \\
\gamma &= 0.85 \\
\epsilon &= 0.5 \\
\delta(rule) &= 0.2/cv(rule)
\end{aligned}
$$

To test the BC the reader is encouraged to use the data defined in Subsection 1.2.

5   Enhancements To the BC

There is a number of improvements that could be made to the above described BC. Some of them are described in the sequel.

1. The BC algorithm previously described could be enhanced by adding a *rule prescanner*. In the BC's current implementation, the premise of each rule (included in the subset of useful rules $\mathbf{R(H)}$) is tested *sequentially* and the verification of each clause in the premise generates a *depth-first* search through the AND/OR tree.

   Therefore, it is possible to devote resources, such as expanding AND/OR subtrees and asking the user, to verify the first clause in the premise, and later realize that another clause in the same premise was already *known to be false*. Unnecessary work was done in attempting to prove a rule that could never be fired.

   This problem can be solved by a *rule prescanner* that will perform a *breath-first* search on the premise clauses when that rule is being evaluated for firing. The breath-first search will insure that the premise will only contain clauses that are either *known true* or *still unknown*.

2. A second improvement could be made by preventing the situation described below. Given a subgoal P, let us assume that the BC's third case is successful (meaning that P is not known nor provable). Let us assume that the user's answer is *UNKNOWN (or cv = 0)*. The system will write P in the *FACT_BASE* with an associated cv of 0. If the same fact P is needed later by the BC, the system will not consider P to be a useful fact (case 1 fails), P is still not provable (the rule base has not changed), and the user will be asked again the same question about P.

   A possible solution is to augment the facts data structure with a *tag* field to be used in this situation (and others as well.)

3. A third improvement could be the compilation of the rule base into a network indicating the logical dependencies of rules, hypotheses, and sub-goals.

4. A fourth improvement could be the replacement of the implicit predicate in the clause of each rule premise (corresponding to "*Is it true?*") with an explicit predicate to be evaluated, e.g., *Is it false?, Is it unknown? Is it true? Is it in this range? Is it in this set?*, etc. This modification would allow to express more interesting tests beside symbol equality.

6 Conclusion

These notes have briefly described the recursive nature of a rule backward chainer (BC). For pedagogical reasons we have focused on the BC for propositional rules and we have finessed a variety of other difficult issues ranging from the use of predicate rules, to the process of rule compilation, to a more correct representation and propagation of uncertainty, to the creation of a development environment to monitor, trace and modify the rules in the KB. The reader is referred to another set of notes, entitled *RUM (Reasoning with Uncertainty Module) and RUMRUNNER (RUM's Run Time System)*, for an elaboration of these topics.

References:

[1] C. Forgy and J. McDermott. OPS, a domain-independent production system language. In *Proc. Fifth Intern. Conf. on Artificial Intelligence*, pages 933–939, 1977.

[2] Lee Brownston, Robert, Elaine Kant, and Nancy Martin. *Programming Expert Systems in OPS5*. Addison-Wesley, Reading, Massachusetts, first edition, 1985.

[3] Inference. *ART Reference Manual*, 1987. Version 3.0.

[4] E.H. Shortliffe and B. Buchanan. A Model of Inexact Reasoning in Medicine. *Mathematical Biosciences*, 23:351–379, 1975.

AN INDUSTRIAL FIRST GENERATION KBS
(EXPERT SYSTEM): DELTA/CATS

Piero P. Bonissone

Artificial Intelligence Program
General Electric Corporate Research and Development
Schenectady, New York 12301, USA

Abstract

Since the early eighties, General Electric Company's Corporate Research and Development has developed expert system technology and applied to a variety of problems. The first application addressed the problem of troubleshooting and repairing diesel electric locomotives in railroad *running repair shops*. In such shops, railroad maintenance personnel had to detect and repair a large variety of faults that partially disabled a diesel electric locomotive.

The expert system used production rules and an inference engine to diagnose multiple problems with the locomotive and to suggest repair procedures to maintenance personnel. In 1982, a prototype system was implemented in FORTH, running on a Digital Equipment PDP 11/73 under RSX-11M. This system contained approximately 620 rules (roughly 420 rules for the Troubleshooting System, and 200 rules for the Help System), partially representing the knowledge of a Senior Field Service Engineer. The inference engine used a mixed-mode configuration, capable of running in either the forward or backward mode. The Help System provided the operator with assistance by displaying textual information, CAD diagrams or repair sequences from a video disk. The rules were written in a representation language consisting of nine predicate functions, eight verbs, and five utility functions. Two copies of the first field prototype expert system, designated CATS (Computer-Aided Troubleshooting System), were delivered in July 1983 for field evaluation.

1 Introduction

During the last decade, numerous applications of expert systems have shown the versatility and usefulness of this emerging technology. The objectives of expert systems are to capture the knowledge of an expert in a particular problem domain, to represent it in a modular, expandable structure, and to transfer it to other users in the same problem domain. Expert systems or knowledge based systems can be subdivided into first and second generation types.

1.1 Characteristics of First Generation Expert Systems

The level of maturity of the first generation expert systems limits its applicability to the class of problems that are well scoped and static, for which there is available expertise, and for which the available information *completely and precisely* describes the knowledge domain. The first generation of expert systems that solve problems with the above characteristics generally exhibit monotonic reasoning, procedurally embedded reasoning control, *ad hoc* representation of uncertainty, shallow explanations, and brittle performance as they approach the frontiers of their knowledge bases.

1.2 Characteristics of Second Generation Expert Systems

More complex domains of expertise involving time-varying, partial and uncertain information cannot be addressed by the techniques common to first generation expert systems. Numerous research issues must be explored and understood, such as: analogues, naive physics, common sense reasoning, meta-knowledge and meta-reasoning, reasoning with uncertain and incomplete information, reasoning from first-principles, compilation and decompilation of know-how, learning, knowledge restructuring, rule breaking, i.e. determination of exceptions, determination of relevance and competence etc. Expert systems exhibiting subsets of the above capabilities are classified as second generation expert systems.

1.3 Classification Problems

The classification problem consists of recognizing a situation from a collection of data and selecting the best action in accordance with some objectives. Examples of classification include diagnosing faulty components, modeling users in terms of goals and beliefs, selecting components from a catalog of items in order to meet certain requirements, performing theoretical analysis, and developing skeletal plans. The classification problem has a recurrent solution structure, as was observed by Clancey [1]. A collection of data, generated from several sources, is interpreted as a predefined pattern. This is illustrated in Figure 1.

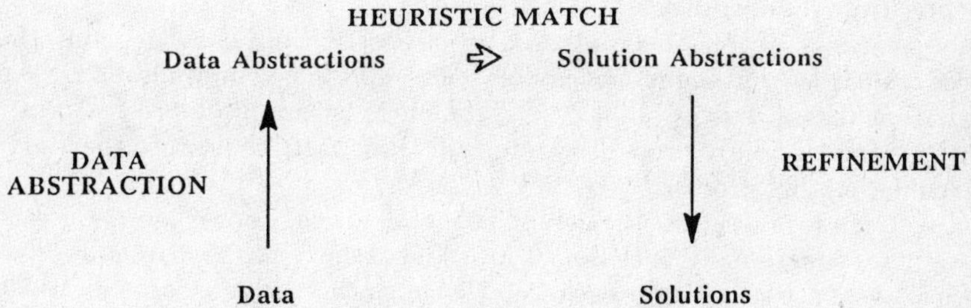

Figure 1: Static Classification Problem

The recognized pattern is mapped into a set of possible solutions, from which one is selected as the most appropriate for the given case. This process is considered a *static* classification problem, since the data are assumed to be invariant over time or at least invariant over the time required to obtain the solution.

The (static) classification problems are typically handled by first generation expert systems. In these notes we will illustrate our experience at addressing one of such problems. We will describe the development of DELTA/CATS, an expert system to aid railroad maintenance technicians in troubleshooting and repairing diesel electric locomotives.

In previous notes we have studied the requirements of dynamic classification problems, and we have discussed the new inference techniques that must be developed to address the increased complexity of reasoning with uncertain and incomplete information.

2 DELTA-CATS: History and Problem Description

Past experience at the General Electric Company's Corporate Research and Development Center [2], [3], [4] has led us in 1982 to the development of DELTA/CATS, an expert system to aid railroad maintenance technicians in troubleshooting and repairing diesel electric locomotives in *running repair shops*. The specific problem for which DELTA/CATS was designed is well-bounded and scoped: railroad maintenance personnel must detect and repair a large variety of faults that have partially disabled a diesel electric locomotive. However, only those problems that can be fixed in a limited and short amount of time, e.g., two hours, are considered. The *a priori* information available to the troubleshooters is the list of *symptoms* reported by the engine crew. More information can be gathered in the repair shop, by taking measurements and performing tests that may consume excessive shop time if performed by inexperienced personnel. The goal is to identify and repair those locomotives that can indeed be fixed within the assigned time-window.

3 Solution Description

The result of this development effort is a rule-based expert system that guides the troubleshooter in his task, enforcing some disciplined troubleshooting procedures that will minimize the cost and time of the corrective maintenance. This system contains approximately 620 rules partially representing the knowledge of a Senior Field Service Engineer. Roughly 420 rules are devoted to the fault diagnosis and repair procedures, i.e., the Troubleshooting System, while about 200 rules form the Help System. The Help System can provide the operator with assistance by displaying textual information, CAD diagrams or repair sequences from a video disk.

A fixed sequence of questions is used to gather the initial facts about the locomotive problem, such as unit number, model year, reported symptoms, etc. An associative information table provides additional facts, such as unit standard features, unit history of failures, model failure propensity, etc. All these facts constitute the starting point for the troubleshooting process.

The set of rules (heuristics) that embeds the empirical knowledge about the diesel electric engine is functionally partitioned into knowledge spaces such as mechanical system, electrical system, etc. Within each knowledge space, the rules are subdivided according to hypotheses (fault areas), such as Operator Error, Engine Unable to Make Power, etc. A set of meta-rules (a smart index of the knowledge partitions) retrieves from the various knowledge spaces the subsets of rules associated with all the hypotheses that could be relevant to the initial symptoms. This collection of hypotheses constitutes a preliminary diagnosis. Working in a backward mode, the interpreter tries to prove or disprove each hypothesis, based on both initial facts and additional facts inferred by the system or asked of the user.

The result of this process is a final diagnosis that indicates the successful hypotheses (faults) and their corresponding corrective actions (repairs).

4 Solution Organization

As many first generation expert systems, DELTA/CATS can be described as the result of programming in an *embedded language*. An embedded language has two components: a representation language, used to describe the problem domain knowledge, usually as production rules, and an inference engine, used to make deductions and derive conclusions to solve instances of the problem domain.

4.1 DELTA/CATS Inference Engine

The inference engine in DELTA/CATS consists of a *backward* and a *forward* chaining rule interpreter. The backward chaining interpreter is activated whenever a new hypothesis set is loaded into the hypothesis stack (dotted box in Figure 2). This may be caused by an external invocation of the backward interpreter, e.g., at the beginning of a session, by a recursive call from the backward chaining interpreter itself, during the evaluation of a sub-goal, or by executing the PROVE verb in the right-hand side of a *meta-rule* being fired by the forward chaining interpreter. The rule search order used by the backward interpreter defaults to the classic depth-first strategy.

Its evaluation of a hypothesis (goal) is a three-step process. First, the system scans the list of facts to verify whether the hypothesis is already known to be true or false. If this is the case, then evaluation terminates. Otherwise, the system scans the conclusion of each rule to determine whether the hypothesis could be proved by at least one rule. In such a case, the system recursively evaluates each clause (sub-goal) in the premise of that rule. Finally, when no hypothesis (or argument) can be directly inferred by a rule, the system requests information from an external source (either the user or a sensor).

The forward interpreter is activated only when a *new fact*[1] is written into the fact set. The new fact is then used by an efficient indexing scheme to retrieve and analyze only those rules (and/or meta-rules) that might have been affected by the new fact. If any of these rules has a completely satisfied premise, the forward interpreter will execute its right-hand side, will cause more new facts to be written and will iterate again.

As indicated in Figure 2, four types of rules are used by the forward chainer: *meta-rules, iff-rules, if-rules* and *when-rules*.

The *meta-rules* are used to control the focus of attention. They evaluate the need of activating and analyzing new knowledge, represented by a new set of hypotheses (dashed box in Figure 2) and their corresponding *iff-rules, if-rules* or *when-rules*.

The *iff-rules* (sufficient and necessary rules) and the *if-rules* (sufficient rules) try to find some new evidence that can be inferred directly, based on the presence of new fact. The new evidence could provide a shorter path in the deduction process. These rules can be accessed by both the backward and forward chainers.

---

[1] A new fact is a 3-tuple that did not previously exist in the fact set, or whose certainty measure had been modified, reflecting a more discriminant value.

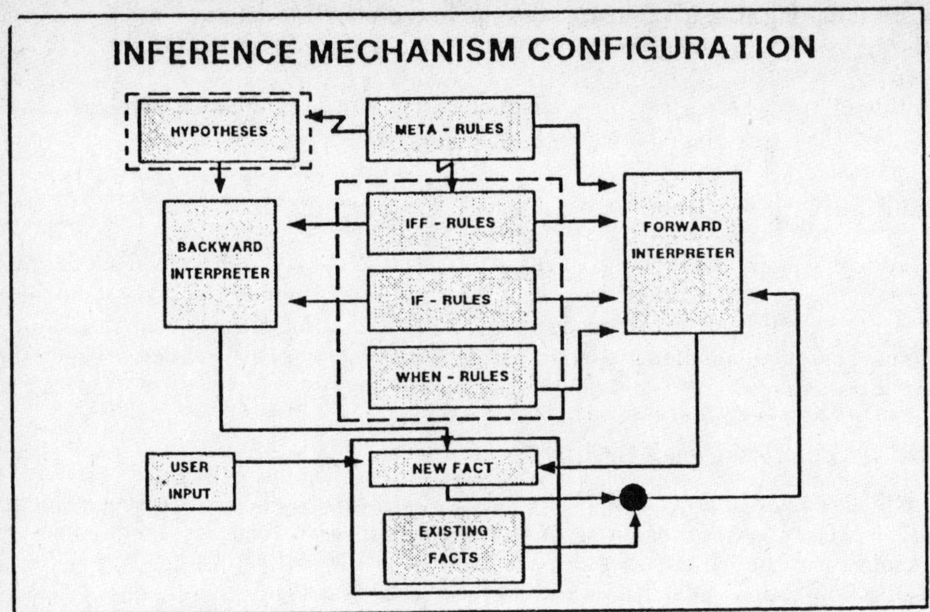

Figure 2: DELTA/CATS Inference Engine Configuration

The *when-rules* attach properties and activate procedures associated with the new fact. These rules are used to propagate the effects of actions taken during the repair procedure. They are also used to propagate dependency constraints to invalidate information that may have changed as a result of such actions. These rules can only be accessed by the forward chainer, thus preventing the backward chainer from using them to establish some goal.

If any of the above rules has a fully satisfied premise (since no explicit rule-chaining or user-prompting is allowed during the evaluation of rules in the forward mode), then the forward interpreter executes that rule, writes another new fact, and iterates again. This forward-chaining process stops when no rule can be executed by the forward interpreter, and control is returned to the backward interpreter.

The backward interpreter will continue its deductive process, until a hypothesis is proven or the entire hypothesis set has been exhaustively evaluated.

4.2 DELTA/CATS Rule Representation Language

The production rules that form the knowledge base of the Troubleshooting System and the Help System are written in a representation language. This user-extensible language currently contains:

- nine predicate functions to describe the conditions of the premise of each rule,

- eight verbs to describe the actions and inferences in the conclusions of each rule,

- five utility functions to interact with the user and display alphanumeric, graphic or pictorial information.

The rules have the traditional format of a conditional statement: IF <premise> THEN <conclusion> <cf>.

The weight "cf" is the certainty factor, a number in the interval [-1, 1], which indicates the strength of such implication. This number is used to control the propagation of uncertainty in rule-chaining, to control the combination of different pieces of evidence supporting the same conclusion, and to evaluate the overall degree to which a premise is satisfied.

Each premise is an intersection of clauses. Therefore, a premise is satisfied if *all* its clauses are also satisfied. In this case, the intersection of clauses corresponds to a boolean AND, or to its legal extensions, triangular norms. The minimum operator was used in DELTA/CATS for this purpose.

Each clause is defined by a *predicate function* and an argument composed of a 3-tuple <object attribute value>. Each clause is satisfied if its predicate function returns a true-value when applied to its argument.

Each conclusion is a sequence of actions that are executed once the premise of the rule has been satisfied. Each action is defined by a *verb* and an argument.

Moreover, five *utility functions* can be present in either the premise or the conclusion of the rule. These functions are transparent to the rule interpreter, in the sense that they do not affect the truth-value of the premise nor they modify the list of facts. The purpose of these functions is to help the user with text, graphics or video-images. These functions form the basis of a rule-driven help system.

4.2.1 Predicate Functions

The set of predicates in the representation language determines the tests that can be used to define a clause within a premise. Typical tests are: *is this argument true, false, or unknown? (using a subset of rules or all the rules), what is the most recent value for this argument?, which entry of this menu represents the observed value?*, etc.

The nine predicates used in the FORTH implementation are:

1. EQ (Equal): evaluates the argument, and returns a true-value if the argument was proven true.

2. EVAL-ALL: forces an exhaustive evaluation of the argument and returns a true value if the argument was proven true at least once during its evaluation.

3. NE (Not Equal): evaluates the argument and returns a true-value if the argument was proven false (Not Equal).

4. NC (Not Confirmed): evaluates the argument and returns a true value if the argument was either proven false or unknown (Not Confirmed).

5. ND (Not Disconfirmed): evaluates the argument and returns a true value if the argument was either proven true or unknown (Not Disconfirmed).

6. ASK-Y: prompts the user with a particular question (the comment associated with the clause), writes the argument as a new fact (with an attached certainty-factor indicating the user's response) and returns a true-value if an affirmative response was input.

7. ASK-N: like ASK-Y, but returns a true-value if a negative response was input.

8. UDO: requires the user to perform a given action (the comment associated with the clause), waits for a confirmation from the user, writes the argument as a new fact and returns a true-value if the action was confirmed.

9. MENU: displays a menu of choices, prompts the user for a specific menu entry selection, writes the selection as a new fact and returns a true value if a legal entry was selected.

4.2.2 Verbs

The set of verbs describes the actions that can be performed as a rule is triggered. Typical actions are: *write a new fact, delete or modify an existing fact, prove a hypothesis, issue a command, ask a question, stop the session,* etc.
The eight verbs used in the FORTH implementation are:

1. WRITE: writes the argument as a fact in the list of facts.

2. CLR: deletes from the list of facts any existing fact which matches the argument.

3. EVAL: activates the backward chaining interpreter, trying to verify the argument.

4. EVAL-ALL: activates the backward chaining interpreter, performing an exhaustive verification of the argument.

5. ASK: prompts the user with a particular question (the comment) and writes the argument (with an an attached certainty factor indicating the user's response) as a fact in the list of facts.

6. UDO: requires the user to perform a given action (the comment), waits for a confirmation and writes the effect of the action (the argument) as a fact in the list of facts.

7. STOP: displays a termination message (the comment) to the user and terminates the session, disregarding any pending tasks.

8. MENU: displays a menu of choices, prompts the user for a specific menu entry selection and writes the selection as a new fact.

4.2.3 Utility Functions

The set of utility functions determines the capability to manage the dialogue with the user by properly displaying text, graphics or video-images. These functions can be present in either premise or conclusion of a rule, and they are transparent to the rule interpreter since they do not affect the truth-value of the premise nor modify the list of facts. These functions form the basis of the rule-driven Help sub-system.
The five utility functions used in the FORTH implementation are:

1. DISPLAY: displays a message to the user.

2. PAUSE: displays a message and waits for an acknowledgment from the user.

3. SHOW: displays a CAD file (graphic picture) or an alphanumeric file on the user's terminal.

4. SCREEN: clears the graphic plane of the user's terminal.

5. DSHOW: displays a video-image (still frame or film sequence) on the auxiliary monitor.

5 DELTA/CATS Implementation and Test

By mid 1983, two copies of the above expert system were successfully deployed in the field. Each unit was implemented in a rugged configuration, packaged by COMARK, containing a PDP 11/73 board (running RSX-11M and an enhanced version of fig-FORTH), a 10 mega-byte Winchester disk, a VT100 terminal and a Selanar graphics board. A SONY laser-video-disk player and an additional color monitor completed the configuration of each field prototype system.

The system showed promising results since its delivery to the General Electric Company's Locomotive Operation in July 1983. The mixed-mode configuration of its inference engine performed very well. The FORTH implementation proved to be easily transportable to small micro-processor-based systems while maintaining fast execution speed. The man-machine interface was very user-friendly and allowed the user to interact with the system via menu selections or simple (single keystroke) answers such as: Yes, No, Unknown , Why?, Help.

As new hardware technology became available in 1984, the system went through a major evolution. DELTA/CATS knowledge base was re-written to run on a simpler inference engine written for IBM-PC. DELTA/CATS inference engine was re-implemented in Lisp to run on 32-bit microprocessor based workstations. The new Lisp based system, called DELPHI, has been used in over a dozen applications in Aerospace and Control [5].

5.1 Lessons Learned

The locomotive troubleshooting problem domain was relatively simple since it was well bounded, the information structure and values were complete and time-invariant, the uncertainty in the information was reduced to a minimum (*true, unknown,* or *false*), and the complexity of the problem was limited by the finite number of symptoms and faults to be considered. For this constrained problem, DELTA/CATS was an adequate tool. Its limitations, typical of the first generation expert systems, were the following:

- The premise of a rule had to be *completely satisfied* to allow the rule to be triggered. If only partial information were available, no inference could be made.

- The search strategy of the inference engine was fixed (depth-first) and the rule order was determined by the order in which the rules were compiled. This procedurally embedded control could not provide any opportunistic control strategy that would react to given facts by changing the priority and order of the rules.

- The time-invariance of the information only required monotonic reasoning. Any change in the truth value of previous statements, e.g., observed parameter values, resulting from actual changes in the physical world, e.g., parts repair or replacement, had small limited consequences that were described by production rules and maintained by the forward chainer.

5.2 Toward the Second Generation Expert Systems

The experience obtained from this application has guided our research efforts in knowledge based reasoning during the subsequent years. Our theoretical work has focused on the theory of reasoning with uncertainty [6],[7], [8], reasoning with incompleteness [9], [10], citeBrown88, reasoning by analogy, and reasoning under time constraints.

Our technology development efforts have been directed toward the integration of uncertainty and incompleteness [11], [12], the automatic compilation of knowledge bases (to improve performance and portability) [13], the development of object-based simulator (to be used in the KB functional validation) [14], [15], and the development of a meta-reasoner (to control the resource allocation for the reasoning system) [16].

Our application efforts have been directed by refinements of the class of problems defined as dynamic classifications [15], such as Pilot's Associate [17], Mergers and Acquisitions, Submarine Commander's Associate, etc.

References:

[1] William J. Clancey. Classification Problem Solving. In *Proceedings Third National Conference on Artificial Intelligence*, pages 49–55. AAAI, August 1984.

[2] P. P. Bonissone. Outline of the design and implementation of a diesel electric engine troubleshooting aid. In *Proceedings of Expert System 82*, pages 68–72. Brunel University, Egham, England, 1982.

[3] P. P. Bonissone. Delta: An expert system to troubleshoot diesel electric locomotives. In *Proceeding of ACM 83*, pages 44–45. New York, 1983.

[4] P. P. Bonissone and H. E. Johnson. Expert system for diesel electric locomotive repair. *Human Systems Management*, 4:255–262, 1984.

[5] J. James, Piero P. Bonissone, D. Frederick, and J. Taylor. A Retrospective View of CACE-III: Considerations in Coordinating Symbolic and Numeric Computation in a Rule-Based Expert System. In *Proceedings of the IEEE Second Conference on Artificial Intelligence Applications*, pages 532–538. IEEE, December 1985.

[6] Piero P. Bonissone and Keith S. Decker. Selecting Uncertainty Calculi and Granularity: An Experiment in Trading-off Precision and Complexity. In L. Kanal and J. Lemmer, editors, *Uncertainty in Artificial Intelligence*, pages 217–247. North-Holland, 1986.

[7] Piero P. Bonissone. Plausible Reasoning: Coping with Uncertainty in Expert Systems. In Stuart Shapiro, editor, *Encyclopedia of Artificial Intelligence*, pages 854–863. John Wiley and Sons Co., New York, 1987.

[8] Piero P. Bonissone. Summarizing and Propagating Uncertain Information with Triangular Norms. *International Journal of Approximate Reasoning*, 1(1):71–101, January 1987.

[9] Piero P. Bonissone and Allen L. Brown. Expanding the Horizons of Expert Systems. In T. Bernold, editor, *Expert Systems and Knowledge Engineering*, pages 267–288. North-Holland, 1986.

[10] Allen L. Brown Jr., Dale E. Gaucas, and Dan Benanav. An Algebraic Foundation for Truth maintenance. In *Proceedings 10th International Joint Conference on Artificial Intelligence*, pages 973–980. AAAI, August 1987.

[11] Piero P. Bonissone, Stephen Gans, and Keith S. Decker. RUM: A Layered Architecture for Reasoning with Uncertainty. In *Proceedings 10th International Joint Conference on Artificial Intelligence*, pages 891–898. AAAI, August 1987.

[12] P.P. Bonissone, D.A. Cyrluk, J.A. Goodwin, and J. Stillman. Uncertainty and incompleteness: Breaking the simmetry of defeasible reasoning. In *Proc. 5th. Workshop Uncertainty in Artificial Intelligence*, Windsor, Ontario, 1989.

[13] Lise M. Pfau. RUMrunner: Real-Time Reasoning with Uncertainty. Master's thesis, Rensselaer Polytechnic Institute, December 1987.

[14] Piero P. Bonissone and James K. Aragones. LOTTA: An Object Based Simulator for Reasoning in Antagonistic Situations. In *Proceedings of the 1988 Summer Computer Simulation Conference*, pages 674–680. SCS, July 1988.

[15] Piero P. Bonissone and Nancy C Wood. Plausible Reasoning in Dynamic Classification Problems. In *Proceedings of the Validation and Testing of Knowledge-Based Systems Workshop*. AAAI, August 1988.

[16] P.P. Bonissone. Now that I have a good theory of uncertainty, what else do I need? In *Proc. 5th. Workshop Uncertainty in Artificial Intelligence*, Windsor, Ontario, 1989.

[17] L. M. Sweet, P. P. Bonissone, A. L. Brown, and S. Gans. Reasoning with Incomplete and Uncertain Information for Improved Situation Assessment in Pilot's Associate. In *Proceedings of the 12th DARPA Strategic Systems Symposium*. DARPA, October 1986.

Specific Issues Relevant for Applications of Expert Systems in Structural Engineering

UNCERTAINTY IN KBS (EXPERT SYSTEMS)

Piero P. Bonissone

Artificial Intelligence Program
General Electric Corporate Research and Development
Schenectady, New York 12301, USA

Abstract

The management of uncertainty in expert systems has usually been left to *ad hoc* representations and combining rules lacking either a sound theory or clear semantics. However, the aggregation of uncertain information (facts) is a recurrent need in the reasoning process of an expert system. Facts must be aggregated to determine the degree to which the premise of a given rule has been satisfied, to verify the extent to which external constraints have been met, to propagate the amount of uncertainty through the triggering of a given rule, to summarize the findings provided by various rules or knowledge sources or experts, to detect possible inconsistencies among the various sources, and to rank different alternatives or different goals.

There is no uniformly accepted approach to solve this issue. A variety of approaches will be described as part of the review of the state of the art in reasoning with uncertainty. We will examine *quantitative* and *qualitative* characterizations of uncertainty. Among the quantitative approaches, we will cover *single-valued* approaches (Bayes Rule, Modified Bayes Rule, Confirmation Theory); *interval-valued* approaches (Dempster-Shafer Theory, Probability Bounds, Evidential Reasoning); and *fuzzy-valued* approaches (Necessity and Possibility Theory). Among the *qualitative* approaches we will exmine formal (Reasoned Assumptions) and heuristic (Endorsements).

These approaches will then be evaluated according to a list of criteria that reflect a crucial set of requirements common to a large variety of problems.

1   Introduction

These notes summarize two papers recently written by the author. The first paper, entitled *"Plausible Reasoning: Coping with Uncertainty in Expert Systems* [1], describes the state of the art of reasoning systems that represent and propagate uncertainty throughout their inferences. The second paper, entitled *"Summarizing and Propagating Uncertain Information with Triangular Norms* [2], defines a set of evaluation criteria for systems that reason with uncertain information.

2 Sources Of Uncertainty

In a survey of reasoning with uncertainty [3], it is noted that the presence of uncertainty in reasoning systems is caused by a variety of sources: the *reliability* of the information, the inherent *imprecision* of the representation language in which the information is conveyed, the *incompleteness* of the information, and the *aggregation* or summarization of information from multiple sources.

The first source type is related to the *reliability of information:* uncertainty can be present in the factual knowledge (i.e., the set of assertions or facts) as a result of ill-defined concepts in the observations (fuzziness), or due to inaccuracy and poor reliability of the instruments used to make the observations (randomness)[1]

Uncertainty can also occur in the knowledge base (i.e., the rule set) as a result of weak implications when the expert or model builder is unable to establish a strong correlation between premise and conclusion. In most expert systems the degree of implication is artificially expressed as a scalar value on an interval (certainty factor, probability, etc.). This value represents the change from the strict implication *for all x, $A(x) \rightarrow B(x)$*, to the weaker statement *for most x*, or *usually, for all x, $A(x) \rightarrow B(x)$*. The latter statement is not categorical and allows the possibility of exceptions to the rule. Thus the logical implication has now been changed into a plausible implication or *disposition* [4], [5]. A natural way to express such a degree of implication is achieved by using fuzzy quantifiers such as *most, almost all*, etc. [6], [7]. [2]

Uncertainty in the data can be compounded by aggregating uncertain data in the premise, by propagating certainty measures to the conclusion, and by consolidating the final certainty measure of conclusions derived from different rules. Triangular norms and conorms [8], [9] can be used to generalize the conjunction and disjunction operators that provide the required aggregation capabilities. A description of their characteristics is provided in reference [10].

The second type of uncertainty is caused by the *inherent imprecision* of the rule representation language. If rules are not expressed in a formal language, their meaning cannot be interpreted exactly. This problem has been partially addressed by the theory of *approximate reasoning* that, in light of imprecise fact and rule descriptions, allows one to make weaker inferences based on a *generalized modus ponens* [11].

The third type of uncertainty is caused by the *incompleteness* of the information. This type of uncertainty has generally been modeled by non-numerical characterizations, such as Doyle's Reasoned Assumptions [12].

The fourth type of uncertainty arises from the *aggregation of information* from different knowledge sources or experts. When unconditional statements (facts) are aggregated, three potential problems can occur: the closure of the representation may no longer be preserved when the facts to be aggregated have *different granularity* (the single-valued certainty measures of the facts may change into an interval-valued certainty measure of the aggregated fact); the aggregation of conflicting statements may generate a *contradiction* that should be detected; the rule of evidence combination may create an over-estimated

---

[1] Randomness and fuzziness reflect different aspects of uncertainty. Randomness deals with the uncertainty of whether a given element belongs or does not belong to a well-defined set (event). Fuzziness deals with the uncertainty derived from the partial membership of a given element to a set whose boundaries are not sharply defined.

[2] A fuzzy quantifier is a fuzzy number representing the relative cardinality of the subset of elements in the universe of discourse that *usually* satisfy the given property, i.e., the implication.

certainty measure of the aggregated fact, if a normalization is used to eliminate or *hide* a contradiction [13], [14]. The first two problems are typical of single-valued numerical approaches, while the last problem is found in the two-valued approach proposed by Dempster [15]. All these approaches will be discussed in the following section.

3 State Of The Art In Reasoning With Uncertainty

The existing approaches to represent uncertainty are divided into two basic categories: *numerical* characterizations and *symbolic (i.e., non-numerical)* characterizations.

Among the numerical approaches, there are three distinct types: the *one-valued*, the *two-valued*, and the *fuzzy-valued* approach. Some of the more traditional techniques found among the one-valued approaches are: Bayes Rule [16], [17], [18], Modified Bayes Rule [19] and Confirmation Theory [20]. A more recent trend is exemplified by the two-valued approaches: Dempster-Shafer Theory [15], [21], [22], Evidential Reasoning [23], [24], [25], Consistency and Plausibility (Probability Bounds) [26], and Evidence Space [27]. The fuzzy-valued approaches include the Necessity and Possibility Theory [28], [29] and the Linguistic Variable Approach [30], [11], [31]. With numerical representations, it is possible to define a *calculus* that provides the mechanism for propagating the uncertainty of the information through the reasoning process. Similarly, the use of aggregation operators provides a summary of the information, which can then be ranked against other such summaries to performed rational decisions. Such a numerical representation, however, cannot provide a clear explanation of the *reasons* that led to a given conclusion.

Models based on symbolic representations, on the other hand, are mostly designed to handle the aspect of uncertainty derived from the *incompleteness* of the information. They are generally inadequate to handle the case of *imprecise* information, since they lack any measure to quantify confidence levels [12]. Among the symbolic characterizations, there are two distinct approaches: the *formal* approach, such as Reasoned Assumptions [12], and Default Reasoning [32] and the *heuristic* approach, such as the Theory of Endorsements [33], [34], [35].

The formal approach has a corresponding logic theory that determines the mechanism by which inferences (theorems) can be *proven* or *believed* to be true. The heuristic approach has a set of context-dependent rules to define the way by which frame-like structures (endorsements) can be combined, added or removed.

The symbolic representations are more suitable for providing a trace from the sources of the information through the various inference paths to the final conclusions. However, no calculus can be defined for the propagation, aggregation, and ranking of such uncertain information. The only available partial solution is the use of context-dependent rules to determine how each piece of evidence can be compared or summarized.

In this paper, selected approaches to the representation of uncertainty will be illustrated and discussed.

3.1 Bayes Rule

Given a set of hypotheses $\mathbf{H} = \{ h_1, h_2, \ldots, h_n \}$ and a sequence of pieces of evidence $\{ c_1, c_2, \ldots, c_m \}$, Bayes rule, derived from the formula of conditional probability, states that the posterior probability $P(h_i \mid c_1, c_2, \ldots, c_m)$ can be derived as a function of the conditional probabilities $P(e_1, e_2, \ldots, e_m \mid h_i)$ and the prior probability $P(h_i)$:

$$P(h_i \mid e_1, e_2, \ldots, e_m) = \frac{P(e_1, e_2, \ldots, e_m \mid h_i) \cdot P(h_i)}{\sum_{i=1}^{n} P(e_1, e_2, \ldots, e_m \mid h_i) \cdot P(h_i)} \tag{1}$$

The Bayesian approach is based on two fundamental assumptions:

- Each hypothesis h_i is mutually exclusive with any other hypothesis in the set $\mathbf{H}$ and the set of hypotheses $\mathbf{H}$ is exhaustive, i.e.:

$$P(h_i, h_j) = 0 \text{ for } i \neq j \tag{2}$$

$$\sum_{i=1}^{n} P(h_i) = 1 \tag{3}$$

- Each piece of evidence e_j is conditionally independent under each hypothesis, i.e.:

$$P(e_1, e_2, \ldots, e_m \mid h_i) = \prod_{j=1}^{m} P(e_j \mid h_i) \tag{4}$$

Note that assumptions 2 and 3 are required to derive Bayes Rule from the formula of conditional probability. Assumption 4, on the other hand, is usually made to alliviate the difficulty of determining the conditional joint probability required by equation 1. Thus, under assumption 4, equation 1 becomes computationally feasible.

This method requires a large amount of data to determine the estimates for the prior and conditional probabilities. Such a requirement becomes manageable when the problem can be represented as a *sparse* Bayesian network that is formed by a hierarchy of small cluster of nodes. In this case the dependencies among variables (nodes in the network) are known and only the explicitly required conditional probabilities must be obtained [17]. The Bayesian approach still exhibits all the same shortcomings discussed in the Modified Bayesian approach.

3.2 Modified Bayes Rule

In addition to assumptions 2 and 3 (for derivational needs) and assumption 4 (for operational convenience) needed by the original Bayes Rule, the Modified Bayesian approach, used in PROSPECTOR, also requires that each piece of evidence e_j be conditionally independent under the *negation* of each hypothesis, i.e.:

$$P(e_1, e_2, \ldots, e_m \mid h_i) = \prod_{j=1}^{m} P(e_j \mid \neg h_i) \tag{5}$$

The Modified Bayesian approach is based on a variation of the odds-likelihood formulation of Bayes rule. When all the pieces of evidence are *certainly true*, this formulation defines the posterior odds as:

$$\begin{aligned}
O(h_i \mid e_1, e_2, \ldots, e_m) &= \frac{P(e_1 \mid h_i)}{P(e_1 \mid \neg h_i)} \frac{P(e_2 \mid h_i)}{P(e_2 \mid \neg h_i)} \cdots \frac{P(e_n \mid h_i)}{P(e_n \mid \neg h_i)} \frac{P(h_i)}{P(\neg h_i)} \\
&= \lambda_{1,i} \lambda_{2,i} \cdots \lambda_{n,i} O(h_i)
\end{aligned} \tag{6}$$

where:
$\lambda{j,i} = \frac{P(e_j|h_i)}{P(e_j|\neg h_i)}$ is the *likelihood ratio* of e_j for hypothesis h_i.

$O(h_i) = \frac{P(h_i)}{P(\neg h_i)}$ is the *odds* on hypothesis h_i.

An analogous odds-likelihood formulation is derived for the case when all the pieces of evidence are *certainly false*:

$$\begin{aligned}
O(h_i \mid \neg e_1, \neg e_2, \ldots, \neg e_m) &= \frac{P(\neg e_1 \mid h_i)}{P(\neg e_1 \mid \neg h_i)} \cdot \frac{P(\neg e_2 \mid h_i)}{P(\neg e_2 \mid \neg h_i)} \cdots \frac{P(\neg e_n \mid h_i)}{P(\neg e_n \mid \neg h_i)} \cdot \frac{P(h_i)}{P(\neg h_i)} \\
&= \lambda_{1,i}^* \cdot \lambda_{2,i}^* \cdot \ldots \cdot \lambda_{n,i}^* \cdot O(h_i)
\end{aligned} \tag{7}$$

The likelihood ratio $\lambda{j,i}$ measures the *sufficiency* of a piece of evidence e_j to prove hypothesis h_i. Similarly, $\lambda_{j,i}^*$ measures the *necessity* of such a piece of evidence to prove the given hypothesis (12).

Formulae 6 and 7 assume that evidence e_j is precise (i.e., $P(e_j) \in \{0,1\}$). This is not the case in most expert system applications. Therefore, the above formulae must be modified to accommodate uncertain evidence. This is accomplished by using a linear interpolation formula. For the case of single evidence, the posterior probability $P(h_i \mid e_j')$ is computed as:

$$P(h_i \mid e_j') = P(h_i \mid e_j) \cdot P(e_j \mid e_j') + P(h_i \mid \neg e_j) \cdot P(\neg e_j \mid e_j') \tag{8}$$

where $P(e_j \mid e_j')$ is the user's assessment of the probability that the evidence e_j is true, given the relevant observation e_j'. An *effective likelihood ratio*, $\lambda_{j,i}'$, is calculated from the posterior odds:

$$\lambda_{j,i}' = \frac{O(h_i \mid e_j')}{O(h_i)} \tag{9}$$

The posterior odds for *all* the evidence is then computed as:

$$O(h_i \mid e_1', e_2', \ldots e_m') = O(h_i) \prod_{j=1}^{m} \lambda_{j,i}' \tag{10}$$

Equation 8, however, requires a modification, because it over-constrains the input requested from the user. In fact, the user must specify:

- $O(h_i)$, the prior odds on h_i from which $P(h_i)$ can be derived

- $\lambda_{j,i}$, the measure of sufficiency from which $P(h_i \mid e_j)$ can be derived

- $\lambda_{j,i}^*$, the measure of necessity from which $P(h_i \mid \neg e_j)$ can be derived

- $O(e_j)$ the prior odds on e_j from which $P(e_j)$ can be derived

These requirements are equivalent to specifying a line in the space $[\, P(e \mid e'), P(h_i \mid e')\,]$ by specifying *three* points:

$$(0, P(h_i \mid \neg e_j)), \; (P(e_j), P(h_i)), \; (1, P(h_i \mid e_j))$$

The modification adopted in this approach to prevent the user's inconsistencies is to change equation 8 into a piece-wise linear function defined by two line segments passing through the above three points [19].

In an analysis of this approach, Pednault, Zucker, and Muresan [36] concluded that for the cases of *more than two* hypotheses, assumptions 4 and 5, requiring conditional independence of the evidence both under the hypotheses and thir negation, were inconsistent with assumptions 2 and 3, requiring an exhaustive and mutually exclusive space of hypotheses. Specifically, Pednault proved that, under these assumptions, no probabilistic update could take place, i.e.:

$$P(e_j \mid h_i) = P(e_j \mid \neg h_i) = P(e_j) \ \forall i, j \tag{11}$$

However, Glymour [37] obtained a pathological counter-example to Pednault's statement (equation 11), finding a fault in the original proof of Hussain's theorem that constituted the basis for Pednault's results. Johnson [38] extended this analysis by first showing that there are also non-pathological counter-examples that refute Pednault's results. However, Jonhson proved that under the same assumptions used in Pednault's work, for every hypothesis h_i there is *at most* one piece of evidence e_j that produces updating for h_i. Further studies done by Cheng and Kashyap [39] have also indicated that there are at least $max[0, (m - \lfloor \frac{n}{2} \rfloor)]$ pieces of evidence that are *irrelevant*[3] to *all* the hypotheses in the system. This lower bound is for a system satisfying assumptions (4) and (5), in which n is the number of mutually exclusive exhaustive hypotheses ($n > 2$), and m is the number of evidence. Their conclusion is that assumption 5 should be dropped.

Pearl has also argued in reference [17] that assumption 5, requiring the conditional independence of the evidence under the negation of the hypotheses, is over-restrictive. By discarding this assumption, Pearl has derived new, more promising results. However, the assumption 4, requiring the conditional independence of the evidence under the hypotheses, was still required.

The Bayesian approach has various shortcomings. The assumptions on which it is based are not easily satisfiable, e.g. if the network contains multiple paths linking a given evidence to the same hypothesis, the independence assumptions 4 and 5 are violated. Similarly, assumptions 2 and 3, requiring the mutually exclusiveness and exhaustiveness of the hypotheses, are not very realistic: assumption 2 would not hold if more than one hypothesis could occur simultaneously and is as restrictive as the single-fault assumption of the simplest diagnosing systems; assumption 3 implies that every possible hypothesis is *a priori* known, and it would be violated if the problem domain were not suitable to a close-world assumption. Perhaps the most restrictive limitation of the Bayesian approach is its inability to represent *ignorance* (i.e., non-commitment) as illustrated by its *two-way betting* interpretation [40]. The two-way betting interpretation of the Bayesian approach consists of regarding the assignment of probability p to event A as the willingness of a rational agent to accept any of the two following bets:

- *If you pay me $ p then I agree to pay you $ 1 if A is true*

- *If you pay me $ (1-p) then I agree to pay you $ 1 if A is false*

The first bet represents the belief that the probability of A is not larger than p, the second bet represents the belief that the probability of A is not smaller than p.

Instead of being explicitly represented, ignorance is *hidden* in prior probabilities. Further shortcomings are represented by the fact that it is impossible to assign any probability to disjunctions, i.e., to non-singletons, which implies the requirement for a uniform

---

[3] An evidence e_j is said to be irrelevant to the hypothesis h_i if $P(h_i \mid e_j) = P(h_i)$.

granularity of evidence. This problem is usually solved with an approximation, using the Maximum Entropy Principle (MEP). According to MEP, the probability assigned to the disjunct (a subset of singletons in the sample space) is equally divided among the singletons in the subset. This approximation, however, creates an interpretation of the original information, which may not always been appropriate. Finally, as was pointed out by Quinlan [26], in this approach conflictive information is not detected but simply propagated through the network.

3.3   Confirmation Theory

The Certainty Factor (CF) approach [20], used in MYCIN, is based on Confirmation Theory. The certainty factor $CF(h,e)$ of a given hypothesis h is the difference between a measure of belief $MB(h,e)$ representing the degree of support of a (favorable) evidence e, and a measure of disbelief $MD(h,e)$ representing the degree of refutation of an (unfavorable) evidence e. MB and MD are monotonically increasing functions that are respectively updated when the new evidence supports or refutes the hypothesis under consideration. The certainty factor $CF(h,e)$ is defined as:

$$CF(h,c) = \begin{cases} 1 & \text{if } P(h) = 1 \\ MB(h,e) & \text{if } P(h \mid e) > P(h) \\ 0 & \text{if } P(h \mid e) = P(h) \\ -MD(h,e) & \text{if } P(h \mid e) < P(h) \\ -1 & \text{if } P(h) = 0 \end{cases} \qquad (12)$$

The measures of belief MB and measure of disbelief MD could be interpreted as a *relative distance* on a bounded interval. Given an interval $[A,B]$ and a reference point R within the interval, the relative distance $d(X,R)$ between any arbitrary point X within the interval and the reference R can be defined as:

$$d(X,R) = \begin{cases} \frac{(X-R)}{(B-R)} & \text{if } X > R \\ 0 & \text{if } X = R \\ \frac{(R-X)}{(R-A)} & \text{if } X < R \end{cases} \qquad (13)$$

By making the following substitutions in equation 13

$$A = 0 \quad B = 1 \quad R = P(h) \quad X = P(h \mid e)$$

the definition of the measure of belief (MB) and measure of disbelief (MD) can be obtained:

$$MB(h,c) = \begin{cases} \frac{P(h|e)-P(h)}{1-P(h)} & \text{if } P(h \mid e) > P(h) \\ 0 & \text{otherwise} \end{cases} \qquad (14)$$

$$MD(h,c) = \begin{cases} \frac{P(h)-P(h|e)}{P(h)} & \text{if } P(h \mid e) < P(h) \\ 0 & \text{otherwise} \end{cases} \qquad (15)$$

The CF was originally interpreted as the relative increase or decrease of probabilities. In fact, from equations 12, 14, and 15, it can be shown that:

$$P(h \mid e) = P(h) + CF(h,e) \cdot [1 - P(h)] \quad \text{for } CF(h,e) \geq 0 \qquad (16)$$

$$P(h \mid e) = P(h) - \mid CF(h,e) \mid \cdot P(h) \quad \text{for } CF(h,e) \leq 0 \qquad (17)$$

Too often the CF paradigm has been incorrectly used in reasoning systems, interpreting the CFs as *absolute* rather than *incremental* probability values. The original interpretation of the CF as a probability ratio, however, can no longer be preserved after the CFs have been aggregated using the heuristic combining functions provided in MYCIN [20].

Ishizuka, Fu, and Yao [41], [42] have shown that these combining functions were an approximation of the classical Bayesian updating procedure, in which a term had been neglected. In their analysis it was concluded that the assumption of mutual independence of evidence was required for the correct use of this approach. The original definition of certainty factor is asymmetric and prevents commutativity. Another source of concern in the use of CFs is caused by the normalization of MBs and MDs before their arithmetic difference is computed. This normalization *hides* the difference between the cardinality of the set of supporting evidence and that of the set of refuting evidence.

Buchanan & Shortliffe [43] have proposed a change to the definition of CF and its rules of combination:

$$CF(h, e) = \frac{MB(h, e) - MD(h, e)}{1 - min(MB(h, e), MD(h, e))} \tag{18}$$

$$CF_{COMBINE}(x, y) = \begin{cases} x + y - xy & \text{for } x > 0, \ y > 0 \\ \frac{x+y}{1-min(|x|,|y|)} & \text{for } x < 0, \ y > 0 \text{ or } x > 0, \ y < 0 \\ -CF_{COMBINE}(-x, -y) & \text{for } x < 0, \ y < 0 \end{cases} \tag{19}$$

where

$$CF(h, e_1) = x \text{ and } CF(h, e_2) = y$$

This new definition avoids the problem of allowing a single piece of negative (positive) evidence to overwhelm several pieces of positive (negative) evidence. However, it has even less theoretical justification or interpretation than the original formulae.

Recently, Heckerman [44] has derived a new definition for the CF that does allow commutativity and has a consistent probabilistic interpretation. The new definition is:

$$CF(h, c) = \frac{P(h \mid e) - P(h)}{P(h \mid c)[1 - P(h)] + P(h)[1 - P(h \mid c)]} \tag{20}$$

There are still numerous serious problems that characterize this approach: the semantics of the CF, i.e., the interpretation of the number (ratio of probability, combination of utility values and probability); the assumptions of independence of the evidence; and the inability of distinguishing between ignorance and conflict, both of which are represented by a zero CF.

This type of representation of uncertainty has also been advocated by Rich [45] as an alternative to default reasoning. In her work, Rich claims that default reasoning could actually better be interpreted as likelihood reasoning, providing a uniform representation for statistical, prototypical and definitional facts.

3.4 Belief Theory

The Belief Theory, proposed by Shafer [22], was developed within the framework of Dempster's work on upper and lower probabilities induced by a multivalued mapping[4]

---

[4] The *one-to-many* nature of the mapping is the fundamental reason for the inability of applying the well-known theorem of probability that determines the probability density of the image of *one-to-one* mappings. In

In this context, the lower probabilities have been identified as epistemic probabilities and associated with a degree of belief. This formalism defines certainty as a function that maps subsets of a space of propositions Θ on the $[0,1]$ scale. The sets of partial beliefs are represented by mass distributions of a unit of belief across the propositions in Θ. This distribution is called *basic probability assignment (bpa)*. The total certainty over the space is 1. A non-zero *bpa* can be given to the entire space Θ to represent the degree of ignorance. Given a space of propositions Θ, referred to as *frame of discernment*, a function $m : 2^\Theta \rightarrow [0,1]$ is called a *basic probability assignment* if it satisfies the following three conditions:

$$m(\phi) = 0 \quad \text{where } \phi \text{ is the empty set} \tag{21}$$

$$0 < m(A) < 1 \tag{22}$$

$$\sum_{A \subseteq \Theta} m(A) = 1 \tag{23}$$

The certainty of any proposition B is then represented by the interval $[Bel(B), P^*(B)]$, where $Bel(B)$ and $P^*(B)$ are defined as:

$$Bel(B) = \sum_{x \subseteq B} m(x) \tag{24}$$

$$P^*(B) = \sum_{A \subseteq \Theta} m(x) \tag{25}$$

From the above definitions the following relation can be derived:

$$Bel(B) = 1 - P^*(\neg B) \tag{26}$$

If m_1 and m_2 are two *bpas* induced from two *independent* sources, a third *bpa*, $m(C)$, expressing the pooling of the evidence from the two sources, can be computed by using Dempster's rule of combination:

$$m(C) = \frac{\sum_{A_i \cap B_j = C} m_1(A_i) \cdot m_2(B_j)}{1 - \sum_{A_i \cap B_j = \phi} m_1(A_i) \cdot m_2(B_j)} \tag{27}$$

Dempster's rule of combination normalizes the intersection of the bodies of evidence from the two sources by the amount of *non-conflictive evidence* between the sources. This amount is represented by the denominator of the formula.

There are two problems with the Belief Theory approach. The first problem stems from *computational* complexity: in the general case, the evaluation of the degree of belief and upper probability requires time exponential in $|Theta|$, the cardinality of the hypothesis

---

fact, given a differentiable strictly-increasing or strictly-decreasing *function* ϕ on an interval I, and a continuous random variable X with a density f, such that $f(x) = 0$ for any x outside I, then the density function g can be computed as:

$$g(y) = f(x) \left| \frac{dx}{dy} \right|, \; y \in \phi(I) \text{ and } x = \phi^{-1}(y)$$

set (frame of discernment). This is caused by the need of (possibly) enumerating all the subset and superset of a given set. Barnett [46] showed that, when the frame of discernment is discrete, the computational-time complexity could be reduced from exponential to linear by combining the belief functions in a simplifying order. Strat [47] proved that the complexity could be reduced to $O(n^2)$, where n is the number of atomic propositions, i.e., intervals of unit length, when the frame of discernment is continuous. In both cases, however, these results were achieved by introducing various assumptions about the type and structure of the evidence to be combined and about the hypotheses to be supported. As a result, in addition to the requirements of mutual exclusive hypotheses and independent evidence which are needed by this approach, the following constraints must be included: for the case of discrete frame of discernment, each piece of evidence is assumed to support only a *singleton* proposition or its negation rather than disjunctions of propositions (i.e., propositions with larger granularity); for the case of continuous frame of discernment, only *contiguous* intervals along the number line can be included in the frame of discernment and thus receive support from the evidence.

The second problem in this approach results from the *normalization* process present in both Dempster's work and Shafer's. Zadeh [13] [14] has argued that this normalization process can lead to incorrect and counter-intuitive results. By removing the conflictive parts of the evidence and normalizing the remaining parts, important information is discarded rather than being dealt with adequately. A proposed solution to this problem is to avoid completely the normalization process by maintaining an explicit measure of the amount of conflict and by allowing the remaining information to be *subnormal* (i.e., $Bel(\Theta) < 1$). Zadeh [14] has proposed a test to determine the conditions of applicability of Dempster's rule of combination. Dubois and Prade [48] have also shown that the normalization process in the rule of evidence combination creates a sensitivity problem, where assigning a zero value or a very small value to a *bpa* causes very different results. Ginsberg [49] has proposed the use of the Dempster-Shafer approach as an alternative to non-monotonic logic. This suggestion is an extension to Rich's idea of interpreting default reasoning as likelihood reasoning [45]. In his work, Ginsberg provides a rule for propagating the lower and upper bounds through a reasoning chain or graph. His result is based on the interpretation of a production rule as a conditional probability rather than as a material implication. Smets [50], [51] has further explained the relations between belief functions, plausibilities, necessities, and possibilities and has extended Dempster's concepts to handle the case when the evidence is a fuzzy set [52].

3.5 Evidential Reasoning

Evidential Reasoning, proposed by Garvey, Lowrance, and Fischler [23], [24], [25] adopts the evidential interpretation of the degrees of belief and upper probabilities. Fundamentally based on Dempster-Shafer's theory, this approach defines the likelihood of a proposition A as a subinterval of the unit interval [0,1]. The lower bound of this interval is the degree of *support* of the proposition, $S(A)$, and the upper bound is its degree of *plausibility*, $Pl(A)$. The likelihood of a proposition A is written as $A_{[S(A),Pl(A)]}$. The following sample of interval-valued likelihoods illustrates the interpretation provided by this approach:

$A_{[0,1]}$	No knowledge at all about A
$A_{[0,0]}$	A is false
$A_{[1,1]}$	A is true
$A_{[.3,1]}$	The evidence partially supports A
$A_{[0,.7]}$	The evidence partially supports $\neg$ A
$A_{[.3,.7]}$	The evidence simultaneously provides partial support for A and $\neg$ A
$A_{[.3,.3]}$	The probability of A is exactly 0.3

Given two statements $A_{[S(A),Pl(A)]}$ and $B_{[S(B),Pl(B)]}$, the set of inference rules corresponding to the logical operations on these statements are defined [23] as:

$$\text{INTERSECTION:}AND(A,B)_{[max(0,S(A)+S(B)-1),min(Pl(A),Pl(B))]} \tag{28}$$

$$\text{UNION:}OR(A,B)_{[max(S(A),S(B)),min(1,Pl(A)+Pl(B))]} \tag{29}$$

$$\text{NEGATION:}NOT(A)_{[1-Pl(A),1-S(A)]} \tag{30}$$

This approach is based on the Dempster-Shafer's (D-S) theory. When distinct bodies of evidence must be pooled, this approach uses the same Dempster-Shafer's techniques, requiring the same normalization process that was criticized by Zadeh.

3.6 Evidence Space

Evidence Space, proposed by Rollinger [27], represents the uncertainty of a statement as a point in a two dimensional space. The (X, Y) coordinates of this space represent the positive or supporting evidence (E+) and the negative or disconfirming evidence (E-) available for any given proposition, respectively. The evidence space is a [0,1]x[0,1] square whose four vertices represent ignorance (0,0), absolute certainty in the support (1,0), absolute certainty in the refutation (0,1), and maximum conflictive evidence (1,1). The diagonal line defined by the equation $x+y-1=0$ represents the locus of probability points, the sum of whose coordinates is one.

It is interesting to note that if the dimensions of the evidence space (E+, E-) represent the *necessary* evidence, i.e., the lower bounds of the degree of support and refutation $(S(E), S(\neg E))$, the evidence space is reduced to the lower left triangle. Its three vertices (0,0), (1,0), and (0,1) represent ignorance, absolute support, and absolute refutation, respectively. The maximum amount of conflict is given by the point (0.5, 0.5). On the other hand, if the dimensions of the evidence space $(E+, E-)$ represent the *possible* evidence, i.e., the upper bounds of the degree of support and refutation $(Pl(E), Pl(\neg E))$, the evidence space is reduced to the upper right triangle. Its three vertices (1,1), (1,0), and (0,1) represent ignorance, absolute support, and absolute refutation, respectively. The maximum amount of conflict is again given by the point (0.5, 0.5). If the lower bounds are equated to the upper bounds, i.e., $(S(E), S(\neg E)) = (Pl(E), Pl(\neg E))$, a new set of coordinates $(P(E), P(\neg E))$, representing Bayesian probability, can be obtained. In this new set of coordinates, the evidence space collapses to the diagonal line $x+y-1=0$ that is the intersection of the two triangles and that indeed represents the probability line. Rollinger suggests the use of a distance to verify the validity of any given premise in a rule. This approach, however, does not suggest any way of aggregating evidence, propagating uncertainty through an inference chain, selecting an appropriate metric of similarity between patterns and data, etc.

3.7 Necessity and Possibility Theory

Necessity and possibility, proposed by Zadeh [29], [28], measure the degree of *entailment* and *intersection* of two fuzzy propositions represented by their normalized possibility distributions. Normal necessity and possibilities correspond to consonant belief and plausibility functions, respectively. Given two fuzzy propositions P and $D \subset U$, characterized by their possibility distributions $\mu_P(x)$ and $\mu_D(x)$, their degree of matching is represented by the interval $[Nec(P \mid D), Poss(P \mid D)]$, where:

$$N(P \mid D) = \bigwedge_x \left(\mu_D(x) \to \mu_P(x) \right)$$
$$= \bigwedge_x \left(max[(1 - \mu_D(x)), \mu_P(x)] \right) = 1 - \bigvee_x \left(min[(1 - \mu_P(x)), \mu_D(x)] \right). \quad (31)$$

$$\Pi(P \mid D) = \bigvee_x \left(min[\mu_P(x), \mu_D(x)] \right) \quad (32)$$

From the above definition, it is possible to derive for necessity and possibility the same duality observed between belief functions and upper probabilities (see formula [22]),

$$Nec(P \mid D) = 1 - Poss(\neg P \mid D) \quad (33)$$

The intersection of necessity measures and the union of possibility measures provide tighter bounds than those obtained by the intersection of belief functions and the union of plausibility functions [53].

3.8 Reasoned Assumptions

Among the non-numerical representations of uncertainty, two approaches typify the characterization of uncertain information in a purely symbolic manner: Reasoned Assumptions and Theory of Endorsements.

In the Reasoned Assumption approach, proposed by Doyle [12], the uncertainty embedded in an implication is (partially) removed by listing all the exceptions to that rule. When this is not possible, assumptions are used to show typicality of a value (default values) and defeasibility of a rule (liability to defeat of a reason). When an assumption used in the deductive process is found to be false, non-monotonic mechanisms are used to keep the integrity of the data base of statements. Assumption based systems can cope with the case of incomplete information, but they are inadequate to handle the case of imprecise information. In particular, they cannot integrate probabilistic information with reasoned assumptions. Furthermore, these systems rely on the precision of the defaulted value. On the other hand, when specific information is missing, the system should be able to use analogous or relevant information inherited from some higher level concept. This surrogate for the missing information is generally fuzzy or imprecise and only provides some elastic constraints on the value of the missing information. It is recognized by Doyle [12] that assumptions based systems lack facilities for computing degrees of belief, which "...may be necessary for summarizing the structure of large sets of admissible extensions as well as for quantifying confidence levels."

3.9 Theory of Endorsements

A different approach to uncertainty representation was recently proposed by Cohen and is based on a purely qualitative Theory of Endorsements [34], [35]. Endorsements are based on the explicit recording of the justifications for a statement, as in a Truth Maintenance System. In addition, endorsements classify the justification according to the type of evidence (for and against a proposition), the possible actions required to solve the uncertainty of that evidence, and other related features. Endorsements provide a good mechanism for explanations, since they create and maintain the entire history of justifications (reasons for believing or disbelieving a proposition) and the relevance of any proposition with respect to a given goal. Endorsements are divided into five classes: *rules, data, task, conclusion,* and *resolution* endorsements. However, combination of endorsements in a premise, propagation of endorsements to a conclusion and ranking of endorsements must be explicitly specified for each particular context, creating potential combinatorial problems.

4 Comparison of Approaches For Reasoning With Uncertainty

From previous reviews of the state of the art of reasoning systems [54], [1], and from previous analysis of applications [2], [55], [56], we have derived a *desiderata* (i.e., a list of requirements to be satisfied by the ideal formalism for representing uncertainty and making inference with uncertainty). We will now compare some of the above approaches against this list of requirements.

4.1 Desiderata for Reasoning with Uncertainty

Quinlan has proposed a list of four requirements to illustrate the shortcomings of the Bayesian and Confirmation theory approaches and to contrast them to INFERNO, his proposed approach to uncertain inference [26]. The requirements proposed by Quinlan were:

- "An inference system should not depend on any assumptions about the probability distributions of the propositions".

- "It should be possible to assert common relationships between propositions ... when the relationships are indeed known".

- "It should be possible to posit information about any set of propositions and observe the consequences for the system as a whole"

- "If the information provided to the system is inconsistent, this fact should be made evident along with some notion of alternative ways that the information could be made consistent".

Quinlan's work has been inspirational in the development of the following *desiderata*, which subsumes and extends Quinlan's initial list. The proposed desiderata describes the requirements to be satisfied by the ideal formalism for representing uncertainty and making inference with uncertain information. To be consistent with the organizing principle described above, the desiderata is subdivided into the same three layers of *Representation, Inference* and *Control.*

Representation Layer

1. There should be an explicit representation of the *amount* of evidence for *supporting* and for *refuting* any given hypothesis.

2. There should be an explicit representation of the information about the evidence, i.e., *meta-information*, such as the evidence source, the reasons for supporting and for refuting a given hypothesis, etc. This meta-information will be used in the control layer to remove conflicting pieces of evidence provided by different sources.

3. The representation should allow the user to describe the uncertainty of information at the available level of detail (i.e., allowing *heterogeneous information granularity*).

4. There should be an explicit representation of *consistency*. Some measure of consistency or compatibility should be available to detect trends of potential conflicts and to identify essential contributing factors in the conflict.

5. There should be an explicit representation of *ignorance* to allow the user to make *non-committing* statements, i.e., to express the user's lack of conviction about the certainty of *any* of the available choices or events. Some measure of ignorance, similar to the concept of entropy, should be available to guide the gathering of discriminant information.

6. The representation must be, or at least must appear to be natural to the user to enable him/her to *describe uncertain input* and to *interpret uncertain output*. The representation must also be natural to the expert to enable him/her to elicit *consistent* weights representing the strength of the implication of each rule.

Inference Layer

7. The combining rules should not be based on global assumptions of *evidence independence*.

8. The combining rules should not be based on global assumptions of *hypotheses exhaustiveness* and *exclusiveness*.

9. The combining rules should maintain the *closure* of the syntax and semantics of the representation of uncertainty.

10. Any function used to propagate and summarize uncertainty should have clear semantics. This is needed both to maintain the semantic closure of the representation and to allow the control layer to *select* the most appropriate combining rules.

Control Layer

11. There should be a clear distinction between a *conflict* in the information (i.e., violation of consistency), and *ignorance* about the information. To solve the conflict, the controller (meta-reasoner) must retract one or more elements of the conflicting set of evidence. To remove the ignorance, the controller must select a (retractable) default value or tag the information with an assumption.

12. The *traccability* of the aggregation and propagation of uncertainty through the reasoning process must be available to resolve conflicts or contradictions, to explain the support of conclusions, and to perform meta-reasoning for control.

13. It should be possible to make pairwise comparisons of uncertainty since the induced *ordinal or cardinal ranking* is needed for performing any kind of decision-making activities.

14. It should be possible to *select* the most appropriate combination rule by using a declarative form of control (i.e., by using a set of context dependent rules that specify the selection policies).

4.2 Evaluation of the Approaches

The above desiderata was used to guide the development of RUM, the Reasoning with Uncertainty Module, which was introduced in [57] and is described in a separate set of notes.

APPROACH	*REPRESENTATION*						*INFERENCE*				*CONTROL*			
	1	2	3	4	5	6	7	8	9	10	11	12	13	14
Modified Bayesian	N	N	N	N	N	Y	N	N	Y	Y	N	N	Y	N
Confirmation	N	N	N	N	Y	N	N	Y	N	N	N	N	N	N
Dempster-Shafer	Y	N	Y	Y	Y	Y	N	N	Y	Y	Y	N	Y	N
Probability Bounds	Y	N	Y	Y	Y	Y	Y	Y	Y	Y	Y	N	Y	N
Fuzzy necessity/posibility	Y	N	Y	Y	Y	Y	Y	Y	Y	Y	N	N	Y	N
Evidence Space	Y	N	N	Y	Y	Y	Y	Y	Y	Y	Y	N	Y	N
RUM	Y	Y	N	Y	Y	Y	Y	Y	Y	Y	Y	Y	Y	Y
Reasoned Assumptions	N	Y	N	Y	N	Y	Y	Y	Y	Y	N	Y	N	Y
Endorsements	N	Y	N	Y	N	Y	Y	Y	Y	Y	N	Y	N	Y

Figure 1: Evaluation of Uncertainty Approaches Against the Desiderata

Table 1 summarizes the evaluation of the formalisms discussed in the previous section against this desiderata. The order in which the formalisms appear in the table reflects their numeric or non-numeric nature: the numeric formalisms are listed above RUM, the non-numeric ones are shown below RUM. RUM itself is considered a hybrid as it uses both numeric and symbolic information.

5 Conclusions

We have described the most common approaches used to represent and propagate uncertainty in deductive reasoning systems. We have analyzed quantitative methods based on point-value representation (e.g., Bayesian), on interval-valued representation (Dempster-Shafer), and fuzzy-value representation (necessity/possibility measures). We have extended our analysis to include to qualitative approaches (reasoned assumptions and endorsements).

We have defined a set of evaluation criteria used to compare these approaches. The evaluation criteria have been used to drive the design and implementation of RUM [57], a Reasoning with Uncertainty Module that relies on numerical and symbolic (structural) information.

References:

[1] Piero P. Bonissone. Plausible Reasoning: Coping with Uncertainty in Expert Systems. In Stuart Shapiro, editor, *Encyclopedia of Artificial Intelligence*, pages 854–863. John Wiley and Sons Co., New York, 1987.

[2] Piero P. Bonissone. Summarizing and Propagating Uncertain Information with Triangular Norms. *International Journal of Approximate Reasoning*, 1(1):71–101, January 1987.

[3] Piero P. Bonissone and Richard M. Tong. Editorial: Reasoning with Uncertainty in Expert Systems. *International Journal of Man-Machine Studies*, 22(3):241–250, 1985.

[4] L.A. Zadeh. Syllogistic reasoning in fuzzy logic and its application to usuality and reasoning with dispositions. *IEEE Trans. Systems, Man and Cybernetics*, SMC-15:754–765, 1985.

[5] L.A. Zadeh. Dispositional logic. *Appl. Math. Letters*, 1(1):95–99, 1988.

[6] L.A. Zadeh. A computational approach to fuzzy quantifiers in natural language. *Computers and Mathematics*, 9:149–184, 1983.

[7] L.A. Zadeh. A computational theory of disposition. In *Proc. 1984 Int. Conf. Computational Linguistics*, pages 312–318, 1984.

[8] B. Schweizer and A. Sklar. Associative Functions and Abstract Semi-Groups. *Publicationes Mathematicae Debrecen*, 10:69–81, 1963.

[9] D. Dubois and H. Prade. Criteria Aggregation and Ranking of Alternatives in the Framework of Fuzzy Set Theory. In H.J. Zimmerman, L.A. Zadeh, and B.R. Gaines, editors, *TIMS/Studies in the Management Science, Vol. 20*, pages 209–240. Elsevier Science Publishers, 1984.

[10] P. Bonissone and K. Decker. Selecting uncertainty calculi and granularity: An experiment in trading-off precision and complexity. In L. N. Kanaal and J.F. Lemmer, editors, *Uncertainty in Artificial Intelligence*. North Holland, Amsterdam, 1986.

[11] L.A. Zadeh. Fuzzy logic and approximate reasoning (in memory of Grigor Moisil). *Synthese*, 30:407–428, 1975.

[12] J. Doyle. Methodological Simplicity in Expert System Construction: The Case of Judgements and Reasoned Assumptions. *The AI Magazine*, 4(2):39–43, 1983.

[13] L.A. Zadeh. Review of Books: A Mathematical Theory of Evidence. *The AI Magazine*, 5(3):81–83, 1984.

[14] L.A. Zadeh. A simple view of the dempster-shafer theory of evidence and its implications for the rule of combinations. Technical Report 33, Berkeley Cognitive Science, Institute of Cognitive Science, University of California, Berkeley,, 1985.

[15] A.P. Dempster. Upper and lower probabilities induced by a multivalued mapping. *Annals of Mathematical Statistics*, 38:325–339, 1967.

[16] J. Pearl. Reverend Bayes on Inference Engines: a Distributed Hierarchical Approach. In *Proceedings Second National Conference on Artificial Intelligence*, pages 133–136. AAAI, August 1982.

[17] J. Pearl. How to Do with Probabilities What People Say You Can't. In *Proceedings Second Conference on Artificial Intelligence Applications*, pages 1–12. IEEE, December 1985.

[18] Judea Pearl. Evidential Reasoning Under Uncertainty. In Howard E. Shrobe, editor, *Exploring Artificial Intelligence*, pages 381–418. Morgan Kaufmann, San Mateo, CA, 1988.

[19] R.O. Duda, P.E. Hart, and N.J. Nilsson. Subjective Bayesian methods for rule-based inference systems. In *Proc. AFIPS 45*, pages 1075–1082, New York, 1976. AFIPS Press.

[20] E.H. Shortliffe and B. Buchanan. A model of inexact reasoning in medicine. *Mathematical Biosciences*, 23:351–379, 1975.

[21] A.P. Dempster. A generalization of Bayesian inference. *J. Roy. Stat. Soc. Ser. B*, 30:205–247, 1968.

[22] G. Shafer. *A Mathematical Theory of Evidence*. Princeton University Press, Princeton, New Jersey, 1976.

[23] T.D. Garvey, J.D. Lowrance, and M.A. Fischler. An inference technique for integrating knowledge from disparate sources. In *Proc. 7th. Intern. Joint Conf. on Artificial Intelligence*, Vancouver, British Columbia, Canada, 1981.

[24] J. Lowrance and T.D. Garvey. Evidential reasoning: an implementation for multisensor integration. Technical Report Technical Note 307, SRI International, Artificial Intelligence Center, Menlo Park, California, 1983.

[25] J.D. Lowrance, T.D. Garvey, and T.M. Strat. A framework for evidential-reasoning systems. In *Proc. National Conference on Artificial Intelligence*, pages 896–903, Menlo Park, California, 1986. AAAI.

[26] J.R. Quinlan. Consistency and Plausible Reasoning. In *Proceedings Eight International Joint Conference on Artificial Intelligence*, pages 137–144. AAAI, August 1983.

[27] C.R. Rollinger. How to Represent Evidence - Aspects of Uncertainty Reasoning. In *Proceedings Eight International Joint Conference on Artificial Intelligence*, pages 358–361. AAAI, August 1983.

[28] L.A. Zadeh. Fuzzy sets as a basis for a theory of possibility. *Fuzzy Sets and Systems*, 1:3–28, 1978.

[29] L.A. Zadeh. Fuzzy Sets and Information Granularity. In M.M. Gupta, R.K. Ragade, and R.R. Yager, editors, *Advances in Fuzzy Set Theory and Applications*, pages 3–18. Elsevier Science Publishers, 1979.

[30] L.A. Zadeh. A theory of approximate reasoning. In P. Hayes, D. Michie, and L.I. Mikulich, editors, *Machine Intelligence*, pages 149–194. Halstead Press, New York, 1979.

[31] L.A. Zadeh. Linguistic variables, approximate reasoning, and dispositions. *Medical Informatics*, 8:173–186, 1983.

[32] R. Reiter. A Logic for Default Reasoning. *Journal of Artificial Intelligence*, 13:81–132, 1980.

[33] P. Cohen. *Heuristic Reasoning about Uncertainty: An Artificial Intelligence Approach*. Pittman, Boston, Massachusetts, 1985.

[34] P.R. Cohen and M.R. Grinberg. A Theory of Heuristics Reasoning about Uncertainty. *The AI Magazine*, pages 17–233, 1983.

[35] P.R. Cohen and M.R. Grinberg. A Framework for Heuristics Reasoning about Uncertainty. In *Proceedings Eight International Joint Conference on Artificial Intelligence*, pages 355–357. AAAI, August 1983.

[36] E.D. Pednault, S.W. Zucker, and L.V. Muresan. On the Independence Assumption Underlying Subjective Bayesian Updating. *Journal of Artificial Intelligence*, 16:213–222, 1981.

[37] C. Glymour. Independence Assumptions and Bayesian Updating. *Journal of Artificial Intelligence*, 25:95–99, 1985.

[38] R.W. Johnson. Independence and Bayesian Updating Methods. *Journal of Artificial Intelligence*, 29:217–222, 1986.

[39] Yizong Cheng and Rangasami Kashyap. Irrelevancy of evidence caused by independence assumptions. Technical Report TR-EE 86-17, School of Electrical Engineering, Purdue University, West Lafayette, Indiana 47907, 1986.

[40] R. Giles. Semantics for Fuzzy Reasoning. *International Journal of Man-Machine Studies*, 17(4):401–415, 1982.

[41] M. Ishizuka, K.S. Fu, and J.T.P. Yao. Inexact Inference for Rule-Based Damage Assessment of Existing Structure. In *Proceedings Seventh International Joint Conference on Artificial Intelligence*, pages 1837–842. AAAI, August 1981.

[42] M. Ishizuka. An extension of dempster-shafer theory to fuzzy sets for constructing expert systems. *Scisan-Kenkyu*, 34:312–315, 1982.

[43] B.G. Buchanan and E.H. Shortliffe. *Rule-Based Expert Systems*. Addison-Wesley, Reading, Massachusetts, 1984.

[44] D. Heckerman. Probabilistic interpretations for MYCIN certainty factors. In L.N. Kanaal and J.F. Lemmer, editors, *Uncertainty in Artificial Intelligence*, pages 167–196. North-Holland, Amsterdam, 1986.

[45] E. Rich. Default Reasoning as Likelihood Reasoning. In *Proceedings Third National Conference on Artificial Intelligence*, pages 348–351. AAAI, August 1983.

[46] J.A. Barnett. Computational methods for a mathematical theory of evidence. In *Proc. 7th. Intern. Joint Conf. on Artificial Intelligence*, Vancouver, British Columbia, Canada,, 1981.

[47] T.M. Strat. Continuous belief functions for evidential reasoning. In *Proc. National Conference on Artificial Intelligence*, pages 308–313, Austin, Texas,, 1984.

[48] D. Dubois and H. Prade. Combination and Propagation of Uncertainty with Belief Functions - A Reexamination. In *Proceedings Ninth International Joint Conference on Artificial Intelligence*, pages 111–113. AAAI, August 1985.

[49] M.L. Ginsberg. Non-Monotonic Reasoning Using Dempster's Rule. In *Proceedings Fourth National Conference on Artificial Intelligence*, pages 126–129. AAAI, August 1984.

[50] P. Smets. The Degree of Belief in a Fuzzy Set. *Information Science*, 25:1–19, 1981.

[51] P. Smets. Belief functions. In P. Smets, A. Mamdani, D. Dubois, and H. Prade, editors, *Non-Standard Logics for Automated Reasoning*. Academic Press, New York, 1988.

[52] L.A. Zadeh. Fuzzy sets. *Information and Control*, 8:338–353, 1965.

[53] H. Prade. A Computational Approach to Approximate Reasoning and Plausible Reasoning with Applications to Expert Systems. *IEEE Transactions on Pattern Analysis and Machine Intelligence*, PAMI-7(3):260–283, 1985.

[54] Piero P. Bonissone and Allen L. Brown. Expanding the Horizons of Expert Systems. In T. Bernold, editor, *Expert Systems and Knowledge Engineering*, pages 267–288. North-Holland, 1986.

[55] Piero P. Bonissone. Using T-norm Based Uncertainty Calculi in a Naval Situation Assessment Application. In *Proceedings of the Third AAAI Workshop on Uncertainty in Artificial Intelligence*, pages 250–261. AAAI, July 1987.

[56] Piero P. Bonissone and Nancy C Wood. Plausible Reasoning in Dynamic Classification Problems. In *Proceedings of the Validation and Testing of Knowledge-Based Systems Workshop*. AAAI, August 1988.

[57] Piero P. Bonissone, Stephen Gans, and Keith S. Decker. RUM: A Layered Architecture for Reasoning with Uncertainty. In *Proceedings 10th International Joint Conference on Artificial Intelligence*, pages 891–898. AAAI, August 1987.

EVIDENCE AND BELIEF IN EXPERT SYSTEMS
(DEMPSTER-SHAFER: A SIMPLIFIED VIEW)

Piero P. Bonissone

Artificial Intelligence Program
General Electric Corporate Research and Development
Schenectady, New York 12301, USA

Abstract

The theory of Dempster-Shafer (DS) is illustrated in relation with traditional probability theory. First we provide some basic definitions of terms used in DS (frame of discernment, evidence sources, basic probability assignments or *bpas*, Dempster's rule of combination, normalization). The theory is extended from crisp data and queries to fuzzy data and queries, introducing the concepts of necessity and possibility measures. A detailed example based on political elections, candidates characteristics, and voters preference is used to illustrate the theory. Finally, we provide a naive version of DS theory, where we show the translation of uncertainty intervals assigned to propositions (à la RUM) into *bpas*.

1 Introduction

In 1967, Dempster proposed a theory of upper and lower probabilities induced by multiple valued mappings of random variables [1]. Subsequently, Glenn Shafer re-proposed the same theory, deriving it from belief functions, rather than probability densities [2]. This theory, which is commonly known as Dempster-Shafer theory, has been recently re-derived by Ruspini [3], [4] from autoepistemic logic (S5) and probability, giving the theory a possible-world semantics interpretation.

The purpose of these notes is to summarize the important aspects of the theory and to illustrate it with some examples. For this purpose we must first introduce some basic definitions.

1.1 Definitions

1.1.1 Frame of Discernment

Let Θ be the universe of discourse of the propositions, whose belief is going to be assessed. Θ is referred to as the *frame of discernment*.

Source	Basic Probability Assignments		
S_i	$m_1^i \to A_1^i$	...	$m_{t_i}^i \to A_{t_i}^i$
S_1	$0.2 \to \{\theta_1, \theta_2\}$	$0.5 \to \{\theta_3, \theta_4, \theta_7, \theta_8\}$	$0.3 \to \{\Theta\}$
S_2	$0.2 \to \{\theta_1\}$	$0.6 \to \{\theta_2, \theta_3, \theta_4, \theta_6, \theta_7, \theta_8\}$	$0.2 \to \{\Theta\}$
S_3	$0.3 \to \{\theta_1, \theta_2, \theta_3\}$	$0.7 \to \{\theta_3, \theta_4, \theta_5, \theta_6, \theta_7, \theta_8\}$	-

Figure 1: Example of bpa Generation from Three Sources

1.2 Evidence Sources

The evidence sources generate the belief assessments for subsets of the frame of discernment. Two evidence sources are *evidentially independent* if the *errors* incurred by one source are independent from the errors incurred by the other one. Let S be a finite set of evidentially independent sources which generate the *basic probability assignments* or *bpas* for various subsets of Θ. Let $n = \mid S \mid$ be the cardinality of S.

Each source will generate a finite set of *bpas*. Source S_i can be interpreted as a mapping from the power set of Θ to the interval [0,1], i.e., $S_i : P(\Theta) \to [0,1]$.

The *bpas* generated by source S_i are defined as: $m_1^i(A_1^i), \ldots, m_{t_i}^i(A_{t_i}^i)$, such that

$$\sum_{k=1}^{t_i} m_k^i(A_k^i) = 1$$

and

$$A_k^i \subseteq \Theta, k = 1, \ldots, t_i$$

1.2.1 Example of bpas Assignments

For illustration purpose, let us assume that Θ is finite, e.g., $\mid \Theta \mid = 10$ and

$$\Theta = \{\theta_1, \theta_2, \ldots \theta_{10}\}$$

Figure 1 illustrates the bpa generation produced by three different sources.

Note that each source can assign its *bpas* to singletons in Θ, to not necessarily disjoint subsets of Θ, or to the entire frame of discernment Θ. The first type of assignment (*bpas* to singletons) is similar to traditional probability theory. The second type of assignment (*bpas* to subsets) represents a major departure from traditional probability: rather than using maximum entropy to uniformly distribute the *bpa* among all the elements of the subset, the *bpa* is directly attached to the subset (heterogeneous granularity assignment). The third type of assignment (*bpa* to the frame of discernment) is an explicit way to express ignorance (or non-commitment).

1.2.2 DS Compared with Traditional Probability Theory

In traditional probability theory, probability assignments are made *only* to singletons in the sample space. For instance, let's us consider the classical example of one unbiased die with six equally likely faces i, $i = 1, \ldots, 6$. The probability of each face is $P(i) = \frac{1}{6}$.

The sample space Ω is the set $\{1, 2, \ldots, 6\}$. Consider now the independent tossing of two unbiased dice: the sample space Ω contains 36 singletons corresponding to all the possible outcomes (i, j). Each singleton (i, j) is assigned a probability $P(i, j) = \frac{1}{36}$. Any (crisp) event A is represented by a (crisp) sub-set defined on this sample space, i.e., $A \subset \Omega$. For instance the event of rolling *any even number less than five*, i.e., $\{2, 4\}$ is defined by $A = \{(i, j) \mid (i + j) < 5, \ (i + j) = even\}$. A is the region containing the singletons $(1, 1), (1, 2), (2, 1), (2, 2)$. The probability $P(A)$ is obtained by counting the probability of the singletons in the sample space, which are *elements* of A, i.e., $P(A) = \sum_{(i,j) \in A} P(i, j)$. The major difference in DS theory is when primitive probabilities are assigned to *subsets* of the sample space, i.e., $S_i \subset \Omega$. Once a region A is defined on Ω once can no longer take the summation of the subsets $S_i \in \Omega$. Thus the symbol $\in$ must be replaced by a suitable set-to-set operator: the most conservative replacement is $\subseteq$, i.e., only evidence *fully contained* in the event may be counted; the most liberal replacement is $\cap$, i.e., any evidence which is *not disjoint* with the event may be counted.

2 Dempster's Rule of Combination

Let $D(S_i)$, for $i = 1, \ldots, n$ be the aggregation of the all sources' assignments. Note that D is associative, i.e.,

$$D(S_1, S_2, \ldots, S_n) = D(S_1, D(S_2, \ldots, S_n))$$

The associativity of D is based on the evidential independence of the sources.

Due to its associativity, we only need to define the aggregation of two sources, i.e., $D(S_i, S_j)$.

Let the *bpas* from the two sources S_i and S_j respectively be:

$$m_1^i(A_1^i), \ldots, m_{t_i}^i(A_{t_i}^i)$$

and

$$m_1^j(A_1^j), \ldots, m_{t_j}^j(A_{t_j}^j)$$

We will now compute the relation $R'_{i,j}$ as the crossproduct $S_i \times S_j$. The element $r'_{k,l}$ in $R'_{i,j}$, where $k = 1, \ldots, t_i$, and $l = 1, \ldots, t_j$, is defined as:

$$m_k^i(A_k^i) \times m_l^j(A_l^j)$$

and represents the combined mass attached to the subset $A_{kl} = A_k^i \cap A_l^j$. Dempster's rule of combination is simply the intersection of random sets under the assumption of uncorrelation.

2.1 Example of Dempster's Rule of Combination from Two Sources

Using the data in Figure 1, we are now going to show the application of the rule of combination to sources S_1 and S_2. Figure 2 illustrates the result of this combination.

$S_1 \times S_2$	S_1		
S_2	$0.04 \to \{\theta_1\}$	$0.1 \to \phi$	$0.06 \to \{\theta_1\}$
	$0.12 \to \{\theta_2\}$	$0.3 \to \{\theta_3,\theta_4,\theta_7,\theta_8\}$	$0.18 \to \{\theta_2,\theta_3,\theta_4,\theta_6,\theta_7,\theta_8\}$
	$0.04 \to \{\theta_1,\theta_2\}$	$0.1 \to \{\theta_3,\theta_4,\theta_7,\theta_8\}$	$0.06 \to \{\Theta\}$

Figure 2: Example of Combination from Two Sources

2.2   Combining More Than Two Sources

If there were other sources to be aggregated, e.g. S_k, we could repeat the same procedure using the computed relation $R'_{i,j}$ and S_k, to obtain a new relation $R''_{i,j,k} = R'_{i,j} \times S_k$. (We are purposely deferring the normalization operations). Once all the sources have been aggregated, the aggregated information is contained in the last computed relation, which we will refer to as R.

The cardinality of R will be $\mid R \mid = t_i \times t_j \ldots \times t_n$. Each element r_s, where $s = 1, \ldots, \mid R \mid$, will contain a combined mass m_s attached to a set $A_s \subseteq \Theta$. Of course, by consolidating assignments to the same subsets into a unique assignment, whose value is the sum of the assignments, the final cardinality of R could be reduced. For instance $R'_{1,2}$, illustrated in Figure 2, has nine computed entries. However, there are only seven different subsets of Θ. The two assignments to $\{\theta_1\}$ can be merged into one, whose value is 0.1, and both assignments to $\{\theta_3,\theta_4,\theta_7,\theta_8\}$ can be consolidated into one, whose value is 0.4.

2.3   Example of Dempster's Rule of Combination for Three Sources

Using the same data from the first example, we are now going to continue the second example by applying the rule of combination to sources S_1, S_2, and S_3.

$S_1 \times S_2 \times S_3$	S_3	
$S_1 \times S_2$	$0.012 \to \{\theta_1\}$	$0.028 \to \phi$
	$0.03 \to \phi$	$0.07 \to \phi$
	$0.018 \to \{\theta_1\}$	$0.042 \to \phi$
	$0.036 \to \{\theta_2\}$	$0.084 \to \phi$
	$0.09 \to \{\theta_3\}$	$0.021 \to \{\theta_3,\theta_4,\theta_7,\theta_8\}$
	$0.054 \to \{\theta_2,\theta_3\}$	$0.126 \to \{\theta_3,\theta_4,\theta_6,\theta_7,\theta_8\}$
	$0.012 \to \{\theta_1,\theta_2\}$	$0.028 \to \phi$
	$0.03 \to \{\theta_3\}$	$0.07 \to \{\theta_3,\theta_4,\theta_7,\theta_8\}$
	$0.018 \to \{\theta_1,\theta_2,\theta_3\}$	$0.042 \to \{\theta_3,\theta_4,\theta_5,\theta_6,\theta_7,\theta_8\}$

Figure 3: Example of Combination from Three Sources

3   Crisp Queries Based on Crisp Data

Let $B \subseteq \Theta$ be an event (query) whose belief must be computed. The answer to the query is represented by a (possibly normalized) lower and upper bound.

Continuing our example, let us define $B = \{\theta_1, \theta_2\}$. We now want to compute the lower and upper bound for this event, given the combined data from $S_1 \times S_2 \times S_3$, illustrated in Figure 3.

3.1   Lower Bound: L(B)

The lower bound of the query represents the amount of confirming evidence provided by the evidence. It is computed as:

$$L(B) = \sum_{A_s \subseteq B} m(A_s)$$

Thus,

$$L(B) = L(\{\theta_1, \theta_2\})$$

$$L(B) = \sum_{i=1}^{2} m_i(\{\theta_1\}) + m(\{\theta_2\}) + m(\{\theta_1, \theta_2\})$$

$$L(B) = 0.078$$

3.2   Upper Bound: U(B)

The upper bound of the query represents the amount to which the evidence fail to refute the query. It is computed as:

$$U(B) = 1 - L(\neg B)$$

$$U(B) = 1 - \sum_{A_s \subseteq \neg B} m(A_s)$$

$$U(B) = \sum_{A_s \cap B \neq \phi} m(A_s)$$

Thus,

$$U(B) = U(\theta_1, \theta_2)$$

$$U(B) = \sum_{i=1}^{2} m_i(\{\theta_1\}) + m(\{\theta_2\}) + m(\{\theta_1, \theta_2\}) + m(\{\theta_2, \theta_3\}) + m(\{\theta_1, \theta_2, \theta_3\})$$

$$U(B) = 0.150$$

3.3 Conflict Measure C and Normalizing Factor K

The amount of conflict C contained in the combined information is determined by the accumulated bpas assigned to the empty set ϕ. Thus,

$$C = L(\phi)$$

$$C = \sum_{A_s = \phi} m(A_s)$$

From Figure 3 $L(\phi) = \sum_{i=1}^{6} m_i(\phi) = 0.282$.
The normalizing factor **K** is defined as:

$$K = \frac{1}{1 - L(\phi)}$$

For this example the normalizing factor is $K = \frac{1}{0.718} = 1.392$
Thus the normalized lower bound $L^*(B)$ and upper bound $U^*(B)$ are:

$$L^*(B) = K \times L(B) = 1.392 \times 0.078 = 0.1086$$

$$U^*(B) = K \times U(B) = 1.392 \times 0.150 = 0.2089$$

4 Fuzzy Queries Based on Crisp Data

In the previous example, the query was defined as a well-defined event, i.e., a crisp set in the frame of discernment. However, it is possible to determine the lower and upper bounds of ill-defined events that will be represented as fuzzy sets in the frame of discernment. Let us assume that the event B is defined by its characteristic function $\mu_B(x)$, for $x \in \Theta$. The lower and upper bounds can be computed as:

4.1 Lower Bound: L(B)

$$L(B) = \sum_{A_s \subseteq \Theta} m(A_s) \times (\bigwedge_{x \in A_s} \mu_B(x))$$

4.2 Upper Bound: U(B)

$$U(B) = \sum_{A_s \subseteq \Theta} m(A_s) \times (\bigvee_{x \in A_s} \mu_B(x))$$

Note that if the event B is crisp, i.e., its characteristic function $\mu_B(x)$ only takes value in 0,1, the above formulae reduce to the crisp case defined in previous section [5].

5 Fuzzy Queries Based on Fuzzy Data

By continuing the generalization process, it is also possible to determine lower and upper bounds of ill-defined events, based on fuzzy data. Such bounds can be computed as:

5.1   Lower Bound: L(B)

$$L(B) = \sum_{A_s \subseteq \Theta} m(A_s) \times N(B \mid A_s)$$

where $N(B \mid A_s)$ is the *necessity measure* of B given A_s.

5.2   Upper Bound: U(B)

$$U(B) = \sum_{A_s \subseteq \Theta} m(A_s) \times \Pi(B \mid A_s)$$

where $\Pi(B \mid A_s)$ is the *possibility measure* of B given A_s.

5.3   Digression: Necessity and Possibility Measures

The necessity measure $N(p \mid d)$ represents the degree of *semantic entailment* of a pattern descriptor p given a datum d. The possibility measure $\Pi(p \mid d)$ represents the degree of *intersection* between the same pattern and datum. Thus, the interval defined by $[N(p \mid d), \Pi(p \mid d)]$ represents the lower and upper bounds of the degree of matching between such pattern and datum.

More specifically, let X be the universe of discourse of the pattern descriptor p, and of the datum d, i.e. X is the collection of all the possible values which can be tested by p or which can be taken by d. Let the meaning of the pattern descriptor p be described by its characteristic function $\mu_P(x)$. Let $\mu_D(x)$ be the characteristic function of the datum d. By definition, we have $\mu_P : X \rightarrow [0, 1]$ and $\mu_D : X \rightarrow [0, 1]$, for all x in X. Then the necessity measure is defined as:

$$
\begin{aligned}
N(p \mid d) &= \bigwedge_x (\mu_D(x) \rightarrow \mu_P(x)) \\
&= \bigwedge_x (max[(1 - \mu_D(x)), \mu_P(x)]) = 1 - \bigvee_x (min[(1 - \mu_P(x)), \mu_D(x)]) .
\end{aligned}
$$

The possibility measure is defined as:

$$\Pi(p \mid d) = \bigvee_x (min[\mu_P(x), \mu_D(x)]) .$$

Notice that $N(p \mid d) = 1 - \Pi(\neg p \mid d)$, i.e., the degree of necessity of a pattern p given a datum d is equivalent to the degree of *impossibility* of the negation of the pattern given the same datum. In the degenerate case in which p and d are crisp sets, defined by boolean characteristic functions, the above is the same relationship that e¿xists between the necessity and the possibility operators in modal logics, i.e., $\Box(p \mid d) = \neg \Diamond \neg (p \mid d)$ [6]. **Note:** *Necessity* and *Possibility* are two of four relative modalities. The other two are *Impossibility* and *Contingency*. The following subsection illustrates their meanings and relationships. Possibility and necessity are duals, in the same way as contingency and impossibility are dual modes. This is illustrated in Figure 4.

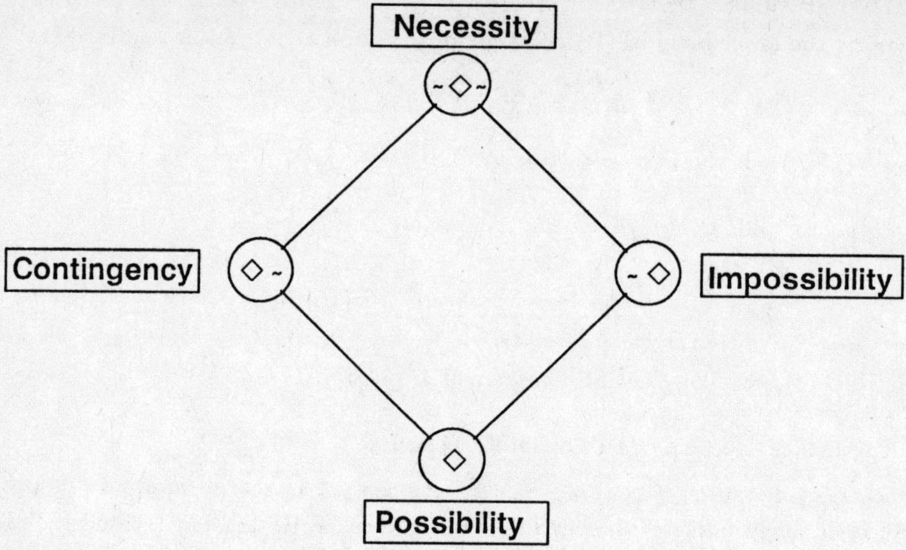

Figure 4: Four Relative Modalities

Necessity : $\Box(B \mid A) = \neg\Diamond\neg(B \mid A)$

$$\forall x[x \in A \rightarrow x \in B]$$

This implies that $A \subseteq B$ or equivalently, $B \cup \neg A = U$, where U is the universe.

Possibility : $\Diamond(B \mid A)$

$$\exists x[x \in A \rightarrow x \in B]$$

This implies that $A \cap B \neq \phi$, where ϕ is the empty set.

Contingency : $\Diamond(\neg B \mid A)$

$$\exists x[x \in A \rightarrow x \in \neg B]$$

This implies that $A \cap \neg B \neq \phi$, where ϕ is the empty set.

Impossibility : $\neg\Diamond(B \mid A)$

$$\forall x[x \in A \rightarrow x \in \neg B]$$

This implies that $A \subseteq \neg B$, or equivalently $A \cap B = \phi$, where ϕ is the empty set.

5.4 Back to the general case

In general the lower bound L(B) and the upper bound U(B) can be expressed as:

$$L(B) = \sum_{A_s \subseteq \Theta} m(A_s) \times I(A_s \subseteq B)$$

and

$$U(B) = \sum_{A_s \subseteq \Theta} m(A_s) \times (1 - I(A_s \subseteq \neg B))$$

$$U(B) = \sum_{A_s \subseteq \Theta} m(A_s) - \sum_{A_s \subseteq \Theta} m(A_s) \times I(A_s \subseteq \neg B)$$

$$U(B) = 1 - \sum_{A_s \subseteq \Theta} m(A_s) \times I(A_s \subseteq \neg B)$$

where $I(A_s \subseteq B)$ indicates the degree of containment of A_s in B, i.e. the degree to which A_s materially implies B. Analogously, $I(A_s \subseteq \neg B)$ is the degree to which A implies $\neg B$. Various authors have proposed different expressions for $I(A_s \subseteq B)$, by using different definitions of material implication in multiple valued-logics.

Yager's definition of the degree of containment [7] is based on extending the boolean definition of material implication $\mu_{(A \to B)}(x) = \mu_{(\neg A \cup B)}(x) = max[1 - \mu_A(x), \mu_B(x)]$:

$$I(A_s \subseteq B) = \bigwedge_x (max[(1 - \mu_{A_s}(x)), \mu_B(x)]) = N(B \mid A_s)$$

Ishizuka's definition of the degree of containment [8] is based on Lukasiewicz definition of material implication $\mu_{(A \to B)}(x) = min[1, 1 - \mu_A(x) + \mu_B(x)]$. Ishizuka's definition takes into account the possible sub-normality of fuzzy set A (denumerator in the expression):

$$I(A_s \subseteq B) = \frac{\bigwedge_x [1, 1 - \mu_A(x) + \mu_B(x)]}{max_x \mu_A(x)}$$

Ogawa's definition of the degree of containment [9] is the ratio of the cardinality of the intersection of A and B, divided by the cardinality of B, i.e.:

$$I(A_s \subseteq B) = \frac{\sum_i min[\mu_A(x_i), \mu_B(x_i)]}{\sum_i \mu_B(x_i)} = \frac{\mid A \cap B \mid}{\mid B \mid}$$

For this general case, Dempster rules of combination must be modified to take into account the degree of intersection of two sets. The above extensions have been studied and described by Yen in [10].

6 A Complete Example

In this example we will consider the case of an election with eight candidates. We will disregard the filtering process caused by party primaries and we will consider all eight candidates at once. First, we will analyze the political positions, affiliations, and characteristics of each candidate, then we will assess the political weights of given positions on issues, as well as the electorate probable reactions to personal characteristics of the candidates. Finally, we will ask various questions about the probable winner of the election. Let the frame of discernment, representing the eight candidates, be the set

$$\Theta = \{A, B, C, D, E, F, G, H\}$$

The political issues, affiliations and personal characteristics considered to be relevant to the electorate are listed in the first column of the following table. The entries in Figure 5 define each candidate relative position, i.e., his/her membership in the set defined by the row.

ISSUE/CHARACTERISTICS	CANDIDATE							
	A	B	C	D	E	F	G	H
Union Endorsement	0	0	0	0	1	0	0	0
Party Affiliation	R	R	R	L	D	D	D	I
Incumbent	1	0	0	0	0	0	0	0
No-new-taxes commitment	1	1	1	1	0	1	1	0
Moderate	0	1	1	0	1	0	1	1
No-scandal involvement	0	1	1	1	1	1	1	1
Charisma/Telegenic	0.8	0.2	1	0.5	1	0.7	0.9	0.5
Campaign Budget	9M	3M	2M	0.5M	6M	2M	4.5M	1M
Age	65	39	55	59	66	58	73	44

Figure 5: Issues, Positions, and Characteristics of the Candidates

Note that the first six rows represent crisp sets (union endorsement, party affiliation, incumbency, no-tax platform, moderate, no-scandals). The seventh row is a fuzzy set representing the degree to which the candidates can be called telegenic. The last two rows are facts about their campaign finances and their own ages. To use this information, we will have to provide the following additional interpretation.

Let's define the fuzzy set *RICH-CAMPAIGN-TREASURE* (on the universe of discourse of mega-dollars) by the fuzzy number (5, 10, 1.5, 10). The four parameter representation of a fuzzy number (A, B, α, β) is interpreted as follows: A is the smallest value of the universe of discourse for which the membership value is 1.0; B is the largest value of the universe of discourse for which the membership value is still 1.0; $(A - \alpha)$ is the value of the universe of discourse where the membership distribution (still at 0.0) begins increasing; $(B + \beta)$ is the value of the universe of discourse where the membership value is going back to 0.0 According to this definition, a rich campaign starts at 3.5M with membership 0, reaches membership 1 at 5M, stays at 1 until 10M, and goes back to membership 0 at 20M, where it is probably *very rich*.

Let's define the following fuzzy sets on the universe of discourse of years:

```
VERY-YOUNG  = (18 30 0 3)
YOUNG = (31 39 3 3)
MIDDLE-AGED = (45 55 4 4)
MATURE  = (55 64 2 3)
OLD = (65 72 2 2)
VERY-OLD = (73 80 2 20)
```

Let us now describe the electorate preferences and typical reactions. We will first assign *brownie points* to the various issues. Later, we will normalize them to add up to one and transform them into *bpas*.

1. About a third of the voters are blue-collar workers with union affiliation. At least half of these blue-collar workers are politically influenced by their union position. Let us translate this into 15 brownie points for whoever has the union endorsement.

2. The mayority of the electorate (60%) is registered democrat. However only one third of the democrats vote in strict accordance with party lines (20 brownie points for Democrats).

3. The power of incumbency is always present, although in this case it has been attenuated by a less then popular incumbent (10 brownie points to him).

4. "No new taxes" is a very popular slogan. The voters clearly like the idea. However, this issue has been inflated and adopted to various degrees by almost all the candidate, thus losing its discriminant power (10 brownie points to the committed ones).

5. Voters no longer trust liberal or extremely conservative candidates. Thus moderate, middle-of-the-road politicians have a better edge (25 brownie points to them).

6. Ethics are playing a stronger role in this election. Candidates who have been tainted by or marginally involved in scandals will be affected in their public image (20 brownie points to all the other guys).

7. As an unfortunate effect of the power of the media, the candidates' personal charisma, amplified by a good telegenic image, is becoming a stronger factor in deciding among them (35 brownie points to the good-looking candidates).

8. Having money never hurts: rich campaign finances allowed for long television ads in large electorate districts (25 brownie points to the riches).

9. Voters are reluctant to elect "green" politicians to important senior positions. However they are also rejecting the idea of electing candidate who are too old and may not have enough stamina and/or mental lucidity to last for the entire mandate. (20 brownie to the middle-age or mature candidates).

Let's summarize the above evaluations:

1. Union Endorsement (15)

2. Democrats (20)

3. Incumbent (10)

4. No new taxes (10)

5. Moderate (25)

6. No scandals (20)

7. Telegenic (35)

8. Rich (25)

9. Middle-aged/mature (20)

Let us first assume that only the first six evaluations represent the assignment of basic probabilities, generated by an expert political scientist. The sum of the brownie points is equal to 100, so we can directly consider them as bpas. This is summarized in Figure 6.

ISSUE/CHARACTERISTICS	CANDIDATE								bpa
	A	B	C	D	E	F	G	H	
Union Endorsment	0	0	0	0	1	0	0	0	$m_1 = .15$
Party Affiliation	0	0	0	0	1	1	1	0	$m_2 = .20$
Incumbent	1	0	0	0	0	0	0	0	$m_3 = .10$
No-new-taxes commitment	1	1	1	1	0	1	1	0	$m_4 = .10$
Moderate	0	1	1	0	1	0	1	1	$m_5 = .25$
No-scandal involvement	0	1	1	1	1	1	1	1	$m_6 = .20$

Figure 6: Issues, Positions, and Characteristics of the Candidates: A simplification

At this point we only have crisp data. Let us first consider a crisp query Q_1, and subsequently a fuzzy query Q_2.

Let Q_1 be the crisp query *"Will a moderate democrat win the election?"*. Q_1 is represented by the crisp set $\{E,G\}$.

The lower bound of Q_1 is $L(Q_1) = m_1 = 0.15$. The upper bound of Q_1 is $U(Q_1) = 1 - m_3 = 1 - 0.1 = 0.9$.

Let Q_2 be the fuzzy query *"Will an OLD and CHARISMATIC candidate win the election?"*. Q_2 is represented by the fuzzy set:

$$OLD \cap CHARISMATIC = \{1/A, 1/E, .5/G\} \cap \{.8/A, .2/B, 1/C, .5/D, 1/E, .7/F, .9/G, .5/H\}$$

$$= \{.8/A, 1/E, .5/G\}$$

The lower bound of Q_2 is $L(Q_2) = m_1 + (m_3 \times .8) = 0.15 + 0.08 = .23$.
The upper bound of Q_2 is $U(Q_2) = m_1 + m_2 + (m_3 \times .8) + (m_4 \times .8) + m_5 + m_6 = 0.96$.

Now we will modify the original description of the set of moderates and the set of politician not tainted by scandals. We will now consider the attribute *political moderate* to be described by a degree of membership. (Some politicians may adopt moderate/conservatives fiscal postures and yet maintain more liberal social postures). Similarly, we will consider the set of ethical politicians not involved in a scandal to be a fuzzy set: some politician have always been above suspicion; others were rumored to be unethical, but were later cleared by the evidence; others were investigated by a grand jury, but were never indicted due to insufficient evidence; finally, others got nailed (and jailed).

We will also evaluate the membership distribution of the candidates in the fuzzy sets *RICH-CAMPAIGN-TREASURE* (5 10 1.5 10) and *FROM MIDDLE AGED to MATURE* (45 64 4 3).

Finally we will normalize the total number of brownie points (180) to transform them into bpas, whose sum adds up to 1.0. This is illustrated in Figure 7.

ISSUE/CHARACTERISTICS	CANDIDATE								bpas
	A	B	C	D	E	F	G	H	
Union Endorsement	0	0	0	0	1	0	0	0	$m_1 = .0834$
Party Affiliation	0	0	0	0	1	1	1	0	$m_2 = .1111$
Incumbent	1	0	0	0	0	0	0	0	$m_3 = .0555$
No-new-taxes commitment	1	1	1	1	0	1	1	0	$m_4 = .0555$
Moderate	0	0.9	1	0	1	0.6	0.9	1	$m_5 = .1389$
No-scandal involvement	0	1	1	1	0.5	1	1	1	$m_6 = .1111$
Charisma/Telegenic	0.8	0.2	1	0.5	1	0.7	0.9	0.5	$m_7 = .1945$
Rich Campaign	1	0	0	0	1	0	0.66	0	$m_8 = .1389$
Middle-aged/Mature	1	0	1	1	0.33	1	0	0.75	$m_9 = .1111$

Figure 7: Issues, Positions, and Characteristics of the Candidates: a Refinement

Let us consider the same two queries $Q_1 = \{E, G\}$ and $Q_2 = \{.8/A, 1/E, .5/G\}$
The lower bound of Q_1 is $L(Q_1) = m_1 = 0.0834$.
The upper bound of Q_1 is $U(Q_1) = m_1 + m_2 + m_4 + m_5 + m_6 + m_7 + 0.666m_8 + 0.333m_9 = 1 - (m_3 + 0.333m_8 + 0.666m_9) = 0.8148$.

The lower bound of Q_2 is $L(Q_2) = m_1 + (1 - 0.2)m_3 + (1 - 0.5)m_8 = 0.19729$. Note that the portion of m_4 included in this lower bound is obtained by computing:

$$N(Q_2 \mid RICH) = 1 - \Pi(\neg Q_2 \mid RICH) = 1 - max[(.2/A, 0/E, .5/G) \cap (1/A, 1/E, 0.66/G)] =$$

$$1 - max[0.2, 0, 0.5] = 0.5$$

The upper bound of Q_2 is $U(Q_2) = m_1 + m_2 + 0.8m_3 + 0.8m_4 + m_5 + 0.5m_6 + m_7 + m_8 + m_9 = 1 - (0.2m_3 + 0.2m_4 + 0.5m_6) = 0.91111$.

Note that the cardinality of the query is roughly comparable or smaller than the cardinality of the bpas. This is the cause for relatively small lower bounds. If we increase the cardinality of the query, we will be able to contain more subsets of the frame of discernment in the query and therefore increase the lower (and upper) bounds. For instance, let us define the query Q_3 *"Will either an INCUMBENT or a CHARISMATIC MODERATE win the election?"*. Q_3 is defined by the fuzzy set $\{1/A, 0.2/B, 1/C, 1/E, 0.6/F, 0.9/G, 0.5/H\}$.

The lower bound of Q_3 is $L(Q_3) = m_1 + 0.6m_2 + m_3 + 0.2m_5 + 0.5m_7 + 0.9m_8 = 0.4306$.
The upper bound of Q_3 is $U(Q_3) = 1$.
So to summarize, using the bpas from Figure 6, we have:

$$Q_1 = [.1, .9] \quad Q_2 = [.23.96]$$

By using the bpas from Figure 7, we have:

$$Q_1 = [.0834, .8148] \quad Q_2 = [.1972.9111] \quad Q_3 = [.43061]$$

7 Going from Intervals to bpas

Let's assume two propositional statements A_1 and A_2. The belief of each statement is represented by an associated certainty interval, i.e., $[L(A_1), U(A_1)]$, $[L(A_2), U(A_2)]$.
It is possible to decompose any interval $[L(A_i), U(A_i)]$ into three *bps* assignments, i.e.:

$$L(A_i) \rightarrow \{A_i\}, 1 - U(A_i) \rightarrow \{\neg A_i\}, U(A_i) - L(A_i) \rightarrow \{\Theta\}$$

Under the assumption of evidential independence, it is possible to combine the belief of propositional statements A_1 and A_2, as it is illustrated in Figure 8

$A_1 \times A_2$	A_2		
A_1	$L(A_1) \times L(A_2)$ $\rightarrow \{\alpha_1 = A_1 \cap A_2\}$	$L(A_1) \times (1 - U(A_2))$ $\rightarrow \{\alpha_2 = A_1 \cap \neg A_2\}$	$L(A_1) \times (U(A_2) - L(A_2))$ $\rightarrow \{\alpha_3 = A_1\}$
	$L(A_2) \times (1 - U(A_1))$ $\rightarrow \{\alpha_4 = \neg A_1 \cap A_2\}$	$(1 - U(A_2)) \times (1 - U(A_1))$ $\rightarrow \{\alpha_5 = \neg A_1 \cap \neg A_2\}$	$(1 - U(A_1)) \times (U(A_2) - L(A_2))$ $\rightarrow \{\alpha_6 = \neg A_1\}$
	$L(A_2) \times (U(A_1) - L(A_1))$ $\rightarrow \{\alpha_7 = A_2\}$	$(1 - U(A_2)) \times (U(A_1) - L(A_1))$ $\rightarrow \{\alpha_8 = \neg A_2\}$	$(U(A_2) - L(A_2)) \times (U(A_1) - L(A_1))$ $\rightarrow \{\alpha_9 = \Theta\}$

Figure 8: Fusion of Two Propositions

Let's eliminate the pathological case of having either propositions being the empty set, i.e., we assume that $A_1 \neq \phi, A_2 \neq \phi$. Then, there are three possible cases:

Case 1: $A_1 \equiv A_2$

Case 2: $A_1 \neq A_2, A_1 \cap A_2 \neq \phi$

Case 3: $A_1 \neq A_2, A_1 \cap A_2 = \phi$

We will now examine each case.

7.1 Case I: Fusing Identical Propositions

Given that $A_1 \equiv A_2$, the entries of Figure 8 change to:
Given the query $Q = A_1 \cup A_2 = A_1 \cap A_2 = A_1$, we can now compute its normalized lower and upper bounds, using the entries of Figure 9.
We first obtain the lower bound $L(Q)$:

$$\begin{aligned} L(Q) &= m(\alpha_1) + m(\alpha_3) + m(\alpha_7) \\ &= [L(A_1) \times L(A_2)] + [L(A_1) \times (U(A_2) - L(A_2))] + [L(A_2) \times (U(A_1) - L(A_1))] \\ &= [L(A_1) \times U(A_2)] + [L(A_2) \times U(A_1)] - [L(A_1) \times L(A_2)] \end{aligned}$$

$A_1 \times A_2$	A_2		
A_1	$L(A_1) \times L(A_2)$ $\to \{\alpha_1 = A_1\}$	$L(A_1) \times (1 - U(A_2))$ $\to \{\alpha_2 = \phi\}$	$L(A_1) \times (U(A_2) - L(A_2))$ $\to \{\alpha_3 = A_1\}$
	$L(A_2) \times (1 - U(A_1))$ $\to \{\alpha_4 = \phi\}$	$(1 - U(A_2)) \times (1 - U(A_1))$ $\to \{\alpha_5 = \neg A_1\}$	$(1 - U(A_1)) \times (U(A_2) - L(A_2))$ $\to \{\alpha_6 = \neg A_1\}$
	$L(A_2) \times (U(A_1) - L(A_1))$ $\to \{\alpha_7 = A_1\}$	$(1 - U(A_2)) \times (U(A_1) - L(A_1))$ $\to \{\alpha_8 = \neg A_1\}$	$(U(A_2) - L(A_2)) \times (U(A_1) - L(A_1))$ $\to \{\alpha_9 = \Theta\}$

Figure 9: Fusion of Two Identical Propositions

The above result can be rewritten as:

$$L(Q) = [L(A_1) + L(A_2) - (L(A_1) \times L(A_2))] - [L(A_2) \times (1 - U(A_1)) + L(A_1) \times (1 - U(A_2))]$$

The above expression corresponds to:

$$L(Q) = S_2(L(A_1), L(A_2)) - L(\phi)$$

where $S_2(x, y) = x + y - xy$ is the triangular conorm corresponding to uncorrelation, and $L(\phi)$ is the conflict measure C, i.e., the lower bound of the empty set.
Then we obtain the upper bound U(Q):

$$
\begin{aligned}
U(Q) &= L(Q) + m(\alpha_9) \\
&= [L(A_1) \times U(A_2)] + [L(A_2) \times U(A_1)] - [L(A_1) \times L(A_2)] \\
&\quad + [U(A_1) \times U(A_2)] - [L(A_1) \times U(A_2)] - [L(A_2) \times U(A_1)] + [L(A_1) \times L(A_2)] \\
&= [U(A_1) \times U(A_2)] \\
&= T_2(U(A_1), U(A_2))
\end{aligned}
$$

where $T_2(x, y) = xy$ is the triangular conorm corresponding to uncorrelation.
The conflict measure C, i.e., the lower bound of the empty set is:

$$L(\phi) = m(\alpha_2) + m(\alpha_4) = [L(A_2) \times (1 - U(A_1)) + L(A_1) \times (1 - U(A_2))]$$

Then the *normalized* lower bound $L^*(Q)$ and upper bound $U^*(Q)$ are:

$$L^*(Q) = \frac{S_2(L(A_1), L(A_2)) - L(\phi)}{1 - L(\phi)}$$

$$U^*(Q) = \frac{T_2(U(A_1), U(A_2))}{1 - L(\phi)}$$

7.2 Case II: Fusing Non-disjoint Propositions

Given that $A_1 \neq A_2$, and $A_1 \cap A_2 \neq \phi$, there are no changes with the entries of Figure 8. Given the query $Q_1 = A_1 \cup A_2$, and $Q_2 = A_1 \cap A_2$, we can now compute their normalized lower and upper bounds.
We first obtain the lower bound $L(Q_1) = L(A_1 \cup A_2)$:

$$\begin{aligned}
L(Q_1) &= m(\alpha_1) + m(\alpha_2) + m(\alpha_3) + m(\alpha_4) + m(\alpha_7) \\
&= L(A_1) + L(A_2) - (L(A_1) \times L(A_2)) \\
&= S_2(L(A_1), L(A_2))
\end{aligned}$$

Then we obtain the upper bound $U(Q_1)$:

$$\begin{aligned}
U(Q_1) &= L(Q_1) + m(\alpha_6) + m(\alpha_8) + m(\alpha_9) \\
&= 1 - m(\alpha_5) \\
&= U(A_1) + U(A_2) - [U(A_1) \times U(A_2)] \\
&= S_2(U(A_1), U(A_2))
\end{aligned}$$

The conflict measure C, i.e., the lower bound of the empty set is:

$$L(\phi) = 0$$

Thus no normalization is required.

For the second query, $Q_2 = A_1 \cap A_2$, an analogous process yields:

$$\begin{aligned}
L(Q_2) &= m(\alpha_1) \\
&= (L(A_1) \times L(A_2)) \\
&= T_2(L(A_1), L(A_2))
\end{aligned}$$

$$\begin{aligned}
U(Q_2) &= L(Q_2) + m(\alpha_3) + m(\alpha_7) + m(\alpha_9) \\
&= U(A_1) \times U(A_2) \\
&= T_2(U(A_1), U(A_2))
\end{aligned}$$

7.3 Case III: Fusing Disjoint Propositions

Given that $A_1 \neq A_2$, and $A_1 \cap A_2 = \phi$, the entries of Figure 8 change to:

$A_1 \times A_2$	A_2		
A_1	$L(A_1) \times L(A_2)$ $\to \{\alpha_1 = \phi\}$	$L(A_1) \times (1 - U(A_2))$ $\to \{\alpha_2 = A_1\}$	$L(A_1) \times (U(A_2) - L(A_2))$ $\to \{\alpha_3 = A_1\}$
	$L(A_2) \times (1 - U(A_1))$ $\to \{\alpha_4 = A_2\}$	$(1 - U(A_2)) \times (1 - U(A_1))$ $\to \{\alpha_5 = \neg A_1 \cap \neg A_2\}$	$(1 - U(A_1)) \times (U(A_2) - L(A_2))$ $\to \{\alpha_6 = \neg A_1\}$
	$L(A_2) \times (U(A_1) - L(A_1))$ $\to \{\alpha_7 = A_2\}$	$(1 - U(A_2)) \times (U(A_1) - L(A_1))$ $\to \{\alpha_8 = \neg A_2\}$	$(U(A_2) - L(A_2)) \times (U(A_1) - L(A_1))$ $\to \{\alpha_9 = \Theta\}$

Figure 10: Fusion of Two Disjoint Propositions

Given the query $Q = A_1 \cup A_2$, we can now compute their normalized lower and upper bounds. (The other query $Q_2 = A_1 \cap A_2$ is equivalent to the empty set, as the propositions are disjoint by definition.)

Using the data shown in Figure 10, let us compute the lower bound $L(Q) = L(A_1 \cup A_2)$:

$$\begin{aligned}
L(Q) &= m(\alpha_2) + m(\alpha_3) + m(\alpha_4) + m(\alpha_7) \\
&= L(A_1) + L(A_2) - 2 \times (L(A_1) \times L(A_2)) \\
&= S_2(L(A_1), L(A_2)) - L(\phi)
\end{aligned}$$

We can then compute the upper bound U(Q):

$$\begin{aligned}
U(Q) &= L(Q) + m(\alpha_6) + m(\alpha_8) + m(\alpha_9) \\
&= 1 - m(\alpha_5) - m(\alpha_1) \\
&= U(A_1) + U(A_2) - [U(A_1) \times U(A_2)] - [(L(A_1) \times L(A_2))] \\
&= S_2(U(A_1), U(A_2)] - L(\phi)
\end{aligned}$$

The conflict measure C, i.e., the lower bound of the empty set is:

$$L(\phi) = m(\alpha_1) = L(A_1) \times L(A_2)$$

Then the *normalized* lower bound $L^*(Q)$ and upper bound $U^*(Q)$ are:

$$L^*(Q) = \frac{S_2(L(A_1), L(A_2)) - L(\phi)}{1 - L(\phi)}$$

$$U^*(Q) = \frac{S_2(U(A_1), U(A_2)) - L(\phi)}{1 - L(\phi)}$$

A subset of these results have been first reported by Baldwin in reference [11].

8 Conclusions

We have reviewed the theory of Dempster-Shafer (DS) and contrasted its less restricted basic probability assignments (*bpa* with the traditional probability assignments to singleton of the sample space. We have extended DS theory to the cases where data and/or queries are ill-defined (fuzzy) and we have illustrated these extensions with an example. Finally, we have shown a way of applying Dempster's rule of combination to propositions qualified by interval-based uncertainty measures (such as those provided by RUM [12], [13], [14]). This application is valid if the reasoning paths used in the derivation of the two propositions (to be combined) are assumed to be uncorrelated.

References:

[1] A.P. Dempster. Upper and lower probabilities induced by a multivalued mapping. *Annals of Mathematical Statistics*, 38:325–339, 1967.

[2] G. Shafer. *A Mathematical Theory of Evidence*. Princeton University Press, Princeton, New Jersey, 1976.

[3] E.H. Ruspini. Epistemic logic, probability, and the calculus of evidence. In *Proc. Tenth Intern. Joint Conf. on Artificial Intelligence*, Milan, Italy, 1987.

[4] E.H. Ruspini. The logical foundations of evidential reasoning. Technical Note 408, Artificial Intelligence Center, SRI International, Menlo Park, California, 1987.

[5] P. Smets. Belief functions. In P. Smets, A. Mamdani, D. Dubois, and H. Prade, editors, *Non-Standard Logics for Automated Reasoning*. Academic Press, New York, 1988.

[6] G.E. Hughes and M.J. Creswell. *An Introduction to Modal Logic*. Methuen, London, England, 1968.

[7] R. Yager. Generalized probabilities of fuzzy events from fuzzy belief structures. *Information Sciences*, 28:45–62, 1982.

[8] M. Ishizuka. An extension of dempster-shafer theory to fuzzy sets for constructing expert systems. *Seisan-Kenkyu*, 34:312–315, 1982.

[9] H. Ogawa and K. S. Fu. An inexact inference for damage assessment of existing structures. *International Journal of Man Machine Studies*, 22:295–306, 1985.

[10] John Yen. Generalizing the dempster-shafer theory to fuzzy sets. In *The Fourth Workshop on Uncertainty in Artificial Intelligence*, pages 382–390. 1988.

[11] J.F. Baldwin. Evidential support logic programming. *Fuzzy Sets and Systems*, 24:1–26, 1987.

[12] Piero P. Bonissone. Summarizing and Propagating Uncertain Information with Triangular Norms. *International Journal of Approximate Reasoning*, 1(1):71–101, January 1987.

[13] Piero P. Bonissone. Using T-norm Based Uncertainty Calculi in a Naval Situation Assessment Application. In *Proceedings of the Third AAAI Workshop on Uncertainty in Artificial Intelligence*, pages 250–261. AAAI, July 1987.

[14] Piero P. Bonissone, Stephen Gans, and Keith S. Decker. RUM: A Layered Architecture for Reasoning with Uncertainty. In *Proceedings 10th International Joint Conference on Artificial Intelligence*, pages 891–898. AAAI, August 1987.

RUM (Reasoning with Uncertainty Module) And RUMRUNNER (RUM's Run Time System)

Piero P. Bonissone

Artificial Intelligence Program
General Electric Corporate Research and Development
Schenectady, New York 12301, USA

Abstract

RUM (Reasoning with Uncertainty Module) is an integrated software tool based on KEE, a frame system implemented in an object oriented language. RUM's architecture is composed of three layers: *representation, inference,* and *control*. The representation layer is based on frame-like data structures that capture the uncertainty information used in the inference layer and the uncertainty meta-information used in the control layer. The inference layer provides a selection of five T-norm based uncertainty calculi with which to perform the intersection, detachment, union, and pooling of information. The control layer uses the meta-information to select the appropriate calculus for each context and to resolve eventual ignorance or conflict in the information. This layer also provides a context mechanism that allows the system to focus on the relevant portion of the knowledge base, and an uncertain-belief revision system that incrementally updates the certainty values of well-formed formulae (*wffs*) in an acyclic directed deduction graph.
RUMrunner, RUM's real-time counterpart, has two major components: a knowledge base compiler and a run-time inference engine. The KB compiler is used to avoid unnecessary run-time checks, searches, and value substitutions. The output of the KB compiler is a RETE-like network, which is interpreted by the run-time inference engine. This engine is capable of asynchronous processing, task scheduling, interrupt handling, rule class scoped forward/backward chaining, and planning to meet time-deadlines.

1   Introduction

The development of RUM and RUMrunner has been documented in a sequence of seven papers written by the author over the past few years. This section provides a brief synopsis of these papers. The reader is referred to the original papers for an in-depth treatment of the topic. The remaining sections will provide a succinct summary of a subset of these papers, by focusing on RUM, the development environment, and RUMrunner, its run-time counterpart.

The first of such papers, entitled *"Selecting Uncertainty Calculi and Granularity"* [1], establishes a theoretical basis for defining the syntax and semantics of a small subset of T-norm based uncertainty calculi, operating on a given term set of linguistic statements of likelihood. Each calculus is defined by specifying a negation, and a conjunction operator. Families of Triangular norms constitute the most general representations of conjunction operators. These families provide us with a formalism for defining an infinite number of different calculi of uncertainty. The term set defines the uncertainty *granularity*, i.e. the finest level of distinction among different quantifications of uncertainty. This granularity limits the ability to differentiate between two similar operators. Therefore, only a small finite subset of the infinite number of calculi produces notably different results. These results have been illustrated by two experiments where nine and eleven different calculi of uncertainty are used with three term sets containing five, nine, and thirteen elements, respectively.

The second paper, entitled *"Summarizing and Propagating Uncertain Information with Triangular Norms"* [2], defines a desiderata that any formalism for reasoning with uncertainty should try to satisfy. The desiderata is subdivided into three layers: *representation, inference* and *control.* The three layers deal with the proper *representation* of information and meta-information, the allowable *inference* paradigms suitable for the representation, and the efficient *control* of such inferences.

The third paper, entitled *"RUM: A Layered Architecture for Reasoning with Uncertainty"* [3], describes the integrated software tool that was built to meet the requirements defined by the desiderata. RUM's representation layer is based on frame-like data structures that capture the uncertainty information used in the inference layer and the uncertainty meta-information used in the control layer. Linguistic probabilities are used to describe the lower and upper bounds of the certainty measure attached to a well formed formula. The source and the conditions under which the information was obtained represent the non-numerical meta-information. RUM's inference layer provides the uncertainty calculi with which to perform the intersection, detachment, union, and pooling of information. Five uncertainty calculi, based on their underlying Triangular norms, are used in this layer. RUM's control layer uses the meta-information to select the appropriate calculus for each context and to resolve eventual ignorance or conflict in the information. This layer enables the programmer to declaratively express the local (context dependent) meta-knowledge that will substitute for the global assumptions traditionally used in uncertain reasoning.

The fourth paper, entitled *"Using T-norm Based Uncertainty Calculi in a Naval Situation Assessment Application"* [4], illustrates the application of the RUM tool to a problem of military situation assessment. RUM has been tested and validated in a sequence of experiments in both naval and aerial situation assessment (SA), consisting of correlating reports and tracks, locating and classifying platforms, and identifying intents and threats. An example of naval situation assessment is illustrated. The testbed environment for developing these experiments has been provided by LOTTA, a symbolic simulator implemented in Flavors. This simulator maintains time-varying situations in a multi-player antagonistic game where players must make decisions in light of uncertain and incomplete data. RUM has been used to assist one of the LOTTA players to perform the SA task.

The fifth paper, entitled *"Plausible Reasoning in Dynamic Classification Problems"* [5] addresses the issues of testing and validating knowledge base and inference techniques. To solve this problem, and to address a broader class of problems, referred to as *dynamic classification problems*, we have implemented a software architecture capable of generating, interpreting, and resolving complex time-varying scenarios. The test-bed architecture is composed of two parts: a simulation environment, *LOTTA*, and a reasoning system, *RUM*. The simulation environment maintains time varying situations in a multiple player antagonistic game where players assess situations and make decisions in light of uncertain and incomplete data.

The sixth paper, entitled *"Uncertainty and Incompleteness: Breaking the Symmetry of Defeasible Reasoning"* [6] describes the integration of defeasible reasoning (based on non-monotonic rules) with plausible reasoning (based on monotonic rules with partial degrees of sufficiency and necessity). Uncertainty measures are propagated through a Doyle-JTMS graph, whose labels are real-valued certainty measures. Unlike other default reasoning languages that only model the incompleteness of the information, our approach uses the presence of numerical certainty values to distinguish *quantitatively* the different admissible labelings and pick an *optimal* one. The key idea is to exploit the information on the monotonic links carrying uncertainty measures. A preference function based on such measures is used to select the extension, i.e., the fixed point of the nonmonotonic loop, which is maximally consistent with the soft constraints imposed by the monotonic links. Thus, instead of minimizing the cardinality of abnormality types [7] or of performing temporal minimizations [8], we maximize an expectation function based on the uncertainty measure. This method breaks the symmetry of the (potentially) multiple extensions in each loop by selecting a *most likely* extension. This idea is currently being implemented in PRIMO (Plausible ReasonIng MOdule), RUM's successor.

The seventh paper, entitled *"Now that I Have a Good Theory of Uncertainty, What Else Do I Need?"* [9], describes the technologies used for knowledge engineering (such as object-based simulator to exercise requirements, and development tools to build the KB and functionally validate it). The paper highlights the difference between development environment and run-time system, and describes the rule cross-compiler, and the real-time inference engine with meta-reasoning capabilities.

2   The Integrated RUM/RUMrunner Technology

2.1   The Rapid Prototyping Paradigm

In previous notes, we have observed that dynamic classification problems are characterized by an evolving set of requirements and need the use of the rapid prototyping methodology for their development [10]. The prototypes are developed in rich and flexible environments in which various AI techniques are used. A knowledge base is generated, debugged, modified, and tested until a "satisficing" solution [11] is obtained from this development phase. Then the prototype is ready for deployment: it is ported to specific platforms and embedded into larger systems. The deployment's success, however, depends on the application performing in *real-time*. If the reasoning system does not provide good timely information, then the application will not be able to react fast enough to its environment. Even after deployment, the prototype cycle must continue, because performance verification can only take place in a real-time environment. Thus, in order to meet the real-time requirements, the knowledge base and algorithms may need additional prototyping.

AI software development is significantly different from the traditional approach. It requires a prototyping cycle which spans two environments: development and target. Usually, instead of having to transition software between these two environments, one environment is eliminated. This approach, however, compromises either the flexibility and richness needed for development, or the speed and efficiency requirements of execution. When both environments are used, a smooth transition of the application between these two environments is essential. If the prototyping cycle cannot completely span the two environments, the knowledge engineer has to re-implement portions of the software.

The development of reasoning systems addressing dynamic classification problems presents another difficulty: testing and validating the knowledge base and inference techniques. To solve this problem, the paper proposes an integrated software architecture capable of generating, interpreting, and resolving complex time-varying scenarios. The test-bed architecture is composed of two parts: a *simulation environment* capable of maintaining the dynamic states of numerous simulated objects; and a *reasoning system* capable of dealing with the uncertain, incomplete, and time-varying information.

2.2 Simulation Environment

The simulation environment is composed of four basic modules: the *window subsystem*, a window based user interface for displaying maps; the *annotation subsystem*, an intelligent database for displaying time varying features; *LOTTA*, the simulator; and a set of tools for interfacing to a reasoning system. LOTTA is a symbolic simulator implemented in an object-oriented language (Symbolics Flavors [14]). LOTTA maintains time varying situations in a multiple player antagonistic game where players assess situations and make decisions in light of uncertain and incomplete data. LOTTA has no reasoning capabilities; these are provided by external reasoning modules, easily interfaced to the LOTTA data structures. LOTTA is further described in [12].

2.3 Reasoning System

The integrated reasoning system is composed of RUM [3], a rich, user-friendly development environment, and RUMrunner, a small and quick run-time system, and translation software to span the two (see Figure 1).

2.3.1 RUM: An Overview

RUM embodies the theory of plausible reasoning described in the previous section. RUM provides a representation of uncertain information, uncertainty calculi for inferencing, and selection of calculi for inference control. Uncertainty is represented in both facts and rules. A fact represents the assignment of a value to a variable. A rule represents the deduction of a new fact (conclusion) from a set of given facts (premises). Facts are qualified by a degree of *confirmation* and a degree of *refutation*. Rules are discounted by *sufficiency*, indicating the strength with which the premise implies the conclusion, and *necessity*, indicating the degree to which a failed premise implies a negated conclusion. The uncertainty present in this deductive process leads to considering several possible values for the same variable. Each value assignment is qualified by different uncertainties, which are combined with T-norm based calculi as described in [2] and [4].

RUM's rule-based system integrates both procedural and declarative knowledge in its representation. The rule-based approach captures expertise gained from experience or "rules

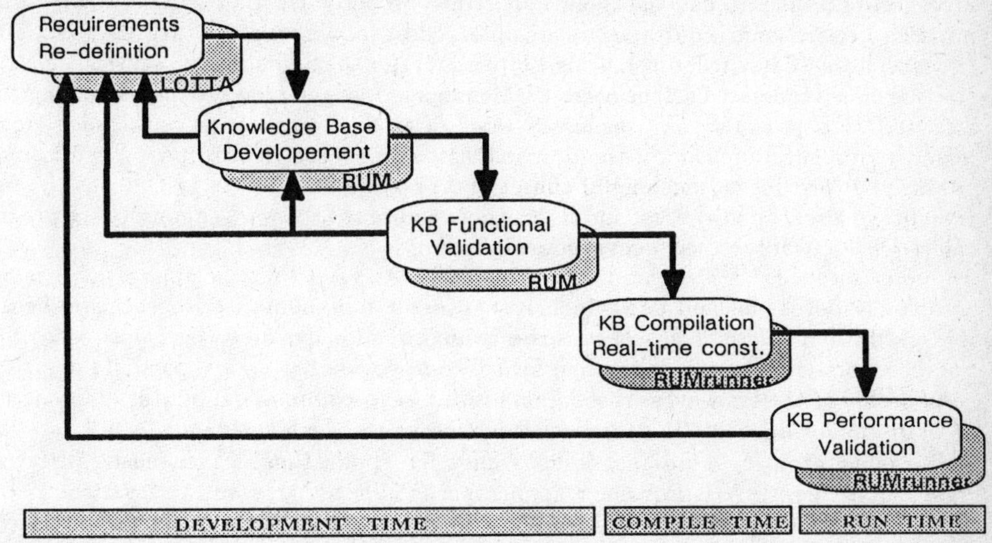

Figure 1: Software Engineering with RUM and RUMrunner

of thumb", thereby codifying heuristic knowledge without any underlying model. In addition, natural expression of procedural knowledge can be smoothly integrated through user-defined predicates in RUM rules. The integration of both techniques is essential to solve situation assessment problems, which involve both heuristic and procedural knowledge.

The expressiveness of RUM is further enhanced by two other functionalities: the *context mechanism*, to determine the rule's applicability to a given situation, and the *belief revision*, to detect changes in the input, keep track of the dependency of intermediate and final conclusions on these inputs, and maintain the validity of these inferences. These features will be described in Section 3 and are further elaborated upon in [3].

These AI capabilities are used to develop a knowledge base, in conjunction with RUM's software engineering facilities, such as flexible editing, error checking, and debugging. Some of these features, however, are no longer necessary once the development cycle is complete. At run-time, applications do not create new knowledge (facts or rules), as their basic structures have been determined at compile-time. The only run-time requirement is the ability to instantiate rules and facts from their pre-determined definitions. By eliminating the development features which are unnecessary at run-time, a real-time AI system can improve upon the algorithms and methodologies used in RUM.

2.3.2 *RUMrunner: An Overview*

The objective of RUMrunner [13] is to provide a software tool that transforms the customized knowledge base generated by the development phase, into a fast and efficient real-time application. RUMrunner provides both the functionality to reason about a broad set of problems, and the speed required to properly use the results of the reasoning process. Performance improvements are obtained by implementing all RUM's functionalities with leaner data structures, using Flavors [14] (for the Symbolics version) or *defstructs* (for the Sun version). Furthermore, RUMrunner no longer requires the use of the KEE software, thus it can be run on any Symbolics or Sun workstation with much smaller memory configurations, and without a KEE software license. RUMrunner has four major qualities: it provides a meaningful subset of AI techniques, it runs fast, it has the functionality of a real-time system, and it does not require the software engineer to re-program the application in the target environment.

To increase speed, RUMrunner takes advantage of the fact that the application has been completely developed and debugged. It provides a minimum of error checking because the application is assumed either to be debugged already, or to be robust enough to handle errors. RUMrunner's time performance in reasoning tasks is partially due to the compilation of the knowledge base. As a result of this compilation, new or different rules or units cannot be created in the knowledge base after the translation.

RUMrunner provides additional functionality for applications which must satisfy real-time requirements. A RUMrunner application is able to carry out and control a set of activities to rapidly respond to its environment. To meet these goals, the interface of RUMrunner with the application program is designed to be asynchronous, allowing the application to avoid unnecessary delays. In addition, the application is able to handle externally or internally driven interrupts. It is also able to prioritize tasks, by using an agenda mechanism [15], so that RUMrunner handles the most important ones first . RUMrunner is performance-conscious by ensuring that tasks execute within a specified amount of time. This is done through planning the execution of a single task as suggested by Durfee and Lesser [16]. Finally, RUMrunner is implemented in Common LISP, thus it can be ported to many machines without requiring any proprietary software. None of this additional functionality takes an unreasonable amount of time, and if not desired, most of it can remain unused without a great time penalty. These features are described in Section 4 and are further elaborated upon in [13].

3 RUM: The Development Environment

RUM is an integrated software tool based on KEE[1], a frame system implemented in an object oriented language [17]. The underlying theory of RUM, centered around the concept of Triangular norms, was described in two previous articles [2], [3]. RUM's architecture is composed of three layers: *representation, inference*, and *control*. A philosophical motivation for RUM's three layer organization can be found in [18]. This section summarizes some of the theoretical results and provides a unified framework for their interpretation and use in RUM's architecture.

3.1 Representation: the Wff System and the Rule Language

The representation layer is based on frame-like data structures that capture the uncertainty information used in the inference layer and the uncertainty meta-information used in the control layer.

---

[1]KEE is a trademark of IntelliCorp

3.1.1 RUM's Wff System

RUM's Wff System modifies KEE's representation of a *wff* (well-formed formula). RUM's *wff* is the pair [<*unit*> <*slot*>], which is the description of a variable in the problem domain. For each *wff* a corresponding uncertainty unit is created. The unit contains a list of the values that were considered for the *wff*. For each value the unit maintains its certainty's *lower* and *upper bounds*, an *ignorance measure*, a *consistency measure*, and the *evidence source*.

RUM's Wff System allows the user to express arbitrary uncertainty granularity by providing the flexibility to mix precise and imprecise measures of certainty in defining the input certainty (points, intervals, fuzzy numbers/intervals, linguistic values) and the rule strengths (categorical and plausible IF/IFF). Various term sets of linguistic probabilities with fuzzy-valued semantics [1] provide a selection of input granularity. The values of the terms can be used as default values or can be modified by the user.

The following table, taken from reference [3], illustrates *L-nine*, one of RUM's four term sets. The meaning of each term in L-nine is represented by a fuzzy number on the [0,1] interval. A four parameter representation (A, B, α, β) [19], [1] is used to characterize each fuzzy number. The first two parameters $[A, B]$ indicate the interval in which the membership value is 1.0. The third and fourth parameters, α and β, indicate the left and right *width* of the distribution.[2] Linear functions are used to define the slopes. The meanings associated with each element of the term set were derived from the results of psychological experiments on the use of linguistic probabilities [20].

Index	Symbol	Meaning
1	impossible	(0 0 0 0)
2	extremely unlikely	(.01 .02 .01 .05)
3	very low chance	(.1 .18 .06 .05)
4	small chance	(.22 .36 .05 .06)
5	it may	(.41 .58 .09 .07)
6	meaningful chance	(.63 .80 .05 .06)
7	most likely	(.78 .92 .06 .05)
8	extremely likely	(.98 .99 .05 .01)
9	certain	(1 1 0 0)

Figure 2: Nine Element Term Set

3.1.2 RUM's Rule System: The Rule Language

RUM's Rule System replaces KEE Rule System-3 capabilities by incorporating uncertainty information in the inference scheme. The uncertain information is described in the uncertainty units of the *wffs*, represented in RUM's Wff System, and in the degrees of

---

[2] A is the smallest value of the universe of discourse for which the membership value is 1.0; B is the largest value of the universe of discourse for which the membership value is still 1.0; $(A - \alpha)$ is the largest value smaller than A, where the membership distribution is still 0.0; $(B + \beta)$ is the smallest value larger than B, where the membership value is back to 0.0.

sufficiency and necessity attached to each rule.[3]
The degree of sufficiency denotes the extent to which one should believe in the rule conclusion, if the rule premise is satisfied. The degree of necessity indicates the confidence with which one can negate the conclusion, if the premise fails.

A rule is internally represented by a frame with several slots. These slots include the name of the rule; the lists of contexts, premises, and conclusions; the rule's sufficiency and necessity; and the T-norm to be used for aggregation. All slots (except the name, premises, and consequences) have default values. The contexts, premises, and conclusions can comprise values, variables, RUM predicates and arbitrary LISP functions. Rules with unbound variables are instantiated with the necessary environment to produce rule instances.

The T-norm specified with each rule is used to aggregate the certainties of the rule premises and to perform detachment (which computes the certainty of the conclusion given the sufficiency and necessity of the rule). It defaults to T_3, which is the MIN function. The associated T-conorm is used to aggregate the certainties of identical conclusions inferred by multiple rule instances derived from the *same* rule. These are often subsumptive, and the value defaults to S_3, the MAX function. Finally, each separate consequence of a rule has a specified T-conorm that will be used to aggregate the consequence with identical consequences derived from *different* rules. (i.e., multiple assignments of the same value to the wff). The negation operator causes the wff to be assigned the complemented value.[4]

3.2 Inference: Triangular norms (T-norms) Based Calculi

The inference layer is built on a set of five Triangular norms (T-norms) based calculi. The T-norms' associativity and truth functionality entail problem decomposition and relatively inexpensive belief revision. The theory of T-norms has been covered in previous articles [1], [18], [2], [4], [3]. A brief review of their definition and their use in RUM is included for the reader's convenience.

3.2.1 Background Information on T-norms

Triangular norms (T-norms) and Triangular conorms (T-conorms) are the most general families of binary functions that satisfy the requirements of the conjunction and disjunction operators, respectively. T-norms and T-conorms are two-place functions from $[0,1] \times [0,1]$ to $[0,1]$ that are monotonic, commutative and associative. Their corresponding boundary conditions, i.e., the evaluation of the T-norms and T-conorms at the extremes of the $[0,1]$ interval, satisfy the truth tables of the logical AND and OR operators.

---

[3] It is important to note that the inference symbol $\rightarrow$ in the production rule $A \overset{s}{\rightarrow} B$ is interpreted as a (weak) *material implication* operator in multiple-valued logics. The value s is the lower bound of the degree of sufficiency of the implication. This is in contrast with the interpretation of *conditioning*, i.e., $s = P(B \mid A)$. The symbol $\leftrightarrow$ in the production rule $A \overset{s,n}{\leftrightarrow} B$ is interpreted as a (weak) *logical equivalence* operator in multiple-valued logics, in which s and n are the lower bounds of sufficiency and necessity, respectively. This (weak) logical equivalence is an *if-and-only-if* (IFF) rule, which can be decomposed into the two rules: $A \overset{s}{\rightarrow} B$ and $B \overset{n}{\rightarrow} A$ (equivalent to $\neg A \overset{n}{\rightarrow} \neg B$). RUM's rules are of the type: $C \rightarrow (A \overset{s,n}{\leftrightarrow} B)$, where C indicates the context of the rule (see Section 3.3.4) and $\rightarrow$ represents the strong material implication.

[4] If a *wff* has a value A with an If the certainty interval attached to a value A is [L(A), U(A)], its complemented value, $\neg A$, has a certainty interval defined by [1-U(A), 1-L(A)].

In a previous paper [1], six parametrized families of T-norms and dual T-conorms were discussed and analyzed by the author. Of the six parametrized families, one family was selected due to its complete coverage of the T-norm space and its numerical stability. This family, originally defined by Schweizer & Sklar [21], was denoted by $T_{Sc}(a, b, \mathbf{p})$, where $\mathbf{p}$ is the parameter that spans the space of T-norms. More specifically:

$$
\begin{aligned}
T_{Sc}(a, b, \mathbf{p}) &= (a^{-p} + b^{-p} - 1)^{-\frac{1}{p}} & \text{if } (a^{-p} + b^{-p}) \geq 1 & \quad \text{when } p < 0 \\
T_{Sc}(a, b, \mathbf{p}) &= 0 & \text{if } (a^{-p} + b^{-p}) < 1 & \quad \text{when } p < 0 \\
T_{Sc}(a, b, \mathbf{0}) &= \lim_{p \to 0} T_{Sc}(a, b, \mathbf{p}) = ab & & \quad \text{when } p \to 0 \\
T_{Sc}(a, b, \mathbf{p}) &= (a^{-p} + b^{-p} - 1)^{-\frac{1}{p}} & & \quad \text{when } p > 0
\end{aligned}
$$

Its corresponding T-conorm, denoted by $S_{Sc}(a, b, \mathbf{p})$, was defined as:

$$
S_{Sc}(a, b, \mathbf{p}) = 1 - T_{Sc}(1 - a, 1 - b, \mathbf{p})
$$

In the same paper it was shown that the use of term sets determines the granularity with which the input certainty is described. This granularity limits the ability to differentiate between two similar calculi; the numerical results obtained by using two calculi whose underlying T-norms are very close in the T-norm space will fall within the same granule in a given term set. Therefore, only a finite, small subset of the infinite number of calculi that can be generated from the parametrized T-norm family produces notably different results. The number of calculi to be considered is a function of the uncertainty granularity. This result was confirmed by an experiment [1] where eleven different calculi of uncertainty, represented by their corresponding T-norms, were analyzed. To generate the eleven T-norms, the parameter $\mathbf{p}$ in Schweizer's family was given the following values:

$$
-1, -0.8, -0.5, -0.3, 0, 0.5, 1, 2, 5, 8, \infty
$$

The experiment showed that five equivalence classes were needed to represent (or reasonably approximate) any T-norm, when term sets with at most thirteen elements were used. The corresponding five uncertainty calculi were defined by the common negation operator $N(a) = 1-a$ and the DeMorgan pair $(T_{Sc}(a, b, \mathbf{p}), S_{Sc}(a, b, \mathbf{p}))$ for the following values of p:

$$
\begin{aligned}
p = -1 \quad & T_1(a, b) = max(0, a + b - 1) \\
& S_1(a, b) = min(1, a + b)
\end{aligned}
$$

$$
\begin{aligned}
p = -0.5 \quad & T_{Sc}(a, b, -\mathbf{0.5}) = max(0, a^{0.5} + b^{0.5} - 1)^2 \\
& S_{Sc}(a, b, -\mathbf{0.5}) = 1 - max(0, (1 - a)^{0.5} + (1 - b)^{0.5} - 1)^2
\end{aligned}
$$

$$
\begin{aligned}
p \to 0 \quad & T_2(a, b) = ab \\
& S_2(a, b) = a + b - ab
\end{aligned}
$$

$$
\begin{aligned}
p = 1 \quad & T_{Sc}(a, b, \mathbf{1}) = (a^{-1} + b^{-1} - 1)^{-1} \\
& S_{Sc}(a, b, \mathbf{1}) = 1 - ((1 - a)^{-1} + (1 - b)^{-1} - 1)^{-1}
\end{aligned}
$$

$$
\begin{aligned}
p \to \infty \quad & T_3(a, b) = min(a, b) \\
& S_3(a, b) = max(a, b)
\end{aligned}
$$

RUM's inference layer provides the user with a selection of the five T-norm based calculi described above. They are referred to as $T_1, T_{1.5}, T_2, T_{2.5}, T_3$, respectively.

3.2.2 Operations in a T-norm Based Calculus

For each calculus, four operations are defined in RUM's Rule System: *premise evaluation, conclusion detachment, conclusion aggregation,* and *source consensus.* Each operation in a calculus can be completely defined by a Triangular norm $T(.,.)$, and a negation operator $N(.)$, just as in classical logic any boolean expression can be rewritten in terms of an intersection and complementation operator. A formal justifications for the following definitions can be found in References [2], [3]. The four operations are defined as follows:
Premise evaluation: The premise evaluation operation determines the degree to which all the clauses in the rule premise have been satisfied by the matching *wffs*. Let b_i and B_i indicate the lower and upper bounds of the certainty of condition i in the premise of a given rule. Then the premise certainty range [b,B] is defined as:

$$[b, B] = [T(b_1, b_2, \ldots, b_m), T(B_1, B_2, \ldots, B_m)]$$

Conclusion Detachment: The conclusion detachment operation indicates the certainty with which the conclusion can be asserted, given the strength and appropriateness of the rule. Let s and n be the lower bounds of the degree of *sufficiency* and *necessity,* respectively, of the given rule, and let [b,B] be the computed premise certainty range. Then the range [c,C], indicating the lower and upper bound for the certainty of the conclusion inferred by such rule, is defined as:

$$\begin{aligned} [c, C] &= [T(s, b), S(N(n), B)] \\ &= [T(s, b), N(T(n, N(B)))] \end{aligned}$$

The degrees of sufficiency and necessity respectively indicate the amount of certainty with which the rule premise implies its conclusion and vice versa. The sufficiency degree is used with *modus ponens* to provide a lower bound of the conclusion. The necessity degree is used with *modus tollens* to obtain a lower bound for the complement of the conclusion (which can be transformed into an upper bound for the conclusion itself).
Conclusion aggregation: The conclusion aggregation operation determines the consolidated degree to which the conclusion is believed if supported by more than one path in the rule deduction graph, i.e., by more than one rule instance. Each group of deductive paths can have a distinct conclusion aggregation operator associated with it. Let the ranges $[c_j, C_j]$ indicate the certainty lower and upper bounds of the *same* conclusion inferred by m rules instances belonging to the same group. Then, for each group of deductive paths, the range [d,D] of the aggregated conclusion is defined as:

$$\begin{aligned} [d, D] &= [S(c_1, c_2, \ldots, c_m), S(C_1, C_2, \ldots, C_m)] \\ &= [N(T(N(c_1), N(c_2), \ldots, N(c_m)), T(N(C_1), N(C_2), \ldots, N(C_m)))] \end{aligned}$$

Source Consensus: The source consensus operation reflects the fusion of the certainty measures of the same evidence A. provided by different sources. The evidence can be an *observed* fact, or a *deduced* fact. In the former case, the fusion occurs before the evidence is used as an input in the deduction process. In the latter case, the fusion occurs after the evidence has been aggregated by each group of deductive paths. The source consensus operation reduces the ignorance about the certainty of A, by producing an interval that is always smaller or equal to the smallest interval provided by any of the information source. If there is an inconsistency among some of the sources, the resulting certainty intervals will be disjoint, thus introducing a conflict in the aggregated result.

Let $[L_1(A), U_1(A)], [L_2(A), U_2(A)], \ldots, [L_n(A), U_n(A)]$ be the certainty lower and upper bounds of the same conclusion provided by n different sources of information. Then, the result $[L_{tot}(A), U_{tot}(A)]$, obtained from *fusing* all the assertions about A, is given by taking the intersection of the certainty intervals:

$$\begin{aligned} [L_{tot}(A), U_{tot}(A)] &= [Max_i(L_i(A)), Min_i(U_i(A))] \\ &= [S_3(L_i(A)), T_3(U_i(A))] \end{aligned}$$

3.3 Control: Calculus selection, Belief Revision, Context Mechanism

3.3.1 Calculi Selection

As it was discussed in the previous section, RUM's Rule System uses a set of five T-norm based calculi. The calculus used by each rule instance is inherited from its rule subclass (the rule before the instantiation). The calculus can be modified through KEE's user interface or programmatically (i.e., by an active value). Class inheritance can also be used to modify the degree of sufficiency and necessity of all the rule members of the same class.

The calculi selection consists of two assignments. The first assignment indicates the T-norm with which the premise evaluation and the conclusion detachment will be computed. Such an assignment is made for each rule, and, through inheritance, is passed to all rule instances derived from the same rule.

The second assignment indicates the T-conorm (represented by its dual T-norm) with which the conclusion aggregation will be computed. This assignment is made for each subset of rule instances generated from *different* rules and asserting the same conclusion.

3.3.2 Rationale for Calculi Selection

The T-norm characteristics will determine the selection choices. For the first assignment, the T-norm assigned to each rule for the premise evaluation and the conclusion detachment will be a function of the decision maker's *attitude toward risk*. The ordering of the T-norms, which is identical to the ordering of parameter p in the Schweizer & Sklar family of T-norms, reflects the ordering from a conservative attitude ($p = -1$ or T_1) to a non-conservative one ($p \to \infty$ or T_3). From the definition of the calculi operations, we can see that T_1 will generate the smallest premise evaluation and the weakest conclusion detachment (i.e., the widest uncertainty interval attached to the rule's conclusion). T-norms generated by larger values of p will exhibit less drastic behaviors and will produce nested intervals with their detachment operations. T_3 will generate the largest premise evaluation and the strongest conclusion detachment (the smallest certainty interval).

For the second assignment, the T-norm assigned to the subsets of rule instances (derived from different rules and asserting the same conclusion) will be a function of the *lack or presence of positive/negative correlation* among the rules in each subset. The ordering of the T-norms reflects the transition from the case of extreme negative correlation, i.e., mutual exclusiveness (T_1), through the case of uncorrelation (T_2), to the case of extreme positive correlation, i.e., subsumption (T_3).

Currently, all calculi assignments are explicitly made and modified through the user interface, to exercise the implemented accessing functions. In the next development phase of RUM control layer, the calculi assignments will be made by a set of selection rules expressing the meta-knowledge about the context. These rules will select the T-norms that better reflect the knowledge engineer's desired attitude toward risk and the perceived amount of correlation among the rules used in such a context.

3.3.3 Uncertain-Belief Revision

A daemon-based implementation of the belief revision of the uncertain information is available in the control layer of RUM's Rule System. For any conclusion made by a rule, the belief revision mechanism monitors the changes in the certainty measures of the *wffs* that constitute the conclusion's support or the changes in the calculus used to compute the conclusion certainty measure. Validity flags are inexpensively propagated through the rule deduction graph. Five types of flag values are used:

Good Guarantees the validity of the cached certainty measure detached by the rule instance and aggregated into the associated wff.

Bad (level i) Indicates that the cached certainty measure detached by the rule instance is no longer reliable, since the support of some of the wff's in the premise of this rule instance has changed. The ith level indicates the correct order of recomputation.

Inconsistent Indicates that the cached certainty measure associated with the wff is conflicting. The inconsistency can be removed by executing a locally defined procedure (differential diagnosis type of experiment, recency of information, split in possible words with subsets of the original sources, etc.)

Not Applicable Indicates that the context of the rule instance is no longer active and the rule instance contribution to the aggregated certainty measure of the wff should be ignored.

Ignorant Indicates that the cached certainty measure detached by the rule instance is too vague to be useful. The default behavior is to ignore the rule instance contribution to the aggregated certainty measure of the wff. Locally defined procedure could be used to remove the ignorance if so specified.

3.3.4 Rule Firing Control via Context Activation

A user-definable threshold can be attached to each rule context, either by local definition or by inheritance from a rule class. A rule context is defined as a conjunction of conditions that must be satisfied before the rule can be considered for premise evaluation. Each condition is described by a predicate on object-level *wffs* (facts in problem domain), or control-level *wffs* (markers asserted by meta-rules). The semantics of a context C attached to an inference rule (establishing the weak logical equivalence between A and B) is given by the following expression:

$$C \rightarrow (A \overset{s,n}{\leftrightarrow} B)$$

where s and n indicate the lower bounds of the degree of sufficiency and necessity that the rule provides; $\rightarrow$ represents the strong material implication; $\leftrightarrow$ denotes the weak logical equivalence.

The context mechanism provides the following features:

- By activating/deactivating subsets of the KB, it limits the number of rules that will be considered relevant at any given time, thus increasing the overall system efficiency.

- By only considering the rules relevant to a given situation, it allows the knowledge engineer to effectively use the necessary conditions in the rule's premise. It is now possible to distinguish between the failure of a necessary test (described in the premise) and the failure of the rule's applicability (traditionally described by other clauses in the same premise and now explicitly represented in the context).

- By using predicates on the control-level *wffs*, it provides the required programmability for defining flexible control strategies, such as causing sequences of rules to be executed, firing default rules, ordering and handling time-dependent information, etc.

- By using hierarchical contexts, it can be used as an organizing principle for the knowledge acquisition task.

3.4   RUM's Constraints: Descriptions and Definitions

The decision procedure for a logic based on real-valued truth values is (potentially) much more computationally expensive than the decision procedure for crisp logic. In crisp logic only one proof is needed to establish the validity of a theorem. In real-valued logic all possible proofs must be explored in order to ensure that the certainty of a proposition has been maximized. RUM deals with.the possible computational explosion through a series of trade-offs:

1. RUM allows the user to create rules with variables (rule-templates). These variables represent complex objects with multiple attributes. The typical problems of first order reasoning are avoided by instantiating the rule templates at run time. The implicit universal quantifier (that determines the scope of the variables in the rules) is replaced by on ongoing enumeration of the instances of the objects encountered by the reasoning system. Thus, a single rule may give rise to many rule instances at run time, but all these rule instances will be propositional.

2. RUM does not allow cyclic (monotonic) rules. Given the rule $P \rightleftharpoons_n^s Q$, RUM uses the amount of confirmation of P, the degree of sufficiency s, and *modus ponens* to derive the amount of confirmation of Q. Similarly, by using the amount of refutation of P, the degree of necessity n and *modus tollens*, RUM derives the amount of refutation of Q. The other two modalities (necessity and *modus ponens*, sufficiency and *modus tollens*) are not used, as they would determine values of P from Q.

    Note that most T-norms are strictly monotonically decreasing. Cyclic rules that propagate their truth values using these T-norms will continually cycle until all the truth values are 0 (unless relaxations, tagging and counting, or other methods were used). Only one T-norm (*min*) satisfies the axioms of a (pseudo-complemented) lattice,[5] and therefore would not exhibit this problem.

---

[5] Idempotency (or equivalently, Non-Archimedean) is the the required T-norm property.

3. P and $\neg P$ are essentially treated independently. The certainty of P is represented by the LB of P. The certainty of $\neg P$ is represented as the negation of the UB of P. With one exception, the LB and UB do not affect each other. That exception is when the LB becomes greater than the UB, which roughly corresponds to the situation when both P and $\neg P$ are believed with high degrees of certainties. When this happens a conflict handler tries to detect the source of the inconsistency.

4. Disjunctions in the conclusions of rules are not allowed.

These restrictions amount to allowing only acyclic quantitative Horn clauses. The following definitions formalize some of the above constraints.

Definitions: A RUM specification is a triple (W, I, J). W is a set of wff's, such that whenever w $\in$ W, $\overline{w} \in$ W. For w $\in$ W, LB(w) $\in [0, 1]$ is the amount of evidence confirming the truth of w. $1 -$ UB(w) $\in [0, 1]$ is the amount of evidence confirming the falsity of w. LB($\overline{w}$) $= 1 -$ UB(w).[6] I $\subset$ W, is a distinguished set of input wffs, that could potentially take on values from the outside world. J is a set of justifications. Each justification is a triple (P, s, c). P $\subset$ W is the premises of the justification. s $\in [0, 1]$ is the sufficiency of the justification. c $\in$ W is the conclusion of the justification.

Definition: A conflict occurs when $\exists$ w $\in$ W s.t. LB(w) $+$ LB($\overline{w}$) > 1.

Definitions: A RUM rule graph is a triple (A, O, E). Where A is the AND nodes in an AND/OR graph, O is the OR nodes, and E is the arcs (n1, n2). The RUM rule graph (A, O, E) of a RUM specification (W, J) is given by A = J, W = O, and (j $\in$ J, w $\in$ W) $\in$ E iff J=(P, s, c) $\wedge$ w = c. (w $\in$ W, j $\in$ J) $\in$ E iff J=(P, s, c) $\wedge$ w $\in$ P. Additionally each arc emanating from a justification is labeled with a real number $\in [0, 1]$ representing its contribution to its conclusion.

Definition: A valid RUM specification is one in which the corresponding rule graph is acyclic.

Definition: A RUM rule graph is admissible iff:

1. the label of each arc leaving a justification equals the T-norm of the arcs entering the justification and the LBs of the premises of the justification and

2. the LB of each wff is the S-conorm of the labels of the arcs entering it.

Due to these restrictions a simple linear time algorithm is able to propagate the correct numeric bounds through a valid RUM rule graph to generate an admissible rule graph. The only possible exponential step is when the conflict handler has to resolve an inconsistency.

4   RUMrunner: The Run-Time System

The approach followed in developing RUMrunner is predicated on the following three steps: *Knowledge Base compilation; execution-time estimation; run-time/real-time execution.* This process is indicated in Figure 3.

---

[6]Note this is the same relationship as that between support and plausibility in Dempster-Shafer theory and that between $\square$ and $\diamond$ in modal logics.

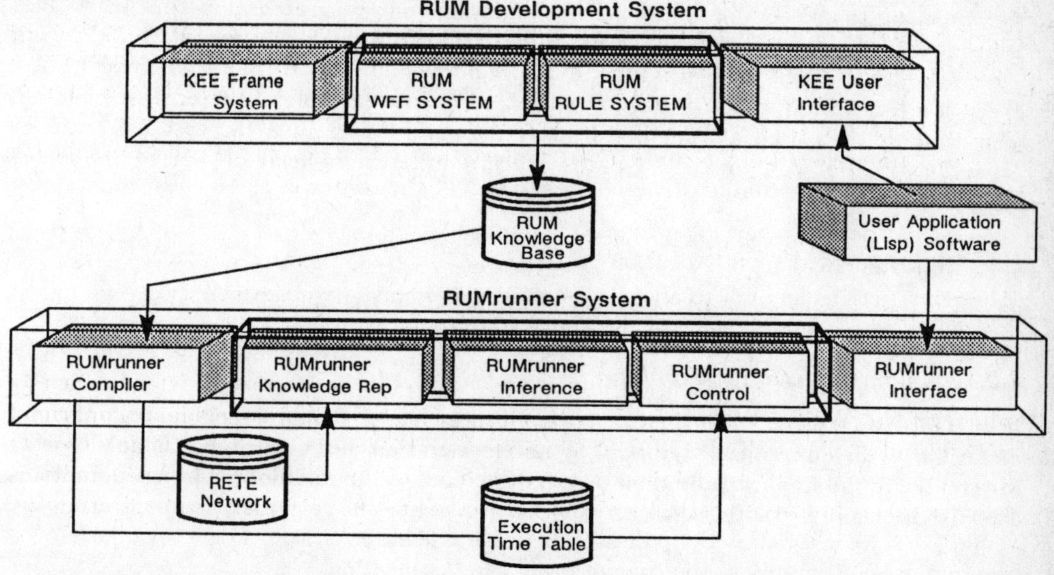

Figure 3: Uncertainty Tool Set Architecture

4.1 Knowledge Base Compilation

One of the most natural steps used to improve the performance of a program is to compile it to avoid unnecessary run-time checks, searches, and value substitutions. In the case of a knowledge base, the compilation is done at the representation language level (beside the traditional compilation from source code to object code, done at the programming language level). Following this philosophy, we have developed a RUM-rule compile, composed of three components:

1. *Reader/Translator* - to read the information stored in the hierarchical rule base, originally written in customized macro-expressions and then expanded into a graph of KEE Units

2. *Analyzer* - to derive and analyze the rule topology (identifying the sharing of common structures), and to create the pointers between the declarative part of the knowledge base (captured by the rule dependency) and the procedural part of the knowledge base (captured by the system or user-defined predicates)

3. *Code Generator* - to write the directed acyclic graph, produced by the Analyzer as a modified RETE network, into an output file to be used by the run-time/real-time inference engine.

4.2 Execution-time Estimation

During the development of the knowledge base, an object oriented simulator (LOTTA) was used to create and run a variety of scenarios representing a substantial set of requirements to be met by the reasoning system. The simulation runs generated numerous track-files, which were then stored. After the rule compilation, the same representative data from the track-files is used to fire each individual predicate and rule. During this process, their execution times are measured and logged into a dedicated table of performance timing. This table is then used at run-time by the planner to determine the cost (in terms of execution time) of using deductive paths which require the re-firing of specific predicates and rules.

4.3 Run-time/Real-time Execution

The real-time execution of RUMrunner is supported by the following features:

4.3.1 Automatic Rule Instantiation

Rule template, written by the knowledge engineer, contain variables scoped by a universal quantifier. This quantifier is replaced by an enumeration of its instances, as new objects are created in the scenario, and their properties bound to the rule slots. The rule templates are automatically instantiated as new objects (linked to the rules) are created during the scenarios

4.3.2 Caching of Node State

Each node in the Direct Acyclic Graph (DAG) generated by the rule compiler, is either a *variable*, a *predicate*, or a *rule*. Each node has its own state: variable-nodes store all their current value-assignments (and their tests on the value distribution of the tested variables; rule-nodes store the results of their detachments (rule conclusions), which are asserting a value assignment for some other variable-node in the network. The local state of each node in the DAG has a validity flag, indicating whether the information in the cache can be fetched, or should be recomputed due to changes or obsolescence. The validity flag are maintained by a belief revision system similar to the one used in RUM.

4.3.3 Asynchronous Processing

An agenda mechanism is used to asynchronously receive any number of input tasks (such as backward-chaining on a goal or forward-chaining on a given piece of evidence) from various sources. The agenda scheduler (see below) sorts the tasks according to their characteristics. The outputs of RUMrunner are isolated from other connecting systems via buffers or streams.

4.3.4 Task Priorities and Deadlines

Each task in the agenda receives a (static) level number, determining the relative priority of the task with respect to the other ones. A time-deadline, expressed in absolute time, is attached to the task to indicate its urgency (i.e., its expiration time). The task are sorted by priority and, within the same priority level, by the shortest deadline.

4.3.5 Individual Processes

Each task in the agenda is assigned a separate process. Such process is then terminated, once the task is completed. This capability requires an underlying operating system supporting multi-tasking.

4.3.6 Interrupts

External or internal interrupts, with re-entrant reasoning, can superseed the current task. Three types of interrupts have been implemented: the internal interrupts caused by queries approaching their assigned time-deadlines; the external interrupts caused by queries with higher priority than the one currently addressed; and the external interrupts caused by new input data characterized by higher priority than the current query.

4.3.7 Scoping the KB by Rule Classes

Design-time partitions of the Knowledge Base, expressed by a hierarchy of rule classes and sub-classes, can be exploited at run-time by adding to the task an optional argument describing the subset of the KB (denoted as a list of rule classes/sub-classes), which is relevant to the task. As a result, forward and backward chaining can be scoped by this optional argument.

4.3.8 Planning to Meet Deadlines

A run-time planning system determines the largest amount of redundancy (parallel proof trees) which the system can afford to use and still meet its time-deadline. At run-time, a graph traversal determines the validity flags of the nodes in the sub-graph used to solve the task. The cost of those nodes whose validity flag is requiring a re-execution are retrieved from the tables. With this information the total estimated time for the path is computed. By taking into account the sharing of common nodes in different paths, the planner maximizes the coverage of the sub-graph that can be executed within the allocated time budget (determined by the task deadline and the current clock-time). A current option for the planner is to use an upper bound of the amount of certainty potentially provided by each path in the compiled network (i.e., the minimum of the sufficiency values attached to the links of the path). This bound is a static measure of information content, which can be used in conjunction with the estimated execution cost to determine the set of reasoning paths to be used.

5 RUM/RUMrunner Applications

Over the last two years, the above technology has been applied to a variety of dynamic classification problems, a subset of which are illustrated in this section.

The first application of this technology has been the Situation Assessment (SA) Module of DARPA's Pilot's Associate (PA) Program. Pilot's Associate aims to improve the combat effectiveness of post-1995 aircraft through the application of mature AI technologies. The Situation Assessment Module is one of six modules within the overall PA system. SA is concerned with analyzing the external environment, such as determining enemy threats to ownship. RUM has been used by GE-CRD and Lockheed Georgia to develop two submodules within the Situation Assessment module. One submodule determines the class and type of enemy aircraft; the second submodule determines the Target Value of an aircraft. Target Value is composed of target importance, target opportunity, target capability and the intent of the bogey. The total KB size is approximately 200 rule-templates.

Another situation assessment module, applied to the submarine problem domain, is currently been developed by the GE-CRD AI Program. The actual Knowledge Base, developed over the last three months, contains about 300 rule-templates to determine contact's knowledge of ownship (counter-detection, location of ownship and threat alert), threat maneuvers of contacts (straight-line drive, turn-away-and-run, turn-to-and-reengage, secondary attack) and the intent of the contact (attack, evade, influence, non-reactive).

RUM and RUMrunner are being used by the GE CRD Control Technology Systems Program to build a guidance and navigation system for the Semi-Autonomous Land Vehicle Project (Mars Roving Vehicle). LOTTA is used to simulate the Martian environment (terrain and related features). RUM is used to implement navigational decisions which will then be communicated to the vehicle control.

Another RUM application, done by GE-CRD Engineering & Business Information Program, is the interpretation of built-in-tests (BIT), such as parity checks and end-to-end processor checks, for in-flight diagnosis of avionics.

All these applications have improved our understanding of the Dynamic Classification Problems and have provided us with valuable feed-back and guidelines to direct our efforts in refining the RUM/RUMrunner technology.

6 Conclusions

We have defined a theory for reasoning with uncertainty based on the semantics of many-valued logics (T-norm operators). We have implemented a subset of this theory, limited to acyclic Horn clauses, and embedded in a software tool. We have integrated this tool with other technologies (object-based simulators, real-time inference engines) into a software architecture designed to address the class of dynamic classification problems. Finally, we have applied this software architecture to a variety of DCP cases: situation assessment for Pilot's Associate and for Submarine Commander's Associate, guidance and navigation for the Mars Roving Vehicle, and in-flight diagnosis of avionics.

References:

[1] Piero P. Bonissone and Keith S. Decker. Selecting Uncertainty Calculi and Granularity: An Experiment in Trading-off Precision and Complexity. In L. Kanal and J. Lemmer, editors, *Uncertainty in Artificial Intelligence*, pages 217–247. North-Holland, 1986.

[2] Piero P. Bonissone. Summarizing and Propagating Uncertain Information with Triangular Norms. *International Journal of Approximate Reasoning*, 1(1):71–101, January 1987.

[3] Piero P. Bonissone, Stephen Gans, and Keith S. Decker. RUM: A Layered Architecture for Reasoning with Uncertainty. In *Proceedings 10th International Joint Conference on Artificial Intelligence*, pages 891–898. AAAI, August 1987.

[4] Piero P. Bonissone. Using T-norm Based Uncertainty Calculi in a Naval Situation Assessment Application. In *Proceedings of the Third AAAI Workshop on Uncertainty in Artificial Intelligence*, pages 250–261. AAAI, July 1987.

[5] Piero P. Bonissone and Nancy C Wood. Plausible Reasoning in Dynamic Classification Problems. In *Proceedings of the Validation and Testing of Knowledge-Based Systems Workshop*. AAAI, August 1988.

[6] Piero P. Bonissone, Dave Cyrluk, John Goodwin, and Jonhatan Stillman. Uncertainty and Incompleteness: Breaking the Symmetry of Defeasible Reasoning. In *Proceeding Fifth AAAI Workshop on Uncertainty in Artificial Intelligence*. AAAI, August 1989.

[7] John McCarthy. Applications of Circumscription to Formalizing Common Sense Knowledge. *Journal of Artificial Intelligence*, 28:89–116, 1986.

[8] Yoav Shoham. Chronological Ignorance: Time, Nonmonotonicity, Necessity and Causal Theories. In *Proceedings Fifth National Conference on Artificial Intelligence*, pages 389–393. AAAI, August 1986.

[9] Piero P. Bonissone. Now that I Have a Good Theory of Uncertainty, What Else Do I Need? In *Proceeding Fifth AAAI Workshop on Uncertainty in Artificial Intelligence*. AAAI, August 1989.

[10] Roger S. Pressman. *Software Engineering: A Practitioner's Approach*. McGraw-Hill, second edition, 1987.

[11] Herbert A. Simon. *The Sciences of the Artificial*. M.I.T. Press, Cambridge, Massachusetts, second edition, 1981.

[12] Piero P. Bonissone and James K. Aragones. LOTTA: An Object Based Simulator for Reasoning in Antagonistic Situations. In *Proceedings of the 1988 Summer Computer Simulation Conference*, pages 674–680. SCS, July 1988.

[13] Lise M. Pfau. RUMrunner: Real-Time Reasoning with Uncertainty. Master's thesis, Rensselaer Polytechnic Institute, December 1987.

[14] Symbolics. *Symbolics Common Lisp - Language Concept (Symbolics Documentation Set, vol. 2A)*, 1986. Genera 7.0.

[15] Lee D. Erman, Frederick Hayes-Roth, Victor R. Lesser, and D.R. Reddy. The Hearsay-II Speach-Understanding System: Integrating Knowledge to Resolve Uncertainty. *Computing Surveys*, 12:213–253, 1980.

[16] Edmund H. Durfee and Victor R. Lesser. Planning to meet deadlines in a blackboard-based problem solver. Technical Report COINS-87-07, COINS at University of Massachusetts, 1987.

[17] Intellicorp. *KEE Software Development System User's Manual*, 1986. Version 3.0.

[18] Piero P. Bonissone. Plausible Reasoning: Coping with Uncertainty in Expert Systems. In Stuart Shapiro, editor, *Encyclopedia of Artificial Intelligence*, pages 854–863. John Wiley and Sons Co., New York, 1987.

[19] Piero P. Bonissone. A Fuzzy Sets Based Linguistic Approach: Theory and Applications. In M.M. Gupta and E. Sanchez, editors, *Approximate Reasoning in Decision Analysis*, pages 329–339. North Holland Publishing Co., New York, 1982.

[20] R. Beyth-Marom. How Probable is Probable? A Numerical Taxonomy Translation of Verbal Probability Expressions. *Journal of Forecasting*, 1:257–269, 1982.

[21] B. Schweizer and A. Sklar. Associative Functions and Abstract Semi-Groups. *Publicationes Mathematicae Debrecen*, 10:69–81, 1963.

UNCERTAINTY REPRESENTATION IN KNOWLEDGE BASED SYSTEMS

H.-J. Zimmermann, B. Werners
RWTH Aachen
Templergraben 55
5100 Aachen (F.R.G.)

1. Introduction

Fuzzy Set Theory as proposed by L. Zadeh in 1965 is adaptable to different problem structures and well suited to model human evaluation and decision making processes. If uncertainty enters into models, it is classically of the stochastic kind which can properly be modelled by using probability theory. There are other fields, however, in which the description of problems contains uncertainties, in which, however, they are of a different kind than randomness. One of the most important areas of this kind is probably that of human problem solving and decision making: here the human factor enters with all its vaguenesses of perception, of subjectivity, of attitudes of goals and conceptions. To use a modeling language which is dichotomous in this area seems far from adequate.

Language is, roughly speaking, the link between the thinking of human beings, the object system, which is to be modeled, and the model itself. Statements in a model are formulated by means of linguistic expressions. That means that the model language semantically and systematically has to be well suited to grasp those aspects of reality which are included in the model; the perspective of a scientific theory is already contained in its linguistic terms. This especially applies to semi-structured problems, which should be handled by expert systems.

2. Types of Uncertainty in Expert Systems

In order to appreciate different types of uncertainty it might be helpful to consider an expert system the knowledge base of which uses rules to represent the knowledge and the inference engine employs some type of logical deduction rules, such as predicate calculus, dual logic or similar. The knowledge base would, for instance, contain the rule "If A then B". From observing the problem situation we would judge "A is true" (the fact), and then conclude "Then B is true" (conclusion).

Usually the following assumptions are being made:

a. A and B are **deterministic statements**.

b. The phenomena contained in A and B are **crisply defined**, i.e. it is clearly defined what belongs to A and what not.

c. The A observed in reality is **identical** to the A contained in the rule.

d. The rule is **correct** (true), not contradictory to a similar rule contained in the knowledge base, and only contains two quantifiers (the universal and the existence quantifier).

Particularly in ill-structured situations, however, these are rather unrealistic assumptions. If the assumptions were valid we would generally not need expert systems, i.e. the expertise of human experts, but the problem solution could be found algorithmically.

Let us consider statements which are not of the assumed type:
"The probability of hitting the target is .6".
This is certainly a probabilistic statement with quantitative probability given. It could properly be modelled using classical probability theory of the Kolmogoroff-type. But how about **"The chances of winning are good"**? Here quantitative probabilities are missing, and it is already more difficult to apply probability theory. In **"It is likely that we will make a good profit"** not only quantitative probabilities are missing, but in addition the term "good profit" is not uniquely defined. If we finally consider a statement of the type **"The experiences of expert A suggest convincingly that B will happen, but

expert C does not believe that B will occur" it becomes obvious that probability theory will no longer be adequate to model these relationships. The above mentioned examples concerned more or less the character of the statements contained in the antecedents and conclusions of the inference.

Another type of uncertainty, **dissimilarity**, occurs whenever the objects of the observed facts are not identical with the components in the rules. Strictly speaking, at least in classical logic, we could not arrive at a conclusion at all. In many real situations, however, a similarity of the observed facts with the components of the rules might still allow a certain conclusion. This similarity is certainly not a dichotomous feature, but rather a matter of degree, which has to be modelled accordingly.

Eventually the inference process itself may contain uncertainties: The degree to which an observation is true, reliable or supported is normally called "material truth". This obviously has some influence on the truth of the conclusion. In addition, however, the rules themselves may not be completely reliable or deterministic. This degree which also has to enter the computation of the degree of truth of the conclusion, may be called "**formal truth**". Closely related is the question of evidence enforcement: How should the reliability of positive independent or dependent evidences be taken into account and combined with that of negative evidences with respect to the reliability or truth of the final conclusion?

Which particular problem will occur when modeling uncertainty depends
- on the type of application or context,
- on the kind of available input,
- on the character of the desired output,
- and on the inference mechanism used.

The context will certainly determine the kinds of uncertainty which have to be taken into account. It will also influence the constraints (time etc.) under which the knowledge and information processing has to occur. Partly it will also determine the input which is available.

Input to an Expert System is knowledge (rules, for instance) and information (facts, observations etc.). Both can be available in numerical, quantitative, crisp or in qualitative, linguistic form. Both types of information might be connected with various kinds of uncertainty (probability, truth, possibility etc.) which again can be expressed either numerically or linguistically. Similar holds for the output: It might be necessary that a crisp conclusion (decision) is required - possibly connected with certain measures of uncertainty, or it can be preferable to obtain the output linguistically, facilitating the communications with human users.

Essentially two types of problems arise:
If the input or the desired output is in another form than the information processed in the inference engine a transformation from linguistic into numerical (crisp) information (defuzzification) or vice versa (fuzzification) has to be performed.
If the input/output is linguistic and the inference engine processes also linguistic information (for instance linguistic variables) then some kind of **matching** or **linguistic approximation** has to be performed whenever the input or output granules are not identical to the granules defined or obtained in the inference engine.

Reasoning under conditions of uncertainty normally creates two major types of problems:
Each type of uncertainty of the antecedents has to be properly carried along to the conclusions. If different types of uncertainty exist, one might want to combine (aggregate) different measures of uncertainty into one measure of overall uncertainty.

Both problems are obviously considerably simplified if one holds the view that there is only one common calculus for uncertainty. We do not share this view, however, because in this case it would not be necessary or meaningful to distinguish between different theories of uncertainty such as probability theory, possibility theory etc. Empirical evidence also does not support this view.

3. Tools to Model Uncertainty

Tools to model uncertainty can roughly be classified into probability based approaches, possibility based and fuzzy set based approaches. Important representatives of probabilistic models are the Bayes approach and the Dempster-Shafer approach. Possibilistic approaches are discussed in more detail in the preceding part of this book. So the consideration here focusses on fuzzy sets. After some basic definitions the most important modeling tools are introduced.

Fuzzy Sets – Basic Definitions

Having recognized the shortcomings of traditional mathematical models in some areas Zadeh suggested in the mid-sixties the notion of a fuzzy set. He defined a fuzzy set as follows (Zadeh 1965):

Definition 1:

If X = {x} is a collection of objects denoted generically by x then a fuzzy set A in X is a set of ordered pairs:

$$A = \{(x, \mu_A(x)), \; x \in X\}$$

$\mu_A(x)$ is the grade of membership of x in A and μ_A is called the membership function, which maps X to the membership-space M. When M contains only the two points 0 and 1, A is nonfuzzy and μ_A is identical with the characteristic function of a nonfuzzy set. In general, M is assumed to be a subset of [0,1].

Example:

Let X = [0,250] be the interval of possible speeds (km/h) at which cars can cruise over long distances. Then the fuzzy set A of "comfortable speeds for long distances" may be defined by a certain individual as

$$A = \{(x, \mu_A(x)), \; x \in [0,250]\}$$

$$\text{with } \mu_A(x) = \begin{cases} 0 & 0 \leq x \leq 90 \\ (x-90)/30 & 90 \leq x \leq 120 \\ 1 & 120 \leq x \leq 140 \\ (160-x)/20 & 140 \leq x \leq 160 \\ 0 & 160 \leq x \leq 250 \end{cases}$$

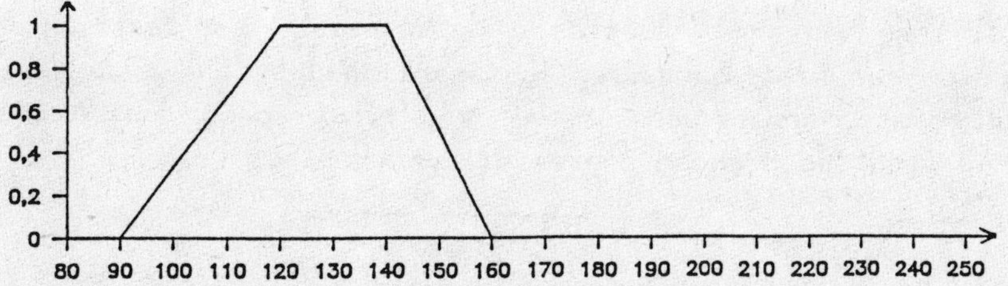

Fig. 1: Membership function "comfortable speed"

A "fast speed" B could, for instance, be described by the following membership function:

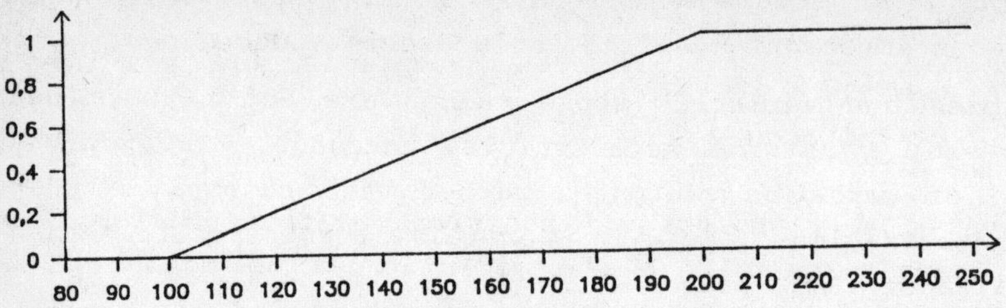

Fig. 2: Membership function "fast speed"

Basic Operations with Fuzzy Sets

Operations on and with fuzzy sets are defined via their membership functions. In this context it should be realized that the set-theoretic intersection corresponds to the logical "and" and the union to the "inclusive or" in logic, respectively.

Intersection of fuzzy sets can be described by the minimum operator:

$$A \cap B = \{(x, \mu_{A \cap B}(x)), \ x \in X\}$$

$$\mu_{A \cap B}(x) = \min \{\mu_A(x), \mu_B(x)\}$$

There exist a lot of alternative definitions of the membership function of A ∩ B which are empirically investigated or axiomatically founded (see Zimmermann, Zysno 1980, Werners 1984, Zimmermann 1985). Some of them are discussed in the following article considering the selection of alternatives.

Example:

If the speed for long distances should be "comfortable and fast", the membership function can, for instance, be established by the intersection of the fuzzy sets "comfortable speed" and "fast speed". Here 150 km/h is the speed with the highest degree of membership.

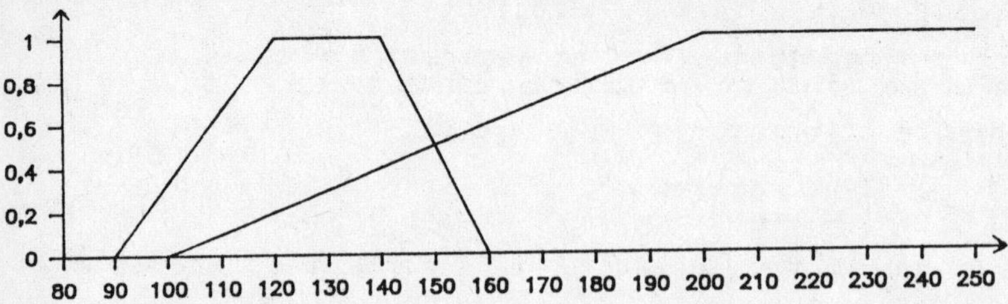

Fig. 3.: Membership function "comfortable and fast speed"

Linguistic Variables

Linguistic variables are variables whose values are not numbers but words or sentences in a natural or artificial language. "The motivation for the use of words or sentences rather than numbers is that linguistic characterizations are, in general, less specific than numerical ones." (Zadeh 1973, p.3)

Definition 2:
A linguistic variable is characterized by a quintuple (X, T(X), U,G,M) in which X is the name of the variable; T(X) (or simply T) denotes the term-set of X, that is, the set of names of **linguistic values of X**, with each value being a fuzzy variable denoted **generically** by x and ranging over a universe of discourse U which is associated with the base variable u; G is a **syntactic rule** (which usually has the form of a grammar) for generating the name, x, of values X; and M is a **semantic rule** for associating with each x its meaning, M(x), which is a fuzzy subset of U. A particular x, that is a name generated by G, is called a **term**. It should be noted that the base variable u can also be vector-valued.

In order to facilitate the symbolism in the following some symbols will have two meanings wherever clarity allows this: X will denote the name of the variable ("the label") and the generic name of its values. The same will be true for x, and M(x).

Example (Zadeh 1973, p.77):
Let X be a linguistic variable with the label "Age" (i.e., the label of this variable is "Age" and the values of it will also be called "Age") with U = [0,100]. Terms of this linguistic variable, which are again fuzzy sets, could be called "old", "young", "very old", and so on. The base-variable U is the age in years of life. M(x) is the rule that assigns a meaning, that is, a fuzzy set to the terms:

$$M(old) = \{(u,\mu_{old}(u)), \ u \in [0,100]\}$$

where

$$\mu_{old}(u) = \begin{cases} 0 & u \in [0,50] \\ \left(1 + \left(\dfrac{u - 50}{5}\right)^{-2}\right)^{-1} & u \in [50,100] \end{cases}$$

T(X) will define the term set of the variable X, for instance in this case:

T(Age) = {old, very old, not so old, more or less young, quite young, very young}

where G(x) is a rule which generates the (labels of) terms in the term set. Figure 4 sketches the above mentioned relationships.

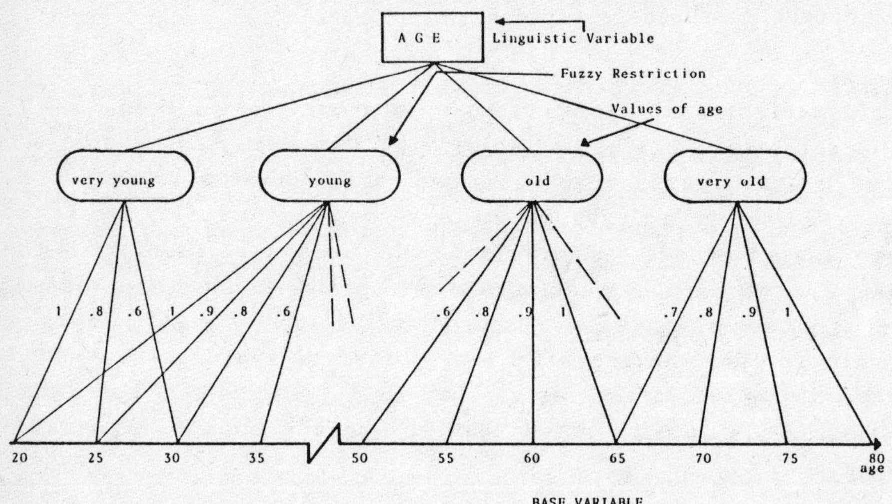

Fig. 4: Linguistic variable age

A fuzzy restriction on the values of the base variable is characterized by a compatibility function which associates with each value of the base variable a number in the interval [0,1].

Fuzzy Matching and Linguistic Approximation

It was already mentioned that "classical" expert systems generally assume that observed facts are identical to the components of knowledge stored in the rules. Obviously complete identity and complete mismatch are only extremes. In general, observed facts will match the components of the rules more or less. A number of methods have been suggested to measure and express the degree to which facts, i.e. observed patterns, match the components predefined in the knowledge base. Cayrol [Cayrol, Farreny and Prade 1982] suggest, for instance, to use the concept of possibility and necessity to express the similarity between observations and defined patterns which are both modelled as fuzzy sets. This min-max approach is certainly not the only way to measure the degree of similarity of patterns. Bonissone [Bonissone, Decker 1986; Bonissone 1982] suggest the power of a fuzzy set (its cardinality), the entropy, the first moment of the membership function and its skewness as characteristic features of fuzzy sets. A quadratic weighted distance in the feature space is then used to evaluate the similarity between two fuzzy sets. Zwick, Carlstein and Budescu [1986] consider nineteen measures to express the similarity of fuzzy sets.

A special type of pattern matching is what Bonissone calls "linguistic approximation": Let us assume that we have defined a linguistic variable "Probability" to express the probability of the output of an expert system. Let the term set of this linguistic variable contain 5 terms. The linguistic probabilities of conclusions emerging from the inference process are characterized by their membership functions. These membership functions will presumably not match any of the terms defined to express the probability of the outputs. Linguistic approximation now tries to find the term of the predefined output term set which is most similar to the computed fuzzy set which is the result of the inference process. The computational efforts of this matching process obviously depends to a large extent on the power of the term set, the restrictions imposed on the shapes of membership functions, and on the similarity measures used in the matching process.

Fuzzy Logic, Approximate Reasoning, Plausible Reasoning

Logics as bases for reasoning can be distinguished essentially by their three topic-neutral (context-independent) items, truth values, vocabulary (operators), and reasoning procedure (tautologies, syllogisms). In Boolean logic truth values can be 0 (false) or 1 (true) and by means of these truth values the vocabularies (operators) are defined via truth tables.

A	B	$A \wedge B$	$A \vee B$	$\neg A$	$A \Rightarrow B$
1	1	1	1	0	1
1	0	0	1	0	0
0	1	0	1	1	1
0	0	0	0	1	1

Assigning meanings (words) to these operators is not difficult: for example the first obviously characterizes the "and", the second the "inclusive or". Here it is assumed that each statement, A and B, could clearly be classified as true or false. If this is not the case, then additional truth values, such as "undecided" or "similar", can and have to be introduced, which leads to the many existing systems of multi-valued logic.

The third topic-neutral item of logical systems is the reasoning procedure itself which generally bases on tautologies such as the well known modus ponens:

$$(A \wedge (A \Rightarrow B)) \Rightarrow B$$

which could be interpreted as "If A is true and if the statement "If A is true then B is true" is also true, then B is true". This tautology obviously depends on the character of the statements, the type of permissible truth values, the definition of the operators. It is assumed that, for instance, all A`s contained in one tautology are identical. The respective assumptions made in dual logic are gradually relaxed in fuzzy logic, approximate reasoning and plausible reasoning.

Fuzzy logic [Zadeh 1973, p. 10] is an extension of the set-theoretic multi-valued logic in which the truth values are linguistic variables or terms of the linguistic variable truth. Since operators, like $\wedge$, $\vee$, $\neg$, $\Rightarrow$ in fuzzy logic are also defined by using truth tables, the extension principle can be applied to derive definitions of the operators. So far possibility theory has primarily been used in order to define operators in fuzzy logic, even though other operators have also been investigated [see, for instance, Mizumoto, Zimmermann 1982] and could also be used. Here we will limit considerations to possibilistic interpretations of linguistic variables and we will also stick to the original proposals of Zadeh [1973]. To the interested reader, however, we suggest to study also alternative approaches such as [Baldwin 1979], [Baldwin and Pilsworth 1980], [Giles 1979, 1980] and others.
It was already mentioned above, that in the truth tables of fuzzy logic the truth values (in dual logic 0 or 1) are terms of the linguistic variable "truth". This variable has been defined differently by different authors.

Approximate and plausible reasoning are ways of drawing conclusions from hypothesis which relax even more assumptions of dual logic than fuzzy logic. On the basis of what has been said so far two quite obvious generalizations of the modus ponens, for example, are:

1. to allow statements which are characterized by fuzzy sets
2. to relax (slightly) the identity of the "A's" and "B's" in
   the implication and the conclusion.

In approximate reasoning the statements in the antecedents are allowed to contain fuzzy sets rather than crisp statements.

In 1973 Zadeh suggested the compositional rule of inference for the above-mentioned type of fuzzy conditional inference. In the meantime other authors (f.i. [Baldwin 1979], [Baldwin and Pilsworth 1980], [Baldwin and Guild 1980], [Mizumoto et al. 1979], [Mizumoto, Zimmermann 1982], [Tsukamoto 1979]) have suggested different methods and investigated also the modus tollens, syllogism, and contraposition. Here, however, we shall restrict considerations to Zadeh`s compositional rule of inference.

Definition 3:

Let $R(x)$ and $R(y)$, $x \in X$, $y \in Y$, be fuzzy sets in X, Y, and $R(x,y)$ a fuzzy relation in XxY, respectively. A and B denote particular fuzzy sets in X and XxY. The **compositional rule of inference** asserts that the solution of the system $R(x) = A$ and $R(x,y) = B$ is given by $R(y) = A \circ B$, where $A \circ B$ is the min-max-composition of A and B.

Plausible reasoning goes one step further and relaxes slightly the identity of the elements in the facts and the rules and tries to still arrive at conclusions on the basis of the generalized modus ponens.

Let A, A`, B, B` be fuzzy statements, then the generalized modus ponens reads:

Premise	x is A`
Implication	If x is A then y is B
Conclusion	y is B`

For instance [Mizumoto, Zimmermann 1982]

Premise	This tomato is very red
Implication	If a tomato is red then the tomato is ripe.
Conclusion	This tomato is very ripe.

It should be mentioned, however, that the generalized modus ponens alone, which characterizes "plausible reasoning", does not allow to obtain conclusions from unequal premises. Such an inference presupposes or necessitates the knowledge about modifications of the premises and their consequences - for instance, that increase in "redness" indicates an increase in "ripeness" (see [Dubois, Prade 1984, p. 325]).

4. Presentation of Uncertainties and Available Tools

One important aspect of uncertainty modeling is the method to process uncertain information. When feeding this information into other information systems or into machines (process control) it might even be rather unimportant in which way the information about uncertainty is presented. If, however, information about uncertainty is obtained from human experts and/or if it is intended to be used by human decision makers, the formal type of representation becomes very important. Due to our limited capability to perceive and process large amounts of information simultaneously, the synchronisation of the granularity with the human capabilities of perceiving information becomes extremely relevant.
The two poles between which we have to find a compromise when deciding about the type of presentation are simplicity (often closely related to acceptance by the user) and perfection or completeness of the types of uncertainty we want to present.

So far there exist only very few empirical results which indicate the influence of the type of presentation of uncertainty on the ease and performance of use by human users [see, f.i., Zwick, Carlstein, Budescu 1987]. We shall, therefore, only indicate possible ways of presenting uncertainty to users. We shall exemplarily assume that one accepts four kinds of uncertainty: stochastic uncertainty (probability), degree of similarity between facts and components of rules, linguistic (lexical) uncertainty and the degree of (formal) truth.
First of all these uncertainties can be represented separately or aggregated as one "over-all-degree" of uncertainty. The former presentation (f.i. as "certainty factors") has probably more appeal for a practitioner, but it has two major disadvantages:

1. The aggregation of different uncertainties presupposes an appropriate, possibly context-dependent aggregation procedure.
2. The user can no longer recognize the "uncertainty mix" which leads to a specific uncertainty factor.

The disaggregated way of presentation, on the other hand, provides more information, but it leaves the user with the problem of aggregation and it might not be very appealing to people not too well trained in uncertainty modeling.

Apart from the degree of aggregation uncertainty can be expressed as a scalar, an interval, a distribution or linguistically. Considering probability we can state a probability as a real number between 0 and 1, as lower and upper probability, as a fuzzy number (possibility distribution) or in terms of the linguistic variable "probability", such as "highly probable", "not likely" etc.

Most results have so far been presented for the combination of (linguistic) fuzziness and (stochastic) probability, focussing on fuzzy random variables, the probability of fuzzy events etc. [see Zwick, Walsten 1987]. They seem to indicate that this combination can be achieved reasonably well. If one interprets, on the other hand, the degree of truth as the degree of consistency (of positive and negative evidence) then a dual measure representing these three fuzzy dimensions seems to be meaningful. The formulation could certainly be linguistic, possibly with underlying numerical scales [see Wallstein et al. 1986].

For the output of an expert system one could obviously also use further alternatives such as simultaneous supply of all disaggregated measures, subsequent or optional presentation or any combination.

In the recent past quite a number of expert systems and -tools have been developed on the basis of fuzzy set theory. As usual some of them have been obtained by emptying expert systems of the context dependent knowledge base. Others have been designed from the very beginning as inference engine. The following table surveys these tools. For a more detailed description see Zimmermann (1987). This table does not include the older area of fuzzy control, the systems of which in a sense could also be considered as expert systems. The last system mentioned in the table - VLSI - is not a software tool, but a fuzzy inference engine on a chip which performs 80000 FLIPS (Fuzzy Logic Inference Per Second).

Name	Application Area	Knowledge Representation Inference	Uncertainty	Reference
ARIES	Tool	Rules, nets	Multiple truth qualification	[Appelbaum, Ruspini 1985]
CADIAC-2	Medical Diagnosis	Rules	Degrees of confidence	[Adlassnig et al. 1985]
FAGOL	Tool	Fuzzy algorithms	Fuzzy probabilities	[Alexeyev 1985]
fINDex	Forecasting	Rules, Ling. Variables	Ling. Variables	[Whalen, Schott 1985]
FLIP	Tool	Fuzzy Logic	Degrees of Belief, Weights	[Giles 1980]
FLOPS	Mgmt. of imprec. Databases	Rules Weakly nonmonotonic	Truth qualific. Ling. Variables	[Buckley et al. 1986]
FRDB	Tool	Fuzzy relations	Qualific. Ling. Var.	[Zemankova, Kandel 1985]
FRIL	Tool	Fuzzy relations	Truth qualific.	[Baldwin 1985, 1979]
Fuzzy Planner	Tool	Fuzzy logic	Truth qualific.	[Kling 1974]
METABOL	Tool	PROLOG-·clauses Relations	Plausibility Credibility	[Ernst 1985]
PROSPECTOR 2	Mineral Exploration	Semantic+Inference Networks	Bayes Fuzzy Logic	[Benson 1986]
PRUF	Tool	Poss. Theory App. Reasoning; Rules	Truth, prob. Poss. Qualific.	[Zadeh 1981]
REVEAL	Tool (Mgmt., Planning)	Rules	Poss. Qual.	[Jones 1986]
SLOP	Tool	Support Logic Programming	"Support Pairs"	[Baldwin 1986]
SPERIL-II	Damage assessment	Rules	Dempster, Shafer	[Ogawa et al. 1985]
SPHINX	Medical Diagnosis	Trees, Rules	Similarity Index, Uncertainty Interval	[Fieschi et al. 1982]
SPII-1	Tool	Rules	Necessity, Poss.	[Martin-Clouaire, Prade 1986]
TAIGER	Tool	Rules	Necessity, Poss.	[Farreny et al. 1986]
VLSI	Hardware Tool	Fuzzy Logic, Rules	Certainty factors	[Togai, Watanabe 1986]

Table 1: Expert systems and - tools using fuzzy set theory

REFERENCES:

Adlassnig, K.-P., Kolarz, G., Scheithauer, W.: Present state of the medical expert systems CADIAC-2, in: Med. Inform. 24 (1985), pp. 13-20

Alexeyev, A.V.: Fuzzy algorithms execution software: The FAGOL-System, in: Kacprzyk, J., Yager, R. (edts.), 1985, pp. 289-300

Appelbaum, L., Ruspini, E.H.: ARIES: An approximate reasoning inference enquire, in: Gupta et al. (edts.), 1985, pp. 745-755

Baldwin, J.F.: A new approach to approximate reasoning using a fuzzy logic, in: Fuzzy Sets and Systems 2 (1979), pp. 309-325

Baldwin, J.F.: Support logic programming, in: Int. Journal of Gen. Systems 1 (1986), pp. 73-104

Baldwin, J.F., Guild, N.C.F.: Feasible algorithms for approximate reasoning using fuzzy logic, in: Fuzzy Sets and Systems 3 (1980), pp. 225-251

Baldwin, J.F., Pilsworth, B.W.: Fuzzy truth definition of possibility measure for decision classification, in: In. J. Man-Machine Studies I (1979), pp. 447-463

Benson, I.: PROSPECTOR: An expert system for mineral exploration, in: Mitra, G. (ed.), 1986, pp. 17-26

Bonissone, P.B.: A fuzzy set based linguistic approach: Theory and applications, in: Gupta, M.M. and Sanchez, E. (edts.): Approximate Reasoning in Decision Analysis, Amsterdam 1982, pp. 329-339

Bonissone, P.P. and Decker, K.S.: Selecting uncertainty calculi and granularity, in: Kanal and Lemmer (edts.), 1986, pp. 217-247

Buckley, J.J., Silver, W., Tucker, D.: A fuzzy expert system, in: Fuzzy Sets and Systems 20 (1986), pp. 1-16

Cayrol, M., Farreny, H., Prade, H.: Fuzzy pattern matching. Kybernetes 11 (1982), pp. 103-116

Dubois, D., Prade, H.: Fuzzy logics and the generalized modus ponens revisited. Cybern. Syst. ISL, pp. 293-331

Ernst, Ch.J.: A logic programming metalanguage for expert systems, in: Kacprzyk, J., Yager, R. (edts.) 1985, pp. 280-288

Farreny, H., Prade, H., Wyss, E.: Approximate reasoning in a rule-based expert system using possibility theory, Proc. 10th IFIP World Congr., Dubl. 1986.

Fieschi, M., Joubert, M., Fieschi, D., Soula, G., Roux, M.: Sphinx: An interactive system for medical diagnoses aids, in: Gupta, Sanchez (edts.) 1982, pp. 269-282.

Giles, R.: A formal system for fuzzy reasoning, in: Fuzzy Sets and Systems 2 (1979), pp. 233-257

Giles, R.: A computer program for fuzzy reasoning, in: Fuzzy Sets and Systems 4 (1980), pp. 221-234.

Jones, P.L.K.: REVEAL: Adressing DSS and expert systems, in: Mitra (ed.) 1986, pp. 49-58.

Kacprzyk, J., Yager, R.R. (edts.): Management decision support systems using fuzzy sets and possibility theory, Köln 1985.

Kling, R.: Fuzzy Planner: Reasoning with inexact concepts in a procedural problem-solving language, in: J. of Cybernetics 4 (1974), pp. 105-122.

Martin-Clouaire, R., Prade, H.: SPII: A simple inference engine capable of accomodating both imprecision and uncertainty, in: Mitra, G. (ed.) 1986, pp. 117-131

Mizumoto, M., Fukami, S., Tanaka, K.: Some methods of fuzzy reasoning, in: Gupta et al. (eds.), Advances in fuzzy set theory and applications, Amsterdam, New York 1979, pp. 117-136

Mizumoto, M., Zimmermann, H.-J.: Comparison of fuzzy reasoning methods, in: Fuzzy Sets and Systems 8 (1982), pp. 253-283

Ogawa, H., Fu, K.S., Yao, J.T.P.: SPERIL-II: An expert system for damage assessment of existing structure, in: Gupta et al. (edts.), 1985, pp. 731-744

Togai, M., Watanabe, H.: A VLSI implementation of a fuzzy inference engine toward an expert system on a chip, in: Inf. Sc. 38 (1986), pp. 147-163)

Tsukamoto, Y.: An Approach to Fuzzy Reasoning Method, in: Gupta et al. (eds.) 1979, pp. 137-149

Wallstein, T., Budescu, D., Rappaport, A., Zweck, R., Forsyth, B.: Measuring the vague meaning of probability terms, in: J. of Exp. Psychology 115 (1986), pp. 348-365

Werners, B.: Interaktive Entscheidungsunterstützung durch ein flexibles mathematisches Programmierungssystem, München 1984

Whalen, Th., Schott, B.: Goal-directed approximate reasoning in a fuzzy production system, in: Gupta et al. (edts.), 1985, pp. 505-518

Zadeh, L.A.: Fuzzy sets, in: Information and Control 8 (1965), pp. 338-353

Zadeh, L.A.: The concept of a linguistic variable and its application to approximate reasoning, Memorandum ERL-M 411, Berkeley, October 1973

Zadeh, L.A.: Test score: Semantics for natural languages and Meaning Representation via PRUF, in: Rieger, B.B. (ed.), Empirical Semantics, Bochum 1981, pp. 281-349

Zemankova-Leech, M., Kandel, A.: Uncertainty propagation to expert systems, in: Gupta et al. (edts.) 1985, pp. 529-540

Zimmermann, H.-J.: Fuzzy set theory and its applications, Boston 1985

Zimmermann, H.-J.: Fuzzy sets, decision making and expert systems, Boston 1987

Zimmermann, H.-J., Zysno, P.: Latent connectives in human decision making, in: Fuzzy Sets and Systems 4 (1980), pp. 37-51

Zwick, R., Carlstein, E., Budescu, D.: Measures of similarity between fuzzy concepts: A comparative analysis, GSIA WP #34/86/87, Pittsburgh 1986

Zwick, R., Walsten, T.: Combining stochastic uncertainty and linguistic inexactness: Theory and experimental evaluation, GSIA Working Paper #37/86/87, Pittsburgh 1987

EVALUATION AND SELECTION OF ALTERNATIVES CONSIDERING MULTIPLE CRITERIA

B. Werners, H.-J. Zimmermann

RWTH Aachen

Templergraben 55

5100 Aachen (F.R.G.)

1. Introduction

In complex decision problems in which a ranking of alternatives, i.e.
projects, objects etc., has to be determined, in general several
criteria have to be considered simultaneously. Often parts of the
description are in addition uncertain. To aid decision makers in such
situations numerous mathematical models and methods have been developed, of which some shall be described in the following. On the one
hand, these models can be used in expert systems to model human decision making behavior properly. On the other hand, they may provide a
benchmark for the quality of decisions in a normative sense.

Let us first consider the basic model of (normative) decision theory:
Given the set of feasible alternatives, X, the set of relevant states,
S, the set of resulting events, E, and a (rational) utility function,
u - which orders the space of events with respect to their desirability - the optimal decision under certainty is the choice of the
alternative that leads to the event with the highest utility.

The set of feasible alternatives can be defined explicitly by enumeration or implicitly by constraints. This model has to be extended to
handle multiple objectives. The decision maker is asked to trade off
the achievement of one objective against another objective. This
requires a subjective judgment of the decision maker. Modeling tools
should allow the handling of such additional information, for example
interactively in close contact with the decision maker (see Werners
1987).

In real world problems - and in particular in semi-structured situations - additional difficulties can arise, for example:
- The feasibility of an alternative cannot be crisply determined, but may be attained to a certain degree.
- The set of relevant states is either probabilistically or nonprobabilistically uncertain.
- The utility function depends on multiple criteria, subjective judgments and risk behavior. The functional dependence is often not specified.
- The problem situation is not formulated mathematically, but is described by linguistic terms of vague concepts.

2. Basic Considerations

Let us now concentrate on decision problems with multiple objectives. In order to facilitate the comprehension of fuzzy models in this area we shall first define the classical (crisp) multi-attribute decision model (MADM):

Let $X = \{x_i, i = 1,...,n\}$ be a (finite) set of decision alternatives and $G = \{g_j, j = 1,...,m\}$ a (finite) set of goals, attributes, or criteria, according to which the desirability of an alternative is to be judged. $g_j(x_i)$ is the consequence of alternative x_i with respect to criterium g_j. The aim of MADM is to determine an alternative x^o with the highest possible degree of overall desirability.

Example: Car selection problem
A person wants to buy a new car. After a market study only five alternatives $X = \{x_1,...,x_5\}$ remain relevant. They differ with respect to the following four criteria which are of main importance to the decision maker:
g_1: The maximum speed should be high.
g_2: The gasoline consumption in town should be low.
g_3: The price should be low.
g_4: Comfort should be high.

The technical details of the different cars are summarized:

	km/h	1/100km	ECU	degree
	$g_1(x_i)$	$g_2(x_i)$	$g_3(x_i)$	$g_4(x_i)$
x_1	170	10,5	17.000	50%
x_2	150	8	15.500	10%
x_3	140	9	8.000	20%
x_4	140	9	9.500	10%
x_5	160	10,5	17.000	40%

Definition 1

Let g_j, $j=1,\ldots,m$, be functions to be maximized.

An element x_i (strictly) dominates x_k <=>

$$g_j(x_i) \geq g_j(x_k) \qquad \forall j = 1,\ldots,m$$
$$\text{and } g_j(x_i) > g_j(x_k) \qquad \text{for some } j \in \{1,\ldots,m\}$$

An alternative $x_i \in X$ is an **efficient solution** of the MADM if there does not exist any $x_k \in X$ which dominates x_i.

Of special importance in multiple criteria models is the set of efficient solutions and a compromise alternative should be an element of this set.

Example: Car selection model

g_2 and g_3 are goals to be minimized. This is equivalent to maximize $-g_2$ and $-g_3$. It is then easy to see that x_1 dominates x_5 and x_3 dominates x_4. The elements x_1, x_2, and x_3 constitute the set of efficient solutions.

Decisions in a Fuzzy Environment

Bellman and Zadeh (1970) have departed from a classical model of a decision and suggested a model for decision making in a fuzzy environment which has served as a point of departure for most of the authors in fuzzy decision theory. They consider a situation of decision making under certainty, in which the objective function as well as the constraint(s) are fuzzy and argue as follows:

The fuzzy objective function is characterized by the membership function of a fuzzy set and so are the constraints. Since we want to satisfy (optimize) the objective function as well as the constraints, a decision in a fuzzy environment is defined by analogy to nonfuzzy environments as the selection of alternatives which simultaneously satisfy objective functions and constraints. According to the above definition and assuming that the constraints are noninteractive the logical "and" corresponds to the intersection. The decision in a fuzzy environment can therefore be viewed as the intersection of fuzzy constraints and fuzzy objective functions. The relationships between constraints and objective function in a fuzzy environment are therefore fully symmetric. This concept is illustrated by the following example.

Example

Objective function: "x should be substantially larger than 10" characterized by the membership function

$$\mu_G(x) = \begin{cases} 0 & x \le 10 \\ (1 + (x-10)^{-2})^{-1} & x > 10 \end{cases}$$

Constraint: "x should be in the vicinity of 11" characterized by the membership function

$$\mu_C(x) = (1 + (x - 11)^4)^{-1}$$

The fuzzy set "decision" is then characterized by its membership function for all $x \in X$

$$\mu_D(x) = \min \{\mu_G(x), \mu_C(x)\}$$

$$= \begin{cases} \min\{(1 + (x - 10)^{-2})^{-1}, (1 + (x - 11)^{-4})^{-1}\} & \text{for } x > 10 \\ 0 & \text{for } x \le 10. \end{cases}$$

This relation is depicted in figure 1.

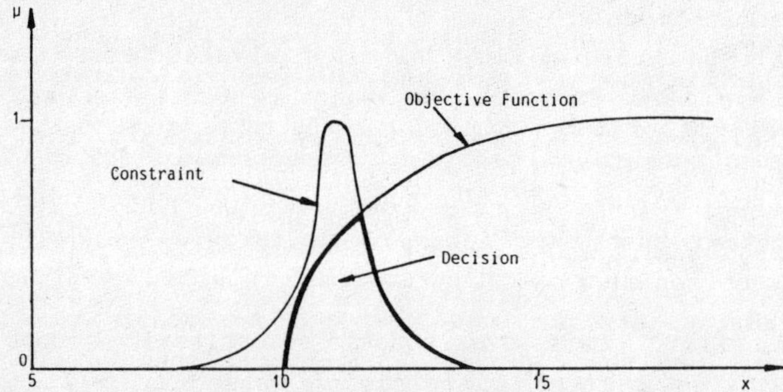

Fig. 1: A fuzzy decision

If the decision maker wants to have a crisp decision proposal, it seems appropriate to suggest to him the value that has the highest degree of membership in the fuzzy set "decision".

Let us call this the **maximizing decision** x_{max} with

$$\mu_D(x_{max}) = \max_x \min \{\mu_G(x), \mu_C(x)\}$$

More generally we can define a decision in a fuzzy environment as follows:

Definition 2

Let μ_{C_i}, $i=1,\ldots,m$, be the membership functions of constraints on X, defining the decision space and μ_{G_j}, $j=1,\ldots,n$, the membership functions of objective (utility) functions or goals on X. A decision D is then defined by its membership function

$$\mu_D = (\mu_{C_1} * \ldots * \mu_{C_m}) * (\mu_{G_1} * \ldots * \mu_{G_n}) = \circledast_i \mu_{C_i} * \circledast_j \mu_{G_j}$$

where $*$, $\circledast$ denote appropriate, possibly context dependent, aggregators (connectives).

Let M be the set of points $x \in X$ for which $\mu_D(x)$ attains its maximum if it exists. Then M is called the **maximizing decision**. If μ_D has a maximum at x_M, then the maximizing decision is a crisp decision which can be interpreted as the action that belongs to all fuzzy sets representing either constraints or goals with the highest possible degree of membership.

Empirical investigations (see Zimmermann, Zysno 1983) lead to the conclusion that the minimum operator and the maximum operator for an aggregation are not always appropriate to model decision making. Often people use averaging operators resulting in membership values between the minimum and the maximum, and the aggregation operator depends on the respective context such that for each decision situation another aggregator might be necessary. This implies a lot of difficulties especially for practical applications. One way to bypass these difficulties is to generalize the classical concept of connectives by combining an and-operator and an or-operator using a parameter $\gamma \in [0,1]$. This is why Zimmermann, Zysno (1983) suggest the γ-operator, a combination of the product and the algebraic sum.

Another disadvantage of the minimum and of the maximum operator is that these operators do not consider non-extreme values. Thus the aggregation of multiple objectives does not guarantee the selection of an efficient alternative. This is shown in the following example:

	μ_1	μ_2	μ_3
x_1	0.1	0.7	0.7
x_2	0.1	0.2	0.7
x_3	0.3	0.3	0.5

Although x_1 strictly dominates x_2 the aggregations min $\mu_j(x_i)$ or max $\mu_j(x_i)$ give the same result for both alternatives, respectively, and so x_1, x_2 are both elements of the set of optimal solutions. This fact is the consequence of the minimum and the maximum not being strictly monotonic and neither is their convex combination.

To avoid this effect and to allow giving higher weight to non-extreme values new operators are suggested (Werners 1988). The idea is to differentiate between the terms "and" and "or", to allow compensation and to get the minimum when expressing the logical "and" and maximum when intending the logical "or":

Fuzzy a$\widetilde{n}$d:

$$\mu_{and} = \min_{j=1}^{m} \mu_j + (1-\gamma) \; 1/m \sum_{j=1}^{m} \mu_j \qquad \gamma \in [0,1]$$

Fuzzy $\widetilde{or}$:

$$\mu_{or} = \max_{j=1}^{m} \mu_j + (1-\gamma) \; 1/m \sum_{j=1}^{m} \mu_j \qquad \gamma \in [0,1]$$

Here γ is the degree of approximating the logical meaning of "and" and "or", respectively. The arithmetic mean gives a compensatory effect. $\gamma = 1$ yields $\mu_{and} = \min$ and $\mu_{or} = \max$. The combination of these two operators leads to very good results with respect to the empirical data of Zimmermann, Zysno (1983) which is shown in (Werners 1984). Another advantage of these operators is that they can be handled computationally in a very efficient way.

3. Multiple Attribute Decision Making

One possible classification of MADM-approaches is into "aggregation approaches" and "order-focussed approaches".

Aggregation approaches generally consist of two steps:

Step 1: The aggregation of the judgments with respect to all goals and per decision alternative.

Step 2: The rank ordering of the decision alternatives according to the aggregated judgments.

In crisp MADM models it is usually assumed that the final judgments of the alternatives are expressed as real numbers. In this case the second stage does not pose any particular problems and suggested algorithms concentrate on the first stage. Fuzzy models are sometimes justified by the argument that the goals g_j themselves or their attainment by the alternatives x_i, respectively, cannot be defined or judged crisply but only as fuzzy sets. In this case the final judgments are also represented by fuzzy sets which have to be ordered to determine the optimal alternative. Then the second stage is, of course, by far not trivial. The aggregation procedure can be direct or hierarchical establishing "consistent" weights for the different criteria.

Order-focussed approaches (outranking) focus attention on the fact that in MADM problems one tries to establish preference orderings of alternatives. If the order relations of alternatives induced by different individual criteria are not consistent with each other or if they do not lead to sufficiently well structured dominance relations then outranking procedures are used to arrive from pairwise dominance statements to the best overall dominance relation which is obtainable. In fuzzy outranking this concept is generalized in the sense that concordance, discordance, and dominance are considered as matters of degree. For a detailed discussion see Zimmermann (1987), for an application see Siskos et al. (1984).

Aggregation Approaches

Let r_{ij} be the (preferability)-ratings of alternative i with respect to criterion j and w_j subjective weights which express the relative importance of the criteria to the decision maker. In crisp MADM-models a frequently used and nonsophisticated way to arrive at overall ratings of alternatives, R_i, is

$$R_i = \sum_{j=1}^{m} w_j r_{ij}$$

Generally the R_i are real numbers according to which the alternatives can easily be ranked.

Example: Car selection problem
With respect to the technical data the decision maker gives (preferability)-ratings r_{ij} for each alternative i with respect to each goal g_j. Additionally he determines his subjective weights for the goals. The result is the following table:

	$g_1(x_i)$	$g_2(x_i)$	$g_3(x_i)$	$g_4(x_i)$
x_1	5	2	2	5
x_2	3	5	3	1
x_3	2	4	5	3
x_4	2	4	4	2
x_5	4	2	2	4
w_j	1	1/3	1/7	1/9

Then for each alternatives x_i a rating R_i as defined above can be computed:

x_i	x_1	x_2	x_3	x_4	x_5
R_i	6.5	5.2	4.4	4.1	5.4

x_1 results as the most preferred solution.

The r_{ij} as well as the w_j, however, will in many cases more appropriately be modelled by fuzzy numbers. This has the following consequences:
In **step 1:** The aggregation procedure of the single criteria ratings will have to be modified.
In **step 2:** R_i will no longer be real numbers but fuzzy sets which have to be ranked.
In the following some approaches to handle fuzziness in MADM-aggregation models shall be described exemplarily:

Hierarchical aggregation using crisp weights (Yager,1978)

Essentially Yager assumes a finite set of alternative actions
$X = \{x_i\}$ and a finite set of goals (attributes) $G = \{\tilde{g}_j\}$, $j=1,\ldots,m$.
The $\tilde{g}_j, \tilde{g}_j = \{(x_i, \mu_{g_j}(x_i))\}$ are fuzzy sets the degrees of membership of

which represent the normalized degrees of attainment of goal j by alternative x_i. The fuzzy set decision, $\tilde{D}$, is then the intersection of all fuzzy goals,

i.e. $\mu_D(x_i) = \min\limits_{j=1}^{m} \mu_{g_j}(x_i)$, $i=1,\ldots,n$,

and the maximizing decision is defined to be the x_o for which

$$\mu_D(x_o) = \max\limits_{i} \min\limits_{j} \mu_{g_j}(x_i).$$

Yager now allows for different importance of the goals and expresses this by eponential weighting of the membership functions of the goals. If w_j are the weights of the goals the weighted membership functions, $\mu_g{}'$, are

$$\mu_{g_j}{}'(x_i) = (\mu_{g_j}(x_i))^{w_j}.$$

For the determination of the w_j Yager suggests the use of Saaty's method, i.e. the determination of the reciprocal matrix by pairwise comparison of the goals with respect to their relative importance (Saaty 1980). The components of the eigenvector of this m x m matrix whose total is m are then used as weights.

The rationale behind using the weights as exponents to express the importance of a goal can be found in the definition of the modifier "very" (Zimmermann 1985). There the modifier "very" was defined as the squaring operation. Thus the higher the importance of a goal the larger should be the exponent of its representing fuzzy set, at least for normalized fuzzy sets and when using the min-operator for the intersection of the fuzzy goals.

The measure of ranking the decision alternatives is obviously the $\mu_D(x_i)$. For the ranking of fuzzy sets in the unit interval Yager suggests another criterion, which bases on properties of the supports of the fuzzy sets rather than on the degree of membership. This is particularly applicable when ranking different (fuzzy) degrees of truth and similar linguistic variables.

Example: Car selection problem

Let X = $\{x_i, i=1,\ldots,5\}$ again be the set of alternative cars and $\{\tilde{g}_j, j=1,\ldots,4\}$ the fuzzy goals, for example described by piecewise linear membership functions

$\tilde{g}_1$:   maximum speed [km/h]

$$\mu_{g_1}(x) = \begin{cases} 0 & x < 100 \\ \dfrac{x-100}{100} & 100 \leq x \leq 200 \\ 1 & 200 < x \end{cases}$$

$\tilde{g}_2$:   consumption in town [1/100 km]

$$\mu_{g_2}(x) = \begin{cases} 0 & 12 < x \\ \dfrac{12-x}{5} & 7 \leq x \leq 12 \\ 1 & x < 7 \end{cases}$$

$\tilde{g}_3$:   price [ECU]

$$\mu_{g_3}(x) = \begin{cases} 0 & 20.000 < x \\ \dfrac{20.000-x}{15.000} & 5.000 \leq x \leq 20.000 \\ 1 & x < 5.000 \end{cases}$$

$\tilde{g}_4$:   comfort [%]

$$\mu_{g_4}(x) = \dfrac{x}{100}$$

The fuzzy goals with respect to alternatives $\{x_i\}$ are then:

$$\tilde{g}_1 = \{(x_1,.7),(x_2,.5),(x_3,.4),(x_4,.4),(x_5,.6)\}$$
$$\tilde{g}_2 = \{(x_1,.3),(x_2,.8),(x_3,.6),(x_4,.6),(x_5,.3)\}$$
$$\tilde{g}_3 = \{(x_1,.2),(x_2,.3),(x_3,.8),(x_4,.7),(x_5,.2)\}$$
$$\tilde{g}_4 = \{(x_1,.5),(x_2,.1),(x_3,.2),(x_4,.1),(x_5,.4)\}$$

By pairwise comparison the following reciprocal matrix has been determined, expressing the relative importance of the goals with respect to each other:

$$W = \begin{array}{c} \\ \tilde{g}_1 \\ \tilde{g}_2 \\ \tilde{g}_3 \\ \tilde{g}_4 \end{array} \begin{array}{cccc} \tilde{g}_1 & \tilde{g}_2 & \tilde{g}_3 & \tilde{g}_4 \\ \begin{pmatrix} 1 & 3 & 7 & 9 \\ 1/3 & 1 & 6 & 7 \\ 1/7 & 1/6 & 1 & 3 \\ 1/9 & 1/7 & 1/3 & 1 \end{pmatrix} \end{array}$$

The eigenvector $w = (w_j)$, $j = 1,\ldots 4$, for which $\sum_{i=1}^{4} w_i = 4$, is

$$w = (2.32, 1.2, .32, .16)$$

Weighting the $\tilde{g}_j$ appropriately yields

$$\tilde{g}_1' = \{(x_1,.44),(x_2,.2),(x_3,.12),(x_4,.12),(x_5,.31)\}$$
$$\tilde{g}_2' = \{(x_1,.24),(x_2,.76),(x_3,.54),(x_4,.54),(x_5,.24)\}$$
$$\tilde{g}_3' = \{(x_1,.6),(x_2,.68),(x_3,.93),(x_4,.89),(x_5,.6)\}$$
$$\tilde{g}_4' = \{(x_1,.9),(x_2,.69),(x_3,.77),(x_4,.69),(x_5,.86)\}$$

Hence,

$$\tilde{D} = \{(x_i, \min_j \mu_{g_j'}(x_i))\} = \{(x_1,.24),(x_2,.2),(x_3,.12),(x_4,.12),(x_5,.24)\}$$

and $D_{max} = \{x_1\}$.

Other models along the same line have been suggested by Laarhoven and Pedrycz (Laarhoven, Pedrycz 1983). They concentrate on step 1 and suggest Saaty's AHP modified for triangular fuzzy numbers.

A different hierarchical aggregation procedure was suggested by Zimmermann and Zysno (Zimmermann, Zysno 1983). This was not intended for solving MADM-problems but it can also be used for step 1. The aggregation of the individual fuzzy criteria is achieved by an appropriate parametrized operation which was tested empirically before. The result of this aggregation procedure is either a fuzzy set or a number (degree of membership) in [0,1] which can both be used for ranking the alternatives in step 2.

Rating and Ranking Multiple-Aspect Alternatives Using Fuzzy Sets

Baas and Kwakernaak suggest the following fuzzy version of the multicriteria model: Let again $X = \{x_i, i=1,\ldots,n\}$ be the set of alternatives and $G = \{g_j, j=1,\ldots,m\}$ the set of goals.

It is assumed that the rating $\tilde{R}_{ij}$ of alternative x_i with respect to goal j is fuzzy and is represented by the membership function $\mu_{R_{ij}}(r)$. Similarly, the weight (relative importance) of goal j is represented by a fuzzy set $\tilde{W}_j$ with membership function $\mu_{w_j}(w_j)$.

All fuzzy sets are assumed to be normalized (i.e. have finite supports and take on the value 1 at least once!).

Step 1: (Determination of ratings for alternatives)
The evaluation of an alternative x_i is assumed to be a fuzzy set $\tilde{R}_i$ which is computed on the basis of the $\tilde{R}_{ij}$ and $\tilde{W}_j$. $\tilde{R}_i = \{(r, \mu_{R_i}(r)\}$ is the final rating of alternative x_i on the basis of which the "ranking ordering" is performed in step 2.

Step 2: Ranking
For the final ranking of the x_i Baas and Kwakernaak start from the observation that if the x_i have received crisp ratings r_i, then a reasonable procedure would be to select those x_i that have received the highest rating, i.e. to determine the set of preferred alternatives. Since here the final ratings are fuzzy sets, the problem is somewhat more complicated. The authors suggest in their model two different fuzzy sets in addition to $\tilde{R}_i$ which give supplementary informations about the preferability of an alternative.

Some authors have criticized this approach for not being sensitive enough, i.e. for not discriminating enough between alternatives with similar ratings.

Jain (1977) concentrates on step 2 and suggests a special concept of a "maximizing set":

Let $S = \bigcup\limits_{i=1}^{r} S(\tilde{R}_i)$ be the union of all supports of ratings. Then the maximizing set $\tilde{M}$ is defined as:

$$\tilde{M} = \{(r, \mu_M(r))\}$$

with $\qquad \mu_M(r) = \left[\dfrac{r}{r_{max}}\right]^n \qquad$ and $r_{max} = \sup\limits_{r \in S} r.$

For each alternative x_i we determine

$$\tilde{R}_i^o = \{(r, \mu_{R_i^o}(r)\}$$

with $\qquad \mu_{R_i^o}(r) = \min\{\mu_M, \mu_{R_i}\}.$

The ranking of alternatives is then performed according to $\arg(\sup \mu_{R_i^o}(r))$. Figure 2 illustrates this procedure. As shown by figure 3, this approach does not either differentiate always between different ratings.

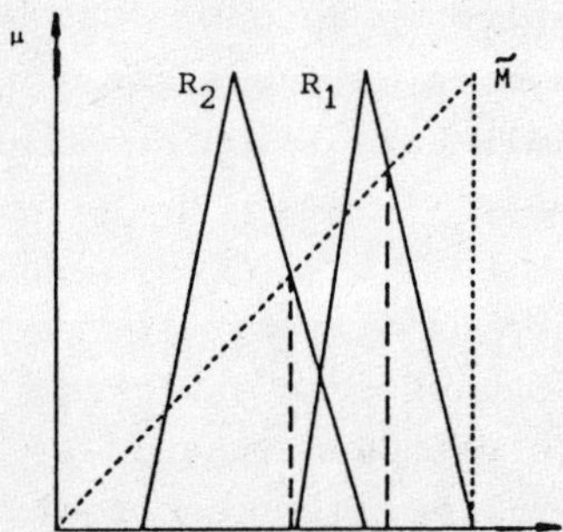

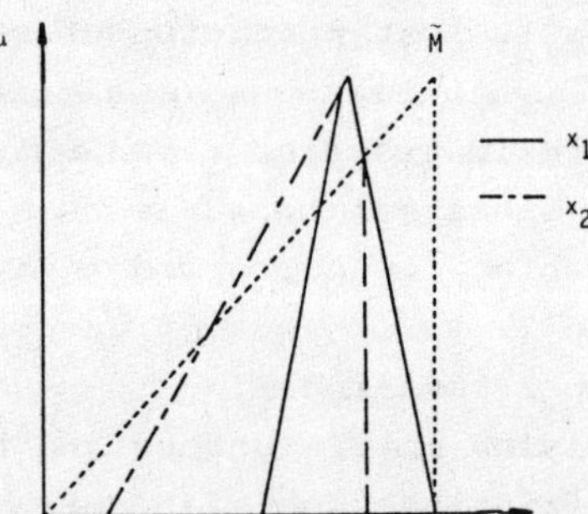

Fig. 2:   Fig. 3:

Baldwin and Guild (1979) tried to increase the sensitivity of this approach. They replace the (crisp) set $(I|\tilde{R})$ of Baas and Kwakernaak by a fuzzy set, use fuzzy relations to rank the fuzzy ratings obtained in step 1 and supply numerically efficient procedures for ratings represented by triangular fuzzy numbers. Chen (1985) eventually supplements Jain's "maximizing set" by a "minimizing set" in order to increase discrimination in step 2.

Dubois and Prade (1984) suggest another approach: They do not even try to determine a unique order of the alternatives in step 2. They rather present to the decision maker four measures of dominance and leave it to him to decide on a final ranking. The four measures or indices they suggest are:
- Possibility of dominance
- Possibility of strict dominance
- Necessity of dominance
- Necessity of strict dominance.

Tong and Bonissone (1984) argue that it is certainly not appropriate to give the decision some artificial precision - therefore, solutions should be linguistic rather than numerical. The authors are only concerned with step 2 of MADM, i.e. the ranking of (normal convex) fuzzy sets, $\tilde{R}_i$, which represent ratings of decision alternatives, x_i, and include already the aggregation of the merits of an alternative with respect to different criteria (goals).

The "ranking" of the final ratings, $\tilde{R}_i$, of the alternatives x_i, $i=1,\ldots,n$ is achieved in two steps:

First a fuzzy preference set is determined, which corresponds essentially to Baas' and Kwakernaak's. In a second step a linguistic approximation and truth qualification of the final decision proposal is generated. We shall consider this step now.

Step 2: (Truth qualification and linguistic approximation)
The rationale behind this second step is the belief that human decision makers understand better a linguistic statement characterizing the decision sets than a (numerical) membership function. The form of the linguistic expressions searched for by the authors is:
"It is very true that x_i is marginally preferred to all other alternatives".
This statement obviously contains three important pieces of information:
- the crisp decision x_i
- the intensity of preference (marginally)
- the degree of truth of the statement (truth qualification).

Formally the statement has the structure:
"x_i is P over all other alternatives is τ".

To translate the decision set $\widetilde{Z}_i$ meaningfully in the above mentioned way the authors interpret $\widetilde{P}_i$ as a fuzzy set on the same universe of discourse as $\widetilde{Z}_i$ and τ as a term of the linguistic variable "truth". They then try to find terms in the term sets of the linguistic variables "Preference" and "Truth" which approximate the unlabelled $\widetilde{Z}_i$ as well as possible and then use the linguistic label of these approximating terms to express $\widetilde{Z}_i$ natural language. As a tool for the linguistic approximation they suggest pattern recognition techniques (Tong, Bonissone 1984) and a context free grammar such as suggested by Zadeh (1977).

4. Comparison and Classification Methods

The methods and approaches presented in the preceding section differ with respect to a number of aspects (for a review see Bartolan, Degani 1985). They generally lead to the same results when applied to rather clear-cut decision problems. When applied to more complex problems, the results, however, differ from each other. We shall compare the methods by Yager, Baas and Kwakernaak, Jain, Baldwin and Guild using the following set of 6 situations:

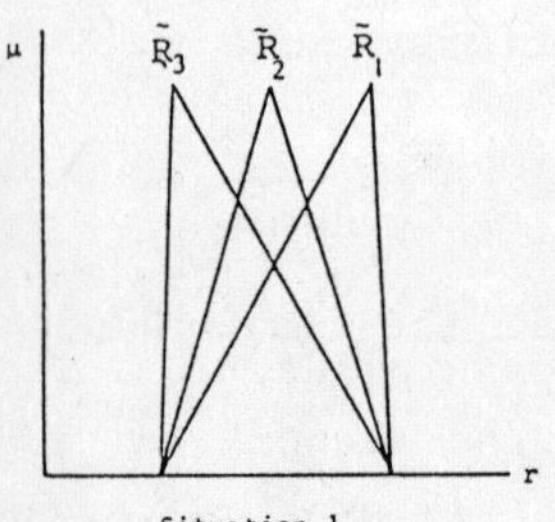

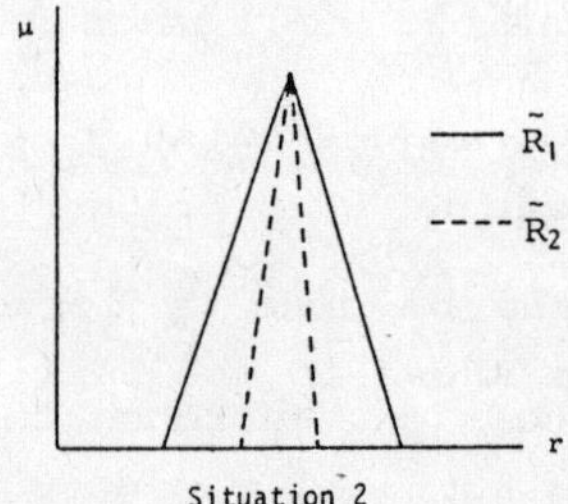

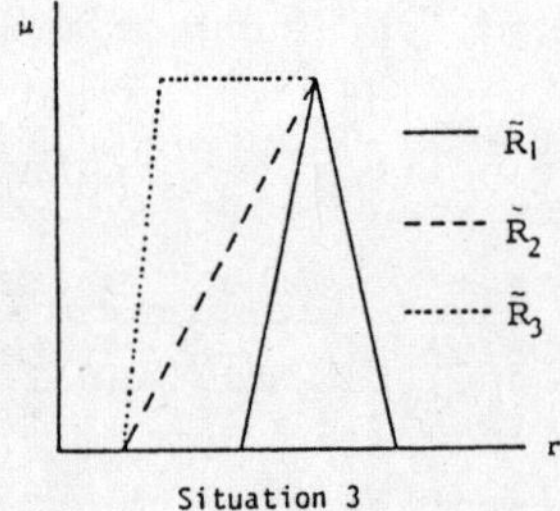

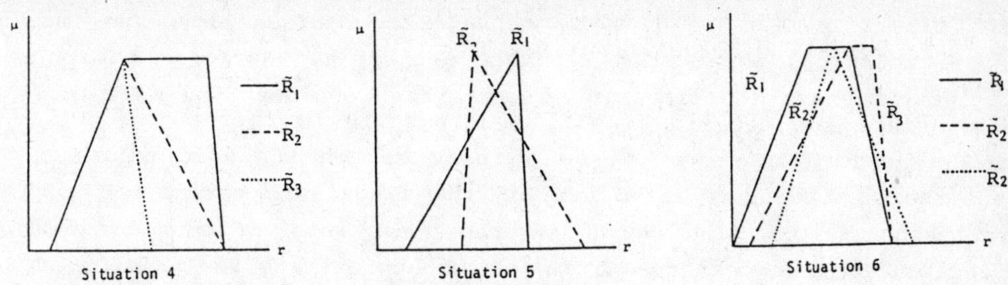

Fig. 4:

Rather than giving the numerical values of the resulting measures, which are not quite comparable, we list the resulting orders of the alternatives in the following table:

Situation	Yager	Baas & Kwakern.	Jain	Baldwin & Guild
1	1 ≻ 2 ≻ 3	1 ≻ 2 ≻ 3	1 ≻ 2 ≻ 3	1 ≻ 2 ≻ 3
2	1 ∼ 2	1 ∼ 2	1 ≻ 2	1 ∼ 2
3	1 ≻ 2 ≻ 3	1 ∼ 2 ∼ 3	1 ∼ 2 ∼ 3	1 ≻ 2 ∼ 3
4	1 ≻ 2 ≻ 3	1 ∼ 2 ∼ 3	1 ≻ 2 ≻ 3	1 ≻ 2 ≻ 3
5	2 ≻ 1	1 ≻ 2	1 ≻ 2	2 ≻ 1
6	3 ≻ 2 ≻ 1	1 ∼ 3 ≻ 2	3 ≻ 2 ≻ 1	2 ∼ 3 ≻ 1

The question arises: "Which is the best MADM-method so far?" This question can certainly not be answered in general, because the answer will depend largely on the situation for which the method is to be used, on subjective evaluations and on other factors. Some help in selecting a suitable method might, however, be provided by a classification of the methods according to a number of criteria.

A first classification could be done by looking at the three aspects scope, process and focus as shown in the next table:

Criterion			
Scope	Step 1	Step 2	Steps 1 and 2
Process	Simultaneous	Hierarchical	Interactive
Focus	Aggregation	Distance	Order-relation

This classification is rather mechanistic. It considers the method from a technical point of view. More appropriate would probably be a multidimensional classification taking into consideration more aspects of the different approaches. The following figure sketches some possible dimensions.

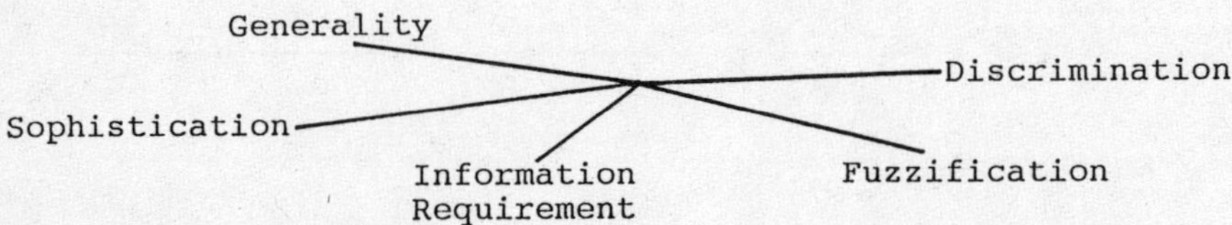

By "generality" we mean the degree of general applicability of the method, i.e. are special types of fuzzy sets assumed or can the ratings have arbitrary forms. Is the method restricted to special operators or can they be adopted to the context etc.

Discrimination refers to the capability of a method to differentiate between alternatives the ratings of which differ only slightly from each other. As mentioned above authors have different views on whether a method should be very discriminatory or rather "stable".

Fuzzification: Obviously different components of the MADM-problem can be represented by fuzzy sets. One extreme would be to only consider the relative weights of the criteria as fuzzy sets. Another extreme could be to consider the ratings, the weights and the alternatives as fuzzy.

Information Requirements: The more standardized the input data the less information has to be processed but the rougher might be the model of the real problem! If, for instance, only triangular fuzzy numbers are allowed each of them can be characterized by three real numbers. If arbitrary fuzzy sets are used much more information has to be provided and processed. The amount of information to be processed would increase even further if Type-2-fuzzy sets are used.

Sophistification refers to the mathematical tools which are being used in steps 1 and/or 2.

An evaluation of the methods according to the above-mentioned criteria would obviously be rather subjective. The following table, therefore, rather describes than evaluates the methods mentioned above.

	Phase	Weights	Criteria	Aggreg(PhaseI)	Crit.f.ranking	Solution
Yager I	I + II	fuzzy	fuzzy sets (norm)	weighted sum	max + min (Phase II)	crisp
Baas & Kwakernaak	I + II	fuzzy	fuzzy sets	max-min	preference sets	crisp
Laarhoven & Pedrycz	I	fuzzy triangular numbers	crisp	hierarchical aggregation	---	fuzzy ratings
Jain	II	--	fuzzy	--	maximizing set	crisp
Baldwin & Guild	II	--	fuzzy (special)	--	relative preference	fuzzy
Chen	II	--.	fuzzy	--	maximizing and minimizing set	crisp
Dubois & Prade	II	--	fuzzy	--	multiple	vectorial
Tong& Bonissone	I + II	fuzzy	fuzzy	max/min	Linguistic approxim.	linguist.

REFERENCES:

Baas, M.S., Kwakernaak, H.: Rating and ranking of multiple-aspect alternatives using fuzzy sets, in: Automatica 13 (1977), pp. 47-58

Baldwin, J.F., Guild, N.C.F.: Comparison of fuzzy sets on the same decision space, in: Fuzzy Sets and Systems 2 (1979), pp. 213-232

Bellman, R.E., Zadeh, L.A.: Decision-making in a fuzzy environment, in: Mgt. Sc. 17 (1970), pp. B141-164

Bortolan, G., Degani, R.: A review of some methods for ranking fuzzy subsets, in: Fuzzy Sets and Systems 15 (1985), pp. 1-20

Chen, S.H.: Ranking fuzzy numbers with maximizing set and minimizing set, in: Fuzzy Sets and Systems 17 (1985), pp. 113-130

Dubois,D., Prade, H.: Criteria aggregation and ranking of alternatives in the framework of fuzzy set theory, in: Zimmermann et al. (1984), pp. 209-240

Jain, R.: Procedure for multi-aspect decision making using fuzzy sets, in: Int. Journal Systems Science 8 (1977), pp. 1-7

Van Laarhoven, P.J.M., Pedrycz, W.: A fuzzy extension of Saaty's priority theory, in: Fuzzy Sets and Systems 11 (1983), pp. 229-241

Siskos, J., Lochard, J., Lombard, J.: A multicriteria decision-making methodology under fuzziness: Appreciation to the evaluation of radiological protection in nuclear power plants, in: Zimmermann et al. (1984), pp. 261-284

Tong, R.M., Bonissone, P.P.: Linguistic solutions to fuzzy decision problems, in: Zimmermann, H.-J. et al. (1984), pp. 323-334

Werners, B.: Interaktive Entscheidungsunterstützung durch ein flexibles mathematisches Programmierungssystem, München 1984

Werners, B.: An interactive fuzzy programming system, in: Fuzzy Sets and Systems 23 (1987), pp. 131-147

Werners, B.: Aggregation models in mathematical programming, in: G. Mitra (ed.), Mathematical models for decision support, Berlin, Heidelberg, New York 1988, pp. 295-305

Yager, R.R.: Fuzzy Decision Making including unequal objectives, in: Fuzzy Sets and Systems 1 (1978), pp. 87-95

Zadeh, L.A.: Linguistic characterization of preference relations as a basis for choice in social systems, Memo UCB/ERL M77/24 Berkeley 1977

Zimmermann, H.-J., Zysno, P.: Decisions and evaluations by hierarchical aggregation of information, in: Fuzzy Sets and Systems 10 (1983), pp. 243-266

Zimmermann, H.-J., Zadeh, L.A., Gaines, B.R. (Eds.): Fuzzy sets and decision analysis, New York 1984

Zimmermann, H.-J.: Fuzzy set theory - and its applications, Boston-Dordrecht-Lancaster 1985

Zimmermann, H.-J.: Fuzzy sets, decision making, and expert systems, Boston-Dordrecht-Lancaster 1987

Practical Building of Expert Systems for Structural Safety and Integrity Assessment

APPLICATION OF EXPERT SYSTEMS TO STRUCTURAL ENGINEERING

Sergio F. Garribba
Cesnef, Politecnico di Milano
Milan, Italy 20133*

Abstract

The purpose of the paper is to describe methodological aspects of expert systems designed to monitor the performance of mechanical structures, identify failures and eventually select appropriate maintenance and repair actions. Three major subject areas are considered. First, is the diagnostic task which requires determining differences between a model of a mechanical structure and the structure itself. The diagnostic procedure should be efficient and incremental, exploiting the iterative nature of the knowledge acquisition. Second is the modelling of the mechanical structure through a process which may allow to identify the various sources of failure and damage propogation the basis of available information including measured and recorded experimental data. Third, is data acquisition to construct a knowledge base, adjust and refine this base through the learning process. This multi-source base should be able to fuse estimates and allow data retrieval taking into account all approximations involved.

1.   Foreword

Expert systems in the area of structural engineering have been developed for

---

*   Present address IEA-OECD, 2 rue André Pascal, Paris, France 75775

many functions: design and planning, monitoring and diagnosis, maintenance and signal interpretation, data collection and analysis. Expert systems have generally worked well along specific goals with well-grounded and organized knowledge. In designing future expert systems more and more cases would be considered where the problem itself needs organization and formalized framing. Automated knowledge acquisition and use may be feasible to the extent that the practical structural engineering domain matches the application domain of the expert system. A machine learning process will rarely provide all of the knowledge for a complete expert system function. Rather, the designer ought to select a piece of the problem that will be handled by the learning process with a view to a number of specified requirements.

Comparing the relative difficulty of recurrent decision- making tasks, namely performance prediction, diagnosis, maintenance and failure repair, it is common understanding that diagnosis is the most difficult task. This consideration bases upon two factors. The first is the ratio of number of values held in working memory to number of values physically visible. The greater this ratio is, the more difficult the task will be. The second factor is the number of mental operations to be performed. Diagnosis is a task that is cognitively very demanding in terms of both factors. Moreover, in many situations, a diagnosis task will include detection of new symptoms and prediction of system behavior.

Monitoring of mechanical components and diagnosis of structural failures are the tasks which will be considered in the following. Reference is generally made to large and complex structural artifacts such as offshore oil platforms, nuclear vessels and piping systems, bridges and structural frames made by a variety of mechanical components with different failure modes. To design the corresponding expert systems the principal requirements must be identified. Although expert systems will progress over time as new applications are prepared and new demands result, an attempt to identify some of the key characteristics can be made.

First, an expert system should be "evolvable", that is, easy to modify and update. This refers to modularity and knowledge acquisition/maintenance. The expert system should follow a top-down, modular design where modules are used to hide design decisions and, through their use, later changes in the design are easy to make. To further ease modification and updating, a function for knowledge acquisition/ maintenance should be incorporated. This is to ensure that knowledge can easily be kept current, complete, and accurate. It would also be useful to have a semi-automatic knowledge acquisition facility, perhaps based on a natural language explanation capability in order to alleviate the knowledge acquisition bottleneck.

Second, the control procedure must be designed so that the expert system can "reason" effectively and quickly within the time constraints of the problem. A strategy might be adopted to develop this control procedure to minimize the amount of work and time required to enter relevant information.

Third, the expert system should have the capability for having both forward and backward reasoning. Forward chaining is a problem-solving paradigm in which chaining starts from a set of conditions and moves toward some (possibly remote) conclusion. Backward chaining entails having a goal or a hypothesis as a starting point, and then "working backwards" along some patterns to see if the conclusion is true. Forward and backward reasoning are both particularly important in the multisensor information integration and some distributed problem solving aspects of the structural engineering environment.

Fourth, it should contain many user-friendly features such as an explanation facility, free-text comments, questioning, help facilities, menus, and some natural language interface. An expert system is not just a set of rules that are processed to determine a solution. An expert system must incorporate user-friendly features in order to make it easier to use, thus creating a smooth implementation.

Fifth, the expert system should be flexible enough to be able to use metaknowledge. Metaknowledge refers to knowledge about the knowledge in the program. Metaknowledge can improve the performance of the program by constraining the search for a solution.

Sixth and last, the expert system should be able to reason about approximation and uncertainty by using probabilistic as well as non-probabilistic measures.

2.   The Human Operator and Fault Diagnosis

2.1   Diagnosis strategies

By focusing on different levels of knowledge representation, we can develop very different diagnostic strategies. The real goal of the diagnostic process is to generate a diagnosis which can "explain" all the observations, especially the deviant. Since a device can have multiple faults, the correct diagnostic answer will often consist of a number of malfunctions, each of which explains some of the data, but together they account for a model which is used. In general, the problem of picking the best model for the mechanical structure and its components is a version of the "set-covering" problem and as such is computationally very complex.

Some faults are known to be more difficult to diagnose than others. A novel fault, which is defined as one that the operator has not experienced in either real system operation or training, is usually more difficult to diagnose than a recurring fault. The difficulty mainly comes from the lack of compiled diagnostic rules. The taxonomy of search strategies suggested by Rasmussen is of great relevance to the question of what makes a novel fault diagnosis so difficult (Rasmussen, 1986). He described two types of search strategies: topographic search and symptomatic search.

Topographic search uses a normal model of the system to select the next field of attention. Each field is then judged to be good or bad through appropriate observation. The next field of attention can be either a subfield of the current field that was judged to be bad, or a field of the same level that is logically or physically adjacent. The advantage of topographic search is its dependence upon a model of normal system operation rather than models of malfunction. Rasmussen pointed out that this advantage makes topographic search suitable for novel fault diagnosis. But, the use of available information by topographic search is rather uneconomical because observations are used only for good/bad judgements. Thus, topographic search by itself may not lead to final diagnosis.

In symptomatic search, the diagnostician searches a mental library of symptom patterns to find a set that matches the observed symptoms. Then, the system state associated with the symptom pattern is identified. When there is ambiguity, more observations need to be collected to resolve it. If the reference symptom patterns were attained by training or experience, the search can be either a parallel, data-driven pattern recognition or a sequential decision table search. Since the symptom patterns are not readily available for novel faults, such symptomatic search may not be used in their diagnosis. A specialization of symptomatic search, however, may be relevant to novel fault diagnosis since the operator can evaluate hypothesis of novel faults as long as he or she can predict the associated symptom patterns through causal understanding. This search strategy by hypothesis and test, generates the symptom patterns on-line according to hypotheses and then compares them to the observed system behavior. Since the symptom patterns are generated via a hypothetical functional model of the system, understanding of the systems dynamics and configuration is more important than in other search strategies.

A view of diagnosis can be maintained which is very general, encompassing defect/failure detection in mechanical devices, troubleshooting in analog and digital circuits, debugging programs, and modeling physical or biological systems. To a large extent, expert system approach to diagnosis is independent of the inference strategy employed to derive predictions from observations. The model of the artifact describes the physical structure of the structural system or mechanical device in terms of its components.

Each type of component obeys certain behavioral rules. A component is a very general concept, including items, processes and even steps in a logical inference. In addition, each component has associated with it a set of one or more possible model-artifact differences which establishes the grain size of the diagnosis. A component is guaranteed to behave according to its model only if none of its associated differences are manifested, i.e., all the component's assumptions hold. If any of these assumptions are false, then the artifact deviates from its model, and the model may no longer apply. Therefore, a diagnosis takes (i) the physical structure, (ii) models for each component, (iii) a set of possible model-artifact differences, and (iv) a set of measurements, and produces a set of candidates, each of which is a set of differences which explains the observations. The diagnostic approach is based on characterizing model-artifact differences as assumption violations. To simplify the problem it can be presumed that the act of taking a measurement (i.e., making an observation) has no affect on the faulty device. Once a quantity is measured to be a certain value, that the quantity remains at the value. This is equivalent to assuming that no component's (correct or faulty) functioning depends on the passage of time. If a component is faulty, the distribution of input-output values contains no information. In general it is not presumed that if a component is faulty, (and exhibiting this faulty behavior) it may exhibit faulty behavior on some other set of inputs.

2.2 <u>Diagnostic Procedure</u>

Model-artifact differences are not directly observable. Instead, all assumption violations must be inferred indirectly from behavioral observations. Intuitively, a symptom is any difference between a prediction made by the inference procedure and an observation. More generally, a symptom is any inconsistency detected by the inference procedure and may occur between two predictions inferred from distinct measurements as well as a measurement and a prediction inferred from some other measurements.

The diagnostic procedure is guided by the symptoms. Each symptom tells about one or more assumptions that are possibly violated (e.g., component that may be faulty). Intuitively, a conflict is a set of assumptions which support a symptom, and thus lead to an inconsistency. In the structural engineering examples, a conflict is a set of structural components which cannot all be functioning correctly.

A case which frequently occurs is detection of multiple failing components. For complex domains any single symptom can give rise to a large set of conflicts, including the set of all components in the structure.To reduce the cominatorics of diagnosis it is essential that the set of conflicts be represented and manipulated concisely. If a set of components is a conflict, then every superset of that set must also be a conflict. Thus the set of conflicts can be represented concisely by only identifying the minimal conflicts, where a conflict is minimal if it has no proper subset which is also a conflict. This observation is central to the performance of our diagnostic procedure. The goal of conflict recognition is to identify the complete set of minimal conflicts (de Kleer & Williams, 1987).

Ultimately, the goal of diagnosis is to identify, and refine, the set of candidates consistent with the observations thus far. A candidate is represented by a set of assumptions. The assumptions explicitly mentioned are false, while the ones not mentioned are true. A candidate which explains the current set of symptoms is a set of assumptions such that if every assumption fails to hold, then every known symptom is explained. Therefore each set representing a candidate must have a nonempty intersection with every conflict.

Diagnosis is an incremental "evolvable" process. As the diagnostician takes measurements he continually refines the candidate space and then uses this to guide further measurements. Within a single diagnostic session the total set of candidates must decrease monotonically. This corresponds to having the minimal candidates move monotonically up through the candidate superset towards the candidate containing all components. Similarly, the total set of conflicts must increase monotonically. This signifies having the minimal conflicts move monotonically down through a conflict superset towards the conflict represented by the empty set. Candidates are generated incrementally, using the new minimal conflict(s) and the old minimal candidate(s) to generate the new minimal candidate(s).

The set of minimal candidates can be incrementally modified. Whenever a new minimal conflict is discovered, any previous minimal candidate which does not explain the new conflict is replaced by one or more superset candidates which are minimal based on this new information. This is accomplished by replacing the old minimal candidate with a set of new tentative minimal candidates each of which contains the old candidate plus one assumption from the new conflict. Any tentative new candidate which is subsumed or duplicated by another is eliminated; the remaining candidates are added to the set of new minimal candidates.

Consider how measuring quantity x_i, could reduce the candidate space. Expert system explicitly represents x_i's values, and their supporting environments:

$$[x_i = v_{ik}, e_{ik1}, ..., e_{ikm}].$$

If x_i is measured to have the value v_{ik}, then the supporting environments e_{ik} of any value distinct from the measurements are necessarily conflicts. If v_{ik} is not equal to any of x_i's predicted values, then every supporting environment for each predicted value of x_i is a conflict. Given expert system database, it is simple to identify useful measurements, their possible outcomes, and the conflicts resulting from each outcome. Furthermore, the resulting reduction of the candidate space is easily computed for each outcome.

The diagnostic process depends critically on manipulating three sets: (i) R_{ik} is the set of candidates that would remain if x_i were measured to be v_{ik}, (ii) S_{ik} is the set of candidates in which x_i must be v_{ik}, (equivalently, the candidates necessarily eliminated if x_i is measured not to be v_{ik}), and (iii) U_i is the set of candidates which do not predict a value for x_i (equivalently, the candidates which would not be eliminated independent of the value measured for x_i). This set R_{ik} is covered by the sets S_{ik} and U_i:

$$R_{ik} = S_{ik} \cup U_i, \quad S_{ik} \cap U_i = \phi.$$

Measurements are expensive, thus not every value at every terminal is known. Instead, some values must be inferred from other values and component models. Intuitively, symptoms are recognized by propagating out locally through components from the measurement points, using the component models to deduce new values. The application of each model is based on the assumption that its corresponding component is working correctly. If two values are deduced from the same quantity in different ways, then a coincidence has occurred. If the two values differ then the coincidence is a symptom. The conflict consists of every component propagated through from the measurement points to the point of coincidence (i.e., the symptom implies that at least one of the components used to deduce the two values is inconsistent). Note however, if the two coinciding values are the same, then it is not necessarily the case that the components involved in the predictions are functioning correctly. Instead, it may be that the symptom simply does not manifest itself at that point. Also, it might be that one of these components is faulty, but does not manifest its fault, given the current set of inputs.

3. <u>Models and Reasoning for Failure Diagnosis of Mechanical Systems</u>

3.1 <u>Fault Diagnosis</u>

Any knowledge-based system for failure analysis must have a passive component - internal models containing knowledge about the fault behavior of the system under analysis - and an active component - reasoning processes that operate on the fault models.

The diagnostic task is to determine why a correctly designed piece of equipment is not functioning as it was intended; the explanation for the faulty behavior being that the particular piece of equipment under consideration is at variance in some way with its design. A sequence of measurements can then be proposed, executed and analyzed to localize this point of variance, or fault. The task for the diagnostician is to use the results of measurements to identify the cause of the variance when possible, and otherwise to determine which additional measurements must be taken.

When the system to diagnose becomes large and complex, the speed, accuracy, and efficiency of diagnosis tends to degrade. There is abundant evidence to this relationship. It has been found that the average length of time the subjects spent solving context-free problems was highly correlated with the degree of functional connection between components. It has also been reported that the accuracy of diagnosis decreased as the number of components in the system increased. The procedural efficiency in terms of the number of tests until correct solution is also increasingly deviated from optimality as the problem size increased. The inverse relationship between problem size and diagnosis speed is the least surprising. The results, in which accuracy and strategic efficiency were degraded by increasing problem size, suggest, however, that the performance decrement was caused by greater mental workload (Fink & Lusth, 1987).

Different approaches have been used for fault diagnosis. In one approach the observed functionality of devices are compared with their predicted functionality which is specified by a quantitative or qualitative model of the structural device (Kuipers, 1984). For a large mechanical system, such as an offshore platform with a number of rapidly changing data values, comparing observed to predicted functionality can be inefficient. In an alternative approach, empirical associations between unusual behavior and faulty components are encoded as heuristic rules. This approach requires extensive debugging of the knowledge base to identify the precise conditions which indicate the presence of a particular fault. When a fault is proposed, and later denied by device models, the heuristic which suggested the fault should be revised so that the hypothesis will not be proposed in future similar cases (Pazzani, 1988). A learning process ensues that can be described as follows.

Let us assume that the goal of diagnosis is to find a hypothesis H which accounts for the abnormal functionality of a device. The process of diagnosis can be viewed as applying the following inference rule to conclude H.

$$\frac{F \text{ and } F \rightarrow H}{H}, \quad \text{if consistent } (H, M, O)$$

where F is a set of features, $F \rightarrow H$ is a fault diagnosis heuristic, H is a fault hypothesis, M is a set of implications which represent a device model, O is the set of observed data (F is a subset of O), and consistent (H, M, O) is true if H is consistent with the device model and the observed data. Note that the above inference rule is similar to "modus ponens" except that the conclusion H must be consistent with the model and the observed data. This is necessary because the implication $F \rightarrow H$ may not be correct. In the proposed approach to learning and fault diagnosis, consistent (H, M, O) corresponds to confirming a hypothesis with models for the structural device. It is computed by first finding M_H, a model of a new device which is identical to M except that a faulty component has a different functionality.

A hypothesis H is not consistent (i.e. consistent(H, M, O) is false) if there is an implication J_i and $M_H \rightarrow P_i$ is false. Here the inconsistency arises because from J_i and M_H - P_i and M_H and J_i it is possible to deduce P_i. However, P_i is known to be false. To summarize, diagnosis is viewed as consisting of two processes: generating a fault hypothesis H from diagnosis heuristics of the form $F \rightarrow H$, and then confirming or denying H by evaluating consistent (H, M, O).

When a diagnosis heuristic proposes a fault hypothesis H which is denied because consistent (H, M, O) is false, the diagnosis heuristic can be revised to not propose the fault in future similar conditions. By similar, is meant those conditions which would result in the same inconsistency. Recall that an inconsistency is detected when P_i and not (P_i) can be derived. The following is an example of an inconsistent theory:

$$F$$
$$F \rightarrow H$$
$$M$$
$$H \text{ and } M \rightarrow M_H$$
$$J_i$$
$$J_i \text{ and } M_H \rightarrow P_i$$
$$\text{not } (P_i).$$

We blame $F \rightarrow H$ for the inconsistency, because we are assuming that the observed data not (P_i) and F, the justification J_i and the implication from the device model J_i and $M_H \rightarrow P_i$ are correct. To avoid this inconsistency the diagnosis heuristic is revised to :

$$F \text{ and } J_i \text{ and } P_i \rightarrow H$$
$$F \text{ and not } (J_i) \rightarrow H.$$

The new theory, no longer allows H and, therefore, P_i to be derived:

$$F$$
$$F \text{ and } J_i \text{ and } P_i \rightarrow H$$
$$F \text{ and not } (J_i) \rightarrow H$$
$$M$$
$$H \text{ and } M \rightarrow M_H$$
$$J_i$$
$$M_H \rightarrow P_i$$
$$\text{not } (P_i).$$

3.2   <u>Fault Sources and Fault Propagation</u>

In case of multiple faults and complex mechanical structures, models which could be used, are fault propagation event-trees (digraphs) that represent the propagative aspects of faults and cause-consequence knowledge bases underlying augmented fault-trees that represent the causal aspects of failures. Failure analysis processes that operate on these models may utilize backtracking with constraint enforcement, implicit problem reduction, and bidirectional chaining of production rules.

Clearly, generating these digraphs and knowledge bases for a specific structural system requires the acquisition and encoding of knowledge about the fault behavior of the mechanical system from human experts. This is a process that incorporates considerable human involvement and judgment. However, the knowledge acquisition process can be systematized to some extent if parameters required by the knowledge representation scheme being used are well-defined and procedures for transforming expertise into these parameter values are available. The parameters that failure models require, such as the failure rates of components, the time for and corresponding measure of uncertainty (e.g. probability) concerning fault transmission between components, and the cause-effect relations among failures are assumed well-defined. Besides, it is admitted that conceptual structures for modeling the fault behavior of systems already exist and techniques for their construction are well-known (McCormick, 1981).

Therefore, reasoning about the causes of failures must involve the reconstruction of fault propagation paths within the structural system to locate the sources of failure and the identification of basic faults which caused the onset of failure at these sources. This two phase approach seems fundamental to the failure analysis methodology. The framework consists of failure models and processes that operate on these models. There is a failure model and a corresponding failure analysis process for each phase. A control task   directs the operation of other processes.

The failure source location process operates on individual subsystems of the hierarchical fault propagation model. It backtracks along all feasible fault propagation paths in a subsystem, starting from the elements indicated to be faulty by sensor's alarms and locates a set of elements which may be the sources of failure in that subsystem. This backtracking is subject to probabilistic (and non-probabilistic) constraints imposed by the failure relations between elements, temporal constraints imposed by the failure relations and real-time alarm information, and feasibility constraints imposed by the subsystem itself.

In one particular model (Narayanan & Viswanadham, 1987) the structural system represented by a hierarchical set of level-structured digraphs. The system S corresponds to a set

$$S = \{s_1, s_2, ..., s_n\},$$

where s_i, $i=1...n$ are the subsystems of S. Then we define a failure relation R on S where $R \subseteq (S \times S)$ and $s_i R s_j$, s_i, $s_j \in S$ implies that it is possible for a failure in subsystem s_i to propagate directly to subsystem s_j. In other words, $s_i R s_j$ if $p(s_i, s_j,)$ a probalistic or non-probabilistic measure of direct failure propagation from s_i to s_j, is nonzero. This failure relation may now be used to specify a fault propagation digraph D for S:

$$D = (V, E).$$

The vertex set $V = \{s_i,...,s_n\}$ and the edge set $E \subseteq (S \times S)$. The vertex set contains two kinds of vertices: monitored vertices and nonmonitored vertices. A monitored vertex corresponds to a subsystem that has an alarm attached to it. Most systems are partially monitored in that monitoring equipment (alarms) are associated only with key units. So the monitored vertex set will usually be a proper subset of V. Each $s_i \in S$ is decomposed into its subsystems, and the corresponding level-structured digraphs are developed. This set of digraphs $\{D_i\}$ forms a level of the hierarchy. The hierarchical fault propagation model may therefore be defined as

$$H = \{L_i\}, \quad i = 1...n$$

where L_i denotes the ith level in the hierarchy.

$$L_i = \{D_{ij}\}, \quad j = 1...m_i.$$

Each D_{ij} is a level-structured digraph (V_{ij}, E_{ij}). Elements of the vertex set V_{ij} correspond to various components of the subsystem, and E_{ij} contains weighted edges that represent the characteristics of fault propagation within the subsystem.

4. Knowledge Acquisition Problems

4.1 Conformity Relations and Inferred Attributes

Use of fuzzy sets allows to represent natural language computations using linguistic variables, and to incorporate the complexity and uncertainty into the model. In effect the desire for perceived precision as found in numeric models is relaxed, in an effort to increase the significance or believability of the results. While the methodology is very general, it is applied to the problem of creating values and updating attributes in the perception of structural components.

The common theme that has emerged with most current models of perception is the notion of mechanisms, such as channels, filters, and detectors, which are selectively sensitive to processing specific sensory qualities of the impinging stimuli, such as stress, size, strain, and other attributes. A general idea underlying the notion of mechanisms is that the sensory data are decomposed into simpler elements concerning various attributes, which are analyzed and subsequently correlated by hierarchical processes of discrimination, recognition, and classification. The outstanding feature of these models of perception is their procedural character.

Human perception ranges widely in complexity, time and space, as well as in dimensionality of inference. Samples of sensory data, which carry measured or assessed values of an attribute, may be represented as propositions of the form: "the attribute of the structural component is value", or, where the object is a conceptual entity and the attribute a characteristic feature of the structural component, which is selected because of its relevance or significance. The value is a scalar quantity or an n-tuple, quantified over some numerical or linguistic scale using fuzzy subsets (Garribba & Servida, 1989).

Consecutive sample values $y(x)$ unfold their attributes $y(x; X)$ over the ordered domain $X = \{..., x_i, x_{i+1},...\}$ in some measurable physical or conceptual space, or in time. The unfolding connections may be represented as a generalized mapping function,

$$y_j(x; X) \in B_X^\infty : X \to Y$$

which relates consecutive elements x of the domain X with elements y of the range Y of possible values of the attribute. B_X^∞ is the infinite set of all possible $y_j \in B_X^\infty$ over the domain X. Note that for simplicity we may denote $y_j(x; X)$ simply as y_j. The various sensory attributes y_j communicate form, meaning, and value to the human faculties of perceptions, interpretation, and evaluation.

To a specific knowledge base the behavior of a structural component is characterized by several, say K, different "prototypical norms", each norm represented by the corresponding reference prototypical attributes

$$\hat{y}^k(x; X_k) \in B_{Xk}^\infty , \quad k= 1, ..., K.$$

Since the prototypical attributes y_k of the same norm (kth) are grossly underdetermined because they conform only to a limited set of characteristics constraints, they form an equivalence class of prototypical attributes, each attribute fully conforming to all the characteristic constraints of this norm.

In formal representation of a conformity distribution relation (CDR) can then be defined (Ligomenides, 1988). CDR is a fuzzy subset over B_x assessing a degree of conformity $\mu_j^k = [0,1]$ to the kth norm, for each sensory pattern $y_j(x; X) \in B_x^\infty$:

$$CDR_k = \{\mu_j^k / y_j(x; X) \,|\, y_j \in B_x^\infty, \mu_j^k = [0,1]\}$$

where μ_j^k, generally a fuzzy number in [0,1], measures overall conformity of $y_j(x; X)$ to the elastic constraints of the kth norm. In this sense, μ_j^k also expresses what we may refer to as the "semantic content" of $y_j(x; X)$, in that it measures the resemblance of $y_j(x; X)$ to $\hat{y}^k(x; X_k)$.

The q-conformity and q-level conformity equivalence classes may also be introduced

$$C_k(q; X) = \{y_j(x; X) \,|\, y \in B_x^\infty, \mu_j^k = q\}$$
$$C_k(\geq q; X) = \{y_j(x; X) \,|\, y_j \in B_x^\infty, \mu_j^k \geq q\} \cdot$$

By extension, the degree of "k-resemblance" between two attributes $y_i(x; X_i)$ and $y_j(x; X_j)$ may be expressed by

$$\mu_{ij}^k = \sup(\min(\mu_i^k, \mu_j^k)) = [0,1]$$

as a measure of the recognized information that one attribute carries about the other, with reference to the characteristic constraints of the kth norm. Then the (fuzzy) K-tuple measure,

$$\mu_{ij} = \{\mu_{ij}^k \,|\, k = 1, ..., K\}$$

may express the overall resemblance between $y_i(x; X_i)$ and $y_j(x; X_j)$. Such conformity/resemblance relations are reflexive, symmetric, but only weakly transitive (i.e., by no means obligatory).

4.2   Data Classification and Recognition

A major part of the knowledge engineer's task involves the acquisition of knowledge from experts in the form of facts and rules and the transformation of this information into a form that can be used by a computer program. The assumption that expert knowledge is mandatory in successful expert systems makes the acquisition and representation of knowledge critical to the design of these systems. There are several problems inherent in the current knowledge engineering methodology that contribute to the knowledge transfer bottleneck. The process of extracting expert knowledge is not well defined, but generally involves interaction between a knowledge engineer and a domain expert. The knowledge elicitation process is often viewed as an art because there are no formal procedures or techniques available. However there are some less formal methods that are commonly employed by knowledge engineers for this purpose. The two most popular of these are interviews and protocol analysis. Whereas both of these methods share many common elements, protocol analysis can be considered a structured interview. Generally, the interview is conducted by a knowledge engineer who poses questions or problems to the expert. The expert in turn, is expected to provide answers or solutions that hopefully reveal some of the facts, rules, and heuristics relevant to the domain in question. In a similar fashion, protocol analysis involves the observation of a domain expert by a knowledge engineer as the expert solves a problem within his domain (Cooke & McDonald, 1986).

Some characteristics of human expertise add to the difficulty of verbally expressing knowledge. Much expert knowledge takes the form of heuristics, rules, and strategies that are procedural in nature. Another problem related to communication has to do with subjectivity on the part of the knowledge engineer. The expert's behavior is observed, interpreted, and transformed into a formalized version by the knowledge engineer. Likewise, whatever the expert says in an interview or protocol is interpreted and transformed by the knowledge engineer.

Again, these subjective procedures can be made more objective by using work-formalized decision-making techniques, based upon fuzzy algebra and multiattribute fuzzy utility functions. But the use of utility theory in practice is complicated by the following difficulties: (i) methods of utility functions construction allowing to take into account fuzzy information about the decision-maker's preference relations hardly exist; (ii) the practical use of multiattribute utility functions construction is based on the assumptions bout preferential independence and decomposition of the mentioned functions; similar methods for fuzzy information are not available.

A more direct approach has been proposed by Driankov (Driankov, 1987). The approach could also apply to query of database, for instance failure data and mechanical properties. Let us assume that proxy-objectives are defined which point at least to some degree in the same direction as the original objective and which can be measured. Then goal G_i is called fuzzy with respect to the decision-maker if it describes the aspiration level L_i^* assigned to the proxy-objective X_i within the context of a certain original objective O, in terms of a fuzzy set, as

$$G_i = (X_i, L_i^*) = \sum_{U_i} \mu_i^*(u)/u$$

where X_i is the name of the i-th linguistic variable employed for the representation of the i-th proxy-objective within the context of the original objective O;

L_i^* is a particular linguistic value of X_i belonging to the term-set T_i of X_i and represents the aspiration level assigned to the proxy-objective X_i within the context of O; $\mu_i^*(u)$ is a membership function defining the fuzzy set labeled L_i^* and is defined on the universe of discourse $U_i = \{u\}$ of X_i.

Since one and the same proxy-objective X_i could be used for the representation of different original objectives and in this case might have different aspiration levels assigned to it within the context of each of these original objectives O_j a comprehensive aspiration level G_{ij} can be introduced with a corresponding degree of attainment α_{ij}. The decision-maker will select options having the highest α (Driankov, 1987).

5. <u>Closing Remarks</u>

Although complex structural engineered systems and their components are generally monitored and diagnosed by human operators, several trends are bringing a growing degree of computerized control and automated data processing. Modern electronics are transforming management of mechanical equipment and structures. Sensors and probes, in the recent years, and many mechanical equipment monitoring apparatuses have become sufficiently rugged and have been accepted by maintenance crews to such an extent that they are now permanent fixtures at critical locations in mechanical structures. Increasingly, sensor readings are routinely rcorded and relayed to the control room or elsewhere, and engineers can look for trends in component degradation for warnings of approaching failure.

As a consequence, significant opportunities surface for the adoption of experts systems in the mechanical performance and fault diagnosis processes. Use of artificial intelligence techniques would allow operators to make quick and reliable inferences which would otherwise be expensive, or impossible within the time constraints imposed by complex engineered systems. Given symptoms of a failure, operators would first try to match from their repertoire of rules. Heuristics might be employed to make this matching process efficient. Once a model is identified operators could exploit the information and data collected to conduct evaluations. If more symptoms are found, operators might use the new evidence to search for better hypotheses. If inferences drawn from models are in conflict with known symptoms, these models are discarded and operators could search in the system or symptom knowledge base for relevant information under the guidance of new accumulating data and experience.

There are of course limitations to the adoption of expert systems for monitoring and diagnosing mechanical structures. Combination of conflicting evidence, representation of operator skills and judgment, decision with multiple objectives, signal recognition are all areas where progress will be needed. The experience which can be accumulated from practical applications may should provide indications for further work and improvement.

References

Cooke, N.M., McDonald, J.E. (1986). A Formal Methodology for Acquiring and Representing Expert Knowledge; Proc. IEEE, 74, 1422-30

de Kleer, I., Williams, B.C. (1987). Diagnosing Multiple Faults; Artificial Intelligence, 32, 97-130

Driankov, D (1987). An Outline of a Fuzzy Sets Approach to Decision Making with Interdependent Goals; Fuzzy Sets and Systems, 21 275-288

Fink, P.K., Lusth, J.C (1987). Expert Systems and Diagnosis Expertise in the Mechanical and Electrical Domains; IEEE Trans. on Systems, Man and Cybernetics, 17, 340-349

Garriba, S., Servida, A. (1989). Fuzzy vs. Probabilistic Methods for Combining Evidence from Expert Judgments; in Advanced Structural reliability II; A.C. Lucia (ed.); Reidel, Dordrecht, The Netherlands

Kuipers, B. (1984). Commonsense Reasoning about Casuality: Deriving Behavior from Structure; Artificial Intelligence, 24, 169-203

Ligomenides, P.A. (1988). Real-Time Capture of Experential Knowledge; IEEE Trans. on Systems, Man and Cybernetics, 18, 542-551

McCormick, N.J. (1981). Reliability and Risk Analysis Methods and Nuclear Power Applications; Academic Press, New York

Narayanan, N. H., Viswanadham N. (1987). Methodology for Knowledge Acquisition and Reasoning in Failure Analysis of Systems; IEEE Trans. on Systems, Man and Cybernetics, 17, 274-288

Pazzani, M. J. (1988). Explanation-Based Learning for Knowledge-Based Systems; pp 217-237 in Knowledge Acquisition for Knowledge-Based Systems, Vol. I, B. R. Gaines, J. H. Boose (eds.); Academic Press, New York

Rasmussen, J. (1986). Information Processing and Human-Machine Interaction; North Holland, New York

PRACTICAL BUILDING OF EXPERT SYSTEMS FOR MATERIALS SELECTION AND CORROSION PROBLEMS

Walter F.L. Bogaerts, Marc J.S. Vancoille
Faculty of Engineering - Dept. MTM
Katholieke Universiteit Leuven
W. de Croylaan 2, B-3030 Leuven, Belgium

Abstract

This paper will illustrate some of the principles of building expert systems. As an example, the expert system PRIME which is one of the most advanced realizations in its field, will be discussed. This expert system has been developed at the University of Leuven and deals with material selection and corrosion problems in the chemical process industry. In its implementation, many of the characteristics of second generation expert systems can be found such as multiple reasoning models, and multiple reasoning strategies. Furthermore, PRIME has a highly advanced, graphics-based and user-friendly interface which takes man-machine interfaces one step further than most present day developments.

Keywords: expert system, material selection, troubleshooting, risk analysis, PRIME, man-machine interface, second generation expert system, multiple models.

1 Introduction

What are the prerequisites for success in industrial or commercial knowledge-based system development? A number of guidelines prevail when starting a project or selecting a potentially successful application of expert systems:

1/ select a *feasible and high-impact* AI application; *i.e.*: consider the following issues:

- what are potential economic or technical benefits ?
- what are the intrinsic features of the AI approach for the problem to be solved ?
- to what extent does AI extend the capabilities of conventional data processing ?
- the area must be technically mature (a KBS is no substitute for knowledge) and the expertise must be available;
- the task (previously) performed by the expert must be frequent, take at least a few hours, and be non-trivial;

2/ develop a strategic plan for *incorporating* AI into the overall engineering environment and carefully determine the components of the knowledge-based system (system *architecture*);

3/ choose appropriate *hardware and software tools* (i.e.: what are available AI technologies and the criteria for their use); allow for smooth incorporation of new technological developments;

4/ put together a "winning" development team;

5/ acquire "sufficient" resources (real-world expert system projects are expensive!);

6/ "manage" knowledge-based system development projects, with regular review points and evaluations;

7/ ... .

In the following paragraphs a brief description is given of how an application in the area of *corrosion* and related *materials performance* has the potential to meet these critical criteria and guidelines.

2 Materials Selection and Corrosion : The Problem

Equipment durability is a major aspect in almost any industrial operation. The issue arises either

- when something has to be built and the question is: will it corrode, creep, crack or deteriorate in some other way; or

- when something has been built and the question is: what is the residual lifetime under the actually observed conditions (and rates) of corrosion or any other materials degradation process; or
- when the equipment has deteriorated and the questions are: what is the failure mode and how can the problem be overcome.

The prediction of corrosion-limited service life, and the evaluation and specification of materials in terms of corrosion resistance, are tasks which require expert knowledge and judgement. However, corrosion is a very wide field, corrosion experts are few in number, and expertise in particular corrosion areas is scattered. The expertise often resides in the presence of a few persons with relevant experience or with access to pertinent information. The absence of a corrosion expert or his inability to locate or interpret relevant information poses difficulties for both plant designers and operators [Westcott 1984]. As a consequence, for example, large financial penalties can be incurred as a result of corrosion induced failures.

Several macro-economic studies have estimated that direct and indirect costs incurred by corrosion related materials failures amount to 4 or 5 percent of the Gross National Product of every industrialized country. Moreover, it is generally accepted that a significant proportion ($\pm$ 1/4 to 1/3) of the costs which result from corrosion failures could have been avoided if only existing know-how would have been used.

A solution to the problem is however not easy to achieve, and any increase of the safety margin applied, when selecting a material for a given application, will almost inevitably increase the overall cost of the structure or component. High costs arise from uncertainties in materials performance and there is an urgent need to explore how the present situation can be improved. An interesting analysis of this issue is described by Edeleanu and Hines [Edeleanu 1983].

3   The Need for Advanced Information Processing Technologies

A great deal of materials information required by engineers is incorporated in codes and related documents. These regulate what is done in engineering and what is, for example, bought and sold (e.g. an "AISI 316L" material, with its well-determined physical and mechanical properties). This is a "well-oiled" existing communication (information exchange) system which works very well because it has evolved over the years. But it still has gaps, and when these are encountered, the system which takes over is cumbersome and expensive. When the engineer is let down by the codes and regular company documentation, additional information is obtained normally from human advisors. This is mostly necessary, since for an "outsider" it is difficult to locate the required information in the technical literature, it is also difficult to assess its validity once located, and there are obvious advantages in sharing the responsibility of doing something which is unusual with a knowledgeable individual [Edeleanu 1983].

Efforts are currently being made to produce on-line accessible data banks in order to improve this (codes- or standards-based) communication system, and some types of information, such as (materials) property data, can and are being incorporated in such data banks. An overview of materials properties data banks is, for example, given in [Hampel 1983].

Valuable as such computer aided services may be, this type of data is the kind of information which is already well documented and which does not normally cause significant hold ups in any engineering task. The greater need is for an aid to help with the problems which are now dealt with by consultation with individual human specialists. Specifically there is a need for on-line assistance on the topics which affect the durability of equipment.

To deal with this problem it is necessary to incorporate information on corrosion, wear, fatigue, creep and other such "processes" [Edeleanu 1983] in the computer-based information system. This is difficult because the rate at which processes occur is a function of the exposure conditions as well as of the material "properties". *Corrosion* and related phenomena can therefore not be treated as conventional materials data but require a more advanced level of information technology in order to be "computer treated".

The provision of an *expert system* seeks to alleviate these difficulties [Bogaerts 1986, 1987]. To achieve this goal fully such expert system should provide assistance at the relevant level to both corrosion experts and plant designers and operators. That such a possibility exists is due to current developments in information technology, more particularly in the area of Expert Systems and Knowledge Engineering.

4   Corrosion Expert System development and A.I. research

Expert systems are already more than simply an exciting laboratory novelty. Major companies and government departments are investing substantial sums to develop demonstrator projects and the first practical systems to come on the market look as if they will save their users the predicted amounts of time and money. Expert systems are the first fruits of many years research into artificial or machine intelligence. The basic concept is that computer software can be written which can sift the professional opinions of experts in a given field and offer 'reasoned' answers to questions put to the system. Expert systems not only provide these answers but they also explain the logical steps taken in coming to a particular conclusion. Their commercial value lies in the fact that they can be used by experts as a time and cost cutting 'second opinion' or by the less expert as a robot adviser.

In the area of corrosion or materials performance, for example, substantial savings may be expected. The topic of materials selection and corrosion control poses a major bottleneck in different engineering disciplines. Because of the multi-disciplinary nature of the required skills an engineer cannot acquire an encyclopedic knowledge of all relevant disciplines and data, and it is generally accepted that significant savings could be obtained if only existing know-how was used and if it could be accessed in a more straightforward manner (see sections 2 and 3 above).

The entire domain of materials performance and corrosion is however vast and complex, and dealing with such knowledge domains requires *stretching the edges of nowadays expert system technology*. Hence, the expert system development work described in this paper involves a significant research component.

To overcome problems associated with high complexity the concept of factorization or restricted responsibility domains has been pursued successfully in many fields. It seems therefore inevitable that effective use of expert system support in the corrosion area will also require a degree of cooperation between different expert systems (each an expert in a specific domain) and between expert systems and non-intelligence based systems. This leads to what is sometimes called Hybrid Integrated Conferring Expert Systems.

The architectural principles for such large complex systems are not fully developed and several important fundamental questions remain to be answered. Progress is needed in areas such as:

- knowledge representation and reasoning,
- cooperation among independent knowledge sources,
- management of very large knowledge and data bases,
- coupling between symbolic and non-symbolic components,
- reasoning with uncertainties,
- a new generation of user interfaces.

The purpose of the present R&D work is to *investigate these questions by experimenting with a number of different corrosion engineering problems* (e.g. materials selection, corrosion troubleshooting and risk analysis, selection of adequate preventive measures, etc.).

One of the results of this effort is a 'second generation' expert system structure, called "**PRIME**" (Process Industries' Materials Expert), which deals with materials selection and corrosion prevention issues in Chemical Process Industries [Vancoille 1988]. The basic programming environment for PRIME is KEE[1] with SUN Common Lisp and Flavors. As a hybrid system, KEE offers several knowledge representation paradigms (rules, frames, and worlds) with their associated inference techniques. Further components of the system include tools for managing the knowledge bases, dialogue processes with the user and explanations of the conclusions that are drawn. The knowledge engineer can use Lisp to supplement the range of functions with individual procedures. The working environment is characterized by diverse graphic functions, mouse-driven menus and window techniques.

KEE is a powerful programming toolkit in its own right but various extensions and modifications have been implemented in this project: integration with Fortran and C programs, a natural language parser to enhance the explanation facilities, etc. PRIME also has no query-based user interface, but employs a rather sophisticated 'active' user interface that makes extensive use of computer graphics features, windowing facilities, soft buttons and mouse-and-menu interface components.

The modular concept of PRIME also makes it easily expandable and allows the contribution of different experts. It is therefore feasible that Prime can be tailor made for specific processes or industries without having to change the whole knowledge representation and inference structure.

Work on the project was mainly carried out in the framework of the joint European ESPRIT programme and has officially started in October 1985, but its development was in fact based on earlier work on expert systems for corrosion technology, i.e. the ESCORT R&D project, which was conceived early 1983 in our laboratories. At present, it mainly deals with so-called CPI (Chemical Process Industries) materials, but it has reached a stage of development where it demonstrates what can be achieved with state-of-the-art A.I.-tools.

A useful prototype is now running in our laboratories. This contains many of the most advanced expert system features existing today and actually contains more than 4000 objects and attributes. The system represents a "computer program" with over 135,000 "lines of code" on top of KEE, accounting for an effort of almost 30 man years and a research & development budget of more than 2 million ECUs. The knowledge bases in PRIME are continuously being extended and a thorough industrial evaluation and (performance) metrication will be the next step in the project. At present, the possibility

---

[1] KEE is a registered trademark of IntelliCorp Inc., Mountain View, CA, USA.

also exists to access external data bases (both bibliographic and factual) and a dedicated 'metaprocessor' is being developed to recognize the type of problem and delegate the processing to the respective knowledge modules (either a knowledge base, a simple data base, a hypertext module [Arents, 1989], or a separate expert system).

5 PRIME: the concept.

5.1 Introduction.

The expert system environment PRIME is intended to assist both experts and novice users in the selection of corrosion resistant materials in chemical process industries and the determination of possible causes of material failure and risk analysis with respect to corrosion.

As pointed out above, expert systems for materials selection and corrosion have specific needs. Until recently, the most widely adopted approach in dealing with automated procedures for materials selection and corrosion control was the development of large data bases and "intelligent" front ends to these data bases. They are, however, limited in what they can offer and always require some human expertise to assist in the interpretation of collected data. *Expert Systems*, which try to reach decisions within specific domains in a way that conforms as closely as current techniques allow to the thought process of human experts, can alleviate part of this problem but we feel there are some requirements the implementation has to obey if the expert system is intended to be successful.

1/   The average user does not want to learn new computer languages, operating systems etc., nor does he wants or has the time to learn complicated commands to access the knowledge bases.

2/   Since most users know what corrosion and materials selection is all about, they fear the "black-box" approach. They want to be able to look at the underlaying knowledge and they want to be able to modify or add knowledge (e.g. company specific or proprietary information). Hence the **explanation, browsing and editing facilities** of an expert system are a major issue.

3/   The underlaying model(s) of the expert system have to conform to expert practices. The models have to be clear and easily understood which adds to the acceptability of the expert system.

4/ It is equally important from an economic and managerial point of view to be able to **incorporate previously developed software** (e.g. simulation software, data bases, ...).

5/ Corrosion and materials selection are complex problems. Every industry has its own very specific problems, practices, solutions and codes. It is impossible to grasp in one system all this knowledge. Therefore, multiple expert systems are necessary and they to be **modular** enabling them to adapt easily to new information. They have to be able to deal with **multiple models** of the same reality and multiple reasoning strategies (see paragraph 7.3). Knowledge sources have to be **independent** so they can easily be added, modified or removed. Moreover, when multiple knowledge sources or experts are available, they have to **cooperate** to solve the problem since none of them has all the necessary competence.

The requirements mentioned above seem to suggest that an expert system for materials selection and corrosion control (the same probably applies to other domains as well) has to be very modular and open. The user has to be able to incorporate his own knowledge although he wants to know as little as possible about artificial intelligence techniques. Practical applications more than theoretical considerations prevail. This also suggests that an expert system, if its is intended to be successful, has to be developed in close cooperation with the interested parties.

For instance, carbon steel is a preferred material in the *chemical process industry* because of its low cost, good weldability and generally good mechanical properties. It is equally well known that in most cases, carbon steel does corrode but at such a slow rate that it is hardly relevant. In the *pharmaceutical industry* on the other hand, carbon steel is almost never used although from a structural point of view it suffices. Possible trace chemicals due to the corrosion of carbon steel are prohibitive.

This example clearly indicates that there are different views of the same reality (see also the discussion on *Second Generation Expert Systems* in paragraph 6.4). Based on the same data, a certain material may be acceptable in one industry, but totally unacceptable in another. It basically comes down to the fact that if an expert system has to be developed, it has to be developed for a specific industry and cannot aim at solving every corrosion problem. However, the underlaying corrosion data may be the same.

From these observations, one could easily conclude that constructing an expert system for materials selection and corrosion control is an almost hopeless task. And it might well be. Still, as much as one industry may differ from another, the requirements of an expert system remain pretty much the same. What is needed is an context specific interpretation of the data. Basically, every problem requires a different view of the same data.

Fortunately, an expert system can be made of modular, independent tools and knowledge bases that interact with each other. These tools and knowledge bases can be reused when constructing expert systems for different applications. Examples of data bases are for instance a data base of all commonly used materials with their chemical, physical and mechanical properties. Examples of reusable tools are for instance the *inference* or *reasoning mechanisms* and the explanation and browsing facilities which can be made independent of the actual knowledge they process.

In the next part of this paper, it will be shown how these different requirements can be implemented. Some properties of *second generation expert systems* will be highlighted together with some implementational details which enable expert system developers to design expert systems for the chemical process industries rather fast.

6   PRIME : the implementation

6.1 Man-Machine-Interaction

As pointed out earlier, the end user does not want to be bothered with (new) programming languages. The use of the expert system has to be as easy and transparent as possible. Furthermore, the interface has to reflect the underlaying model thus enhancing the acceptance of the expert system since the user will recognize his way of analyzing the problem.

Most of today's expert systems' user interaction is (almost) completely text oriented. These query-based systems require a lot of effort from the user to understand completely what the developer means. Subtle details in the phrasing of a question may be completely overlooked by the user. This of course bears an enormous risk: the user may not be aware that he has accepted certain default built-in assumptions. These in term may cause the answer to be completely irrelevant for the user's application. Furthermore, many of these questions may be the same for every session. There is a substantial risk that the user might get bored and won't use the expert system again. The amount of textual information / clarification which can be -usefully- presented on a single screen is also very limited: typically 5 to 9 lines of (new) text is memorized by the user at every screen display.

These are all significant disadvantages of conventional query-based user interfaces, but what is even more worrying in conventional interface technology is that the expert system might not ask the right questions. The user has a lot of information he considers important for his problem. If the system fails to ask for these, the user will mistrust the answer. This problem can only be overcome through the design of an interface where the user is, or at least gets the feeling to be, in control of the interaction process (*'active' man-machine interface*).

All these considerations have led us to believe that text oriented interfaces are out. The success and ease of use of the Apple Macintosh have been largely attributed to its highly graphical interface, its use of icons and windows, and a sound and consistent underlaying philosophy of how the user prefers to interact with a computer. A completely new kind of expert system interface, of which we believe it can alleviate much of the problems sketched above, has therefore been developed for PRIME along the same lines. The PRIME-interface is completely graphics based with windows that correspond to delimited parts of the problem at hand. It has been implemented in a consistent way so the user very quickly understands how to interact with the system. The user is also very much in control of the interaction through the provision of so-called *Engineering Worksheets*. Every engineering worksheet can be "filled-out" by the user to any level of detail to which he has his information available. Input of specific details is possible through a series of default or context-sensitive pop-up menus.

An example of this interface is represented in Figure 1. It depicts the **First Corrosion Engineering Worksheet** (see paragraph 6.3). It is conceived much the same way the user interacts with a human expert. At the start of the consulting session he will describe to the expert from which industry he comes and what they are working with (equipment, environment). It is only at a later stage that the expert will ask him specific questions such as the operating conditions (temperature, pressure, what specific chemicals they are dealing with, contamination, user preferences,...).

6.2 The Tool Level

If at all possible, it is certainly prohibitively expensive to build an expert system that will solve "all" imaginable problems for materials selection and corrosion control. Most problems are so complex, that if an expert system aspires to be successful in a company it has more or less to be tailor made. An acceptable approach is to construct a **generic** system that incorporates most commonly accepted corrosion knowledge (the way it can be found in engineering handbooks). This will, however, only enable to do a screening of the problem, leaving some important gaps that can only be filled by experimental data and rules of thumb which are problem specific.

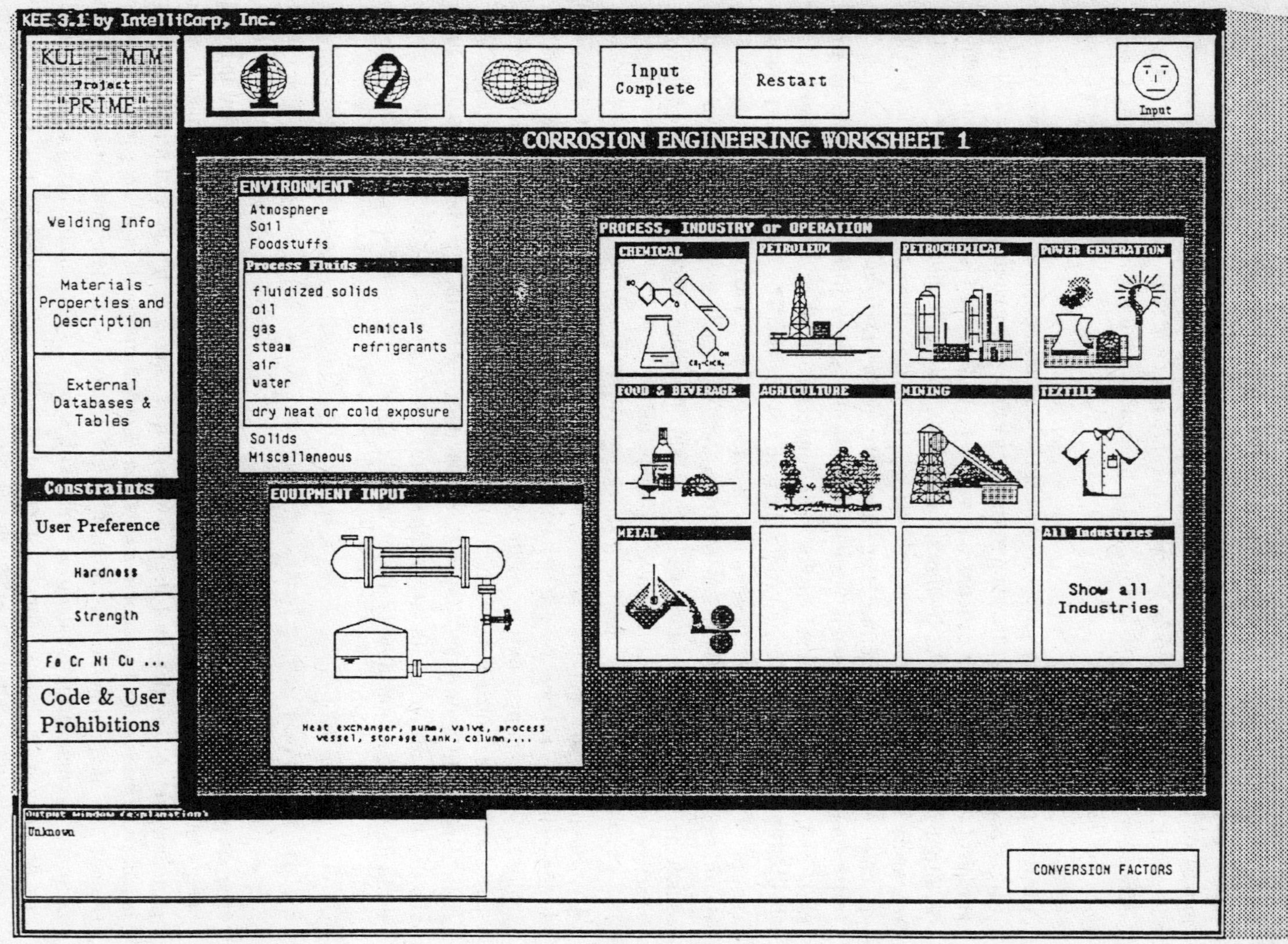

Figure 1: First Corrosion Engineering Worksheet.

It is this philosophy that has dominated the conception and implementation of PRIME. As a basis, it incorporates a vast amount of generally accepted corrosion knowledge. It furthermore incorporates a number of tools which allow the user to build problem specific knowledge bases and interfaces.

6.2.1 Software layers

PRIME has been implemented by means of the expert system building environment "KEE" (Knowledge Engineering Environment). KEE itself consists of a number of tools. The *KEEpicture* package for instance, is a facility that allows the creation of object-oriented graphics by combining a set of predefined graphics primitives such as lines, circles and rectangles. *Active Images* is a graphics package that is built on top of the KEEpictures package and provides displays for viewing and displaying attributes of KEE *objects* which are the conceptual entities in a knowledge-based application and which are typically used to refer to units. Units are the objects or entities of an application domain.

On top of these basic tools, which require a thorough understanding of the LISP programming language and the KEE system, we have built our own domain specific tools and tool components that allow for quick development of expert system modules in the area of corrosion control and materials (selection) problems. These tools provide high level functionalities to easily extend the basic PRIME system and allow for seamless integration of new modules thus allowing for an easily expandable and custom tailored expert system.

The different software layers that implement this kind of approach are represented in Figure 2. The operating system of the workstation constitutes the lowest level (e.g. UNIX[2]). On top of that comes the programming language Common Lisp[3] in which KEE is implemented. With the different facilities already available in KEE specific tool components have been constructed such as a customized windowing system, browsing facilities, complete interface building tools and tailor made *inference engines* (the software that processes the knowledge contained in the knowledge bases). These different tool components can then be combined to form higher-level tools such as the explanation facility *HyperExplain* (including future enhancements such as a *knowledge* and *reasoning browser*) [Arents 1989]. The *HyperExplain* module, for instance, is a combination of an inference engine component and the windowing package.

---

[2] UNIX is a registered trademark of AT&T Bell Laboratories.
[3] Lucid and Lucid Common Lisp are registered trademarks of Lucid Inc.

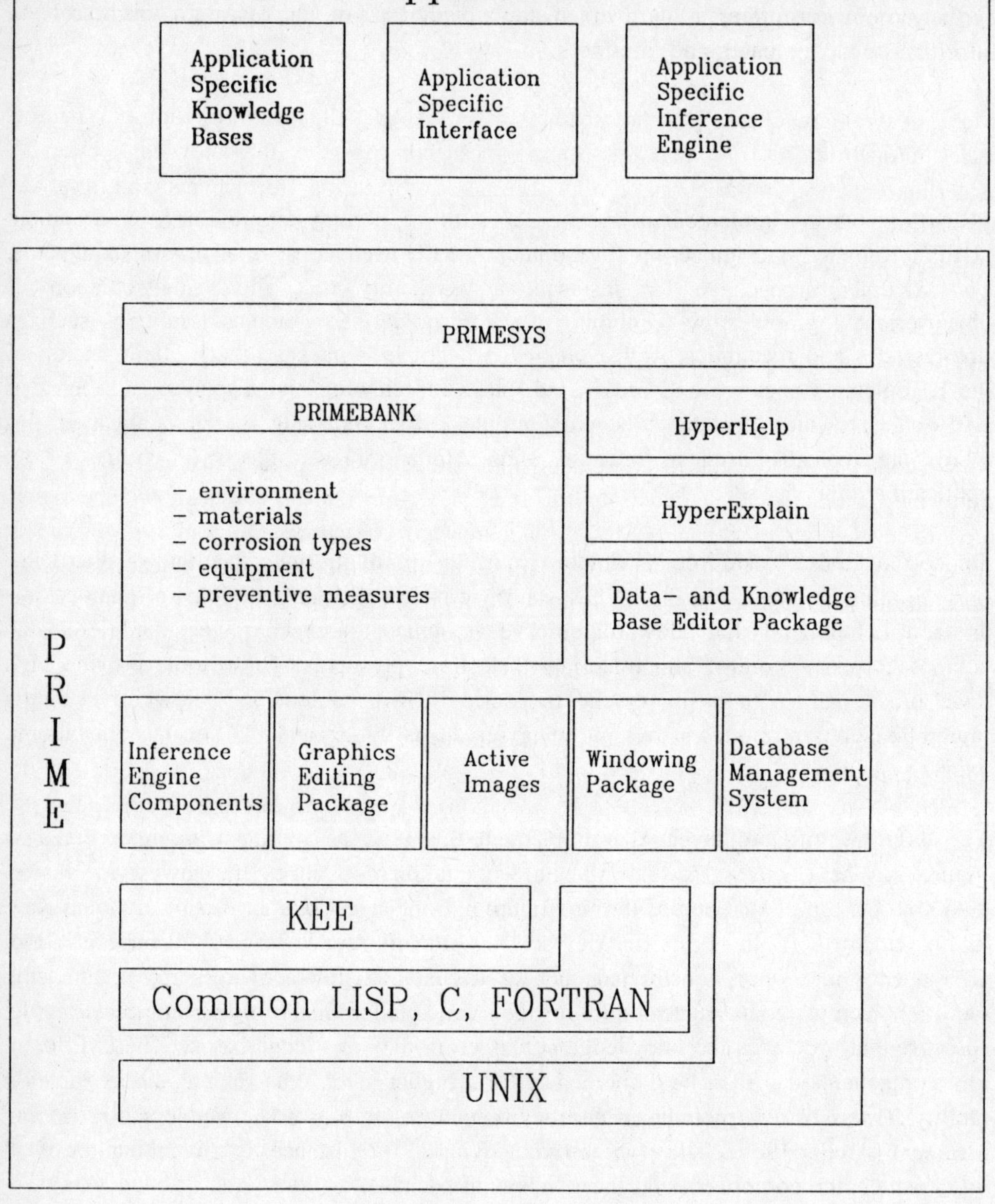

Figure 2: The different implementation layers of PRIME.

The different components and tools can be reused for very different applications than those for which they were originally developed. Therefore, PRIME can also be seen as an expert system assembling toolkit which has already part of the corrosion and materials information implemented.

Many of the knowledge bases that are present in PRIME contain information that can be used in other kinds of expert system modules. For instance, the knowledge base on materials contains information on the chemical composition, mechanical and physical properties, ... for a vast amount of materials, both metals and polymers. The information contained in this knowledge base is completely independent of the application domain.

6.2.2 Overview of the implementation of PRIME

PRIME basically consists of a number of independent knowledge bases (e.g. **equipment**, which contains some generic equipment parts such as pumps, vessels, valves, etc., or **materials**, which contains material properties and description). These knowledge bases are linked together by means of a working knowledge base called 'primesys'. Figure 2 shows these different modules and their hierarchical dependencies. All objects and *slots* (the attributes or characteristics of a unit) are represented in the interface. Figure 3, for instance, shows part of the **Second Corrosion Engineering Worksheet** together with the corresponding representation of this worksheet in the knowledge base 'primesys'.

The way these different knowledge bases and the underlaying software have been implemented allows for the incorporation of a "multiple worlds" concept. A world can be defined as one state of a knowledge base; *multiple worlds* enable the expert system to deal with multiple, different states of the same knowledge bases at once. For our applications the multiple worlds concept implemented in KEE has some major shortcomings. Therefore, we implemented our own domain specific version. All software has been designed generally enough to allow for multiple worlds. In principle, the number of simultaneous worlds is unlimited. In PRIME, this has been implemented by the creation of multiple *working knowledge bases* which connect to the same underlaying knowledge bases.

The usefulness of this principle can easily be shown by a simple example. Consider the design of a tube and shell heat exchanger. Materials can be selected for either the tube side or the shell side of the heat exchanger tubes. In most cases the optimal materials will be different. The same applies for materials for the shell, baffles etc. All these selection problems constitute different sessions with PRIME or could be handled by the multiple worlds concept. The different working knowledge bases will then contain the complete problem specification and solution. These different problems could either be

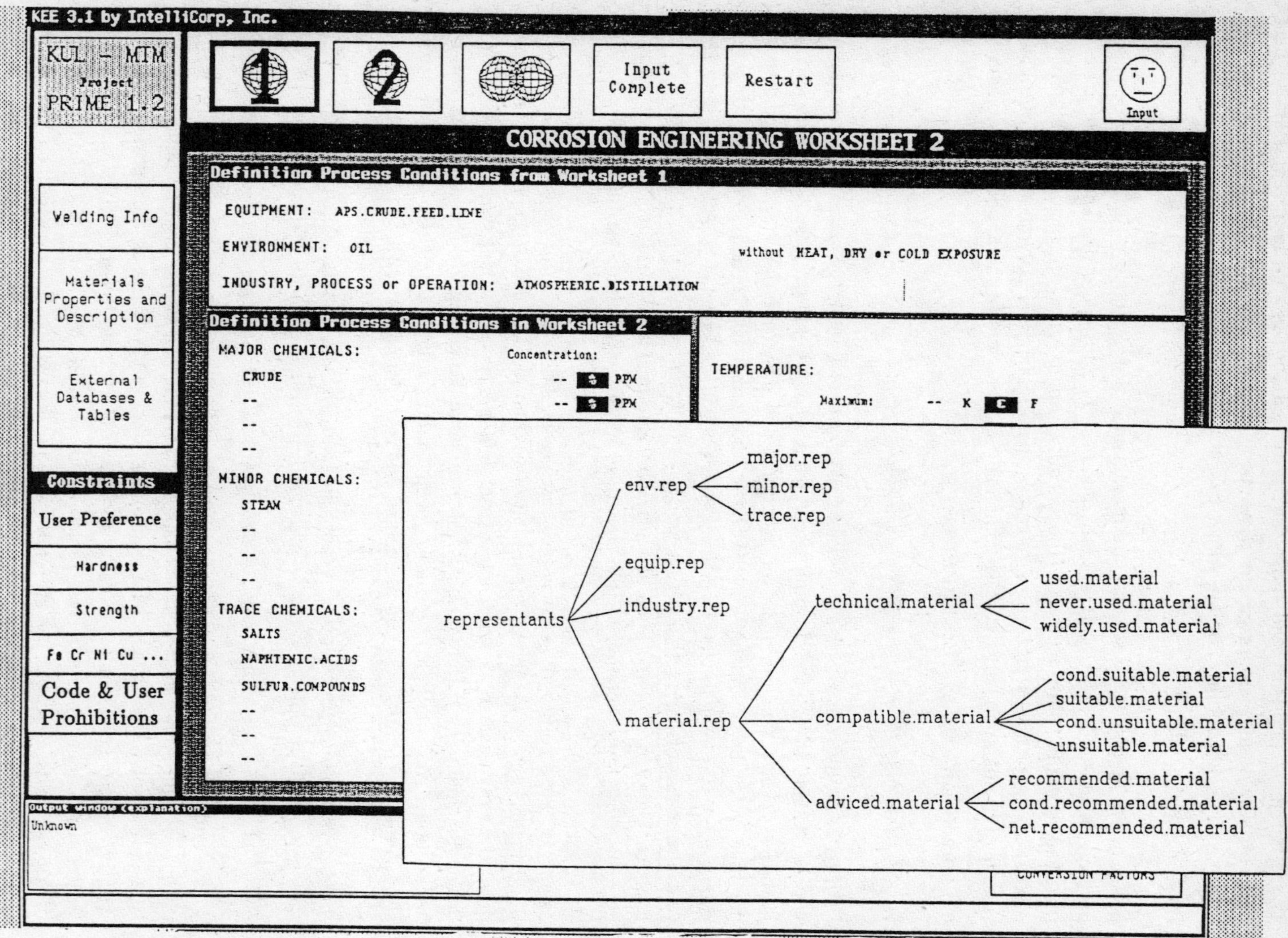

Figure 3: Second Corrosion Engineering Worksheet with corresponding representation in 'primesys'

solved in parallel (on a parallel computer or on different workstations at once) or in sequence.

Information can be exchanged between these knowledge bases which allows to *merge* different worlds. This is an important feature. In merging different worlds for a heat exchanger for instance, one can apply knowledge concerning galvanic corrosion, to come up which a materials selection that is compatible with the different fluid streams and in which the selected materials are compatible from a galvanic point of view.

6.3 Functional characteristics of PRIME

A complete functional description of PRIME can, for instance, be found in "PRIME - The European Esprit project on expert systems for materials selection" [Vancoille 1988].

6.4 PRIME and Second Generation Expert Systems

Second generation expert systems are characterized by their ability to solve problems by combining different types of reasoning. Such systems often use multiple representations of the problem to develop different problem solving strategies. They combine representations and reasoning techniques that complement one another in some way, such as heuristic and model-based, or qualitative and quantitative [Steels 1985]. A central problem in building expert systems is deciding how to model the system and which reasoning techniques should be used to exploit the model(s) chosen. A major difficulty in multi-model based reasoning is deciding how to control the use of strategies with different viewpoints on how to solve the problem.

PRIME incorporates some of the characteristics of second generation expert systems. For materials selection it employs two distinctly different models. It consults a theoretical corrosion model, based on fundamental corrosion information the way it can be found in engineering handbooks. This knowledge usually does not suffice to solve real world problems since it does not address technical problems such as weldability, availability, price and user practices which may differ significantly from the (more conservative) theoretical approach. Therefore, PRIME uses both models (theoretical and industrial practices) and then uses a third model to combine both.

One of the applications developed in the framework of the PRIME R&D work is a CAD/AI-integration feasibility study (see section 7.3.3). This involves the design (both selection of suitable materials and structural design) of heat exchangers. The hybrid knowledge based system developed for this purpose combines a numerical model for the structural design (written in FORTRAN) and symbolic reasoning techniques for such non-numerical reasoning problems as the selection of the appropriate heat exchanger type, rear- and front-end head types, etc. Both models work in an iterative way to yield an optimal design.

The development of Second Generation Expert Systems is still in a very early phase. It is difficult to say whether it will be possible to realize its potential for solving complex, real-world problems. There is still much theoretical work to do to solve difficult issues such as those sketched above.

7 Application building

7.1 General

As pointed out earlier, it is impossible to grasp in one system all necessary knowledge to solve all imaginable (corrosion or materials) problems. This was one of the main reasons to specify the design of PRIME as it stands now. This modular approach allows for the swift construction of *application specific expert system modules*. If they are designed according to the specification provided by PRIME, they can be simply integrated with PRIME itself and with other modules, thus taking advantage of all the knowledge that is already available in PRIME.

Every user application normally consists of four subcomponents: the PRIME 'core', the **application specific knowledge bases**, the **application specific interface** and, eventually, an **application specific inference engine**. The most important task in the construction of a new expert system is, of course, the generation of a set of dedicated knowledge bases. They contain all the knowledge (mostly in the form of if-then rules). Nevertheless, many (tedious) tasks of a typical application building process will be eliminated, since the PRIME core already provides such standard corrosion and materials information as: definition of objects and their attributes, definition of relationships between these objects, "deep" corrosion knowledge and non-application specific corrosion related materials information, etc.

Also the other application specific subcomponents (interface, inference engine) can of course take advantage of existing PRIME modules. The interface can either be a dedicated interface (e.g. specific graphics that are only useful for the particular problem), or can be the standard PRIME interface if this suffices for the application.

If necessary, a specific inference procedure can be defined. Normally, the standard inference procedures such as the built-in KEE forward and backward chaining, or the modified PRIME inference engine which provides an additional high-level explanation facility [Arents, 1989], are suitable enough.

7.2 Specific Industries

The prime 'core' has been designed essentially to serve a "generic" chemical process industry. Its user-input interface, for example, contains very general equipment or equipment component representations as shown in Figure 4. For a specific industry, however, such generic user input facility can be replaced by a dedicated system, either a concrete process flowsheet or, for example, a more detailed equipment representation. Some examples, stemming from feasibility studies carried out in our laboratories, are shown in Figures 5 to 7. Figure 5 represents the **atmospheric distillation of an oil refinery**. A "zooming" function allows the display of more equipment details, e.g. distillation tower components (Figure 6). Figure 7 illustrates another example, i.e. a detailed analysis -and concomitant representation- of **different types of shell-and-tube heat exchangers**.

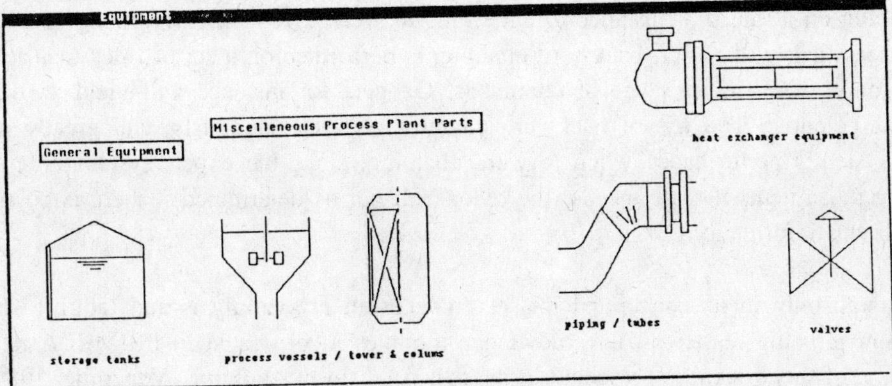

Figure 4: Generic Equipment Input Interface

7.3 Specific Applications

7.3.1 *Troubleshooting Assistance and Risk Determination : ' dual scope ' expert systems and the power of backward and forward chaining*

One of the most interesting features of the KEE expert system building toolkit is the identical syntax for both forward and backward chaining rules. Rules are structures in first-order predicate logic. They are typically expressed in the form: IF <CONDITIONS> THEN <CONSEQUENCES>. With *forward chaining* or *data- or event driven* rules, the inference engine matches rule premises to the new information, finding all possible implications of the new fact based on the present state of the system. In *backward chaining*, the system tries to prove goals or hypotheses. A goal is matched to a rule conclusion, and if the premise or premises of that rule are true in the knowledge base, the conclusion(s) become true. Otherwise, the chainer goes backward, poses each premise as a new goal and tries to match it with conclusions of other rules until the backward links are exhausted.

Both reasoning strategies have different applications. Forward chaining can be typically applied to a selection problem, finding all consequences of the user input. This is also the way PRIME reasons mostly when solving material selection problems, although also backward chaining is employed. Backward chaining is suitable for trouble shooting and risk analysis. From an initial state of the knowledge base, it hypothesizes what problems rise or may have arisen. Since KEE supports both reasoning paradigms and since the rule syntax is identical for both of them, the same knowledge and knowledge representation schemes can be used, thus eliminating duplicate work.

PRIME sometimes uses backward chaining to prove certain premises when forward chaining on a rule. For instance to determine the redox (reducing, oxidizing or neutral) if not specified by the user, backward chaining is performed on a set of rules that determine the redox from the presence of chemicals. Oxygen, for instance will yield an oxidizing redox potential. The use of backward chaining, if used cautiously, can greatly enhance the system's performance when forward chaining, since the expert system only tries to prove those items it really needs (the redox will not be determined if there is no rule that needs this information).

Although only in its conceptual phase, a **corrosion risk analysis** and **trouble shooting system** is being built with the knowledge modules incorporated in PRIME. A dedicated version of this system is presented in another paper in this volume [Vancoille, 1989].

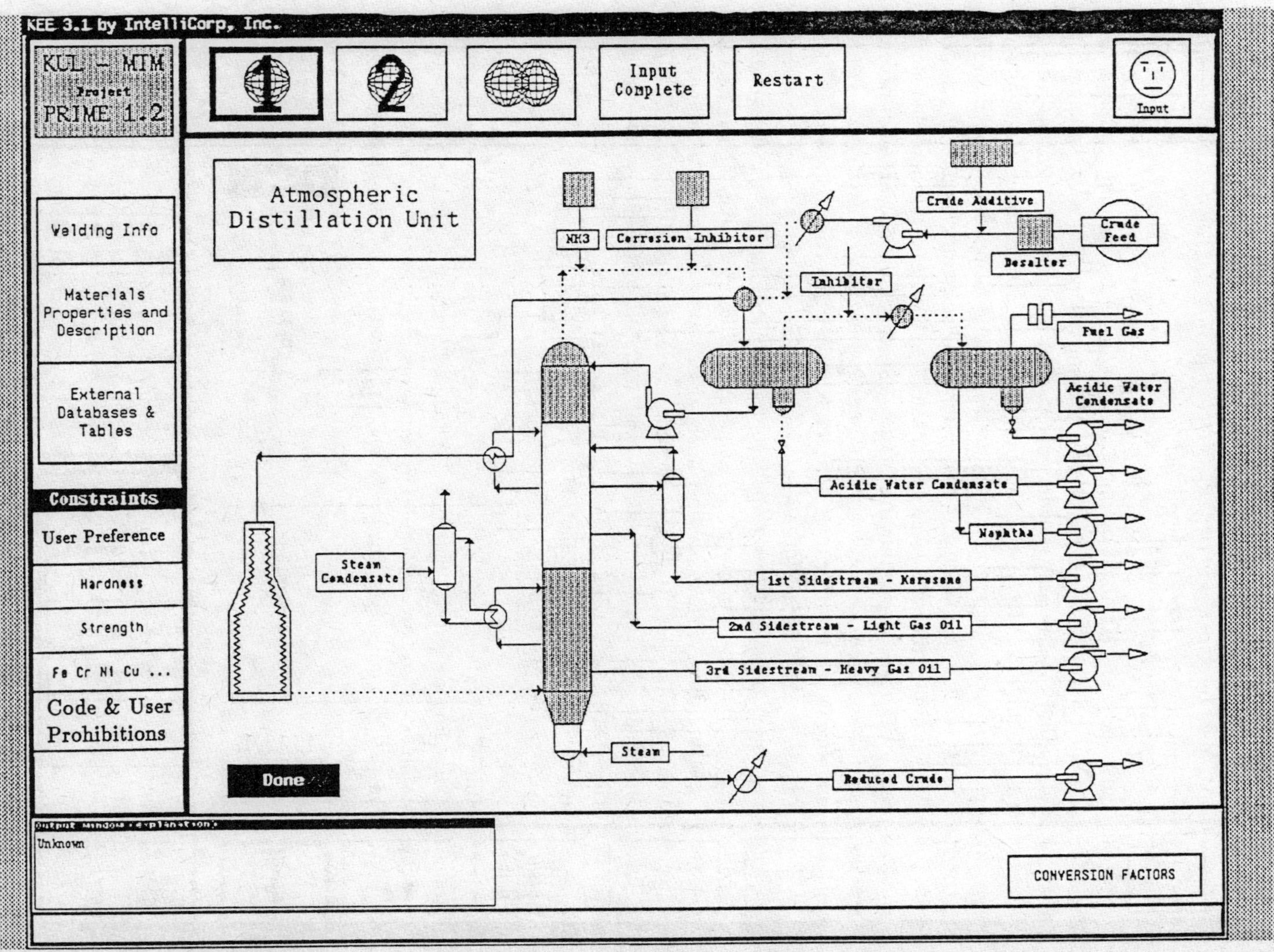

Figure 5: Atmospheric Distillation Flowsheet.

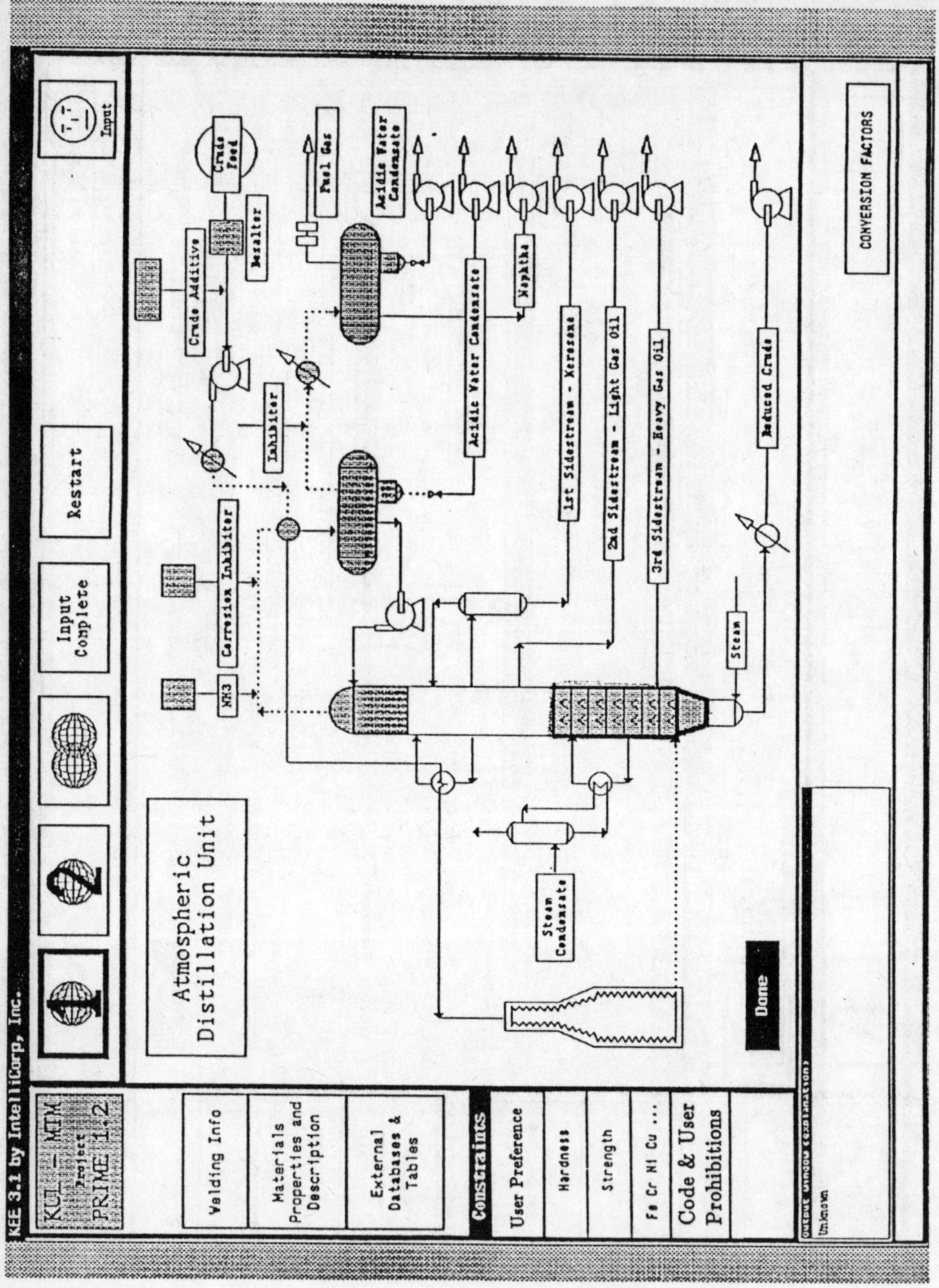

Figure 6: Distillation Tower Components.

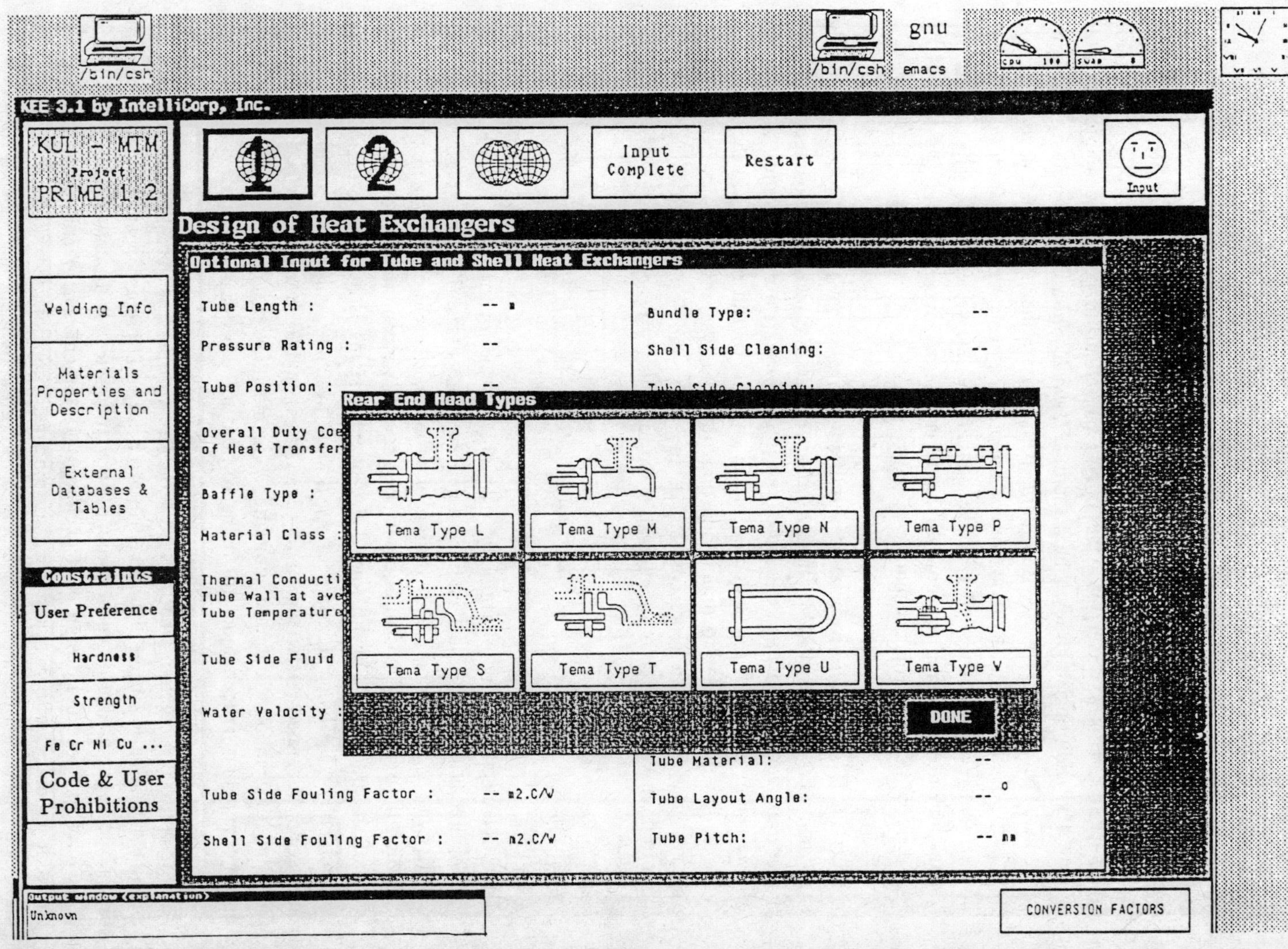

Figure 7: Specific Heat Exchanger Type input.

7.3.2 *Training & Education* (adapted from [Arents 1989])

The same modular design of PRIME can be used for quite different purposes. The available knowledge incorporated in PRIME yields a valuable source of information for *training and educational purposes*. A context sensitive help facility is already incorporated, allowing the user to get help on the explanation of the different functionalities. This however, provides only a partial aid in getting acquainted with the system. The additional high-level help facility, *HyperExplain*, can be used by the expert to validate the system-proposed reasoning path and by the non-expert user to explore only parts of the proof path he thinks important , which enables him to gradually increase and refine his knowledge of the domain and the system itself. This results in a system-supported **tutoring system** which nevertheless leaves absolute freedom to the user, thereby increasing his acceptance of the system.

In order for *HyperExplain* to become a fullfledged intelligent tutoring tool, assisting the user when working with PRIME, future enhancements are under development such as a *glossary*, explaining terms and concepts used in the formulation of the rules; a *knowledge browser*, giving a visual representation of the contents of and relations in the knowledge base(s), allowing the user to understand and trust the system's knowledge; a *reasoning browser*, giving a visual representation of the entire proof path, allowing the user to zoom in on a particular point of interest in the reasoning if necessary.

7.3.3 *Computer Aided design*

Materials selection, which originally was one of the major goals of PRIME, is a significant bottleneck is most engineering design processes. The link of a system like PRIME with a (computer aided) design environment therefore seems a natural way of extending the scope and possibilities of nowadays CAD equipment [Bogaerts 1988]. The issue of integrating CAD with knowledge-based systems is however still in its (early) research phase. Advanced engineering design systems will have to make use of large-scale integrated knowledge and data bases, added to today's CAD equipment. Coupling of symbolic and non-symbolic components is essential to achieve practical useful results but the techniques for doing so need to be better developed.

An attempt to demonstrate what can be achieved with present state-of-the-art technology and understanding has been undertaken in a recent R&D project in our research team. In this project knowledge-based techniques are integrated with conventional numerical procedures to select and **design heat exchange equipment**. The overall system comprises different modules, either expert system modules (implemented in Lisp & KEE) or numerical calculation modules (implemented in Fortran), which interact with each other

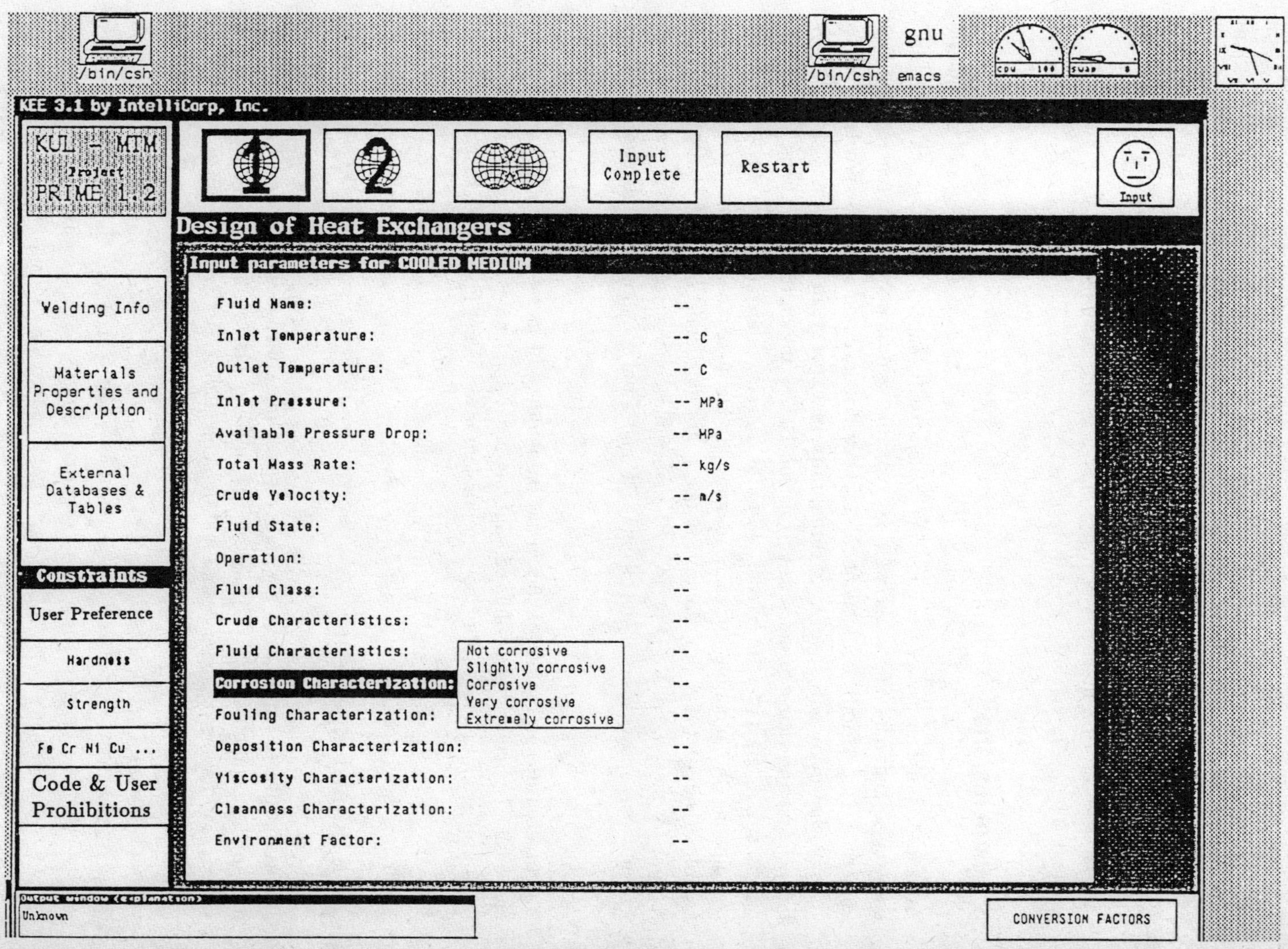

Figure 8: Computer Aided Heat Exchanger Design Interface.

in an "iterative" manner as to achieve an "optimized" heat exchanger design. Some of the essential modules are: Type Selection of the Heat Exchange Equipment (expert system), Calculation of total Heat Exchange Surface Area (numerical), Detailed Calculation of Exchanger Components, e.g. tubes, shell, etc. (numerical, default rules), Material Selection for Exchanger Components (PRIME-based expert system). An illustration of the system's interface is shown in Figure 8.

8   Conclusions

Advanced computer-assisted engineering systems will make use of large-scale integrated knowledge and data bases, as well as symbolic and non-symbolic (i.e. conventional numerical calculation) components.   Large-scale developments of knowledge based systems, however, will involve a **community of experts**. In these cases different and independent knowledge sources can be established by the current knowledge engineering practices but, in order to create a useful system, integration has to be done by coordinating the functions of each source. The study of cooperation of such different modules, and of the 'computational models' for this cooperation is one of the main reasons for the present R&D project. The emerging concepts of second generation expert systems may provide a sound theoretical and implemental foundation for future developments.

The **layered design** of PRIME facilitates the building of user specific applications which can easily be incorporated, thus simplifying the implementation by providing a set of high-level routines and a lot of built-in problem independent knowledge bases.

9   Acknowledgement

We would like to express our gratitude towards our colleagues *H. Arents*, for the design and implementation of the explanation facility and his ongoing programming support, *O. Erickson* for the database design and implementation, and *H. Smets* for the knowledge engineering contribution on engineering plastics.

10 References

[Arents 1989]   Arents, H.C., Bogaerts W.F. *Use of Hypertext for Corrosion Analysis*, this volume.

[Bogaerts 1986]   Bogaerts, W.F.L., Rijckaert, M.J., Bettendorf, C. (1986). *Artificial Intelligence, Expert Systems and Computer Aided Engineering in Corrosion Control*, in Computers in Corrosion Control, ed. NACE, pp.98-109.

[Bogaerts 1987]   W. Bogaerts, M. Rijckaert and C. Bettendorf (1987). *Expert systems as a tool for corrosion technology transfer*, Proc. Eurocorr '87, p.403-409, ed. Dechema, Karlsruhe.

[Bogaerts 1988]   W.F. Bogaerts and M.J.S. Vancoille (1988). *Knowledge based systems for engineering-design suppport. - A test bed: Materials selection and corrosion prevention*, Proc. Avignon'88, "Les systemes experts et leurs applications", Vol. 3, p.567-571.

[Edeleanu 1983]   Edeleanu, C., Hines, J.G. (1983). *Modelling approach to corrosion prediction*, Br. Corrosion J., Vol.18, p. 6-9.

[Hampel 1983]   Hampel, V.E., Hilsenrath J., Westbrook J.H., Gaynor C.A., Johnson P.S. (1983). *A directory of databases for material properties*, ASME Int. Computers in Engineering Conf., Chicago, August 1983.

[Steels 1985]   Steels, L. (1985). *Second Generation Expert Systems*, in Future Generation Expert Systems, North-Holland Pub.,Amsterdam.

[Vancoille 1988]   Vancoille, M.J.S., Bogaerts , W.F.L., Rijckaert, M.J. (1988). *PRIME, The European Project on Expert Systems for Materials Selection*, Proc. Corrosion'88, ed. NACE, Houston, Texas, USA.

[Vancoille 1989]   Vancoille, M.J.S., Bogaerts, W.F.L., Perdieus, F. (1989) *AI-based Corrosion Risk Analysis in Oil Refinery Operation*, this volume.

[Westcott 1984]   Westcott C., Williams D.E., Wilkins N.J.M., Marsh G.P., Wanklyn J.N., Croall I.F., (1984). *The application of expert systems to corrosion problems*. AERE Publication M3445.

Applications

DEVELOPING EXPERT SYSTEMS FOR STRUCTURAL DIAGNOSTICS AND RELIABILITY ASSESSMENT AT J.R.C .

A. C. Lucia

Commission of the European Communities, Joint Research Centre,
Ispra Establishment, 21020 ISPRA (VA), Italy

1. INTRODUCTION

The on-going activities at JRC-Ispra on structural diagnostics and reliability tackle the problem of the assessment of the degree of deterioration of the integrity of structures and of the quantification of their residual safety margins during service life or, in other words, of the remnant lifetime.

The lifetime distribution of a structure depends on several factors and processes :

- material properties
- past and future loads
- environmental conditions
- fabrication accuracy
- inspections, maintenance, repairs
- microscopic degradation of material for ageing and damage accumulation (embrittlement, irreversible strain accumulation, nucleation of microcracks, etc.)
- macroscopic damage (fatigue or corrosion crack formation and propagation, local buckling, etc.)
- human errors

Not all of the related variables are fully independent of each one another.

The failure or end of life condition can be thought of as a situation of level-crossing between two multidimensional stochastic processes representing resistance (capacity) and load (demand) and which are not statistically independent.

Unfortunately, a general analytical solution of the problem is not possible; simplifying assumptions, approximations, reduction of the space of the input variables etc. are necessary.

The mathematical models available for representing damage processes and failure modes are only approximate representations of physical processes, having peculiar, but often ill-defined, characteristics of precision, sensitivity and range of validity. Furthermore they do not constitute an exhaustive representation of reality.

The information we can collect is not perfect but affected by imprecision, uncertainty, vagueness and is not homogeneous because it comes from many and quite different fields of knowledge.

The knowledge relevant to the various domains (e.g. fracture mechanics, non destructive inspection, cumulative damage, material science, etc) is not fully representable by algorithms or mathematical tools but contains also qualitative and heuristic parts.

Any *a priori* estimates of the life span distribution of a structure show therefore quite a large scattering which can be progressively reduced by using proper up-dating techniques exploiting any fresh data obtained during service.

General problems to be tackled are consequently:

- diagnostic procedures : inspection techniques, inspection policy, data analysis and interpretation, effect of repair and maintenance
- uncertainty modelling and treatment
- knowledge representation and combination
- relevant up-dating procedures and their effectiveness in presence of few fresh data.

Traditional algorithmic approaches are unable to cope with such a complex context, to integrate heuristics and qualitative knowledge and to handle situations that were not foreseen. Expert systems are, potentially, the breakthrough.

Expert systems, roughly consisting of a procedure for inferring intermediate or definitive conclusions on structural damage and remnant lifetime, using the domain knowledge and the accumulating input data, can deal with real world problems by properly incorporating all the knowledge which may become available.

Through this process, prediction capability may improve and specialize in time.

2. DIAGNOSTICS AND RELIABILITY

The approach to the problem followed by DLL (Diagnostic and Lifetime Prediction Laboratories) in Ispra J.R.C. is systemic, meaning that we try and investigate the different main fields necessary to achieve a reliability assessment.

In particular:

a)   Inspection and monitoring:
- development of special techniques: (laser holographic interferometry; thermal emission measurement; use of optical fibres; etc)
- data analysis (pattern recognition; non linear filtering; image processing)

b)   Analytical tools:
- mathematical modelling
- treatment of uncertainties
- use of non standard logic
- expert systems

c)   Experimentation:
essentially aimed at developing, testing and validating techniques and methods, it includes :
- tests on specimens
- tests on small components
- automatic facilities for long running tests on large components

Special software packages have been implemented and tested which can be incorporated in expert systems or used as stand alone tools, in particular:

COVASTOL, based on probabilistic fracture mechanics, estimating fatigue crack propagation and failure probability of pressure components

RELIEF, representing the damage accumulation process as a discrete semi-Markovian process and estimating crack propagation and lifetime distribution

CRACK, evaluating the stress intensity factor at the contour of a crack for any shapes

COVAL, for the combination of random variables under the form of histograms

MUP, for the combination of random variables, based on Montecarlo method

APP, a pre-processor for the application of statistical pattern recognition techniques to random signals

ELISA, for the digital analysis of interferograms and the evaluation of strain; efforts are also devoted to the application of fuzzy logic and the development of response surface methodologies.

3. PROCEDURE FOR STRUCTURAL RELIABILITY ASSESSMENT

An expert system for structural reliability assessment must have the

- ability to analyse and interpret large quantities of information coming from different sources and to obtain from them a realistic representation of the state of the structure and of its future behaviour and to suggest appropriate actions, i.e. the following goals have to be achieved:

- identification of the actual state of the structure and of the damaging process which is currently taking place

- prevision of the future

- decision and planning

The bone of the expert system can be thought of as a coordinator and manager of "operators" which mutually collaborate and supply the pieces of information the system needs, e.g.:

    crack and defect population
    geometry
    material properties selection
    past and future load definition
    stress calculation
    damage processes identification and modeling
    failure criteria
    reliability/safety/dependability assessment methods
    risk and cost evaluation
    external constraints and decision criteria
    decision taking and action planning

Beside these operators, we define two "distributed" operators which interact with almost everyone of the aforementioned operators. These two "distributed" operators deal with two crucial problems:

> modelling of uncertainties
> updating procedures

A third distributed operator could account for consequences of gross and human errors.

The problem of the representation and treatment of "imperfect" (uncertain, imprecise, vague) information is approached not only by the use of the probability theory, but also by non standard logic, namely, fuzzy set and possibility theory can provide a powerful and more suitable mathematical tool for handling non-crisp data sets.

Furthermore, fuzzy set theory allows the use of fuzzy linguistic variables which seem to be the most appropriate instruments for dealing with human expert judgements.

As far as up-dating procedures are concerned, both Bayesian and non-Bayesian methods are used in order to update the knowledge base without demanding its complete restructuring.

Updating may be particularly difficult when, as in many real structures, the fresh data coming from monitoring and inspection techniques, applied during service, are very few.

This leads to the practical impossibility to find a soundly based likelihood function and, consequently, to correctly apply Bayesian techniques.

In this context it is very important to look at the time-structure of data and to take advantage of it in order to improve progressively the reliability assessment and to optimize the decision.

In fact, although in the prevailing models and theories of decision making only marginal attention has been given to the time dimensions, the decision is intimately linked with time.

Time is information and plays quite a central role in cognitive processes of human beings.

The concepts of time and decision lead us to another challenging and important aspect of structural diagnostics: decision making under pressing time constraints.

It is proved that human beings are information processing systems whose behaviour may be strongly affected by stressful situations (e.g. emergency, urgency, etc).

This may cause a reduction of the attention and processing capacity, altering either the perception of the temporal structure of information or the attention to other non temporal properties of the information.

These considerations clearly single out the two types of structural reliability oriented expert systems on which the J.R.C. is focusing its efforts, namely:

a)   ES for combination and elaboration of large amounts of non homogeneous pieces of information, data sets and knowledge

b)   ES for real time decision making under pressing time constraints.

At present, most of our work is devoted to type a) ES's, while the activity on type b) is just being started.

Going back to the above listed operators, it is evident that robust diagnostics, prediction and decision can be achieved only if the operators contain and use also the extensive knowledge and deeper understanding of the physical processes that human expert use to judge and decide.

In the following, the main operators are listed with some comments indicating for which of their parameters the uncertainties are worth to be taken into account and heuristic or qualitative knowledge to be represented.

OPERATOR	UNCERTAINTY	KNOWLEDGE
Crack population	Dimensions Position History	Type of crack/defect History Crack modelling
Geometry	Dimensions	
Material Properties Selection	F.M.Parameters Mechanical Parameters Damage Models Parameters Failure Criteria Parameter	Qualification of data Material Choice Validity Limits
Load Definition	Load Intensity Frequency Min/Max Highest Load	Load Combination Dynamic vs Static

OPERATOR	UNCERTAINTY	KNOWLEDGE
Stress Analysis	From:crack population, geometry, material properties, load	Validity Limits Linear Elastic vs Elastic-Plastic Crack/No crack Local vs Global Analysis Residual Stresses
Damage modelling	From:crack population, material properties, load	Crack type Crack history Residual Stresses Model choice and effectiveness
Failure criteria	From:crack population, material properties, load, stress & damage	Validity Limits LEFM vs EPFM Local vs Global Criterion choice and effectiveness
Reliability Assessment	From all preceding Operators	Validity Limits Local vs Global Model Choice
Risk and Cost Evaluation	From preceding Operators	Social Constraints Political Constraints Model Choice

The user can exploit interactively the functions performed by the operators. Rules and decision criteria can be modified under a set of meta-rules. Since rules and decision criteria are made manifest, comparisons become possible with assessments made by other analysts.

The modular array allows an easier representation of the base of knowledge and an incremental construction of the system.

More specifically:

i)   the number of operators can be changed without requiring modifications in the overall architecture and procedures for information management

ii)   single operators can possibly find separate usage

iii)   the system can be easily analysed to make its content understandable to the user

iv)   the knowledge paradigms can be varied and adapted locally according to the nature of the problem.

4. TWO ON GOING PROJECTS AT J.R.C.: ARTIC AND RAMINO

ARTIC (Assessment of Residual lifeTIme for Creep damaged components) is a knowledge based system being developed in the frame of an Interest Club set up by the collaborating partners : CEC Joint Research Centre (Ispra and Petten Establishments), CEGB (U.K.), LABORELEC (Belgium), ENEL (Italy), MPA (Germany).

This expert system, aiming at the damage assessment of pressurized steam headers, is envisaged as a "procedural expert system" mainly based on expert knowledge and operation experience whose representation in algorithmic form is impossible or not effective.

ARTIC carries out the assessment of the residual lifetime of steam headers according to the procedures commonly used by human experts while assuring a rational framework for knowledge representation and merging and can be seen as an answer to the pitfalls and drawbacks of the conventional approach.

In the conventional approach, standards and internal procedures define models for the approximate estimation of expired life and give the operator rather generic guidelines.

Qualitative knowledge and heuristics, constituting quite often the basis of human expert assessment, are not accounted for in the conventional procedure. Uncertainties and imprecisions are not considered: the approach is fully deterministic.

Furthermore, for creep damage assessment, a list of suitable inspection techniques is given, but problems such as data analysis, combination of information from different techniques, qualification of the techniques and their reliability, are completely ignored.

All these aspects are considered in the expert system approach and taken into consideration in a coherent and rational framework.

A schematic block diagram of ARTIC is shown in Figure 1.

RAMINO (Reliability Assessment for Maintenance and INspection Optimization) is an expert system under development in the context of the BRITE Project 2124 : Enhancement of Inspection and Maintenance of Industrial Structures using Reliability based Methods and Expert Systems.

The partners of the project are : Elf Aquitaine (France), Framatome (France), CEC-Joint Research Centre (Ispra Establishment, Italy), Synthesis SaS (Italy), Politecnico di Milano (Italy), University of Pavia (Italy), Technical University of Munich (Germany), Siemens-KWU (Germany), CSR/OC (Denmark), Swansea University College (U.K.).

They agreed on the statement that, although some industries have developed probabilistic methods and tools to estimate the lifetime of structures, yet these studies, which can usually be used at the design stage, do not correctly take into account the effect of time (new information, degradation, etc), uncertainties and heuristics.

Studies aiming at accounting more effectively for these factors, are expected to lead to an optimization of the inspection and maintenance strategies on any industrial particular structure. These studies would be most useful in industries involving large multi-element structures, and particularly in the off-shore, nuclear and petrochemical fields where inspections can represent a large percentage of the exploitation cost, while not always resulting in significant risk reduction costs.

The work foreseen for the development of RAMINO is organized in four tasks:

1.   data collection and organization of data bases

2.   stochastic modelling of the time dependent structural processes at element as well as system level

3.   optimization of strategies for inspection and maintenance of structures

4.   development of a relevant expert system

Two application areas are foreseen:

- offshore structures

- nuclear pressure vessels

RAMINO can be considered a "stage" forward with respect to ARTIC, because it also allows integration of existing software packages (e.g. codes for damage accumulation, failure probability assessment, estimation of reliability index, stress analysis, identification of response surfaces, etc.)

The general objective of the project is to collect and organise data and knowledge, to improve analysis methods and to develop an intelligent system in order to:

- identify the processes by which the structure may be lead to failure and predict the progress of structural integrity degradation during service life

- ensure a given level of reliability at the lowest cost

The architecture of the expert system entails the intelligent interaction of four levels of units : data bases, operators, intelligent interface modules (IIM) and one Overviewing Supermodule.

The IIM, one for each operator, manages the relevant data set for the operator and acts as a local supervisor and a communicator of the classical informatic tools when engineering calculations have to be made (communication between inference part and external packages). It assures also the communication and interlinks between operators.

The supermodule controls and activates IIMs, elaborates meta-rules, coordinates the sequence of inference actions and assures also the communication and interlinks between operators.

A block diagram of RAMINO configuration is shown in Figure 2.

In both projects the target is a portable tool, open to available software packages in real domain environment and runnable on PC and 386 based systems.

The chosen hardware is the IBM PS/2 80 system with 115 MB of hard disk. The choice is based on:

- consortium wide availability and compatibility

- ability to use existing software/code where appropriate

- availability of distributors to provide local support

Several expert system shells have been reviewed, and GoldWorks has been thought to be the most suitable, because:

- runnable on PC and 386 based systems

- relatively low cost

- a proven product with good documentation

- interacts with DBase III Plus (chosen for data bases)

- good knowledge representation and inference capabilities

- extendable on the common lisp level.

REFERENCES

D.Basile; V.Khong; A.C.Lucia " General Architecture of RAMINO Expert System" BRITE 2124, Sp 4 Rp, March 1989

S.Garribba; A.C.Lucia; A.Servida; G.Volta " Fuzzy measures of uncertainty for evaluating non destructive crack inspection" Structural Safety, 5 (1988) 187-204

S.Garribba; A.C.Lucia; G.Volta " Architecture and knowledge paradigm of an expert system for residual lifetime assessment of mechanical structures " Workshop on Intelligent Decision Support Systems for Plant Operation, Ispra, Nov.1986

S.Garribba; A.C.Lucia; A.Servida; J.Bressers " Damage assessment and residual lifetime estimation : an artificial intelligence approach to structural safety analysis " 5th International Conference on Structural Safety and Reliability, San Francisco, August 1989

G.Lebas " BRITE Project P2124 Milestone Report " March 1989

A.C.Lucia " Probabilistic structural reliability of PWR pressure vessels " Nuclear Eng. and Design, 87, 35-49

A.C.Lucia " Stochastic methods for lifetime prediction ", 8th SMiRT Conf., Brussels, August 1985

A.C.Lucia " Treatment of uncertainties in structural safety analysis ", 11th MPA Seminar, Stuttgart, Oct.1985

A.Michon; J.L.Lackson (Editors) " Time, Mind and Behavior " Springer-Verlag, 1985

G.Volta " Time and decision ", NATO Advanced Study Institute of Intelligent Decision Aids, San Miniato, 1985

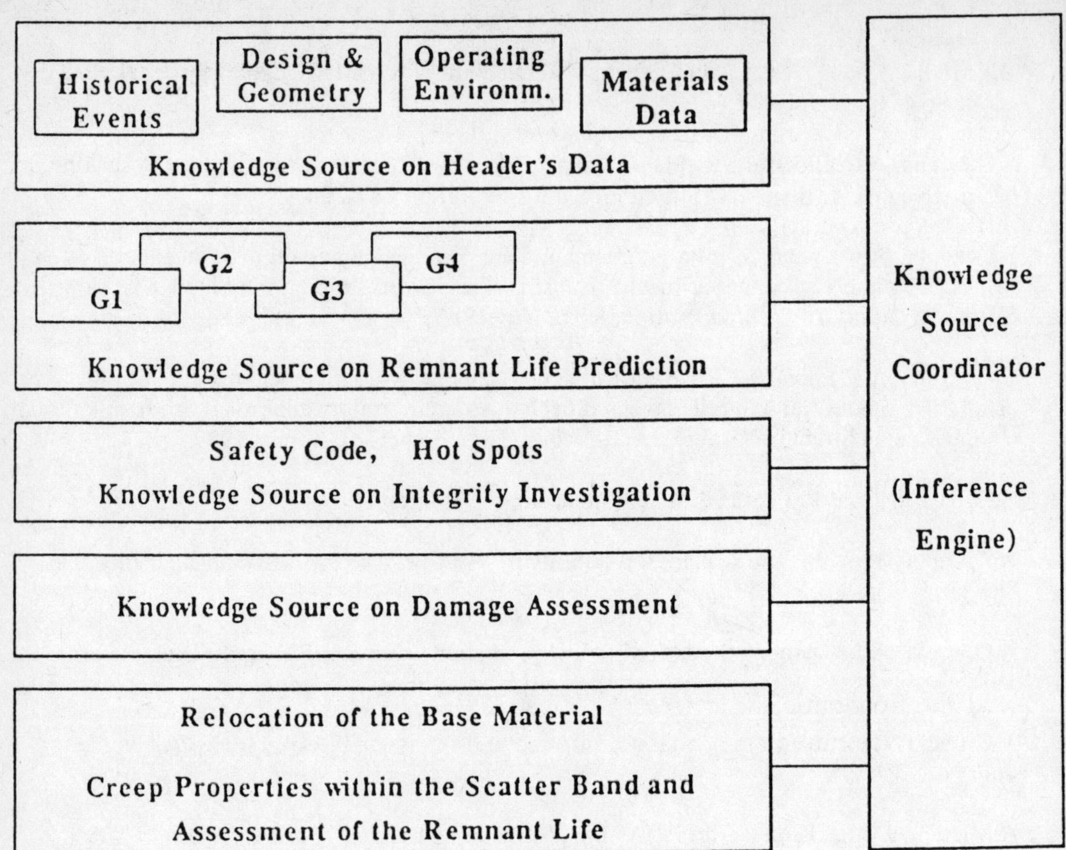

Figure 1: Schematic view of the ARTIC functionality

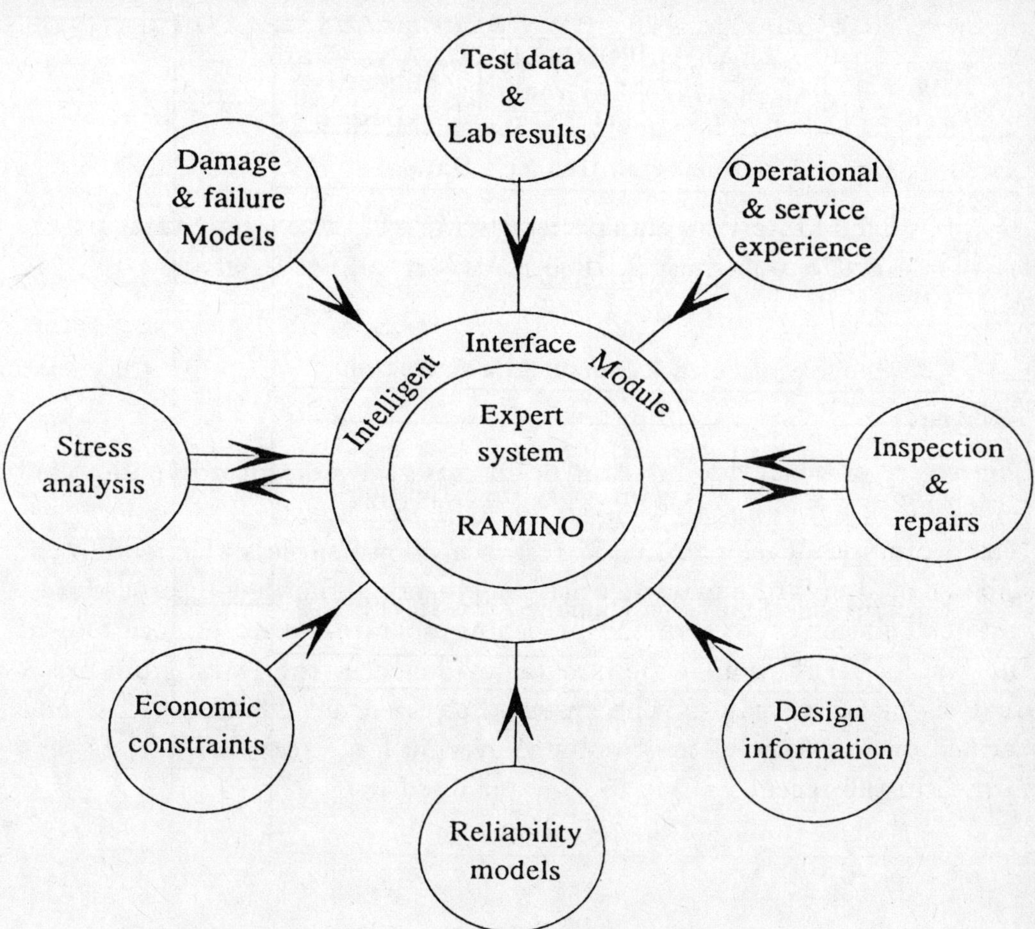

Figure 2: Schematic view of the RAMINO functionality

DEVELOPMENT OF AN EXPERT SYSTEM FOR
LEAK-BEFORE-BREAK ANALYSIS

A. Jovanović, D. Sturm, M. Hassler

Staatliche Materialprüfungsanstalt (MPA) Universität Stuttgart
Pfaffenwaldring 32, 7000 Stuttgart 80, FR Germany

Abstract

The paper presents development of an expert system prototype applied for the leak-before-break analysis at MPA Stuttgart. The expert system prototype developed in the research allows successful introduction and use of standard and/or heuristic engineering knowledge (procedures, code calculations, etc.), of the preceding operational experience and of the testing results and experimental evidence in the leak-before-break analysis of future cases. The system has been developed, applied and verified on the basis of the results of over 50 large scale MPA-tests performed in the recent years.

1. INTRODUCTION

In the domain of fracture mechanics, as well as in many of the domains related to it, there is a number of issues entailing strong involvement of heuristics, such as: interpretation of nondestructive examination and material testing results, pressurized thermal shock analysis (Okamura, Yagawa, 1987), treatment of uncertainties, corrosion influence assessment, leak-before-break (LBB) analysis, etc. Numerical analysis alone, even when very complex, provides just a part of the total information necessary for reaching a valuable expertise, judgment and decision in the application domain. The expertise is necessarily linked to active role of humans, which cannot be avoided. However, current state of the art in the field of knowledge engineering (KE) and expert systems (ES), can offer a substantial complement to the "conventional" decision basis when dealing with some of the issues listed above.

The expert system described in this paper tackles the LBB-analysis, where the term "leak-before-break" describes behaviour of a pressurized component (pipe, vessel, etc.) during an actual or hypothesized failure caused by fatigue crack growth. Namely, LBB means that, once the crack has grown enough to penetrate the wall, only a leak occurs, leaving, under all possible circumstances, a substantial margin (e.g. in terms of safety and time for the component replacement/repair) between the leak occurrence and final rupture of the component caused by the fact that the crack has reached the critical size ("break" - Fig.1). LBB is essential for inherent safety of pressurized systems and components (e.g. those in power plants), because hazards and technical remedies related to leak usually differ drastically from those of a break (Fig.2).

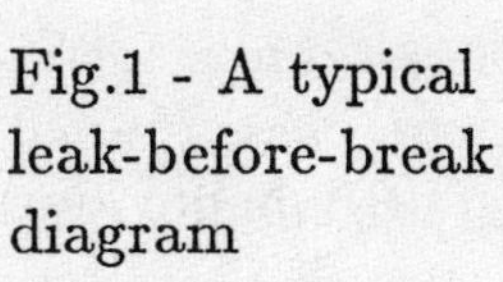

Fig.1 - A typical leak-before-break diagram

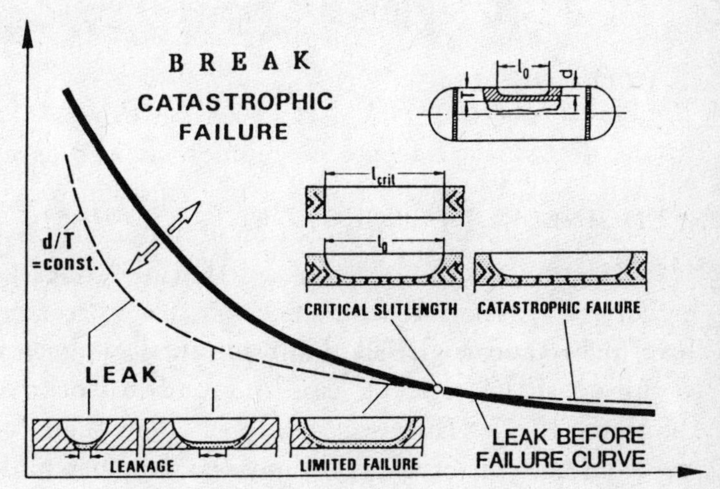

Design and practical development of the expert system for the LBB analysis has been related to other research efforts in the field of ES/KE applications in structural safety and integrity assessment, at the State Materials Testing Institute - MPA Stuttgart (see "Deep Knowledge ...", 1988). The issues of knowledge elicitation, knowledge representation and reasoning in the LBB expert system resulted to be of a great importance, so the work on these issues lead to development and application of a new and original concept (Jovanovic, 1989).

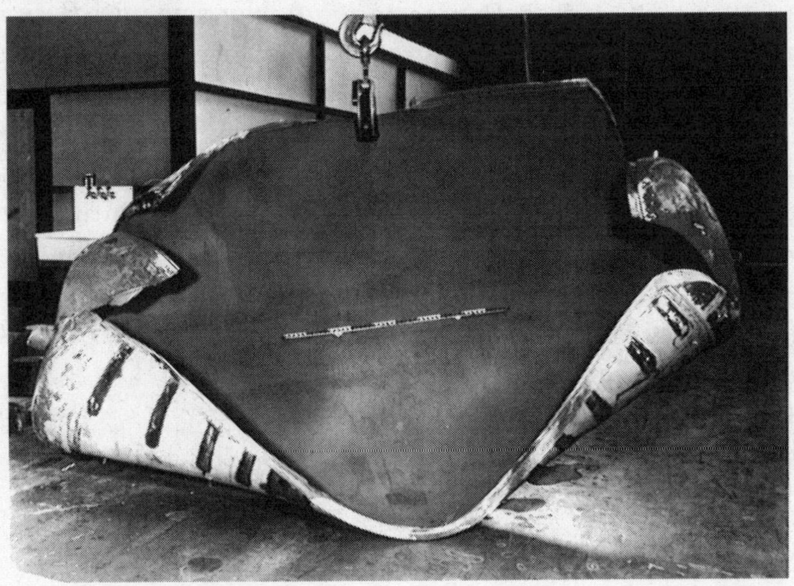

Fig.2: A pressure vessel after break - MPA tests (Sturm, Stoppler, 1985)

2. ENGINEERING BACKGROUND AND BASIC PARADIGMS OF THE SYSTEM

Several methodologies enable numerical analysis of the LBB behaviour (a survey of the methods can be found in works of Sturm and Stoppler, 1985 and 1987). However, many of them tend to be very conservative, or, when not conservative, to have large uncertainty margins. E.g., when applied to the same data base as the one used for the development of the LBB expert system (results of the MPA tests performed in the last years), the methods and the data with which they have been applied, resulted in imprecision ranging between -30 and +40%, when predicting the critical bending moment or the critical internal pressure for cracks in large pipes.

These results, for the MPA experiments (a summary of experiment related data is given in Tables 1 to 3), are shown in Tables 4 and 6 (longitudinal and circumferencial cracks, respectively). They are obtained using methods of Kiefner (Folias), of Kitching (shell parameters), of Dowling and Thownley, of Newman and Raju, of Dufresne, of Delale and Erdogan, of Zahor and Kanninen and of Hasegawa (complete references are given in the work of Hassler, 1989). The results show also a high degree of uncertainty in the qualitative prediction of test outcomes (break/leak), confirming thus that the numerical methods cannot encompass all the factors relevant for the structural failure, such as material toughness, corrosion influence, stiffness of piping, nature of crack, temperature influence, etc.

| | | Material | | | |
| | | 20 MnMoNi 55 | | NiMoCr-Melt | |
	Unit	Value	Standard Deviation	Value	Standard Deviation
Guarantee Values					
Yield Strength $R_{p0,2}$	MPa	$\geq$392 *)	–	–	–
Ultimate Strength R_m	MPa	$\geq$530 *)	–	–	–
Yield Strength $R_{p0,2}$	MPa	428 *)	10	417 *)	43
Ultimate Strength R_m	MPa	605 *)	17	622 *)	25
Outer Diameter	mm	797,9	3,8	703,9	4,5
Wall Thickness	mm	47,2	0,4	47,2	0,4
Notch Depth	mm	–	0,1	–	0,1

*) at 300 °C

Table 1: Material Data

Experiment Number	Dimensions mm	Flaw			Temperature °C	Internal Pressure MPa	Maxim. Bending Moment kNm
		Type	Angle grd	Depth mm			
BVZ 090	Outer-Dia-meter 800	0	–	–	20	15,2	12.110
BVZ 161		S	20	47,2	130	0,0	13.910
BVZ 091		S	60	47,2	20	0,0	9.650
BVZ 100		S	60	47,2	5	15,0	8.970
BVZ 160		A	20	26,0	233	14,6	13.110
BVZ 110		A	60	39,4	220	13,4	7.050
BVZ 120		A	60	20,0	195	13,3	11.030
BVZ 130	Wall Thick-47,2	I	60	20,0	235	15,2	11.300
BVZ 150		A	90	20,0	267	15,7	9.300
BVZ 140		A	120	20,0	248	15,2	8.300
BVS 090	Length 5000	0	–	–	250	14,8	11.500
BVS 060		S	60	47,2	140	15,6	5.510
BVS 110		A	20	36,0	247	15,3	8.500
BVS 102		A	20	20,0	235	15,1	11.240
BVS 070		A	60	20,0	232	16,0	6.600
BVS 080		A	120	20,0	253	14,8	5.550

```
BVZ: Material 20 MnMoNi 55 (CVN Upper Shelf Energy C_v > 150 J)
BVS: NiMoCr-Melt          (CVN Upper Shelf Energy C_v ~  50 J)

Flaw Type A: Outer Surface Crack
          I: Inner Surface Crack
          0: Without Crack
          S: Through-Wall Crack
```

Table 2: Experiments with circumferential cracks

Experiment Number	Dimensions mm	Flaw			Temperature °C	Internal Pressure MPa	Nominal Stress MPa
		Type	Length mm	Depth mm			
BVZ 010	Outer Dia-meter 800	S	650	–	20	23,8	187
BVZ 011		S	1102	–	20	14,8	117
BVZ 012		S	1105	–	20	14,4	113
BVZ 022		A	782	38,3	305	21,9	173
BVZ 030		A	1500	36,2	300	19,5	155
BVZ 080	Wall Thick-ness 47,2	A	1500	36,2	17	20,4	164
BVZ 070		I	700	38,2	265	22,4	177
BVZ 060		A	1500	36,0	305	18,0	143
BVS 010	Length 2500/5000	S	800	–	155	17,5	137
BVS 020		A	709	37,3	320	14,8	117
BVS 030		A	1100	35	305	13,1	103
BVS 042		A	709	38,3	245	16,8	132

```
BVZ: Material 20 MnMoNi 55 (CVN Upper Shelf Energy C_v > 150 J)
BVS: NiMoCr-Melt          (CVN Upper Shelf Energy C_v ~  50 J)

Flaw Type A: Outer Surface Crack
          I: Inner Surface Crack
          S: Through-Wall Crack
```

Table 3: Experiments with longitudional cracks

experiment code	failure pressure in experiment [MPa]	the most conservative calculation [MPa]	prediction error [percent]	the most optimistic calculation [MPa]	prediction error [percent]
V051.	22.1	16.6	25	23.3	-5
V052	19	15.3	19	21.4	-13
V053	15.7	13.6	13	19.4	-23
V054	17.23	13.7	20	20.8	-21
V055	18	15.6	13	21.9	-22
V056	28.3	26.4	7	36.4	-29
V057	93/5	21.5	16	30.2	-62
V058	19.3	14.5	25	21.9	-13
V059	9.4	8.0	15	11.8	-26
V060	18.4	16.4	15	23.0	-19
V061	10.1	7.7	24	10.8	-7
V062	14.3	12.7	11	16.7	-17
V063	7.4	5.8	22	8.2	-11
V064	10.2	8.4	20	11.4	-16
V065	10.8	8.8	19	12.6	-17
V066	10.2	8.8	14	12.6	-24
V067	20.7	16.4	21	23.1	-12
V068	21.7	16.3	25	23.0	-6
V069	18.3	18.1	1	21.2	-16
V070	19.1	16.8	12	19.6	-3
V071	37.8	38.8	3	49.6	-31
V072	11.2	9.8	12	16.4	-46
V073	29.6	23.9	19	36.9	-25
V074	13.5	11.4	16	19.3	-43
V075	3.7	2.8	21	4.3	-20
V076	21.4	20.8	3	28.6	-34
V077	13.3	12.1	10	17.1	-28
V078	22.2	17.3	22	26.6	-20
V079	28	21.7	22	33.2	-19
V080	24.6	21.3	13	31.3	-27

Table 4: Verification of the computational methods for longitudional cracks implemented in the expert system

experiment code	failure moment in experiment [kNm]	The Bending Moment method [kNm]	prediction error [percent]	The MPA method [kNm]	prediction error [percent]
BVZ110	7400	5090	31	6940	6
BVZ120	10500	6610	37	8970	15
BVZ130	10600	6615	37	8918	16
BVZ140	8650	5040	41	7167	17
BVZ150	9000	5685	37	7920	12
BVZ040	–	–	–	578	–
BVZ050	–	–	–	1050	–
BVS060	5500	5176	6	6920	-26
BVS070	6500	7294	-12	9703	-50
BVS080	5700	5713	-0.2	7937	-40
BVS102	11000	8993	18	11037	-0.3
BVS050	–	–	–	1294	–

Table 5: Verification of the computational methods for circumferential cracks implemented in the expert system

A way to analyze influence of the above mentioned factors, is to include them as the "deep knowledge" (see "Deep Knowledge...", 1988) in the expert system, but this way has been replaced by a quicker and simpler one, described below.

The adopted concept (Jovanovic, 1989) is based on analysis of analogy between the given case and the cases in the knowledge base (KB). Numerical and non-numerical information from, currently, about 50 tests (Sturm and Stoppler 1985; Sturm and Stoppler 1987; Stoppler and coworkers, 1987, Sturm and Stoppler, 1989) is stored in the KB. Analogy in single factors (temperature, loading conditions, material toughness, etc.) is combined with results of the numerical analysis. Then, the relevant cases (if any!) are extracted form the knowledge base. "Relevant" are either those cases showing very high analogy with the analyzed one, or those being totally different. The prediction *leak* or *break* is made accordingly. In this respect, the major difficulty is the definition of effective measures of similarity and proximity between the stored cases and the case which is to be analyzed. The definition of this measure concerns the generic, still opened, KE issue of characterizing real entities by analogy.

3. ELICITATION OF ENGINEERING KNOWLEDGE IN THE DOMAIN

Three methods have been used in knowledge elicitation: organized inquiry, analysis of test reports and direct interviewing of the domain experts. The inquiry, containing a list of questions presented in form of tables (Fig.3), has been performed among the engineers of MPA and other institutions and companies (46 participants, see Jovanovic, Hassler, 1989), and it provided information regarding possibility distributions of numerical and linguistic variables used in the LBB analysis. Analysis of test reports provided the basic information (both numerical and qualitative) about the tests. Interviewing has been done with the MPA experts only, and its results have been used mainly for confirmation/redefinition of the heuristic rules (used in the ES) derived from analysis of reports.

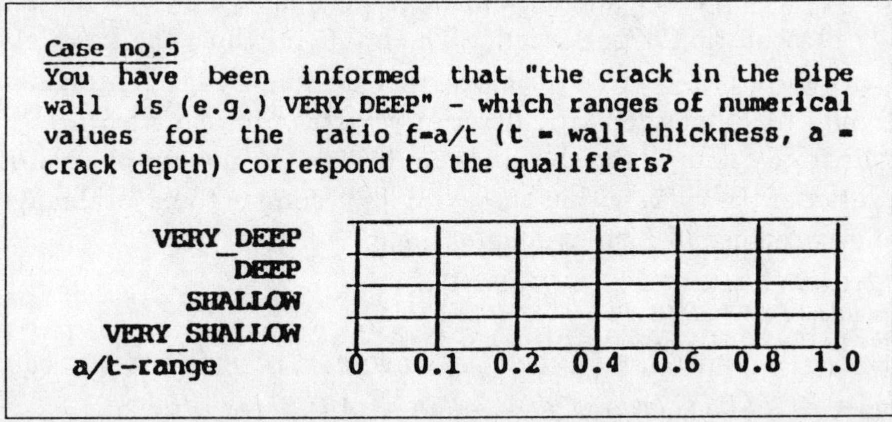

Fig.3: A sample question from the inquiry (Jovanovic, Hassler, 1989)

4. KNOWLEDGE REPRESENTATION

In the Lisp version of the ES, the knowledge base is organized as a series of lists, containing both numerical and qualitative data. Numerical data taken into account are, e.g., geometry of the component, crack dimensions, working temperature, loads, material characteristics, etc. Qualitative information regards type of the crack (e.g. inside/outside, fatigue crack, welding defect,etc.), stiffness of the construction, etc. Rules, in this version of ES, are purely heuristic, e.g. *"when the crack is deep, then the chance of the failure mode leak is high"*. The terms like *deep* and *very high chance* are defined as trapezoidal fuzzy numbers (Jovanovic, Servida, Sauter, 1988), the parameters of which are derived from the inquiry - Fig.4. Technically, the solution is designed as an open architecture, shell free, intelligent module, interacting with the user in the search of the appropriate representation of various pieces of knowledge and in determination of its importance for the final conclusion.

In the second version of the LBB expert system, the KEE^{TM}-shell was used. KEE^{TM} is a hybrid tool which joins the KE-methods (object-oriented programming, frame-based knowledge representation, rule-based reasoning, use of Lisp functions), with graphic oriented interactive possibilities for editing and browsing the knowledge bases as well as for a user-friendly presentation of the results. The frame in which the knowledge is stored, is an object-centered construction for knowledge representation, which puts together the static-descriptive attributes (i.e. features of an object) with dynamic-procedural attributes (i.e. behaviour of an object). The syntax of the rules within these frames is similar to natural language and allows the use of Lisp expressions, e.g.:

```
(if    (find (MIN.SIMILAR is in SIMILAR.MIN))
       (cant.find (MAX.SIMILAR is in SIMILAR.MAX))
       (the FAILURE.MIN of MIN.SIMILAR is ?X)
then do   (lisp (cond ((equal ?X 'LEAK) (unitmsg 'BREAK.ANSWER 'ANSWER.50))
       ((equal ?X 'BREAK) (unitmsg 'LEAK.ANSWER 'ANSWER.50)))))
```

5. REASONING BY ANALOGY

The main task of the reasoning process is the intelligent recognition of analogy/similarity between the analyzed case and the cases in the knowledge base. To each piece of knowledge regarding the influencing factors, is attributed a weighting factor, the initial value of which is provided by the domain expert, but the value of which can be reviewed interactively. Analogy between numerical data is defined as a combination of the weighting factors and the algebraic ratio/difference. Obviously, the operating temperature of 120°C is more similar to 150°C than to 30°C . Analogy in qualitative data is more difficult to define. So far, the issue in the LBB expert system is tackled either by direct specification of the granularity of the descriptors (*very corrosive* is more similar to *corrosive*, than to *neutral*), or by assigning membership functions to the descriptors (Jovanovic, Hassler, 1989) - Fig.5. Thus, the analogy/similarity can be "quantified" and, later on, compared. The search for analog cases is however, still "unintelligent", as the limited number of cases in the knowledge base still allows to compare the actual case with each of the stored cases. This issue, however, has to be improved further on, as an increase of the number of cases in the knowledge base is expected.

Reasoning has two levels: the one of the pre-established production rules and the one of the self-generated and "implicit" rules. Its main result is the diagnosis/prediction of the structural state, in terms of *leak, break, no failure* and *prediction impossible* (not any). In addition, semantic (linguistic) probability estimators (e.g. *very high chance, meaningful chance, it may be*, etc.) are assigned to ES outcomes, accordingly to how strong the analogy between the analyzed and the reference case(s) was. Transition between the semantic and numerical probabilities is obtained by means of fuzzy algebra (Jovanovic, Servida, Sauter, 1988; Zimmermann, 1987). Precision of the system can be tuned through interactive definition of the analogy significance limits. Thus, if a high certainty of answers is required only "sure" analogies will be identified, with the consequence that the probability of not finding the significant analogues increases. Practical, software, coupling between the engineering numerical calculations and the symbolic analysis used in reasoning is in such a way that the numerical programs and subroutines are invoked from the Lisp level directly.

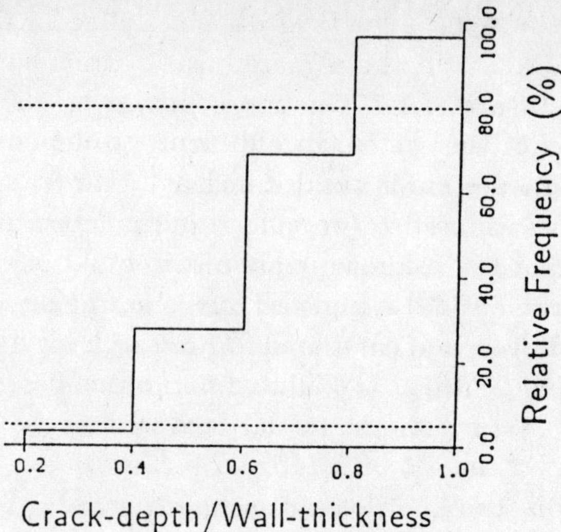

Fig.4 - Results of the inquiry showing the link between term "very deep crack" and the ratio crack-depth/wall-thickness

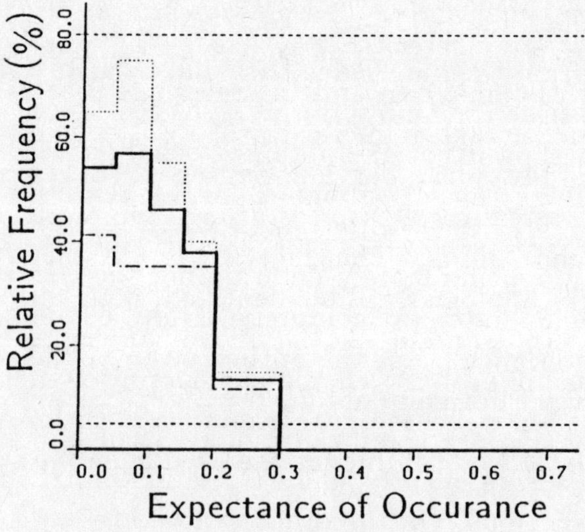

Fig.5 - Results of the inquiry showing the link between the estimator "unlikely" and the expected probability of occurrence (thick line)

6. PRACTICAL IMPLEMENTATION OF THE SYSTEM
LISP VERSION

The Lisp-version of the system has been programmed in the frames of a small project (Hassler, 1988) performed at MPA Stuttgart. In this version, the user is supposed to provide standard data regarding the component (loads, materials, known state of defects), as well as the evidence (description of the current situation), directly. The user is also supposed to provide inputs relevant for the definition of the representation frame, granularity and semantics, as well as for the definition of uncertainty in the information given as input to the system.

The results of the system verification, performed on the MPA LBB full-scale tests internally pressurized pipes, $\phi = 800\ mm$, exposed to bending and containing longintudinal i.e. circumferential defects, respectively (see Stoppler and co-workers, 1985, 1987 and 1989), are shown in Tables 6 to 12.

Verification is done in such a way, that each time one of the tests has been taken out from the knowledge base and then analyzed by the ES on the basis of the tests remaining in the knowledge base.
The verification shows that the obtained reliability of the system's prediction is quite high (usually about 70-90 %), reaching 100% in some of the cases (Tables 11 and 12). Two factors obviously play a very significant role: the weighting factors determining the relative importance of similarity in single factors (a "compressed" and compiled form of the domain expert experience), and the optimization of the weighting factors. The best results are normally obtained with optimized weighting factors. Narrowing of the analogy significance limits (which determine what is to be considered as "significantly similar", i.e dissimilar), leads to a higher precision of predictions. On the other hand, this can contemporary lead also to the reduction of number of cases in which the system provides an answer at all (Table 11).

experiment code	failure mode of the experiment	expert system prediction	"subjective certainty" of the prediction
V051	LEAK	BREAK	70%
V052	LEAK	BREAK	70%
V053	LEAK	BREAK	70%
V054	LEAK	BREAK	70%
V055	BREAK	LEAK	70%
V056	LEAK	BREAK	70%
V057	BREAK	BREAK	65%
V058	BREAK	LEAK	70%
V059	LEAK	LEAK	70%
V060	BREAK	LEAK	70%
V061	BREAK	BREAK	70%
V062	BREAK	LEAK	70%
V063	BREAK	BREAK	80%
V064	LEAK	LEAK	65%
V065	LEAK	LEAK	75%
V066	LEAK	LEAK	75%
V067	LEAK	BREAK	70%
V068	BREAK	LEAK	70%
V069	LEAK	LEAK	70%
V070	BREAK	LEAK	70%
V071	BREAK	BREAK	70%
V072	LEAK	LEAK	70%
V073	BREAK	BREAK	60%
V074	LEAK	LEAK	70%
V075	BREAK	BREAK	51%
V076	LEAK	BREAK	70%
V077	LEAK	LEAK	80%
V078	LEAK	BREAK	70%
V079	LEAK	LEAK	65%
V080	BREAK	LEAK	70%

number of predictions vs. number of analyzed cases   30/30 (100%)
exact predictions: 16 (53%)

Table 6: LBB expert system results, for *unoptimized* weighting factors and the analogy significance limits 0.8 and 0.2 for the application to longitudional cracks

experiment code	failure mode of the experiment	expert system prediction	"subjective certainty" of the prediction
V051	LEAK	LEAK	70%
V052	LEAK	LEAK	70%
V053	LEAK	LEAK	80%
V054	LEAK	LEAK	70%
V055	BREAK	BREAK	70%
V056	LEAK	LEAK	70%
V057	BREAK	BREAK	65%
V058	BREAK	BREAK	70%
V059	LEAK	LEAK	70%
V060	BREAK	BREAK	85%
V061	BREAK	BREAK	70%
V062	BREAK	BREAK	70%
V063	BREAK	BREAK	80%
V064	LEAK	LEAK	65%
V065	LEAK	LEAK	75%
V066	LEAK	LEAK	75%
V067	LEAK	LEAK	70%
V068	BREAK	LEAK	70%
V069	LEAK	LEAK	70%
V070	BREAK	LEAK	70%
V071	BREAK	BREAK	70%
V072	LEAK	LEAK	70%
V073	BREAK	BREAK	60%
V074	LEAK	LEAK	70%
V075	BREAK	BREAK	51%
V076	LEAK	BREAK	70%
V077	LEAK	LEAK	80%
V078	LEAK	LEAK	70%
V079	LEAK	LEAK	70%
V080	BREAK	BREAK	70%

number of predictions vs. number of analyzed cases 30/30 (100%)
exact predictions: 25 (83%)

Table 7: LBB expert system results, for *optimized* weighting factors and the analogy significance limits 0.8 and 0.2 for the application to longitudional cracks

experiment code	failure mode of the experiment	expert system prediction	"subjective certainty" of the prediction
V051	LEAK	LEAK	70%
V052	LEAK	BREAK	70%
V053	LEAK	LEAK	70%
V054	LEAK	LEAK	70%
V055	BREAK	LEAK	70%
V056	LEAK	LEAK	70%
V057	BREAK	BREAK	65%
V058	BREAK	BREAK	70%
V059	LEAK	LEAK	70%
V060	BREAK	BREAK	85%
V061	BREAK	BREAK	70%
V062	BREAK	BREAK	70%
V063	BREAK	BREAK	70%
V064	LEAK	LEAK	70%
V065	LEAK	LEAK	75%
V066	LEAK	LEAK	65%
V067	LEAK	no prediction	–
V068	BREAK	LEAK	70%
V069	LEAK	LEAK	70%
V070	BREAK	BREAK	51%
V071	BREAK	BREAK	70%
V072	LEAK	LEAK	70%
V073	BREAK	BREAK	60%
V074	LEAK	LEAK	51%
V075	BREAK	BREAK	51%
V076	LEAK	no prediction	–
V077	LEAK	LEAK	51%
V078	LEAK	no prediction	–
V079	LEAK	no prediction	–
V080	BREAK	BREAK	51%
number of predictions vs. number of analyzed cases			26/30 (87%)
exact predictions:			23 (88%)

Table 8: LBB expert system results, for *optimized* weighting factors and the analogy significance limits 0.95 and 0.05 for the application to longitudional cracks

experiment code	failure mode in the experiment	expert system prediction	"subjective certainty" of the prediction
BVZ110	LEAK	LEAK	70%
BVZ120	LEAK	LEAK	70%
BVZ130	LEAK	LEAK	70%
BVZ140	LEAK	BREAK	60%
BVZ150	BREAK	LEAK	70%
BVZ040	BREAK	BREAK	70%
BVZ050	BREAK	BREAK	70%
BVS060	BREAK	BREAK	70%
BVS070	LEAK	BREAK	60%
BVS080	BREAK	BREAK	70%
BVS102	BREAK	LEAK	60%
BVS050	BREAK	BREAK	70%
number of predictions/ total number of experiments		12/12 (100%)	
exact predictions:		8 (67%)	

Table 9: LBB expert system results, for *unoptimized* weighting factors and the analogy significance limits 0.8 and 0.2 for the application to circumferential cracks

experiment code	failure mode in the experiment	expert system prediction	"subjective certainty" of the prediction
BVZ110	LEAK	LEAK	70%
BVZ120	LEAK	LEAK	70%
BVZ130	LEAK	LEAK	70%
BVZ140	LEAK	LEAK	70%
BVZ150	BREAK	LEAK	70%
BVZ040	BREAK	BREAK	70%
BVZ050	BREAK	BREAK	70%
BVS060	BREAK	BREAK	70%
BVS070	LEAK	LEAK	65%
BVS080	BREAK	BREAK	70%
BVS102	BREAK	BREAK	65%
BVS050	BREAK	BREAK	70%
number of predictions vs. number of analyzed cases		12/12 (100%)	
exact predictions:		9 (75%)	

Table 10: LBB expert system results, for *optimized* weighting factors and the analogy significance limits 0.8 and 0.2 for the application to circumferential cracks

experiment code	failure mode in the experiment	expert system prediction	"subjective certainty" of the prediction
BVZ110	LEAK	no prediction	–
BVZ120	LEAK	LEAK	70%
BVZ130	LEAK	LEAK	70%
BVZ140	LEAK	no prediction	–
BVZ150	BREAK	no prediction	–
BVZ040	BREAK	BREAK	70%
BVZ050	BREAK	BREAK	70%
BVS060	BREAK	no prediction	–
BVS070	LEAK	no prediction	–
BVS080	BREAK	no prediction	–
BVS102	BREAK	no prediction	–
BVS050	BREAK	no prediction	–
number of predictions vs. number of analyzed cases			4/12(33%)
exact predictions:			4(100%)

Table 11: LBB expert system results, for *unoptimized* weighting factors and the analogy significance limits 0.95 and 0.05 for the application to longitudional cracks

experiment code	failure mode of the experiment	expert system prediction	"subjective certainty" of the prediction
BVZ110	LEAK	LEAK	70%
BVZ120	LEAK	LEAK	70%
BVZ130	LEAK	LEAK	70%
BVZ140	LEAK	LEAK	70%
BVZ150	BREAK	BREAK	51%
BVZ040	BREAK	BREAK	70%
BVZ050	BREAK	BREAK	70%
BVS060	BREAK	BREAK	65%
BVS070	LEAK	LEAK	60%
BVS080	BREAK	BREAK	51%
BVS102	BREAK	BREAK	51%
BVS050	BREAK	BREAK	51%
number of predictions vs. number of analyzed cases			12/12(100%)
exact predictions:			12(100%)

Table 12: LBB expert system results, for *optimized* weighting factors and the analogy significance limits 0.95 and 0.05 for the application to longitudional cracks

7. PRACTICAL IMPLEMENTATION OF THE SYSTEM — KEE^{TM}-VERSION

So far, the practical implementation in KEE^{TM} differs from the one in Lisp, mainly due to the introduction of more sophisticated graphic oriented, user interface (mouse, menues and so-called active images), Fig.6. Also, in addition to the failure mode prediction (*leak/break*) the system, in case of a leak, tells the threshold values where a break is to be expected, i.e. under which loading conditions will a catastrophic failure occur.

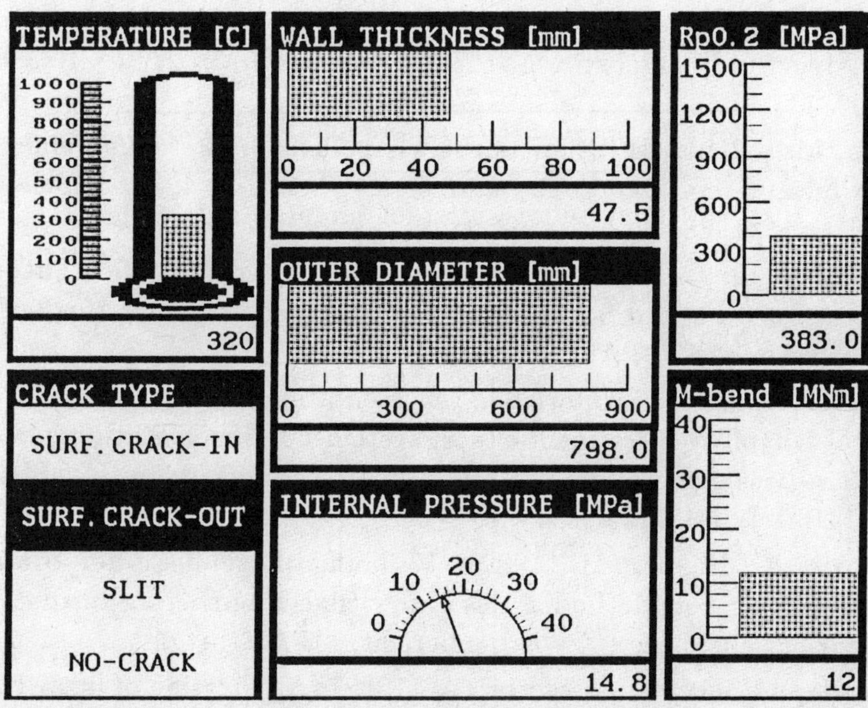

Fig. 6: One of the screens in the KEE^{TM}-version in the expert system

8. CONCLUSIONS

The LBB expert system has been designed and developed as a practice oriented, KE-based engineering tool. Although simple, its first results appear to be good, and allow to expect that, with future developments, will be further improved. Its practical use should help in enhancing the structural integrity assessment and life extension analysis for pressurized components. The efficiency of the system is partly tested in the experiments of MPA, while further verification in practice should be done in collaboration with industry.

REFERENCES

Deep knowledge expert system using coupled symbolic numerical analysis for structural diagnostic, *ESPRIT-II Proposal Nr. 2224 (Area II.2.3)* (1988)

Hassler, M. (1988). HF-Studienarbeit Nr. 754938: Erstellung eines Lisp-Moduls für ein Expertensystem zum Thema Leck-vor-Bruch, MPA Stuttgart, 1988

Hassler, M. (1989). Diplomarbeit: Anwendung der Fuzzy Set Theory für Wissenserwerb und Schließen in einem Expertensystem zür Leck-vor-Bruch Analyse, MPA Stuttgart

Jovanovic, A. (1989). Knowledge representation and reasoning in structural reliability assessment expert systems: Implementation in a Leak-before-Break prototype expert system, submitted for ICOSSAR-89 (R14B-07), San Francisco

Jovanovic, A., Hassler, M. (1989). Vorläufige Ergebnisse der Umfrage - Anwendung der Methoden des Knowledge Engineering in der Struktursicherheitsanalyse, Zwischenbericht, MPA Stuttgart

Jovanovic, A., Servida, A., Sauter, A. (1988). Application of fuzzy algebra for treatment of uncertainties in the structural analysis of pressure vessels exposed to thermal shock. Proc. 4th SAS-World Conf. - FEMCAD, Vol. 2, Paris

KEE^{TM} 3.1 (1988). Users' Manual, IntelliCorp, W. Mountain View, CA.

Okamura,h., Yagawa, G. (1987). An expert-interactive algorithm of structural integrity evaluation and its application to thermal shock, in SMiRT - Advances 1987, ed. F.H.Wittmann, Balkema, Rotterdam

Stoppler, W., Schiedermeier, J., Hippelein, K., Sturm, D., SHEN, S. (1987). Bruchverhalten von Rohren unter Innendruck mit gleichzeitig wirkenden äußerem Biegemoment, 13th MPA Seminar, MPA Stuttgart

Sturm, D., Stoppler, W. (1985). Forschungsvorhaben "Phänomenologische Behälterberstversuche" – Traglast- und Bruchverhalten von Rohren mit Längsfehlern – 150279, Abschlußbericht Phase I, Juli 1985, MPA Stuttgart

Sturm, D., Stoppler, W. (1987). Forschungsvorhaben "Phänomenologische Behälterberstversuche" – Traglast- und Bruchverhalten von Rohren mit Umfangsfehlern – 150279, Abschlußbericht Phase II, Dezember 1987, MPA Stuttgart

Sturm, D., Stoppler, W. (1989). Strength behaviour of flawed pipes under internal pressure and external bending moment - Comparison between experiment and calculation. Specialist Meeting Leak-Before-Break in Water Reactor in Water Reactor Pipipng and Vessels, Toronto, October 24-27

Zimmermann, H.J. (1987). Fuzzy sets, decision making and expert systems, Kluwer Academic Publishers, Boston - Dodrecht - Lancaster

ARTIC: A KNOWLEDGE ENGINEERING APPROACH TO DAMAGE ASSESSMENT OF

HIGH TEMPERATURE PRESSURIZED COMPONENTS

Andrea Servida

Commission of the European Communities, Joint Research Centre

Ispra Establishment, 21020 ISPRA (VA), Italy

and

Petten Establishment, 1755 ZG PETTEN, The Netherlands

Abstract:

A research project for the Assessment of Residual life TIme for Creep damaged components (ARTIC) has been launched. It aims at building an intelligent system for dealing with the assessment of steam headers. The paper describes how the complexity of the problem domain can be handled by the capabilities which the proposed technology exploits. Particularly, the knowledge engineering approach to damage assessment of high temperature pressurized components is addressed.

The expert system ARTIC is a "procedural expert system". The term "procedural" refers to the fact that the expert system architecture is based on a generalized procedure obtained by merging a set of existing assessment procedures. The procedural features of the expert system are investigated.

The architecture of the system and the structure of the problem solving methodology are shown in the paper, as they result from the knowledge engineering approach to the problem domain.

Moreover, the difficulties of the elicitation stage are presented. Particularly, we address the problem of combination of experts from both the problem domain as well as the knowledge engineering viewpoints.

Lastly, is the part devoted to what can be called the "side effects" of this approach to the problem domain. These side effects mainly concern the methodology adopted to provide a structure to the domain information and the treatment of uncertainty and imperfection pervading the data. The envisaged representation techniques and the results achieved are presented.

KEYWORDS

expert system, approximate reasoning, creep damage assessment, residual life assessment, knowledge elicitation.

INTRODUCTION

The ARTIC project is the result of a multinational collaboration between some European Electric Utilities (CEGB-UK, ENEL-I, LABORELEC-B), the German university institute MPA and the CEC-Joint Research Centre. The partners involved in the project have set up an Interest Club which regulates and coordinates the project activities. Aim is at developing an expert system to deal with damage assessment of pressurized steam headers of conventional power plants. The growing interest in life extension of power plant components [1] and the need to enhance accuracy and reliability of damage assessment urge the exploitation of innovative computer techniques. Artificial intelligence and particularly expert system technology may be the breakthrough. However, the effective exploitation of these tools hardly depends on the domain problem chosen as target.

The assessment of the residual life time of steam headers, which have normally been operated over 100.000 hours, is very critical because of both their intrinsic worth and the extremely high outage costs they induce [2] [3]. Moreover, these components require an extremely flexible procedure as they can be characterized by a wide range of geometrical and mechanical configurations, and they perform in many different operational environments. To date, the existing european standards (e.g. TRD 508, T 0102 V, T 0203 V, T 0204 V) or internal procedures (e.g. CEGB-GOM 101) ensure flexibility by establishing a set of recommendations which, a part from the numerical calculations, are quite general. Thus, a number of decisions should be taken by the operator according to his/her competence and experience. Moreover, although the numerical calculation of the residual life is straightforward and it provides extremely pessimistic results, the final assessment of the component, that must take in account both the structural damage as well as the materials creep damage, is a very complex task. Complexity mainly emerges when dealing with the structure of data and information and when characterizing the significance and relevance of each piece of information with regard to the problem environment. A skilled operator retains in his/her mind the knowledge and the expertise which allows him/her to consistently and effectively handle this complexity. Nevertheless, one can say that the accuracy and reliability of the final result depend heavily on the competence of the human operator. In this respect, a direct improvement of the final result, in terms of consistency and repeatability, can be achieved by coupling of knowledge based and standard assessment techniques.

The technology of knowledge based systems [4] is envisaged as suitable for representing and managing all the sources of uncertainty and imperfection which pervade the problem domain. It is argued that an expert system can overcome all the drawbacks which usually emerge when coping with the above problem by means of standard computer programming. On the one hand, powerful computer systems enlarge the scope of the analysis as they can easily collect and process numerical data (e.g. temperature and pressure records, strain readings) and finally they may permanently back-up the calculation. On the other hand, the expert system technology allows one to build (to represent) a consistent (in terms of representation paradigms) framework to better understand the connection and relationships among the different sources of information that cooperate in achieving the solution. Moreover, the representation of human operator knowledge and expertise in a computer system, which ensues the knowledge based approach to the problem, provides one with an effective tool to optimise and to handle the investigation and eventually the exploitation of the component.

The suitability of this technique for dealing with the residual life assessment of headers comes from the fact that often the assessment involves reasoning about images appearing under a microscope, qualitative data, interpretations and heuristic. That is, it mainly deals with the so called 'rules of thumb' based on experience, whose representation in terms of rigid numeric-like information is not effective. Mostly, the reasoning is symbolic, that is, it processes labels which "translate" very complex pieces of information. Thus, the modelling and representation techniques should take in account the imprecision and the uncertainty pervading each step of the inference process.

The paper is divided in three parts. First, the strategy common to the procedures usually adopted to assess headers is detailed, pointing out why the knowledge based approach is envisaged. Second, the architecture of the expert system ARTIC is presented and the models for dealing with uncertainty described. Lastly, the elicitation technique and the difficulties faced in dealing with multiple experts in the domain are explained.

1.THE ASSESSMENT OF STEAM HEADERS AND THE KNOWLEDGE-BASED APPROACH

The residual life time assessment of any component requires [6] knowledge of

     i) the degree of damage currently in the component

     ii) the rate of damage accumulation

     iii) the degree of damage needed to cause failure.

Hence, the first goal of the assessment is to characterize the possible damage mechanism, in order to acquire the knowledge as above. In case of steam headers operating at temperature over 500 C the damage is due mainly to creep and structural degradation. Whenever the header is thermally cycled, low cycle fatigue damage may occur. In the existing European standards(e.g. TRD 508, T 0102 V, T 0203 V, T 0204 V) attention is paid to all these possible aspects of damage. However, in these codes as in the internal procedures adopted by some utilities (e.g. CEGB-GOM 101) only the numerical part of the assessment is thoroughly detailed. Usually, the section devoted to the non destructive examination and destructive testing supplies only some general recommendations and guidelines. Thus, especially during the evaluation of results, large effort is required to the operator implement them. For the evaluation of results yielded by some techniques, reference is often made to specialized teams, whose availability might be limited.

Moreover, a resources saving, unified and repeatable procedure for dealing quantitatively with uncertainty in this contest does not exist yet. The scientific and the technological communities devoted many resources in developing tools which could overcome the above difficulties.

To understand why and to what extent it is worth adopting knowledge-based techniques and how they can enhance the assessment capability, let us point out some critical structural aspects related to the assessment. Here are some features common to most of the available residual life assessment procedures. It must be stressed that each procedure does differentiate from the other, because of different pragmatical, political and economical motivations. However, the goal that they pursue and the strategy they follow will be, in equivalent conditions, the same.

The general strategy of these recommendations and how they are actually implemented are here briefly reviewed. Looking at the typical layout of a power plant one would notice that a plant is usually compound of a number of units. Each unit has usually more then one boiler and in each boiler there are different headers. Thus, even if we are concerned with a single unit the number of components to be assessed is large. Hence, a previous screening must be carried out in order to reduce the burden of the assessment without possibly affecting safety. The very first action aims at locating the most heavily loaded component to be inspected. This preliminary part is considered either as fully belonging to the assessment or as a stage preceding the assessment. The difference is somehow just formal as the named location is in any case carried out.

Generally, three different steps towards the complete assessment are common to most of the procedures usually adopted by the electric Utilities. First, a structural investigation is performed in order to check whether there is any particular macroscopic problem with any headers, by means of visual checks, thickness and dimensional measurements, the study of the past failure history and failure analysis reports, historical data. Once this preliminary operation has been done, the following levels of investigation are envisaged:

i)   numerical assessment of the expired and/or the residual life time based on steam temperature records or operational, minimum material properties, design or nominal dimensions;

ii) numerical assessment of the expired and/or the residual life time based on measured metal temperatures, minimum material properties, measured or nominal dimensions;

iii) direct access to the component aiming at performing those experimental inspection techniques which evaluate the structural degradation and other forms of creep damage.

Steps 1 and 2 use the linear creep accumulation Robinson's rule [7] [8] [9] to calculate the expired percentage of the theoretical life of the header material, as it is defined by the standards valid at the operating temperatures and stresses. If a given threshold value is achieved or crossed, one has to step into a new stage of the procedure. The cost of the assessment increases as we proceed from level 1 to level 3. In fact, apart the increasing cost of the tests themselves, considerable is the induced cost due to the outage that they demand. Thus, the conditions to be satisfied when passing from one step to another are fixed according to both economical and safety reasons. In this regard, we can highlight how the degree of approximation of the assessed state of damage to the real one decreases going from step 1 to 3 as the accuracy increases. However, the results produced by step 1 and 2 are, in terms of residual life time, so pessimistic that whenever a header passes them then there is no point in worrying about it. This is true as long as all the input data in the above calculation are really the worst one with regards to the life time estimation. As a general remark we can say that only in level 3 properly one gives a direct indication of the state of damage of the component, as the component will be characterized metallurgically and mechanically.

The complexity of the assessment surfaces as soon as one focuses the attention on each single action of the approach. For instance, even in the very straightforward stage of numerical calculation of the expired life time one has to describe the thermal and the stress fields in the component in order to pick up the worst combination of input data. Thus, whenever direct data are not available, the identification of the so called hot spots is based on the analysis of geometrical factors (i.e. geometry of the component and its location into the boiler), environmental aspects (i.e. boiler firing system, stress system acting on the component), historical data and so forth. The reasoning involved is mainly symbolic as it is based both on the accumulated experience as well as on the evaluation of information whose pattern is informative to the operator. Moreover, the information are quite uncertain and imprecise and the reasoning path is highly problem dependent, as it can not be exactly stated a priori.

In level 3, the set of information and data to be merged is variegate. On the one hand, are the results of creep damage testings (cavitation replica and extraction replica) which are pictures of a very small area. On the other hand, are the results of a variety of different techniques which provide direct or indirect indication on degradation of the material due to creep ageing (i.e. detailed 3D analysis, strain measurements, metallographic techniques). Hence, the data to be processed can be either numbers, either pictures reporting some modifications which happened in the structure, either physical parameters readings, or damage parameter estimation and so forth. Each type of data retains a peculiar expressiveness, as it points out a particular aspect of the problem. A human operator faces the burden of looking at all these sets of information successfully and consistently.

The technology of knowledge based systems has been envisaged as suitable for representing and managing the whole variety of data and all the sources of uncertainty and imperfection which pervade this domain. It is argued that an expert system approach can overcome all the drawbacks which usually emerge when trying to treat the above problem with standard computerised tools. In fact, the residual life assessment of headers requires the representation of information and data whose complexity is related to the knowledge on how to use and to merge them effectively. Furthermore, the lack of homogeneity among the data is unavoidable and can not be disregarded in the system implementation.

Mainly, this knowledge is not algorithmic but heuristic. It is based on acquired expertise, whose representation in term of a rigid numerical-like scheme is not worthwhile. This knowledge, commonly couched in terms of IF-THEN or WHEN-THEN rules, is fired by the problem contest according to a strategy which has been chosen as the most appropriate to deal with the problem itself. In expert systems the knowledge representation paradigms are suitable to coherently handle this variety of the domain knowledge. Furthermore, they also naturally embed the treatment of uncertainty both at representation level as well as at propagation level. The explicit representation of knowledge, in terms of either rules or facts, makes the maintenance and updating of the knowledge easier and effective. The architecture and the main features of the knowledge paradigms adopted in the system ARTIC are shown in the following chapter.

2. ARTIC: ARCHITECTURE AND REPRESENTATION PARADIGMS

The expert system ARTIC is a "procedural" expert system. The complete exploitation of its capabilities will allow the development, in a later stage, of a distributed intelligence expert system whose modules will either cooperate in searching for the solution or will perform independently. The term "procedural" refers to the fact that the expert system architecture is based on a generalized procedure obtained by merging together a set of existing assessment procedures. The definition of generalized procedure does reflect on the system as abstract characterization of the assessment strategy. Thus, one can look at this generalized procedure as a kind of meta-procedure that states, according to what it has been acquired so far from the standard approaches, what are the assessment goals or intermediate goals to be achieved by the expert system. Moreover, being the concern of this meta-procedure mainly related to the assessment strategy and not to the inference path, it is clear that it affects only the actual planning of the assessment actions being taken, without overlapping with what is called the control strategy of the system which overwievs the inference in the expert system. This skeleton has provided both the structural design elements, upon which the system is developed, as well as the general pattern for the representation paradigms being used. The resulting architecture and knowledge representation paradigms are presented in this section, pointing out the advantages of the proposed approach. Moreover, the uncertainty modelling technique will be introduced and detailed.

The assessment of the residual life of the steam headers is usually carried out according to some guidelines which precisely define only the strategy to be followed. In this regard, we argue that the approach to the problem of remaining life assessment of headers can be divided into various different tasks. Each of them is not isolated as, together with the others, it defines the **Assessment Strategy** to be followed. Precisely we may have

1. The **complete characterization** of the header, which aims to thoroughly describe the component in terms of maintenance history, operating environment, temperature distribution, stress distribution and material properties.

2. The **calculation of the assessed header life (AHL)** and of the residual life, which gives the very first relative indication of the severity (in terms of creep damage) of the operating conditions.

3. The **structural integrity** investigation which concerns with possible structural damage of header critical parts (e.g. main butt welds, end cap welds, large branch fillet welds, stub to header fillet welds, and so forth);

4. The **damage assessment**, aiming to investigate the degree of creep damage in the header materials.

These tasks form as a rough partition of the problem solving process that may be identified in the experts' approach to the problem. Within each task is a number of subtasks whose meaning is that of further partition of the knowledge of the problem.

In the expert system, each task is represented by a particular knowledge set-up of the blackboard model [5]. In fact, in order to perform each particular task the different knowledge sources (KBs) are not all required. The activation of the different KBs is carried out by the **control structure**, which defines the active knowledge set-up according to both the problem state as well as the particular task being undertaken. In this regard, each particular task can be considered as an intermediate goal on the path of problem solution. This particular control strategy is achieved by designing the expert system as composed of three main parts:

1. The **Overviewer module**, which contains the <u>control structure</u> of the system and the <u>planning space</u> partitioned roughly like the above mentioned tasks.

2. The **Internal Knowledge bases** (IKBs), which retain the knowledge about the specific header. The IKBs are partially built-in and are completed by interaction with the user.

3. The **External Knowledge bases** (EKBs), which retain the knowledge about the models and techniques which are of concern with the assessment of headers.

The Overviewer module is an open blackboard which exploits the different KBs according to the goal defined by the schedule in the planning space. The schedule is defined as result of the partial construction of the IKBs. In fact, these KBs specifically concern the being assessed header for what regards both the typicality of the header (design, manufacturing, structural behaviour, etc.) as well as the peculiarity of the header (historical data, maintenance record, temperature pattern, etc.).

The control structure activates the KBs which are relevant to the actual goal in the planning space. Thus, for each goal a different knowledge set-up is accessible to the open blackboard. The correspondences between tasks (in the assessment strategy) and goals (in the planning space) ensure that the problem solving in the expert system is fully consistent. It is worth underlining that for each knowledge set-up the expert system behaves as it was a single module, where all the active knowledge bases cooperate to perform the particular task. Moreover, the correspondences tasks/modules and subtasks/knowledge bases make easier both the elicitation of the domain knowledge and the construction of the various parts of the expert system.

Here is a sketch which outlines the different parts of the expert system and shows how the IKBs and EKBs are linked by the overviewer module.

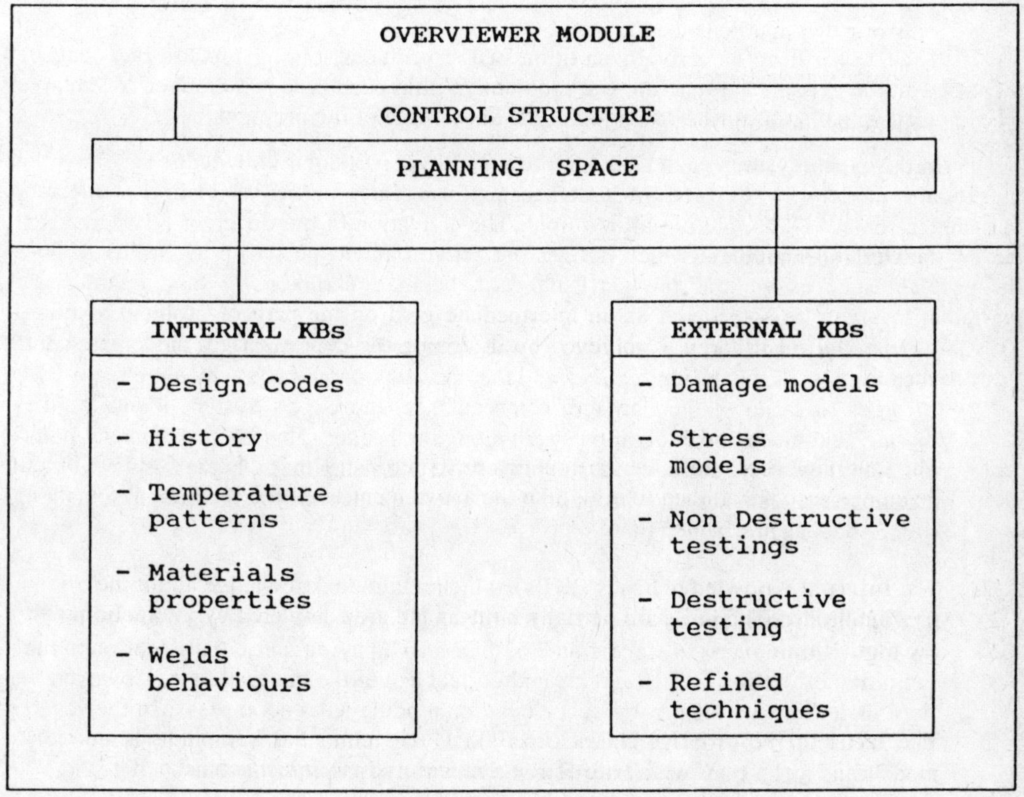

Fig. 1 Architecture of the expert system ARTIC

2.1. THE DIFFERENT MODULES OF ARTIC

It is argued that the approach to the problem could be divided in some different tasks. Each of them is not isolated as it contributes with the others to define the **Assessment Strategy** to be pursued. Thus, these tasks represent a rough partition of the problem domain knowledge that somehow is present in the experts' approach to the problem. The above partition points out the capabilities of the different modules of the expert system. Each module is a "blackboard" where all the different knowledge bases (subtasks in the partition) will cooperate to perform the task for which the module has been designed. Moreover, the activation of the different modules would be carried out by an Overviewer module, whose capabilities are related to the implementation of the assessment strategy.

The process of structuring the problem domain knowledge, for the expert system implementation, may also lead, as side effect, to the definition of a **common format** to be implemented in collecting, presenting and evaluating data and information for the assessment. Let us briefly describe the capabilities of the different modules.

1. The **complete characterization** of the header aims to assess the component in terms of

   - historic and maintenance records;

   - geometry and design;

   - operating environment;

   - temperature distribution;

   - stress distribution;

   - material behaviour.

   Hence, the result of this module is the complete assessment of the component according to all the design features and the operating parameters which are required in order to evaluate the damage. Consequently, the expert system identifies the thermal and mechanical loading pattern along the header. This characterization points out both the critical areas where the loads have been maximum, higher, ass well as all those parameters and data which should be used to calculate the theoretical remaining life. It must be underlined that

2. The **calculation of the assessed header life (AHL)** regards the implementation of the **algorithmic parts** of stage 1 an 2 of GOM 101, paying particular attention to the definition of the correct histograms and correct use of the material data. However, we must underline that the theoretical calculation module will encompass also the very first steps of the problem characterization. Thus, aim of this module is possibly twofold: i) <u>complete</u> characterization of the headers in terms of subtasks of 1. ; ii) <u>actual</u> calculation of the theoretical remnant life.

The <u>format</u> of the calculation is enhanced by embedding an uncertainty quantification of both the data and the results in terms of possibility measure. In fact, the applicability and the rule, on the one side, and the quality assurance of both data and information regarding the material, on the other, may be modelled in a unique representation frame.

3. In the **structural integrity** investigation module, a complete exploitation is done of the headers characterization provided by modules 1 and 2. Target of the module is the **structural integrity assessment** of the header. Thus, the following NDT examinations data are considered in the evaluation

    - visual control;

    - thickness and diameter measurements (vs. strain   measurement);

    - magnetic particle inspection;

    - dye penetrant;

    - ultrasonic examination;

    - Eddy Current;

    - endoscopic examinations.

    The analysis is performed using only the NDT data that either may be required by the problem or can be collected because of limited resources, time and economical constraints.

    Each technique is represented in the expert system by means of a pattern which retains all those information which might be relevant to the successive analysis (i.e. areas of inspection, applicability of the examination, methodology implemented for the examination, format used in reporting the results, and so forth).

    It is worth underlining that the nature of the data produced by the techniques is different from one to another. In this respect, the expert system permits to exploit these different character of the available knowledge.

4. The module for the **damage assessment** investigates the creep damage of the header materials. The aim of the module is twofold. On the one hand, the module is to assess of the cavitation damage by evaluating the replica examination results on the headers girth welds, on the welding area of some inlet stubs, as well as on the base material of the hottest ligament and the branches (special pieces), and so forth. Hence, the state of microcavitation is sensitive index of creep damage. On the other hand, the module is to evaluate the data from the analysis by transmission electronic microscopy of extraction replica, as they may provide information on abnormal evolutions of the microstructure induced by high temperature operation.

The pattern use to represent the above techniques is similar to the one adopted in the previous module. However, the evaluation scheme is quite different as it is supposed to provide a global assessment of the materials by means of hot spots data.

2.2 ARTIC KNOWLEDGE PARADIGMS

The knowledge elicitation of the human experts highlighted that the complexity of the problem domain can be resolved only by embedding in the expert system a formal tool for dealing with uncertainty. In this respect, we argue that non standard logics are crucial. In particular, the fuzzy uncertainty measures are deployed to represent and handle the imperfection affecting both the data as well as the inference path. These measures are suitable as they can treat, within a unique formal scheme, very different types of imperfections. The treatment of uncertainties, optimised for the designed architecture, enhances the capabilities of the expert system as it would coherently justify any computed result.

The knowledge is represented according to the needs which were pointed out during the knowledge elicitation phase. Thus, the format and the internal structure of the objects are as close as possible to the original data even if the necessity of the structurised representation is not been disregarded. The frame representation scheme provided by the shell GoldWorks, which is used for the prototype development, has been adapted to our needs. Particularly, the send-message utility is used to make the inference uncertainty driven.

3. MULTIPLE EXPERTS KNOWLEDGE ELICITATION

The knowledge elicitation is an important step in any expert system process. In the ARTIC project we have been dealing with multiple experts in order to achieve a common approach acceptable by all the partners. Although the background and the capabilities of the experts were comparable, it should be pointed out that for each expert the mental representation of the assessment process is different and peculiar. Hence, a great effort was devoted in trying to formally establish what might be called the cross-links among the experts. Commonly, we experienced that the expert is urged on reacting to an external domain stimulus (like an assertion or a statement collected by an other expert) by tuning is mental representation and eventually by independently clarifying the hidden commonalities or discrepancies. This effect is extremely helpful. It provides the knowledge engineer with an unexpected tool for maintaining and tuning the so far acquired knowledge. Moreover, the variety of "points of view" which associated to multiple expert enhances also the in depth investigation and understanding of the assessment process. In fact, in order to support a certain achievement or evidence, with regard to some other that might be argued by some other expert, one has to gain a deeper understanding of his/her own knowledge.

The multiple experts elicitation has produced as side effect what may be called a "collective perception" of the uncertainty affecting the assessment process. In fact, as result of the above mentioned cross-links established during the interview it was possible to gather different perception of what is the uncertainty affecting the assessment process. Hence, a homogeneous typicality of uncertainty has been reached and modelled in the system. This typicality is characterized by the highlighted absence either of precise and reliable models for the quantitative interpretation of available evidences or of a pattern or format to represent the evidence as a whole. Consequently, the partial description of the reality of the domain problem induces the need for an uncertainty modelling.

ACKNOWLEDGMENTS

The author is indebted to all the partners in the Interest Club project which have been contributing so far. Special thanks go to the domain experts who patiently allow him to acquire a deeper understanding of the remaining life assessment problem.

REFERENCES

[1]   Proceedings of the Int. Conf. "Refurbishment and Life Extension of Steam Plant", London, Inst. Mech. Eng. London,1988.

[2]   FIELDING,P.J., et alt.: "An Assessment of the remanent creep life of a superheater header", as ref. [1].

[3]   ANGELI,F., et alt.: "Methods for Residual Life Assessment of Elevated Temperature Components", Proceedings of the Int. Conf. on "Life Extension and Assessment", The Hague, The Netherlands, Vol. 1, 190-199, 1988.

[4]   BOUCHON,B., et alt.(Eds): "Uncertainty and Intelligent Systems system", Springer Verlag, berlin, 1988.

[5]   HAYES-ROTH,B. "A blackboard architecture for control". Artificial Intelligence, Vol. 26, 251-321, (1985).

[6]   "Generic Guidelines for the Life Extension of Fossil Fuel Power Plants", EPRI CS-4778, November 1986, Palo Alto, California.

[7]   ENEL private communication.

[8]   LABORELEC private communication.

[9]   CEGB private communication.

E S R PROJECT: EXPERT SYSTEM AIDED ASSESSMENT AND MANAGEMENT OF REMAINING LIFE OF PRESSURIZED HIGH TEMPERATURE COMPONENTS

A. Jovanović

Staatliche Materialprüfungsanstalt (MPA) Universität Stuttgart
Pfaffenwaldring 32, 7000 Stuttgart 80, FR Germany

Abstract

The paper presents a project aimed at development of an expert system for assessment and management of remaining life of the high temperature high pressurized components of power and process plants (pipework and headers, mainly). The project has been prepared in collaboration with other European partners (industry and research institutions), as a build up of their current activities in the domain of expert system applications in the structural assessment of high temperature components and systems. In that case of MPA Stuttgart, these activities are performed in collaboration with the German utility companies and the within the Interest Club of European Utilities. The background of the project, its objectives and its tasks, leading to the definition of the architecture of the expert system, are described more in detail in the paper.

1. INTRODUCTION

Possibility of failure of high temperature pressurized components (pipework, headers, etc.) is one of the principal considerations in the safety and economy of power and other industrial plants. A catastrophic failure

of such a component could both endanger the industrial plant and put
at risk the population of the surrounding areas. The economic effects of
improved/reduced availability are also considerable: the cost of one day
of outage in a modern 500 MWe power plant can be as much as 150,000
ECU. In other industrial plants, the outage costs are usually much smaller
(e. g. $\sim$ 8,000 ECU/day, for a 10 MW plant), but the indirect costs due
to disturbances in production can be much greater.

Improving methods and procedures used for assessment and management of remaining life of the high temperature pressurized components is
therefore extremely important. Development of new, more sophisticated
numerical engineering methods and tools is often coupled with severe
limitations in practical application. The limitations are in most cases
imposed through unavailability of time, resources and highly qualified
personnel which is required for a sophisticated analysis. In addition, it
has become evident that the numerical analysis alone, cannot cover all the
factors influencing assessment and management of the target components.

The development proposed in the ESR project should significantly improve the current practice by making it:

- *more comprehensive and reliable* (a direct coupling of heuristics and
  specialists' knowledge with standard and/or advanced
  engineering calculi);
- *quicker and more easily available* (expertise resides in computer
  systems at the plants, available round-the-clock); and
- *more uncertainty independent* (systematic treatment of uncertainties
  involved).

To reach these goals, it is proposed:

- to develop the "reference procedure", and
- to base the assessment and the management on the expert system
  technology.

The proposed development will thus *improve safety* (reduced risk of wrong
assessment), *reliability and availability* (reduced outages caused by wrong

management), and *economy* (see the figures) of the target components and plants. It is estimated that there are more than 25,000 components operated in the 5 European countries (Belgium, Germany, Finland, Italy, and UK) involved in the project. Moreover, the *Reference Procedure* would be a contribution to a unified European approach in the field (year 1992).

The project has been prepared in collaboration with nine European partners (industry and research institutions). The project is a build up of the current activities of MPA in the same domain, performed in collaboration with the German utility companies and the so called Interest Club of European Utilities. The particular goals or applications cases are slightly different within each of the mentioned collaborations: piping problems and the current state assessment are of the primary interest in the collaboration with German utilities, the expert system prototype developed within the Interest Club is oriented versus headers, the *Reference Procedure* is an additional goal in the project. But the main goal - development of an expert system for high temperature pressurized components, is common in every of the collaborations. Common are also the technical problems to be resolved: knowledge acquisition and representation, design and development of the expert system, its verification and deployment.

2. STATE OF THE ART

Operation of today's power, process and other industrial plants is characterized by the steadily increasing requirements to enhance their safety, and availability, and, if possible, to extend their life. The target components selected for this project (high temperature pressurized components, headers and pipework in particular), are essential for the fulfillment of the above mentioned requirements. Safety, availability and life extension, both of the components and the plants in which they are operated, depend heavily on the engineering assessment of remaining life (also referred to as residual or remanent life) of these components. The life assessment, on its own, as performed currently, is based mainly on:

- numerical calculation of the remaining creep and fatigue life

- destructive (material) testing, nondestructive measurements and strain measurements, and
- human expertise and experience involved in the preceding steps.

The existing methods/procedures of calculation are either relatively crude or they are applicable only to a particular item under particular conditions. The basic methods in the current European standards are in principle similar, but their practical applications exhibits differences. Furthermore, their application to pipework and headers meets serious difficulties, mainly due to:

a) frequent operation outside the range of the design parameters,
b) local variations of stress and temperature, and
c) complex geometry.

On the other hand, an alternative overall advanced method does not exist. Use of alternative, more advanced theoretical concepts, not implemented in standards, is often restricted by the lack of the necessary input data, by the uncertainty connected to transferability of the specimen test results to the components, by the uncertainty regarding applicability of the chosen concept in the given conditions, etc.

The examination and/or test data are in the real-life conditions often incomplete and uncertain, and their proper interpretation and use in the analysis require a high level of expertise, as the available methods only indicate that the damage has occurred — they do not give the amount of accumulated damage or the remaining life of the component.

The human expertise of different specialists mentioned above, is essential for the assessment, but it is often unavailable at the plants in the very moment when it is needed. In its absence, the plant engineers, are thus often saddled with the decision on what to do with the component and/or plant (e.g. *stop and re-inspect, reduce load*, etc), or how to deal with a possibility of an unnecessary outage or with the increased risk of the component failure.

In addition, the target components are expensive and difficult to replace, requiring extensive outage times and long delivery times (up to 1-2 years), while their failures by creep and/or fatigue rupture during service (catastrophic accidents) have been recorded.

All these difficulties are well known in the engineering community (cf. Int.Conf. on Life Assessment and Extension, Den Haag, 1988, and some preceding BRITE projects, e.g. 1209-1-85, 1290-1-85, 2341-1-87) and new ways to tackle the problem are continuously examined.

On the other side, the expert system technology and tools appear nowadays to be sufficiently developed to offer a significant contribution to the solution of the above described problems (e.g. Ashley, Duggan and Shor, 1988; Singh and co-workers, 1987; Wollenberg and Sakaguchi, 1987). Personal computers are widely used by plant engineers and extending such a well known tool with an expert system would be a great benefit. It coincides also with the European interests to use and develop expert systems in related fields (e.g BRITE 2124-1-87, ESPRIT-P-857-990) and even more with those in Japan and USA (cf. EPRI's or Battelle's programs).

3. OBJECTIVES OF THE PROJECT

Main objectives of this project are:

1. Develop the *Reference Procedure* (RP) for the remaining life assessment and management of the high temperature pressurized components, headers and pipework in particular. The procedure should merge and summarize "good and desirable practice" at the European level and thus provide a technical support towards the development of a unified European approach (year 1992).

2. Develop the *ESR* expert system aided assessment and management of the remaining life of the target components (based on the *Reference Procedure*).

3. Substantially and qualitatively improve assessment and the short- and the longterm management (by reaching the two preceding objectives), and thus improve safety, reliability, availability and economy

both of the component and of the plants in which they are operated.

4. Extend the results of the earlier established collaboration started among some of the partners in the Consortium (the expert system prototype).

In order to reach the main objectives, a series of sub-objectives (such as development of the *ESR/RP* library of computer codes and development of the generic demonstrator) is specified. Realization of the *ESR* project will also introduce several innovative elements in the current practice, namely:

- The assessment and management will be expert system based (*ESR*) will be a computerized system, exploiting the engineering and heuristic knowledge for remaining life prediction of pressurized, high temperature components, the capabilities of which go well beyond the software used at present in this domain);

- Integration of the numerical analysis with heuristic and qualitative knowledge, within the same system, thus allowing a more comprehensive and reliable prediction;

- Exploitation of uncertain, incomplete and imprecise information by providing a framework for the systematic treatment of uncertainties in numerical and symbolic analysis in the expert system;

- Guided (by the expert system) use of advanced analysis methods beyond the normal engineering practice;

- Extensive use of knowledge- and databases for material data, design data, operational history, experiences and maintenance data and evidence, rules, etc.;

- Making the high-level expertise available round-the-clock, in situ (e. g. power plants) where it is to be used, on the computers normally available there, i. e. on PC/PS's;

- Use of Hypertext for the *Reference Procedure* consultancy.

Reference Procedure on which it will be based, on the other hand, is much more than just a collection of rules for lifetime prediction, reliability and maintainability of components - it's a new, European approach to this problem.

4. A SURVEY OF TASKS IN THE PROJECT

To reach the specified objectives, the following tasks in the project has to be performed:

a) In-depth comparative analysis of current practice and experience of the partners, in 5 European countries, in order to establish the *Reference Procedure*.

b) Analysis of the case studies, supplied by the industry partners, resulting in a "matrix" of cases with typical combinations of component types, materials, applications, etc.

c) Interviewing of domain experts (knowledge elicitation), aimed in particular to the extraction and formulation of heuristic rules and reconstruction of the expert reasoning in sub-domains like non-destructive testing, material testing, component disposition decision making, future operation regime definition, etc.

d) Development of a set of computer codes (*ESR/RP Library*) covering numerical analysis in the *Reference Procedure* and chosen alternative algorithms (life assessment in terms of creep and fatigue). The tasks includes codes for elaboration of numerical data from service records.

e) Definition and preparation of the software and hardware environment. Detailed architecture of the *ESR* expert system (modules, interactions, etc.).

f) Development of modules of the *ESR* system: interfacing with the user and use of stored numerical data (dialogue module), decision optimization (optimization module), module for the interactive treatment of uncertainties, the system data base, and the Hypertext-based "reference procedure" module.

g) Development of interfaces and final assembly of the *ESR* system.

h) Development of the *ESR - generic demonstrator*, handling a generic case of a pressurized high temperature component.

i) Testing/verification/modification loop, including the establishment of the testing/validation procedure and testing of the generic demonstrator.

j) Development of the *ESR Specific Demonstrators* corresponding to the case studies from Task 2, and their testing.

k) Evaluation of the project and its results. Documentation.

5. DEVELOPMENT OF THE *Reference Procedure*

An in-depth comparative analysis of current practice and preceding experience of the partners has been foreseen, with a goal to establish the *Reference Procedure* for assessment and management of remaining life of the target components (high temperature pressurized components in power and process plants). The procedure will include the regulatory requirements and the task will result in a summary of the "good practice" in assessment of high temperature pressurized components. The structure, i.e. the "skeleton", and the principal elements of the procedure were agreed upon by the partners during the preparation of this proposal and are shown in the flowchart in figure 1.

In order to perform the task it is first necessary to define and provide a set of typical input information, including all the information relevant for the remaining life assessment and management of the target components (design and as-built-in data, fabrication data, operating history, economy related data, etc.), including also the non-numerical and the heuristics-based information. The typical uncertainties related to the information and assess their relevance/importance qualitatively, from the domain expert's point of view must be identified.

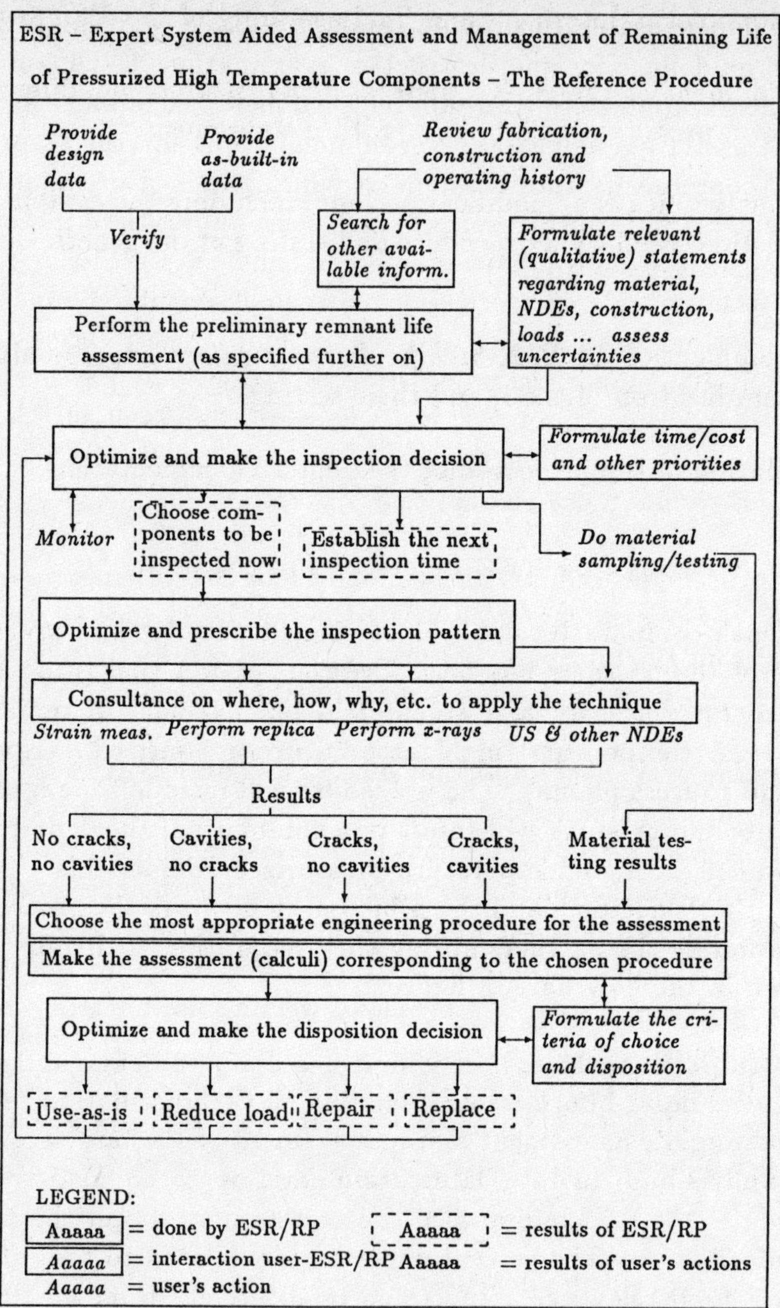

Fig.1 - Flowchart of the *Reference Procedure*

It is also indispensable to provide (make available) and analyze procedures and guidelines for non-destructive examinations (NDE) and strain measurements applied by the different partners, including the written and/or heuristic guidelines and recommendations for their application (choice of components and locations at which the measurements/examinations will be performed, choice of the most appropriate methods and techniques, choice of equipment and personnel, etc.). Principal NDE methods to be considered are: x-rays, ultrasonic testing, dye penetrants, magnetic particle examination and replication. Uncertainties are to be evaluated and a comparison of experiences gathered by partners participating in the subtask analyzed. Reliability and limits of the short-term prediction of the remaining life, using fracture mechanics and the obtained NDE results must be analyzed.

Furthermore, the procedures and guidelines for the destructive testing of the target component materials has to be analyzed too, as well as the information for the interpretation of the testing results (in terms of measured values only, not in terms of their relevance for the remaining life).

The typical and the most important repair procedures, necessary to maintain the component in service, have to be identified and specified. The specification must include the limits for repairs, depending on the degree of creep damage of the component and the extent of the necessary repairs. The written and/or heuristic guidelines (a summary of "good practice") must cover the fundamental aspects like selection of the welding material, heat treatment conditions, scattering of welding parameters, etc. They are to be specified for the most relevant steels used for the high temperature components. Economic aspects of the repair have to be specified both at the level of the component (system) and of the plant. The same is to be done for the safety aspects. Both safety and economy criteria have to be specified in explicit form(s).

The *Reference Procedure* has to include all the regulatory requirements identified as relevant and an overall synthesis/summary of good practice, both for the total of the procedure (from the input data to the disposition decision) and for the related subdomains, like operation, inspections, material testing, etc. Particular attention will be devoted to the issue of formulation and formatting of semantic, qualitative and/or heuristic statements in the procedure in order to obtain their clear and doubtless applicability, both by the human user and by the knowledge engineers developing the ESR expert system. Furthermore, an additional analysis of the disposition decision ("what to do with the component" – repair, use-as-it-is, use under reduced loading, re-inspect, replace, etc.) will be done.

6. KNOWLEDGE ELICITATION

6.1. Analysis of case studies

In order to give a sound practical (technical) base to the Reference Procedure it is necessary to identify and analyze a series of "case studies". The case studies will provide further evidence for the development of the Reference Procedure and of the ESR expert system demonstrators, as well as provide a basis for its validation. A "matrix" of cases will be analyzed, with typical combinations of component types, materials, applications, etc. The matrix for component-material combination is shown figure 2.

The analysis of case studies should distinguish between the issues relevant for specific cases only and those issues of a generic importance. The generic issues will be "extracted" and brought into a form directly usable in the formulation and in the application of the Reference Procedure. Particular attention will be devoted to the designers' and analysts' actions and decisions (choices) based on practical experience, rather than on codified guidelines and standards, and significantly influencing the final result.

6.2. Interviewing of domain experts

Interviewing of domain experts (knowledge elicitation), is necessary for elicitation of the knowledge which is regularly/occasionally used by the experts when dealing with remaining life assessment and related subdomains, *in addition* to the knowledge already encoded in standards, other national and/or internal regulatory requirements and in the numerical

Component	Material				
Type	A	B	C	D	E
headers	*3*	*6*	*1.1*		
pipework	*2 , 5*	*5 , 6*	*1.1*	*4*	*2*

Material codes:
A = $2\frac{1}{4}$Cr1Mo, 10CrMo9.10;
B = 1Cr$\frac{1}{2}$Mo, 13CrMo44;
C = $\frac{1}{2}$Cr$\frac{1}{2}$Mo$\frac{1}{4}$V, 14MoV63;
D = 12Cr1Mo, X20CrMoV12 1;
E = Austenitic.
Numbers in the table indicate the partner providing the case study.

Fig. 2 - Matrix of the case studies

algorithms. This also includes the knowledge necessary for a correct and consistent application of these requirements and algorithms. Particular attention during the interviewing will be devoted to the extraction and formulation of heuristic rules and to the reconstruction of the expert's reasoning process in sub-domains such as input formulation, non-destructive testing, material testing and component disposition decision making, which correspond to the first four subtasks in the task.

The knowledge necessary for the input of the RP has several components: the design data, the as-built-in data, data from fabrication, data from the construction, data regarding the operation history and other, usually less relevant data. Fabrication and construction data, if available, are almost entirely qualitative, while data from the operational history are often only assumptions and/or cover just a small part of the actual history of the component. Advanced monitoring systems should improve this situation, but the current situation is as described above. Applicable codes and standards require normally only a very restricted set of the above quoted data, resorting to overconservatism whenever uncertainty and/or a lack of information appears. On the other hand, an "experienced engineer" can often overcome many of the above mentioned difficulties by exploiting his experience, consisting usually a large set of "rules of thumb", allowing him to avoid excessive conservatism and at the same time satisfy the regulatory and the safety requirements.

The knowledge to be elicitated concerns a series of quite typical situations, appearing during the planned maintenances and inspections. Costs, technical resources, available time and common sense, in practice to limit the analysis, NDE and other investigations to reasonably small number of components. Decisions, i.e. selection, are rarely possible starting just one of the criteria, it is rather "a search for good compromise". The search, on the other hand, is again based mainly on the experience of the expert (e.g. operator), who can in many cases "smell" a possible problem, even in situations when the available data do not indicate some particular problems. Extracting this kind of knowledge is a difficult task which requires an extremely high degree of cooperation between the domain and the knowledge engineers, pushing probably both of them to learn more about each other's job.

A (multidimensional) problems-solutions "matrix" will be used as a tenchical basis for the consultancy which the *ESR* - expert system is supposed to provide (the matrix should actually consist of rules of the type: *"if the given situation is characterized by A, B, C, ... , than the proposed pattern consists of M, N, O, ..."*).

The disposition decision is a typical multi-criteria decision problem. The decision to be made consists practically of finding, or choosing the best alternative from a given subset of possible alternatives. In other words, at the end of the remaining life analysis, when inspections and testing have been carried out, a decision has to be made, taking into account not only these results but also factors like the relevance of the component for the safety and availability of the plant, economic and/or operation time schedule impact, etc. While the regulatory and numerical analysis related criteria are usually very clear ("crisp"), other criteria are often very ill-structured. Therefore the final decision also is often ruled mainly by the experience of the domain expert. The interviewers should therefore insist on elicitating these additional, experience-based guidelines applicable in the disposition decision making, and formulate them in a format that can be used later on in the *ESR* expert system.

At the end of interviewing, set of standard formats and a list of heuristic rules related to application of the RP has to be established.

7. LIBRARY OF ENGINEERING COMPUTER CODES

The *ESR* expert system should encompass all currently available engineering numerical alternatives allowing an "on-line" (in terms of the user's session with the expert system) assessment of the remaining life of the target components. In order to have them in the expert system, these alternatives will be first codified as "classical" engineering computer codes, assembled in the *ESR/RP Library*, covering the numerical analysis in the *Reference Procedure* and the chosen alternative algorithms (life assessment in terms of creep and fatigue). The library will be used as a basis for the *ESR* module for numerical calculi, but also a selfstanding software package. The task includes also codes for elaboration of numerical data from service records.

A review of algorithms for the remaining life assessment, will cover all published information regarding the algorithms for the remaining life assessment of the *ESR* target components, both in terms of creep and fatigue, as well as in terms of their combination (the latter is however not foreseen to be directly applied in the *ESR*). In addition, the information on internal procedures applied by the partners, made available by the participating partners will be reviewed too. The subtask review will also include the heuristic guidelines regarding application of different alternatives in different assumable conditions in practice.

Further on, for the *ESR/RP* library of engineering numerical codes for the remaining life assessment of the target components will be developed. The library will thus represent the principal numerical engineering software tool in the project, based on the *Reference Procedure* and used in the numerical calculation module of the *ESR* expert system.

8. ARCHITECTURE OF THE SYSTEM

The architecture of the system, as defined at the stage of the proposal, has to be reviewed and elaborated in detail, in order to establish the specific functionalities of the modules constituting the system, and their interrelationships The software structure will allow for a decoupled development of the various system modules.

It has been foreseen that the *ESR* system should be developed on a PS/2 personal computer and run on a PC-386, both under the DOS operating system. This will make the system easier available to the final users, at power and process plants, where PCs are predominant in the computer environment. In order to speed up the development, the system will be built on top of GoldWorks-II^{TM}, a commercial expert system development toolkit, available for PCs. The PS/2 hardware configuration will be properly dimensioned to run the GoldWorks-II^{TM}. Within the *ESR* system, the development in GoldWorks-II^{TM} (modules of the expert system) will include the programs for numerical analysis, written in conventional languages (C, FORTRAN), and it will be interfaced with the data base and Hypertext modules.

The design of the expert system foresees the following modules:

a) Module for dialog interface with the user and for the interpretation of history data from files.

b) Module for numerical analysis, performing the engineering calculations concerning stresses, fracture mechanics parameters and damage accumulation.

c) Module for treatment of uncertainties and overviewing, having the capability to "elaborate" the information containing imprecision, uncertainty, fuzziness, etc. Its second function is overviewing (supervision), i.e. activation and coordination of the other system modules.

d) Module for decision optimization, providing consultancy related to

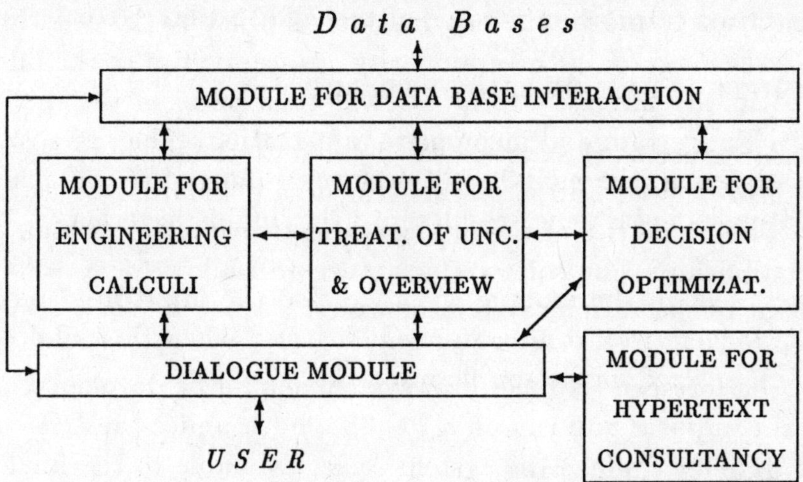

Fig. 3 - Architecture of the ESR system

the decisions such as which component to chose for detailed examination/nalysis and which testing technique to adopt on it, or what to do with the component (the "disposition decisions") once the inspection results and the results of numerical analysis programs are obtained.

e) Module for data base interaction/interfacing.

f) Module for hypertext based consultancy.

The architecture of the system is shown in figure 3.

9. MODULES OF THE SYSTEM

9.1. The dialogue/input management module

The module performs the dialogue with the user. It searches to obtain a maximum available information, both the numerical and the non-numerical one. Its "duty" is therefore to put all the question (to the user) which may lead to this goal, in other words it will solicitate the user. In the interaction with the user, the module should also assess the correct representation frame for the information supplied by the user, in particular the issues regarding the granularity and semantics of the information provided in the linguistic form. It will thus perform the first elaboration of the uncertain and incomplete information. The same role, but in oposite direction, the module will perform when giving explanations and "translating" the inference results into the linguistic terms.

With user's help, the module will take also the input data recorded by the data logger (from the component history, when this kind of data is available), entered as files on floppy disks.

9.2. Module for engineering calculi

The module is based on the *ESR/RP Library* of engineering computer codes. The module will guide the choice of the "right" algorithm and procedure, according to the given situation, and perform the necessary numerical calculations (e.g. to calculate stresses, fracture mechanics parameters, creep analysis parameters, creep life, etc.). In the most simple case, the module will perform a complete guidance of an inexperienced user through the calculations prescribed in standards. For more sophisticated, alternative methods of analysis, the module will require a higher level of the user's understanding.

9.3. Module for treatment of uncertainties and overviewing

Module must have the capability to elaborate the uncertain and incomplete information to the level corresponding to the capabilities of the inference machine of the knowledge engineering tool (GoldWorks-II^{TM}) used for development of the expert system. It will also build up these capabilities in the direction of a more coherent representation of the linguistic terms (variables), mainly by the means of fuzzy algebra. MPA will provide here its data from the inquiries among the structural engineers. Second role of the module is to activate and coordinate other system modules, i.e. to overview (supervise) the whole system. The supervising function will be performed on the basis of the current goal to be achieved and on the status of the problem solving activity.

9.4. Module for decision optimization

The main decision problem to be handled are the decisions related to choice of the components to be inspected in detail and to the choice of NDE techniques to applied, and the decision on what to do with the component ("disposition decision"). The module can be described as "single-person, multi-criteria (multi-attribute)" decision making aid, acting in an ill-structured situation. The module will operate on a finite set of decision alternatives, i.e. it shall be assumed that *decision* is equal to *choice* of the "best" alternative or the of the subset of "good" alternatives. E.g., decision at the end of o structural integrity assessment analysis might be *"replace"*, *"repair"*, *"use-as-is"*, *"use-under-reduced-load"*, etc. The desirability of an alternative is judged with respect to a finite set of goals, attributes or criteria, and the "best" alternative(s) is/are chosen. Goals and constraints will be given by the user, partly as crisp, partly as fuzzy. E.g., limits imposed by codes, standards, regulatory requirements, etc. are usually crisp, but economic and operational criteria are usually fuzzy. Final rating/ranking of alternatives is obtained applying the basic concepts extended in the direction of application of linguistic variables. Experiences and data gathered in the preceding work of the MPA will be exploited. The user shall weight the importance of consumed life, safety relevance, economy/availability relevance and importance of uncertainty in the data available for each component. Importance of each of the criteria is specified, e.g. in terms like *very-important*, *moderately-important*, etc. The result supplied by the module is the fuzzy sets based rating of alternatives:

(a) - according to each of the criteria,
(c) - according to a given subset of the criteria, and
(c) - according to all the criteria taken together.

9.5. Module for data base interaction/interfacing

The data bases foreseen to be used in the project are three: the SQL-based ones (e.g. ORACLE), the dBase-III and -IV based ones and the data base developed in the ESPRIT project PRIME (see Chapter 1 for the complete reference). Development of a specific data base, in this project, is not foreseen.

9.6. Module for the Hypertext-based consultancy

Hypertext is a very innovative element of the project. It can be defined as a hybrid environment for storage and retrieval of information in form of texts, factual and numerical data and images. It is, at the most basic level, a software tool which lets the user connect screen of information using associative links. At its more sophisticated levels, it is a collaborative environment, producing mimic of the human expert ability to store and retrieve information by referential links for quick and intuitive access.

As such, Hypertext is an ideal representation tool for such a document as the *Reference Procedure*. It will allow a quick and well directed access not only to the relevant parts of the procedure itself, but also to the documents allowing better and more comprehensive explanation of relevant details (e.g. other procedures used in derivation of the RP, codes and standards, electronic flowcharts, scanned design drawings, etc.).

10. APPLICATION AND VERIFICATION OF THE SYSTEM

Once the *ESR* expert system is assembled, it will not be possible to test and evaluate its performance without developing first a demonstrator, i.e. evolve the first working version of the *ESR* expert system into the first performing version of the system. In order to reach this goal, a *typical (generic) case* will be defined first and the *Generic Demonstrator* specified. In order to evaluate the *ESR* system it is necessary to specify precisely what are the cases and what are the practical, theoretical and engineering problems related to them, which the demonstrator must be able to deal with.

The level to be achieved by the demonstrator must be high enough to prove that the advocated approach and the developed software are indeed valid and adapted to the problem characteristics.

In order to prove the validity and the usability of the concept of the remaining life assessment based on the *Reference procedure* and the *ESR* expert system, it it necessary to test and validate the system, and according to the results of validation, make the necessary modifications.

The validation will be based on the validation procedure developed. The validation procedure will be developed in order to foresee the steps to done in order to obtain a reliable and consistent evaluation of the *ESR* expert system. The procedure will cover:

- Review of the system requirements

- Review of the hardware and specification
- Review of the preliminary software and hardware design
- "Test Bed" Requirements
- Coding and Debugging
- Integration
- Validation tests program and test results
- Field installation and field results
- Evaluation report

Quality of decision and advice, the human-computer interaction, the system efficiency and the cost effectivness will be imposed as the main evaluation criteria.

The system will be first tested on the generic case, in the environment of the partners participating directly in the development of the system. The second phase of testing foresees tests in the industry, by personnel non directly involved in the system development. At the beginning of the task a detailed plan of testing will be developed, precising what, under which conditions and by whom will be tested.

11. ECONOMIC AND TECHNICAL BENEFITS

Application of the Reference Procedure and of the ESR expert system is expected to bring improvements in the remaining life assessment and management (including the life extension) of the target components and plants, leading to the following economic and technical benefits:

- reduced component replacement costs (typical value of a large component is over 500,000 ECU, raising up to 4 mio ECU - more than 500 large components, in power industry alone, will have to be replaced up to the year 2010 in the countries participating in the project) improved availability of systems and plants in which the components are operated
- shorter and better exploited maintenance periods in power and process plants (see above for economic effects)
- improved possibilities for life extension of the plants (capital costs for large plants often over 2,000 mio ECU - life extension reduces needs for the new plants)
- better utilization of the existing plants, due to improved possibilities of the long term prediction of component behaviour
- reduced costs of daily operation (specialists called only when necessary)
- reduced costs of scheduled inspections due to the optimized inspection strategy
- reduced unplanned costs, thanks to avoiding the unforeseen outages, damage costs, indemnities caused by failures or leaks to the environment, etc.
- the Reference Procedure will reduce costs and times in the European transfer of technology, goods and services in its application domain (the "unification/standardization" effect)
- more available and less dependent on human experts (the expertise resides at the plant, in the computers already available there (PC/PS's) - low equipment/investment costs)

12. CONCLUSIONS

This outline of the ESR allows to see the relationship between the domain problem and the offered solution, based on the expert system technology. The scope and the volume of the research work necessary to reach the specified goals are remarkable, but the project must be seen as e long term goal. Seen in such a way, the project gives a sort of a frame to the activities performed currently by the partners in this field. It defines final goals and allows better coupling, better complementarity and avoiding of pitfalls in the long term collaboration on the European and national levels.

Acknowledgments

Preparation of the ESR project is a results of a joint effort of several institutions. A particular contribution has been supplied by Messrs. *H. Arents (KUL, Belgium), W. Bogaerts (KUL, Belgium), P. Bulens (Laborelec, Belgium), M. De Witte (Laborelec, Belgium), S. Garribba (Politecnico di Milano, Italy), M. Gallanti (CISE, Italy), A. Lucia (JRC Ispra), K. Maile (MPA, FR Germany), T. Mikkola (VTT, Finland), K. Penttilä (IVO, Finland), J. Phillips (CEGB, UK), A. Servida (JRC Ispra, Italy) and G. Sluze (Solvay, Belgium).* Their efforts and precious help are gratefully acknowledge here.

REFERENCES

Ashley,G. K., Duggan, P. E. and Shor, S. W. W. (1988) Expert systems provide help in life extension, availability improvement. Power Engineering.

BATELLE-COLUMBUS Research Programme No. 779-P-1820: Residual Life of Materials Used in Elevated-Temperature Process Service

BRITE Project P-1209-1: Lifetime Prediction and Extrapolation Methodologies for Computer aided Assessment of Component Service Behaviour under Stress at High Temperature

BRITE Project P-1290-1: Condition Assessment of Industrial Pressure Parts operating at elevated Temperatures

BRITE Project P-2341-1: Performance and Reliability Evaluation of Weldments in elevated Temperature Service

BRITE Project P-1145-1: Prediction of Fatigue Crack Initiation on Defects in Welds and Cast Components

BRITE Project P-1436-1: Methods for Predicting the Effect of Surface Degradation on Fatigue and Fracture Behaviour

BRITE Project P-2147-1: Creep Behaviour of Full Size Weld Joints: Design Improvements and Defect Acceptance Criteria

Deep knowledge expert system using coupled symbolic numerical analysis for structural diagnostic, *ESPRIT-II Proposal Nr. 2224 (Area II.2.3)* (1988)

EPRI Research Project 2596-7 (CS-4774): Guidelines for the Evaluation of Seam-Welded Steam Pipes

ESPRIT Project 857-990 PRIME: The European Esprit Project on Expert Systems for Material Selection

Proceedings of the International Conference "Life Assessment and Extension" (1988). June 13 to 15, 1989, The Hague, The Netherlands

Singh, G. P., Divakaruni, S. M., Gehl, S. M., Scheibel, J. R. (1987). Plant Availibility Improvement: Expert System Approach for Determining Boiler-Tube Failure Mechanisms. Joint ASME/IEEE Power Generation Conf., Pap 87-JPGC-Pwr-14

Wollenberg, B. F., Sakaguchi, T. (1987). Artificial Intelligence in Power System Operations Proc IEEE, Vol 75, No. 12, pp 1678-1685

LEARNING EXPERT SYSTEMS IN NUMERICAL ANALYSIS OF STRUCTURES

J. ZARKA, Directeur de Recherches au CNRS
J.M. HABLOT, Ingénieur de Recherches Peugeot
Laboratoire de Mécanique des Solides
Centre commun X-Mines-ENPC-CNRS
Ecole Polytechnique
91128 Palaiseau Cedex, France

ABSTRACT : *In this paper, we present a method to use a new class of expert systems, the Learning Expert Systems. In numerical analysis of elastoplastic structures, an application of this expert system may be for example the evaluation of the error or the search of optimal discretizations.*

1. INTRODUCTION

Numerical computations of elastoplastic structures are very costly; moreover, we do not know the error which has been reached during such computations.
No general answer exists to the following problems :
- find the error between the exact solution of a problem, and the result of the numerical computation;
- find rules which can give discretizations with a prescribed error or a prescribed cost.

The use of the artificial intelligence may be considered but some difficulties appear, no expert can give rules valid for such problems.

In this paper, we study the application of a non-classical expert system, the Adaptive Learning Systems (the mathematical software was developped by M. Schoenauer and M. Sebag (1988) and is called SEA). It can find rules from an examples base and these rules can be used for any new case.

2. THE LEARNING EXPERT SYSTEM (SEA)

2.1. Principles

The SEA finds rules from an examples base. Our aim is not to describe the mathematical algorithms of this software (M. Sebag,1989) but to present the approach, valid for any examples base.
The base contains examples of a problem. They may be found by observation of real cases or from the experience of an expert (breakdown of machines...). The examples need to be in sufficient number and representative of the problem, conveniently distributed.
The problem has the form (hypotheses or parameters $\Rightarrow$ conclusions), with in general known hypotheses and unknown conclusions. Each example of the base has the same form, but hypotheses and conclusions are known. The SEA finds rules from the examples and the application of these rules to the hypotheses of any unknown case gives the conclusions of this case.
Hypotheses and conclusions are described by components, which have the following properties :
- they must contain all the features of the example
- the cost of their computation must be low and in order to reduce the cost of their treatment,

their number must be reasonable

- they must be dependent only on the intrinsic properties of the example.

The forms of components are usually boolean terms, multivaluate terms, numerical terms or character strings.

The choice of a good system of components is important.

Real examples bases may contain noise. This noise may be due to the precision or the dispersion of physical observations, approximate computation of components, or other reasons as typing errors... The SEA eliminates a part of the noise by acceptance of an error rate, i.e. some examples of the base do not verify the rules.

Practically, the examples base is splitted into two parts, the learning base and the test base. The determination of the rules is made only on the learning base and the rules are qualified on the test base. Usually the quality of the rules is evaluated by the proportion of good responses in the classification of the examples of the test base. A response is good when the conclusion given by the rules base is in the same class as the real conclusion; otherwise the result is bad. In some cases the conclusion remains unknown; the result is taken null. The global result of the test base has the form

$$\alpha \ \% \ \text{good}; \ \beta \ \% \ \text{bad}; \ \gamma \ \% \ \text{null}.$$

Of course, a good rules base gives a high value of α.

2.2. Application to the numerical analysis of structures

We must first define what we want to evaluate with an expert system.

The most ambitious objective would be the obtention of a quasi-optimal spatial and temporal discretization which would give, for a given structure, a prescribed error at the lowest cost, or the lowest error at a prescribed cost.

An unpretentious objective would be the obtention of the error and the cost from a given structure and its discretization.

These objectives need to create an examples base in analysis of elastoplastic structures. Each example contains the full description of a structure, its loading, its material, its spatial and temporal discretizations, the cost of the numerical computation and the errors between the numerical results and the corresponding exact values.

2.3. Description of an example

The hypotheses of an example may be divided into

• "intrinsic" hypotheses, independent of any discretization i.e. geometry, loading, material constitutive law;

• other hypotheses and the conclusions, linked to the discretizations in space and time, the error and the cost of the computation.

The loading is the most difficult hypothese to describe; it can have multiple forms as concentrated or distributed loads, thermal gradients, assumed displacements, pressures, dynamical loadings... We propose to describe a stress field computed from this loading and the geometry of the structure, but with a material assumed elastic. The cost of this computation is low and the singular effects of both the geometry and the loading are taken into account by the elastic computation. The amplitude of the stress field may be characterized by the field of the Mises equivalent stresses; it can be written as

$$\phi = f(x,y)$$

and the components have to describe its shape.

- Description of the shape of a 2-D domain

The description of a plane domain is similar to the "pattern recognition" but we shall not use the mathematical techniques of this theory. We propose to describe a plane shape by non-dimensional geometrical invariants. For example the ratio

$$\frac{\text{perimeter of the domain}}{\text{perimeter of a circle which has the same area}} \quad \text{or the ratio} \quad \frac{S^2}{I_x + I_y}$$

may be used. Invariants of shape tensors of second, third and fourth order are also used; they are similar to the moments, "skewness", "curtosis" used in statistics or signal analysis. These components may be generalized to 3-D volumes.

- Description of the shape of a field $\phi = f(x,y)$
This field is defined on a plane domain; in our case this domain is of course the geometry of the structure.
We shall describe the shape of the surface by taking only the invariants of the volume contained between the base domain and the surface; the principle of these invariants is the same as presented for a 2-D domain and also may be extended to an hypersurface $\phi = f(x,y,z)$ for 3-D structures.
A description of these codes can be found in (J.M. Hablot, 1989).

- Description of a mesh
As for the loading we use a field $\phi = f(x,y)$ where ϕ is the element size, and the components are the invariants of this field.

- Other parameters
The last parameters of an example are described by
• the elastic and plastic coefficients for the behaviour;
• the mean and extreme values of elastic stress increments for the time discretization;
• the maximal global errors at the end of the loading for the errors;
• the ratio between elastoplastic and elastic computation times and CPU times for the cost.

3. EXAMPLES BASE OF ELASTOPLASTIC STRUCTURES

We need an example base of elastoplastic structures, with errors and costs associated to discretizations. To compute the error (defined as the difference between an exact field and the approximated corresponding field), we need exact solutions to various problems. In elastoplasticity, a few analytical solutions of problems are known. Their geometries are generally equivalent to one-dimensional problems (thick cylinders or spheres under pressure (R. Hill, 1950 or J. Mandel, 1966)). These solutions are not sufficient to build an examples base.

So we have been obliged to build one by using an *inverse method*.

3.1. Hypotheses
- General hypotheses
We use the hypotheses of the mechanics of continuous solid media. We limit the study to the case of small strains and purely mechanical problems.
- Mechanical hypotheses
We assume that the materials are homogeneous and isotropic. The constitutive laws are elastoplastic. We limit our study to the standard plasticity. We shall assume the Mises criterion and perfectly plastic or kinematic hardening mathematical representation. The elastoplastic material is virgin at the beginning of the loading.
The elastic behaviour has the form

$$\sigma = \mathbf{L}\,\epsilon \quad \text{or} \quad \epsilon = \mathbf{M}\,\sigma. \tag{1 a and b}$$

with σ the stress tensor, ϵ the strain tensor, $\mathbf{L}$ and $\mathbf{M}$ the elastic tensors computed from the Lamé coefficients λ and μ, or Young's modulus E and Poisson's ratio ν.
The Mises criterion is written

$$f(\sigma, \epsilon^p) = \frac{1}{2}(S - C\epsilon^p) : (S - C\epsilon^p) - y_0 \leq 0 \tag{2}$$

where we have $S = \text{dev}(\sigma) = $ deviatoric part of the stress tensor σ, $S_{ij} = \sigma_{ij} - \delta_{ij}\,\sigma_{kk}/3$, C is the hardening modulus, σ_0 the yield limit in uniaxial tension, $y_0 = \sigma_0^2/3$ and $\mathbf{a{:}b} = $ inner product of the tensors $\mathbf{a}$ and $\mathbf{b}$.
Equations of standard plasticity are
- stress/strain relation : $\sigma = \mathbf{L}\,(\epsilon - \epsilon^p)$ $\tag{3}$
or for the deviatoric part : $S = 2\mu\,(\epsilon' - \epsilon^p)$ $\tag{4}$

- with the flow rule : $\dot{\epsilon}^P = \lambda_p \dfrac{\partial f}{\partial \sigma}$ (5)

with $\lambda_p \geq 0$ only when $f(\sigma, \epsilon^P) = 0$.

3.2. Principle of the Inverse Method
- Exact fields

- Displacements

We start by taking a displacement field $u_i = f_u(x_j,t)$ with analytical functions f_u. These functions are continuous and have defined second derivatives with respect to the space and time variables.

- Strains

Strains are computed from displacements

$$\epsilon_{ij} = \frac{1}{2}\left[\frac{\partial u_i}{\partial x_j} + \frac{\partial u_j}{\partial x_i}\right]$$ (6)

- Stresses

Stresses are computed from strains using the material constitutive law. If in elasticity the relations are straightforward, in elastoplasticity, we need to perform incremental computations with projections onthe yield surfaces.

- Body forces

At last body forces X_i are obtained from the equilibrium equation

$$X_i = \quad -\frac{\partial \sigma_{ij}}{\partial x_j} \text{ (statical case) or}$$

$$-\frac{\partial \sigma_{ij}}{\partial x_j} + \rho \frac{\partial^2 u_i}{\partial t^2} \text{ (dynamical case)}$$ (7a and b)

We have defined 4 exact fields $\mathbf{u}$, ϵ, σ and $\mathbf{X}$ in $\mathbf{R}^2$ or $\mathbf{R}^3$.

- Particular solution

Now, in $\mathbf{R}^2$ or $\mathbf{R}^3$, we can choose any domain Ω of arbitrary shape and with the boundary $\partial\Omega$. This boundary may be splitted into $\partial_u\Omega$, where the displacements are taken from the previous displacement field and $\partial_F\Omega$, where we compute the surface forces associated to the previous stress field

$$F_s = \sigma_{ij}n_j$$ (8)

where n_j is the external normal on $\partial_F\Omega$ at the given point. In Ω the body forces are taken from the previous body forces field.

So, we have defined a classical initial boundary values problem, but for which, of course, we know the exact solution.

- Numerical solution

We give this problem to any finite element code to evaluate the numerical solution.

3.3. Stress and Strain Increments in Elastoplasticity

The computation of σ from ϵ is made by increments in elastoplasticity.

At a point P, at the time t_k, we have the strain tensor ϵ_k, the plastic strain tensor ϵ_k^P, the stress tensor σ_k and the inequality $f\left(\sigma_k, \epsilon_k^P\right) \leq 0$.

During the time increment Δt, we add a strain increment $\Delta\epsilon$ and we search the stress tensor σ_{k+1} which verifies (2) at time $t_{k+1} = t_k + \Delta t$.

The trial stress is $\sigma^* = \sigma_k + \Delta\sigma^{el}$; the trial stress increment is $\Delta\sigma^{el} = L.\Delta\epsilon$.

If $f\left(\sigma^*, \epsilon_k^P\right) \leq 0$, we write $\sigma_{k+1} = \sigma^*$; the problem is solved.

If $f\left(\sigma^*, \epsilon_k^P\right) > 0$, we need to find $\Delta\epsilon^P$ which verifies

$$f\left[\sigma_k + L.(\Delta\epsilon - \Delta\epsilon^P), \ \epsilon_k^P + \Delta\epsilon^P\right] = 0 \tag{9}$$

Two methods may be used to solve these equations.

- First method : radial projection

The principle is to use a first-order development of the evolution equations for the computation of $\Delta\epsilon^P$:

$$\Delta\epsilon^P = \Delta\lambda_p.\left(\frac{\partial f}{\partial\sigma}\right)_{\sigma \ = \ \sigma_{k+1}} \tag{10}$$

In the deviatoric stress space, we project the point σ^* on the convex set of plasticity at time t_{k+1} (implicit algorithm) in order to verify the criterion (9) (see Q. S. Nguyen, 1973 or J. Zarka, 1978).

In standard plasticity, the spheric part of the stress tensor is not changed

$$\sigma_{k+1} - S_{k+1} = \sigma^* - S^*$$

and the exact value of the deviatoric stress increment is

$$\Delta S^{ex} = \frac{1}{C_2}\left[\Delta S^{el}\left(\frac{C}{2\mu} + \eta\right) + \left(S_k - C\epsilon_k^P\right)(\eta - 1)\right] \tag{11}$$

with $C_2 = 1 + C/2\mu$; $C_1 = C_2/2\mu$; $\eta = \sqrt{(2y_0)/\xi}$ and $\xi = \left(S^* - C\epsilon_k^P\right) : \left(S^* - C\epsilon_k^P\right)$.

For a numerical use with reasonable accuracy, the strain increments $\Delta\epsilon$ must be "not too far" from the convex set of plasticity, except during a radial loading.

- Second method : analytical integration.

For the Mises criterion and for a finite strain increment $\Delta\epsilon$ with the form $\Delta\epsilon = \dot{\epsilon}\ \Delta t$ the analytical integration of the evolution equations is possible; thus the exact value of the plastic strain and stress increments can be computed (R. D. Krieg & D. B. Krieg, 1977 or P. J. Yoder & R. G. Whirley, 1984).

The complete formulation of this method can be found in (J. M. Hablot, 1989).

3.4. Computation of the Body Forces in Elastoplasticity

The computation of the body forces needs the derivatives of the strains with regard to spatial coordinates; of course the plasticity will modify their values. These values are associated to the method used to compute the stress increments.

Here, we show only the idea for the radial projection.

The derivative of (11) with respect to the space variables is

$$\frac{\partial(\Delta S^{ex})}{\partial x} = \frac{1}{C_2}\left[\left(\frac{C}{2\mu} + \eta\right)\frac{\partial(\Delta S^{el})}{\partial x} + (\eta - 1)\frac{\partial S_1}{\partial x} + S_2\ \frac{\partial\eta}{\partial x}\right] \tag{12}$$

with $\sigma_1 = \sigma_k - C\epsilon_k^P$, $\sigma_2 = \sigma^* - C\epsilon_k^P$ and S_1 and S_2 their deviatoric parts.

This derivative presents a discontinuity at the boundary between the elastic zone and the plastic zone; the body forces will have also a jump.

Let us assume that the point σ_k is a contact point, $f(\sigma_k, \epsilon_k^P) = 0^-$ (the point reaches the boundary of the convex set, without crossing it). We add a strain increment $\Delta\epsilon$ such that

$f(\sigma_k + L.\ \Delta\epsilon, \epsilon_k^P) > 0$ and such that $\|\ \Delta\epsilon\ \|$ tends towards zero.

The expression (12) becomes

$$\frac{\partial(\Delta S^{ex})}{\partial x} = -\frac{1}{2y_0 C_2}\left[S_1 : \frac{\partial S_1}{\partial x}\right] S_1 \tag{13}$$

and generally does not tend towards zero with $\|\ \Delta\epsilon\ \|$. This residual value gives the jump of the body forces, at the crossing of the boundary between elastic and plastic zones.

An expression exists also for the analytical integration; we refer to (J. M. Hablot, 1989) for more details.

3.5. Practical Way to build an Element of the Examples Base

The exact solution i.e. its 4 fields $\mathbf{u}$, ϵ, σ et $\mathbf{X}$ are known in the space and time. The domain Ω and its boundaries $\partial_F\Omega$ and $\partial_u\Omega$ are given. An example is built by the following scheme :
i) Choice of one or more spatial and temporal discretizations
ii) For each discretization

> a) data preprocessing for the numerical computation
> discretization, loading, material properties with a classical input or a specialized pre-processor
> b) numerical elastoplastic analysis with these data
> c) computation of local and global errors by difference between the approximate field given by the code, and the exact corresponding field
> d) evaluation of the cost of the computation
> e) computation of the components of the example and storage of them in the examples base.

4. USING OF THE EXAMPLES BASE

4.1. Benchmarks

Exact solutions can be used for testing of computer codes, for finding formulation or programmation errors or comparing the performances of some algorithms.

4.2. Training of Users

Didactic examples may be chosen for the training of new users of an industrial code.

4.3. Verification of some Rules of Optimal Mesh Generation

Exact solutions allow the test of some rules of optimal mesh generation, given by several authors in the literature. Other parameters of the computations are kept constant. With the help of an automatic mesh generator (G. Coffignal, 1987), kindly given by this author, we build series of meshes respecting these rules. The rule which gives for a majority of examples the lowest error for a constant number of DOF, is taken as the best rule. This procedure does not need the use of the SEA but does not give the error.

For example, D. J. Turcke and G. M. McNeice (1974) proposed to place the nodes on iso-SED lines, where SED is strain energy density.

We generalize this rule by taking

$$\frac{h}{r_0} = \frac{k}{(r_0 \parallel \mathrm{grad}\ \phi \parallel)^\alpha} \tag{14}$$

where r_0 is a caracteristical length of the structure, ϕ a quantity evaluated on a coarse mesh of the structure, k an unknown constant and α an unknown exponent. In the Turcke and McNeice case, ϕ is the SED and $\alpha = 1$.

Other rules may be searched in this way, with other values of ϕ than the SED. In elastoplasticity, the evaluation of such quantities may be costly; thus we use only quantities computed with the structure assumed elastic.

Application

We consider an elementary example in plane strain with a rectangular domain Ω and a polynomial displacement function

$$u = [\ -0.032\ x^2y + 0.16\ xy\ ]\ \zeta(t)$$
$$v = [\ 0.032\ xy^2 - 0.08y^2\ ]\ \zeta(t)$$

where $\zeta(t)$ is a piecewise linear function with $\zeta(t = 0) = 0$, $\zeta(t = 20) = 0.004$, $\zeta(t = 60) = 0.16$ and $\zeta(t = 100) = -0.016$.

The mechanical constants are E = 216000 MPa, C = 7200 MPa, $\nu = 0.2$ and $\sigma_0 = 400$ MPa. The exact stress tensor computed from these functions is purely deviatoric in all the space. The loading on the domain Ω is elastic then elastoplastic between times 20 and 60; an elastic unloading then an elastoplastic reloading occur between times 60 and 100.

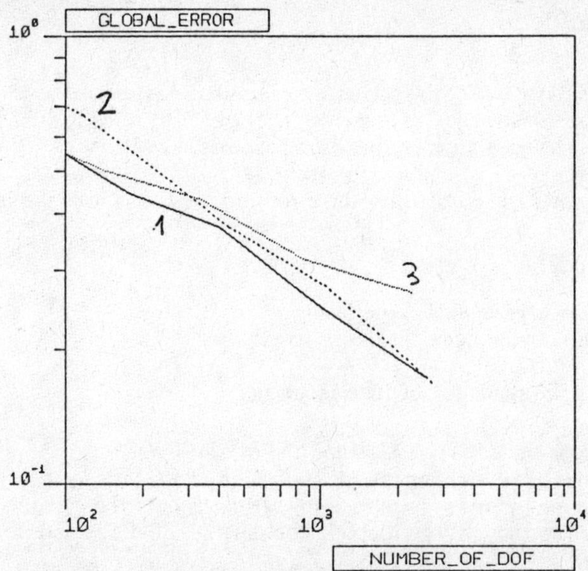

Fig. 1. Errors versus the number of degrees of freedom (DOF)

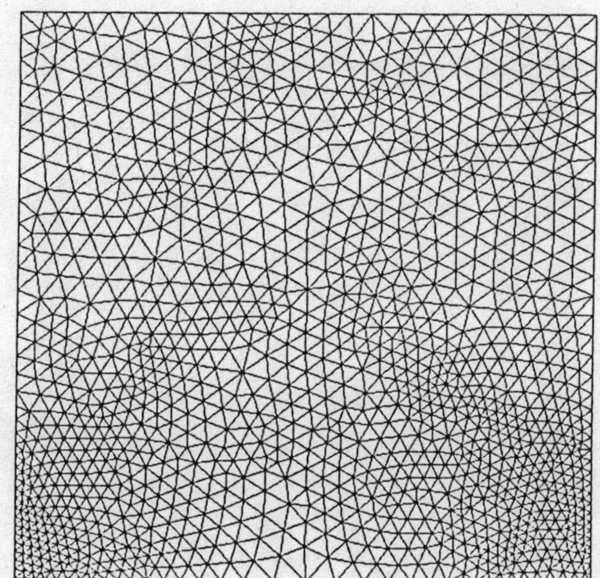

Mesh with 2505 elements, 1319 nodes and an error equal to 24.5%

Fig. 2. An example of mesh found with (14) and $\alpha = 0.5$

The errors described in Fig. 1 for three meshing rules are computed for the quantity σ. Fig. 2 shows an example of mesh given by the best rule, which is (14) with ϕ = Mises equivalent stress and $\alpha = 0.5$.

4.4. Use of the examples base by SEA

4.4.1. General case

- Obtention of a rule base

The examples base is given. Its analysis with SEA allows a generation of clear rules. The rules link components with relations. The construction of the rules may be costly but it is made only once; after finding the rules we do not need anymore the examples base.

- Practical use for an unknown case

We assume that the rules base is created. The rules give the error and the cost versus the intrinsic components and the discretization.

i) Before the use of the rules base

      a) Data preparation

            - the structure : geometry, loading, material

            - computation of the intrinsic components

            - error or cost.

      b) For a given discretization, computation of its components.

ii) Using the rules base

      a) computation of the error or the cost with SEA and the previous rules

      b) if the results are not correct, try with another discretization and return to i) b).

The cost of the evaluation of the error by this procedure is low. It needs only the computation of the components, approximately close to the cost of an elastic computation of the structure.

4.4.2. One example

We present the first results obtained with the SEA. We limit this case to the search of the error versus the mesh, all other features are kept constant : geometry, loading, material, time discretization. This gives 49 components for the hypotheses and 1 for the conclusion.

Geometry, loading, material properties are the same as the Application of §4.3. The time step has the value 5.

The examples base contains 42 meshes and is splitted into 30 examples in the learning base and 12 in the test base.

The repartition of the errors versus the number of DOF for all the elements of the examples base is shown on Fig. 3; one can see that all qualities of meshes are present.

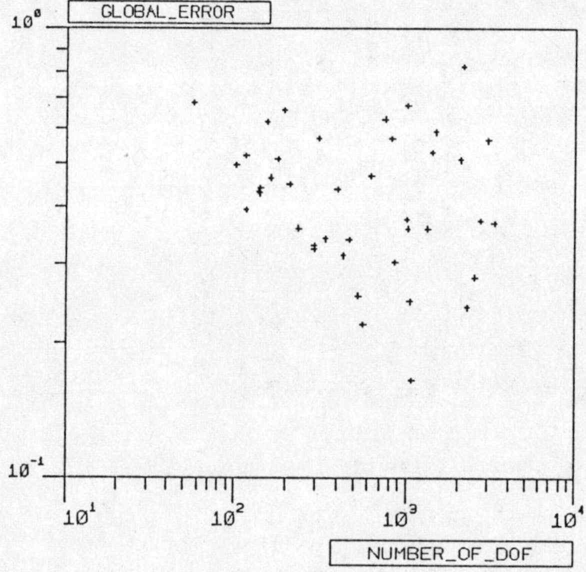

Fig. 3. Errors versus number of DOF for the meshes of the examples base

A rules base may be generated for several classifications of the conclusion, i.e. number of classes and values of the borders of the classes. Of course a learning base of only 30 examples allows not to separate the conclusion into a too big number of classes; we try with 3 classes : smaller than 25%, between 25% and 45% or greater than 45%.
The "statistical" results of the learning and test bases are on Table 1

Base	good	bad	null
Test base	84	8	8
Learning base	100	0	0

Table 1: *Classification of the bases*

But an engineer needs not statistical results but the value of the error. So we compute the intervals of error found for some examples of the test base :

Example	Number of DOF	Real error %	Estimated error %
1	398	37.8	25 to 45
2	858	34.2	25 to 45
3	3290	24	< 25
4	308	47	undefined
5	154	49.5	> 45
6	2082	25.5	< 25 §
7	3024	53	> 45

undefined = no conclusion is given by SEA for this example
§ = false conclusion

Table 2 : *Errors given by SEA for some examples of the test base*

We can see that it is possible to foresee rather well the level of the error. We recall that, of course, other parameters and not only tne number of degrees of freedom are taking part in this a priori evaluation. Of course, other cases will be tested.

ACKNOWLEDGEMENTS

This study was supported in part by PSA Etudes et Recherches and by ANRT through a CIFRE contract. The numerical computer code used as an example is Dynaflow, from Princeton University.

REFERENCES

Coffignal, G., 1987, "Optimisation et Fiabilité des Calculs Eléments Finis en Elastoplasticité", thèse de doctorat ès sciences, Université Paris VI, Paris.

Hablot, J.M., 1989, "Solutions Analytiques Exactes en Elastoplasticité", thèse Ecole Nationale des Ponts et Chaussées, Paris, to appear.

Hill, R., 1950, *Mathematical Theory of Plasticity*, Clarendon Press, Oxford.

Krieg, R.D. and Krieg, D.B., 1977, "Accuracy of Numerical Solution Methods for the Elastic-Perfectly Plastic Model", *J. of Pressure Vessel Technology*, November, pp. 510-515.

Mandel, J., 1966, *Mécanique des Milieux Continus*, Gauthier-Villars, Paris.

Nguyen, Q.S., 1973, "Contribution à la Théorie Macroscopique de l'Elastoplasticité avec Ecrouissage", thesis, Paris.

Schoenauer, M. and Sebag, M., 1988, "Generation of Rules with Certainty and Confidence Factors from Incomplete and Incoherent Learning Bases", *Proceedings of the European Knowledge Acquisition Workshop*, Bonn.

Sebag, M., 1989, "Apprentissage Multicouche", Paris Dauphine thesis, to appear.

Turcke, D.J. and McNeice, G.M., 1974, "Guidelines for Selecting Finite Element Grids based on an Optimization Study", *Computers and Structures*, vol.4, pp. 499-519.

Yoder, P.J. and Whirley, R.G., 1984, "On the Numerical Implementation of Elastoplastic Models", *J. of Applied Mechanics*, vol. 51, June, pp. 283-288.

Zarka, J., 1978, *Calcul des Structures*, course.

Expert Systems within the German HDR-programm for the
Assessment of the Integrity of Nuclear Components-
Correlation of NDT-Results with Acoustic Emission
Measurements and Fracture Mechanical Evaluations

G. Deuster, F. Walte
Fraunhofer-Institut für zerstörungsfreie Prüfverfahren,
Universität des Saarlandes, Gebäude 37, D-6600 Saarbrücken

1. INTRODUCTION

Within the German HDR-programme simulations of different accident
conditions are performed in a real reactor plant out of operation. Pressurized thermal shock tests on a saturated steam
outlet nozzle have been conducted under simulated realistic
operation conditions (t = 300° C, p = 11 MPa) for the past
three years. On the one hand the tests are to improve our understanding of crack formations and crack growth under thermal
shock conditions and on the other hand to evaluate and to improve
the suitability of non-destructive test methods for detection
and quantitative description of natural cracks and crack fields
both in on-lineapplication and off-line.

To facilitate the handling of the large amount of data out
of the material investigations, fracture mechanical and NDE-
evaluations an integrated central evaluation system (CES) was
developed enabling to work in parallel with all available technical information.

2. GENERAL FRAMEWORK OF THE CES-EXPERT-SYSTEM

The main block of an expert system consists of a inference
machine and the knowledge base (Fig. 1). In the inference machine
a strategy for the solution searching and a regulation interpretation are available. The data base represents the technical
and scientific knowledge as well the description of the problem
contexts and delivers indications for decisions "if - then"
on the basis of expert knowledge. Via an expert dialogue the
knowledge input and the examination on contradictions is performed, the dialogue component gives information about background
values, decision guidelines and intermediate results for the
user of the expert system.

3. HDR-CENTRAL-EVALUATION-SYSTEM (CES-EXPERT-SYSTEM)

Fig. 2 shows the block diagram of the HDR central evaluation
system (CES). The software components for the management, manipulation, processing and representation of the test and the
analytical data are detailed.

Firstly in the management phase the different data will be
identified and standardized. Then an administration scheme
is developed to store the different test and analytical data,
also giving the possibility for sorting and merging in a logi-
stical way.

For corrections and adaptions of the data a manipulation step
is possible.

In the processing phase the measured quantities out of the
different investigations can be referred to the different areas
of the primary circuit of this real reactor plant. In the software system also the functional relationships in between the
recorded data are available. Comparisons between test and analytical results can be performed. The most important step is
to create correlations and draw conclusions on the basis of
an integrated consideration of all available results.

Moreover the results can be represented in corresponding drawings, time and cyclic representations for the different loadings
are possible, also multidimensional graphics.

The central evaluation system (CES) described above is implemented in a data flow system of the total HDR-safety-programme.
(Fig. 3) This evaluation and documentation system with input
and output paths consists of a data acquisition system for
the different measurement quantities delivering also analytic
results of external institutions like non-destructive and destructive test data and material properties. The CES-output
is completed by plots, lists, tapes, test reports, drawings,
static and dynamic colour graphics, if necessary, after a plausibility control.

In Fig. 4 the specific measured quantities for the evaluation
and documentation system within the thermal shock experiments
are listed. Domain and consultation knowledge are the basis
for a strategy on the highest decision level (inference machine).

At the present time the optimization of the strategic searching
for solutions is still running. Special shell programmes and
refined symbolic programming languages are introduced. The
working data, the specific expert (domain) knowledge as well
as the consultation knowledge are complete.

4. THERMAL SHOCK EXPERIMENTS

On the saturated steam outlet nozzle for measurements thermo-
couples, acoustic emission transducers,stationary ultrasonic
probes, strain gauges, displacement transducers, clip gauges
and a potential drop array were installed (Fig. 5) on-line.

For <u>off-line</u>-measurements after the different trial phases
advanced US-inspection and analysis systems also in comparision
to convential US-testing were applied. Besides ALOK, SAFT and
special ultrasonic surface waves, potential drop and dye penetrant tests were performed (Fig. 6).

Now some examples to the representation capability of the CES-
system: Fig. 7 shows a roll-out of the inside nozzle area after
different cycle numbers with the dye penetration test; one
thermal shock was consisting out of 120 seconds cooling and
a following 240 seconds heating up time. It is shown that already
after 1752 thermal cycles a relevant damage was produced in
the surface near cladding area. After about 3000 thermal cycles
already a maximum crack depth of 22 mm was reached. By zooming
(Fig. 8) the expert system allows now to find out detailed
information about the damaging, for instance for the 270° angular
position. In this area already after 551 thermal shock cycles
the first surface breaking crack could be observed. With the
CES-system it is possible to follow the evaluation of this
specific crack up to about 3.000 cycles.

What concerns the crack front profile, side views in the different angular positions in the nozzle area are possible so that
one can compare the crack front profile development according
to the thermal shock loadings (Fig. 9). This result was found
by potential drop measurements.
In addition comparison between potential drop and fractographi-
cally recorded crack depth is possible after having taken out
a boat sample (Fig. 10). Depending on the angular position
also the results of other techniques as is shown in Fig. 11,
the profile measurement using ultrasonic i.d. SAFT-technique
from opposite directions can be represented. Satellites and
crack closure effects are disturbing the real crack depth results.

With a special software-programme also the acoustic emission
results-related to the dye penetrant and potential drop results -
after a location procedure can be compared graphically (Fig.
12). In order to understand the acoustic emission phenomenas
also the temperature in the cooling and heating phase at the
crack tip can be plotted out to study the crack propagation
behaviour. These curves were created after special finite element
calculations (Fig. 13), which are running within the CES-system.

Time distribution of acoustic emission events within the thermal
shock cycles can be plotted as a function of temperature and
strain conditions. Fig. 14 shows in contradiction with normal
fracture mechanical consideration (Paris-law) main crack growth
in the reheating phase. By destructive examination it could
be shown that time dependant corrosion crack growth was domina-
ting the normal cyclic crack growth.

For the understanding of the AE-crack type signals in this
context a finite element procedure beginning with the crack
starter was initiated. In the Fig. 15 the real NDT-result and
the calculations according to the different crack rate models
are shown what demonstrates the conservativity of the theoretical
algorithms.

The superposition of AE-results with the da/dt-curve considering
the temperature at the crack tip indicates the main crack propagation in the heating up phase as a time dependent corrosion
damage (Fig. 16, 16a).

Another CES-presentation (Fig. 17) delivers a three-dimensional-
diagram of the acoustic emission results versus time and angular
position; in the upper part all located signals in the first
10 seconds after start of cooling are shown. In the middle
the acoustic emission signals of friction type as expected
according to crack opening displacement measurements and calculations are in agreement with fracture mechanics. In the lower
picture acoustic emission-type crack growth is recorded and
again it can be shown, that during the cooling phase no normal
crack propagation according to Paris-law is occuring.

Out of an overall diagram which is showing the development
of 9 dominant cracks during the thermal shock loading can be
stated that in the austenitic cladding the first crack growth
runs comparably fast and after reaching the ferritic base material there is a flattening in the corresponding curve. What
concerns the crack depth in relation to the crack growth rate
again the expert system delivers on the basis of measurements
and calculations the rather high crack growth rate for the
cladding and the decreasing crack growth rate in the base material (Fig. 18).

5. SUMMARY

The described "Multiparameter computerized surveillance and
monitoring system" is handling the on-line and off-line measurement values, the theoretical calculation data of the fracture mechanical investigations, the material properties and
the NDE-results via functional relationship. The CES-software
programmes enable to call for a defined area in the primary
circuit and to let print out plots, lists or 3-dimensional
graphic diagrams for different measurement values. With the
integrated system it is possible to interprete the results
under all thinkable marginal conditions, in the sense of a
rather well equipped expert system.

The CES-expert system is universally applicable for surveillance
and monitoring of real plants. It can fullfill in the sense
of fit for purpose the requests for lifetime prediction especially for monitoring critical defects in complicated geometries,
which have normally unfavourable stress conditions. In addition
what concerns the availability aspect, the system will give
the basis for preventive maintenance and a long-term repair
planning and in-service inspection planning. Using the CES-system
the operation conditions can be steered and controlled dependent
on the system conditions. This means, it represents an intelligent operation control system which can be finally integrated
in a regulation circuit.

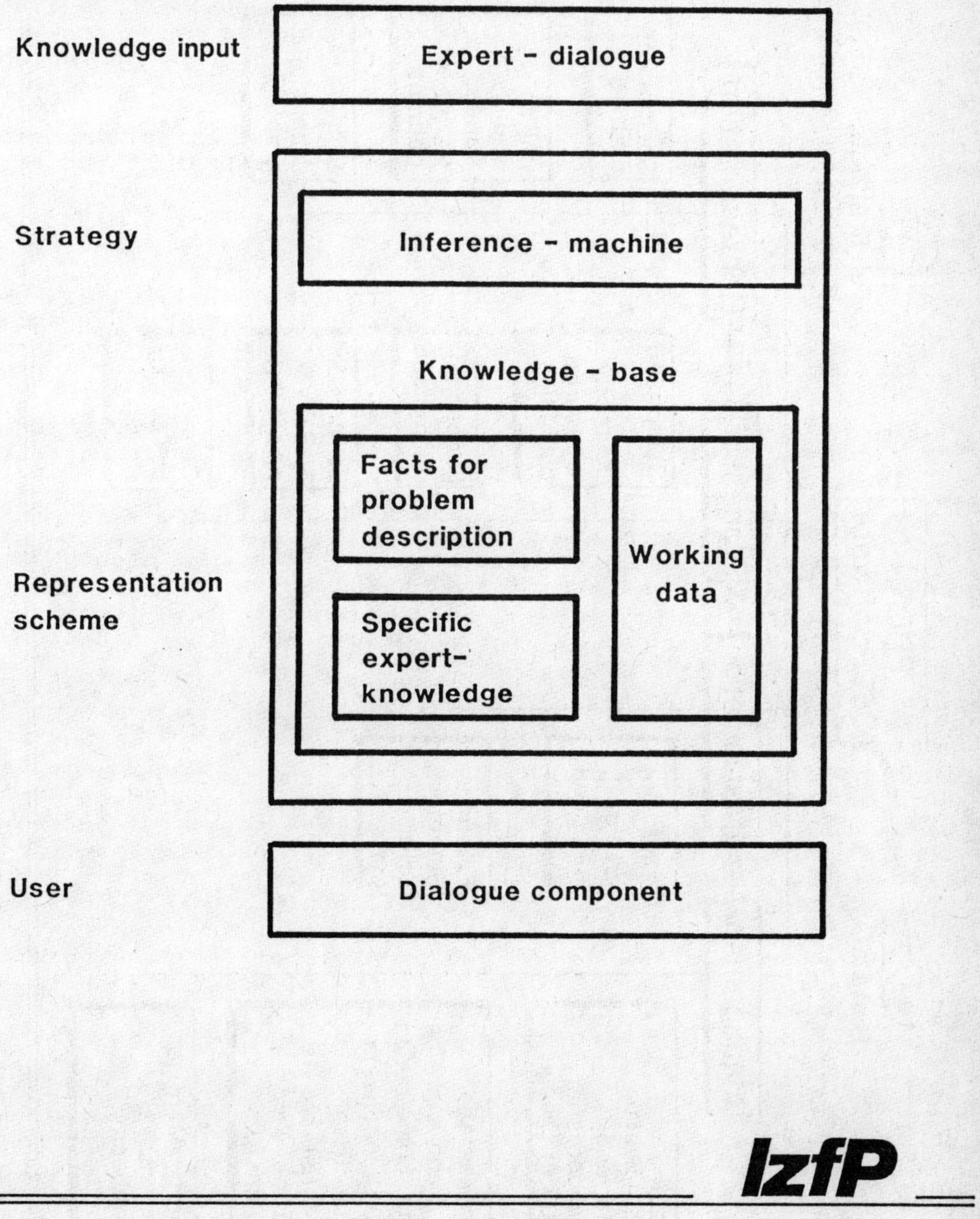

Expert – system   Fig. 1

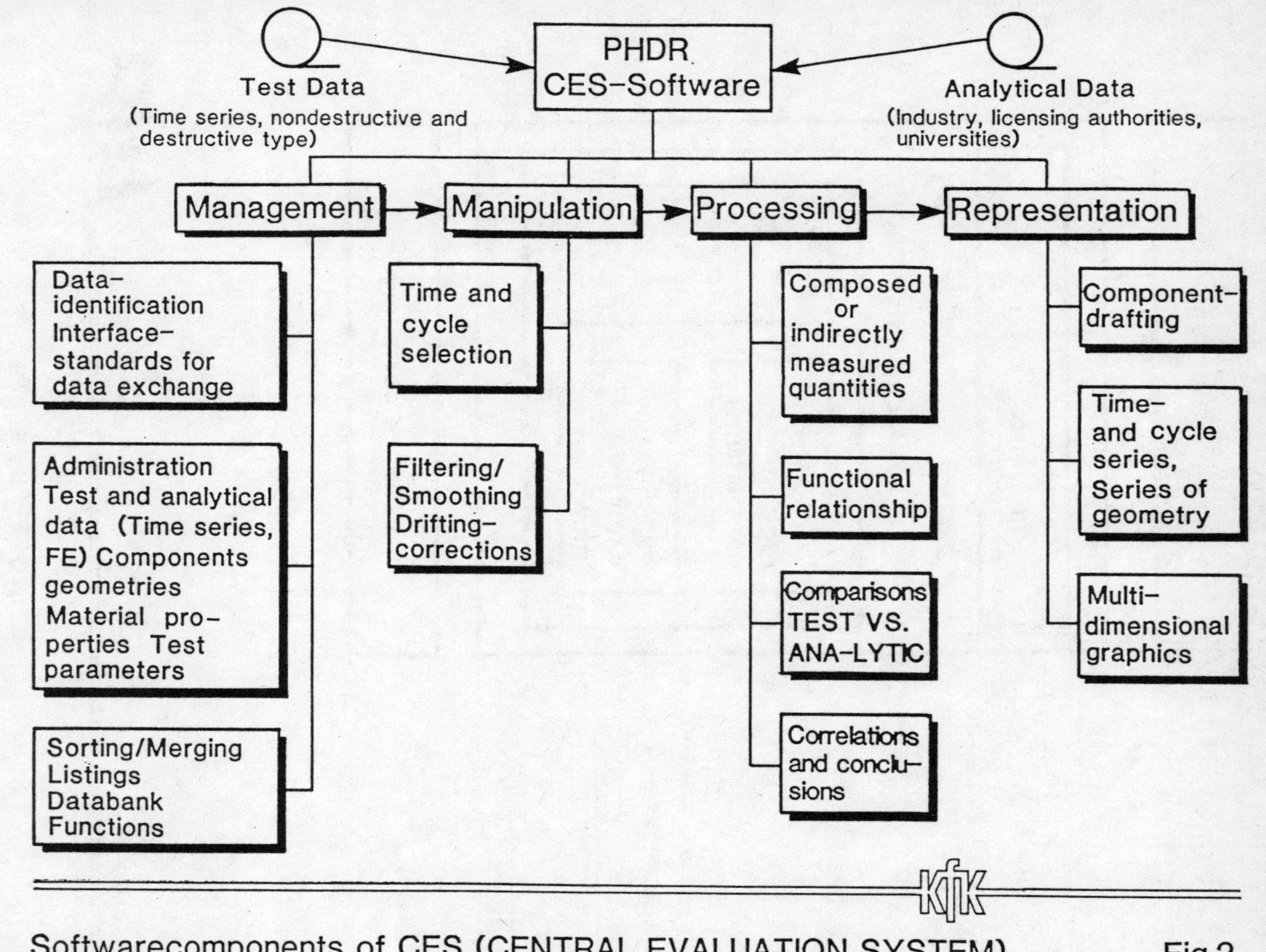

Softwarecomponents of CES (CENTRAL EVALUATION SYSTEM)   Fig.2

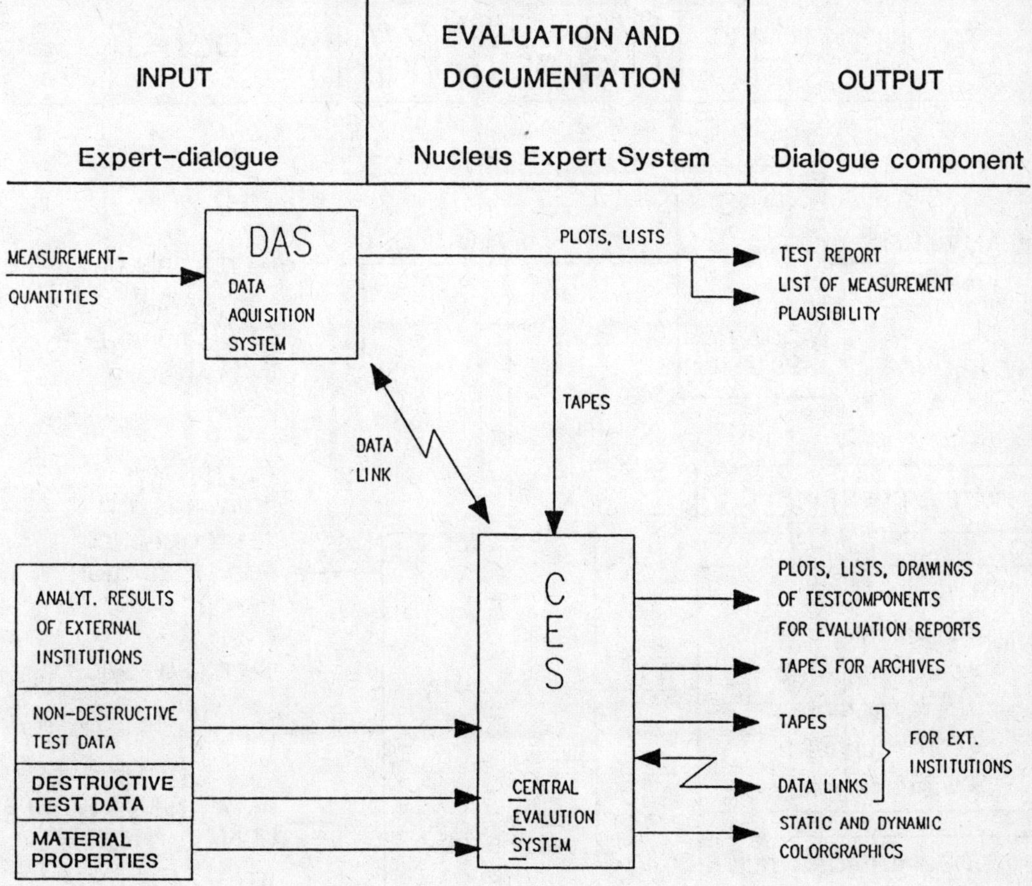

Data flow within the HDR-safety program   Fig.3

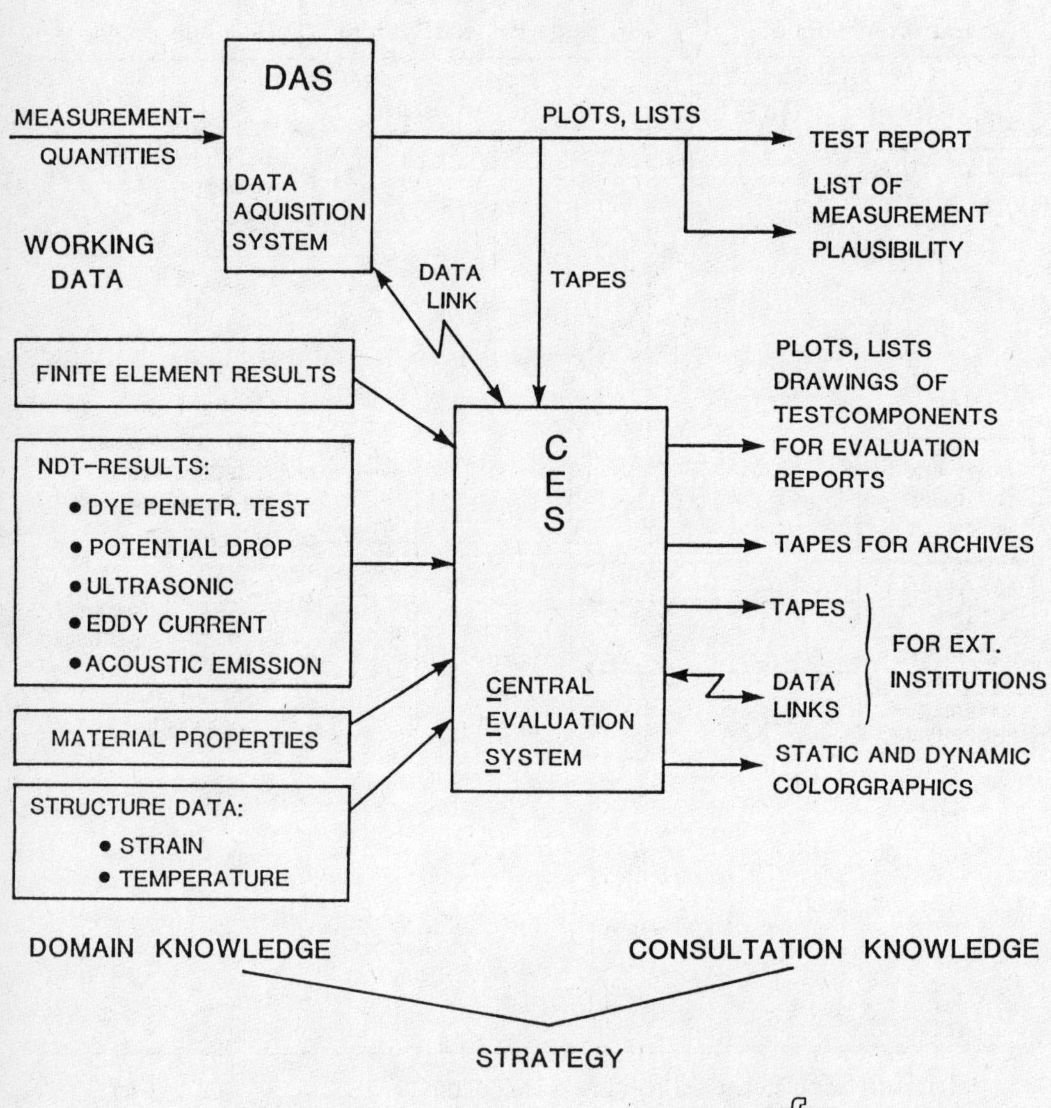

Data flow within the thermal shock experiment   Fig.4
(Reactor pressure vessel and nozzel)

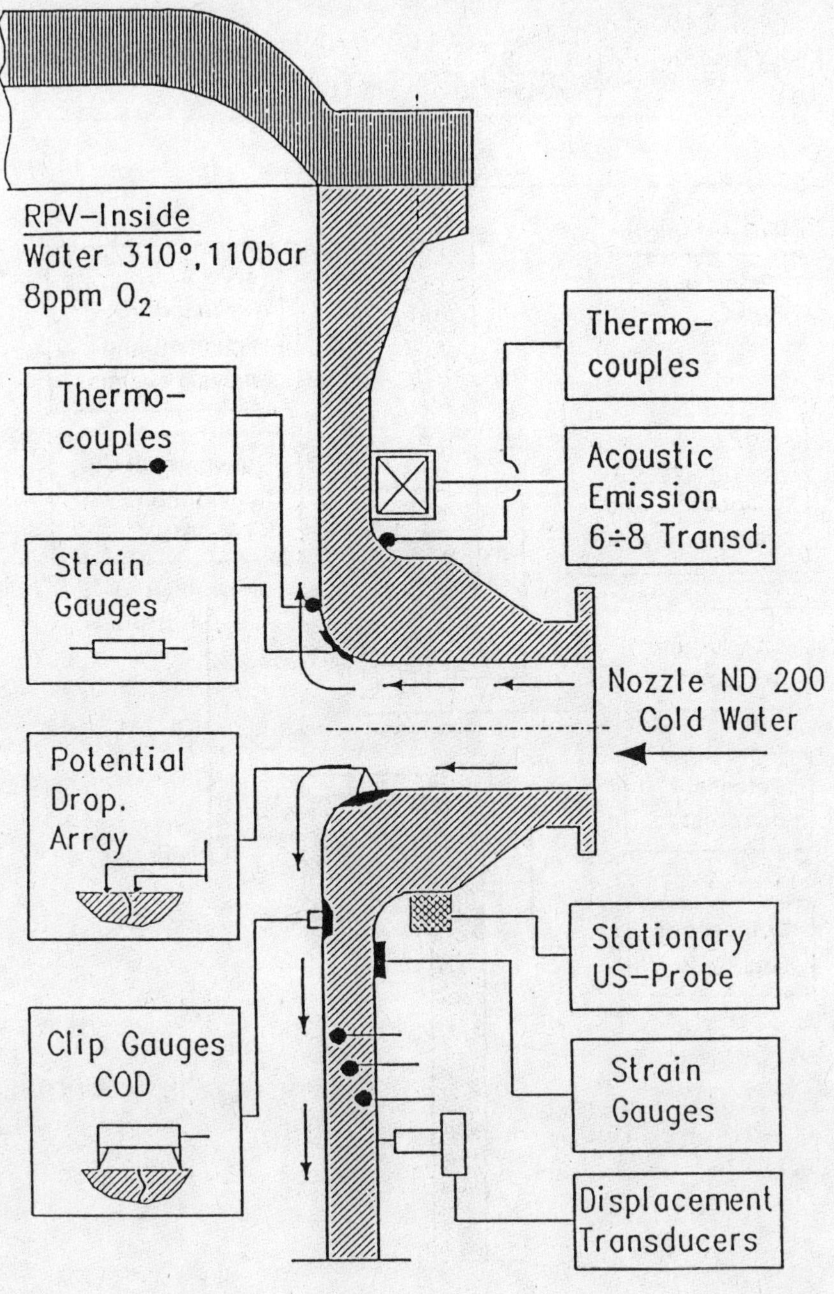

Survey of online measurements on HDR-Vessel   Fig.5

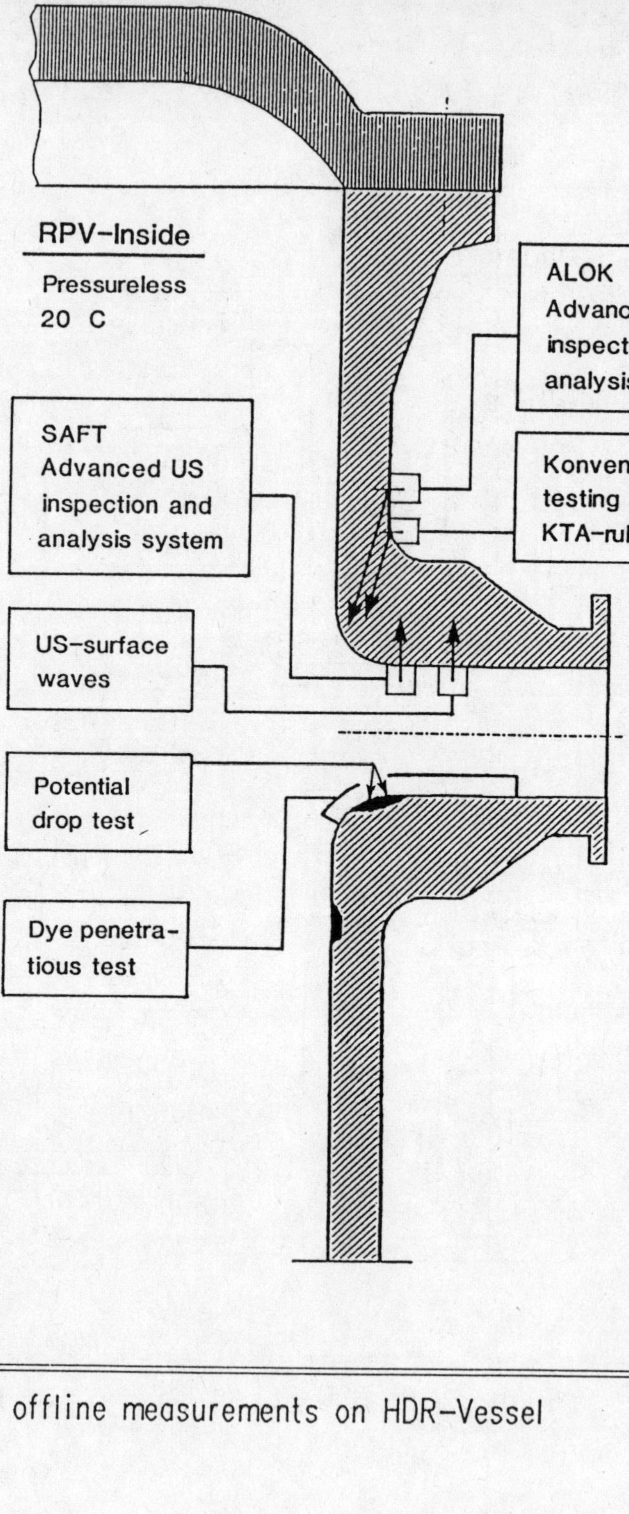

Survey of offline measurements on HDR-Vessel   Fig.6

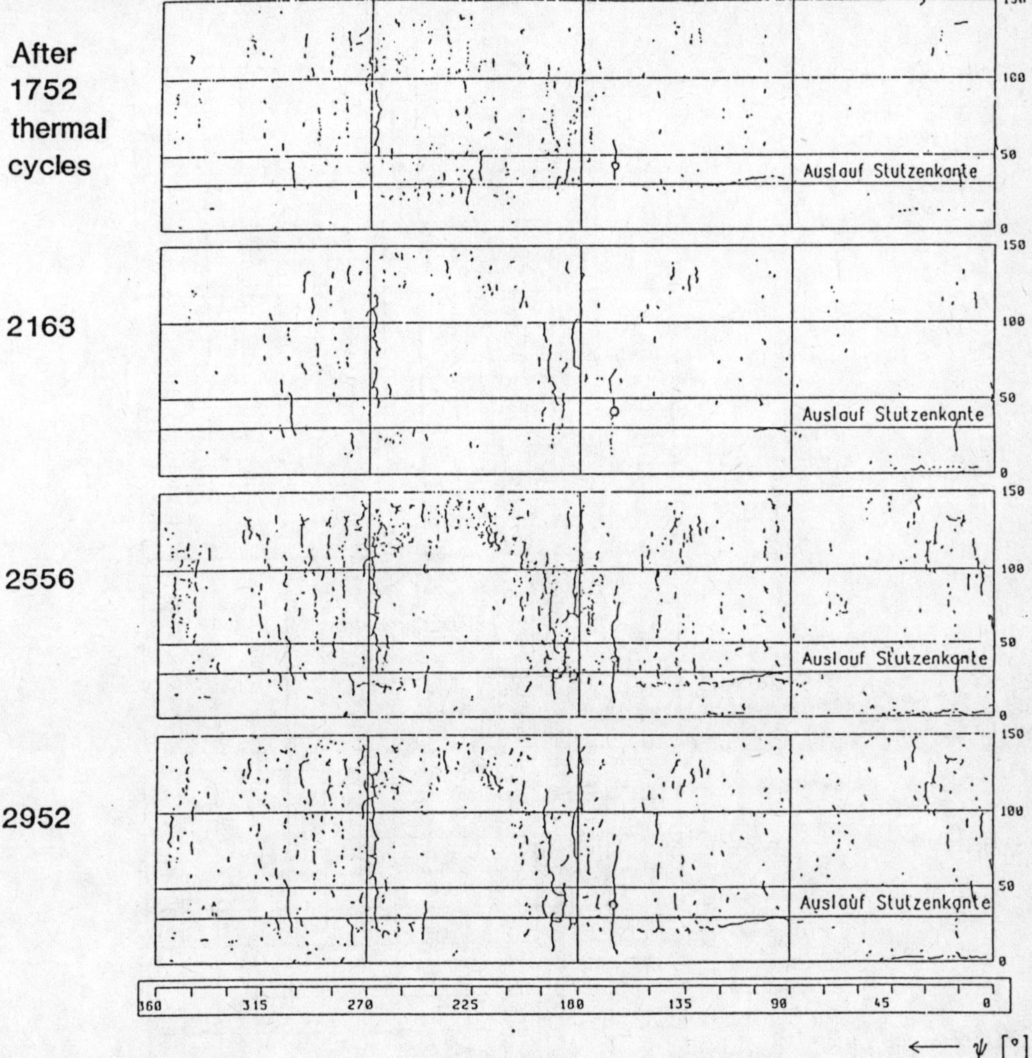

Results of dye penetration test (after different thermal shock cycles) on the inner nozzel surface

Fig.7

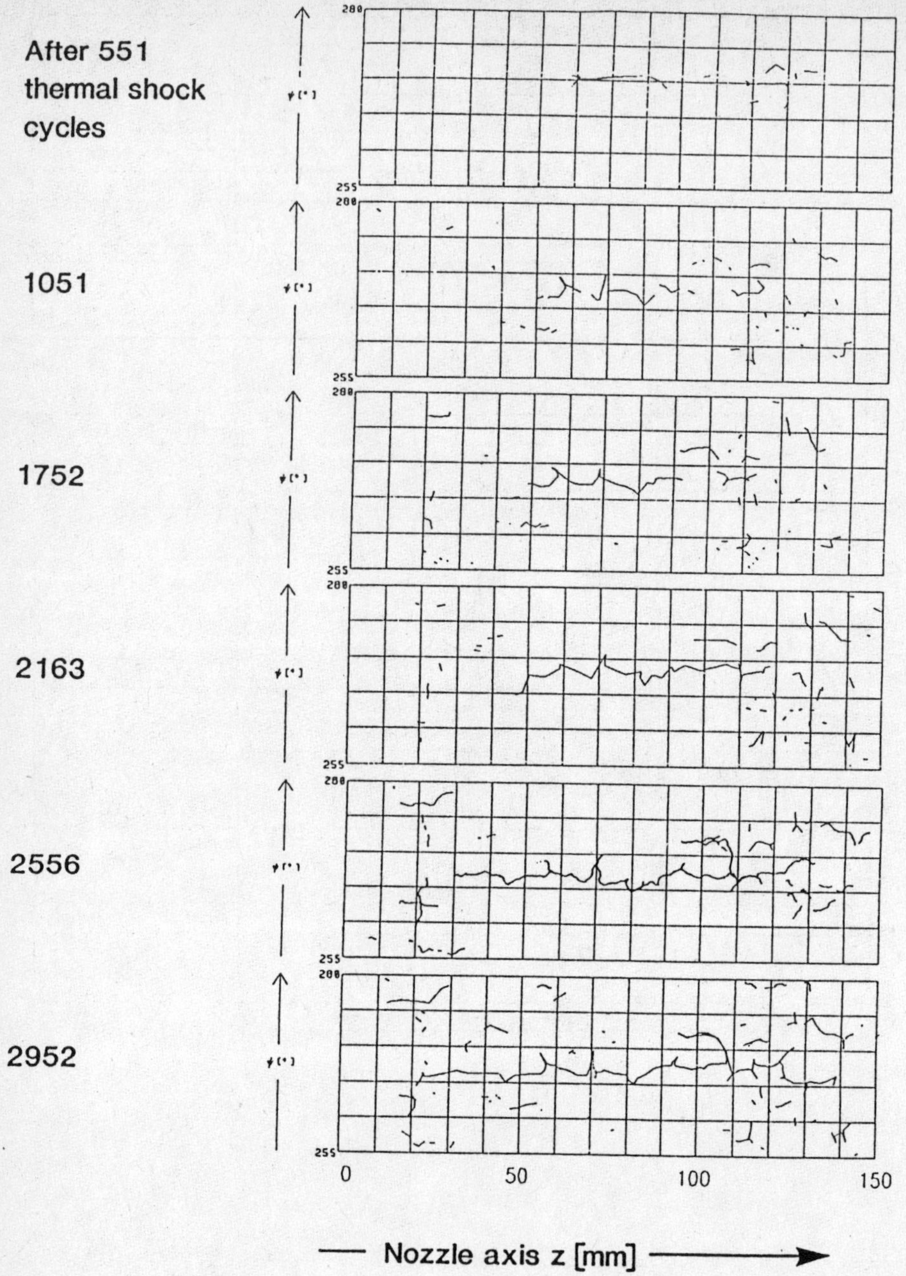

Results of dye penetration test (after different thermal shock cycles) on a part of inner nozzel surface ($\Psi = 270°$) using a zoom function

Fig.8

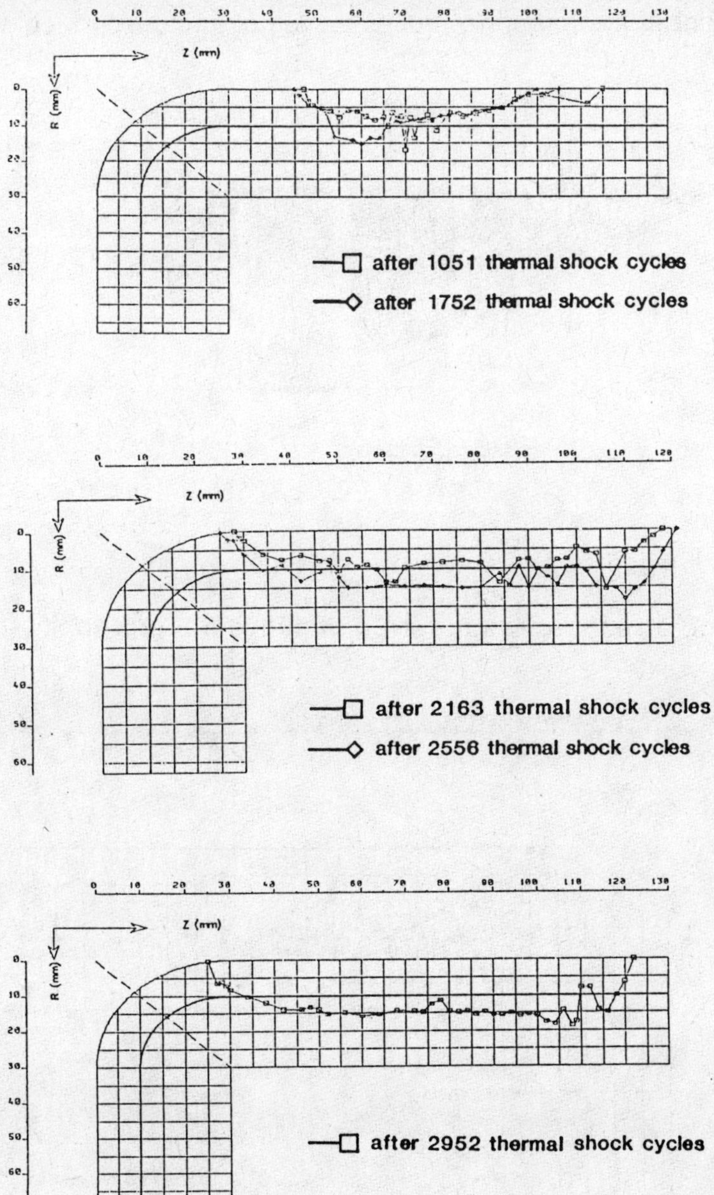

Crack front profil measurement using potential drop test after different thermal shock loadings (crack position : 270° circumference angle)

Fig.9

Original fractographie reconstruction of fracture surface

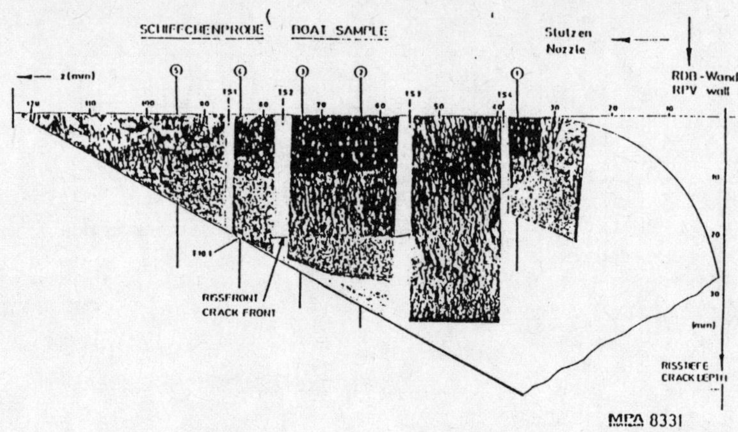

Crack arrest profiles received of above fractographie

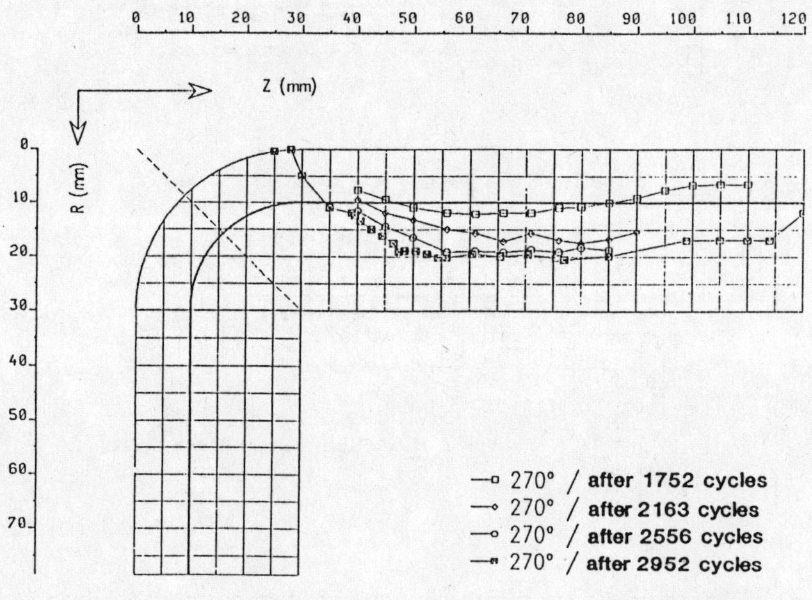

Storage of destructive test results

Fig. 10

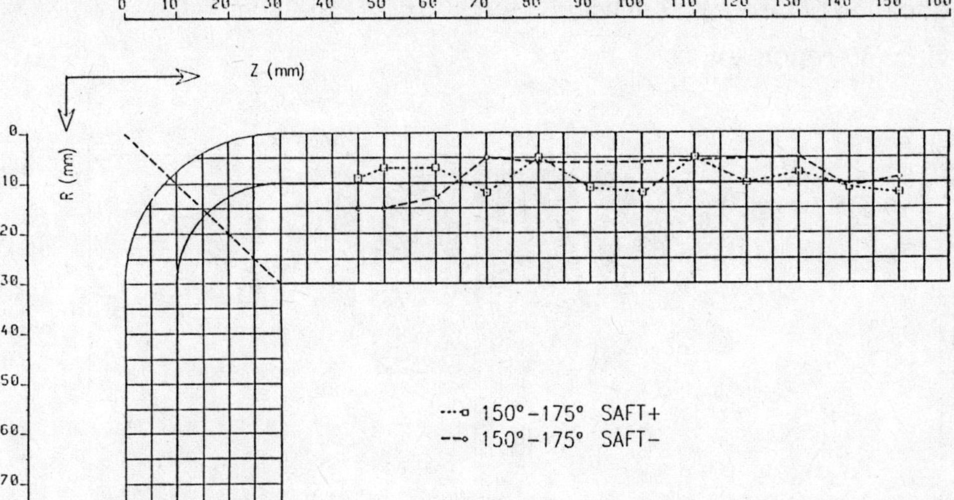

Crack profile measurement using ultrasonic
SAFT technique
(crack position : 170° circumference angle)

Fig.11

Locating of acoustic emission sources about the nozzle circumference angle

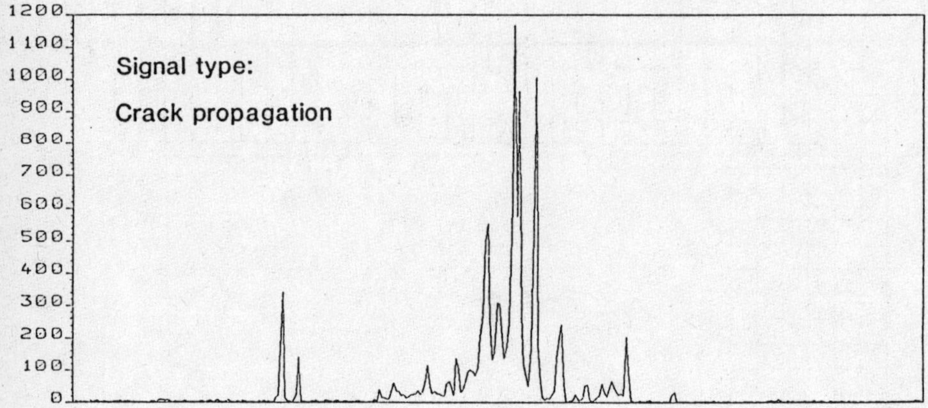

Result of dye penetratious test

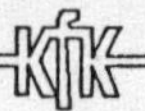

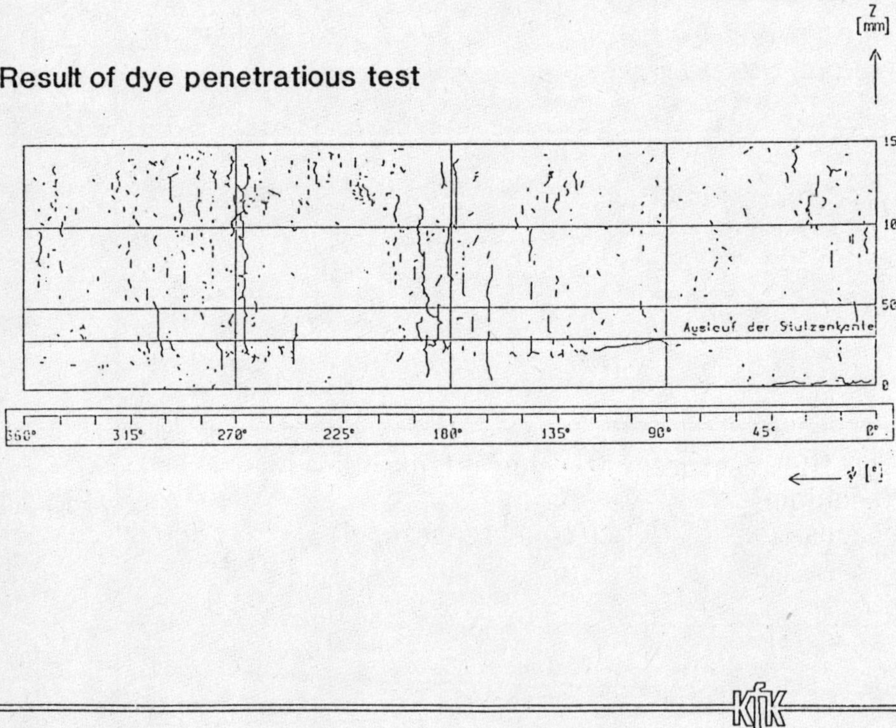

KfK

Locating results of acoustic emission inspection compared with dye penetration test

Fig.12

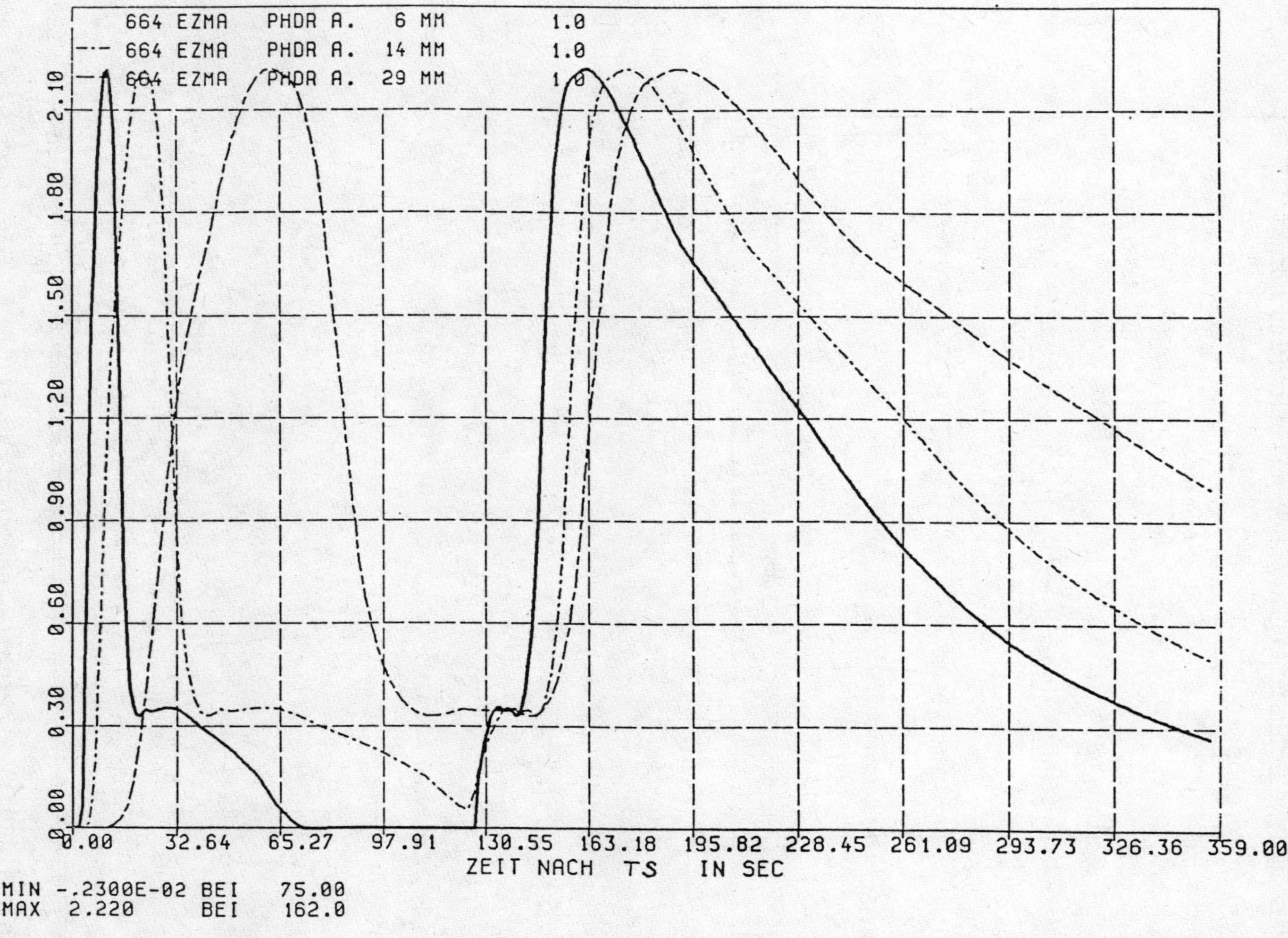

Temperature at the crack tip during one thermal cycle   Fig.13

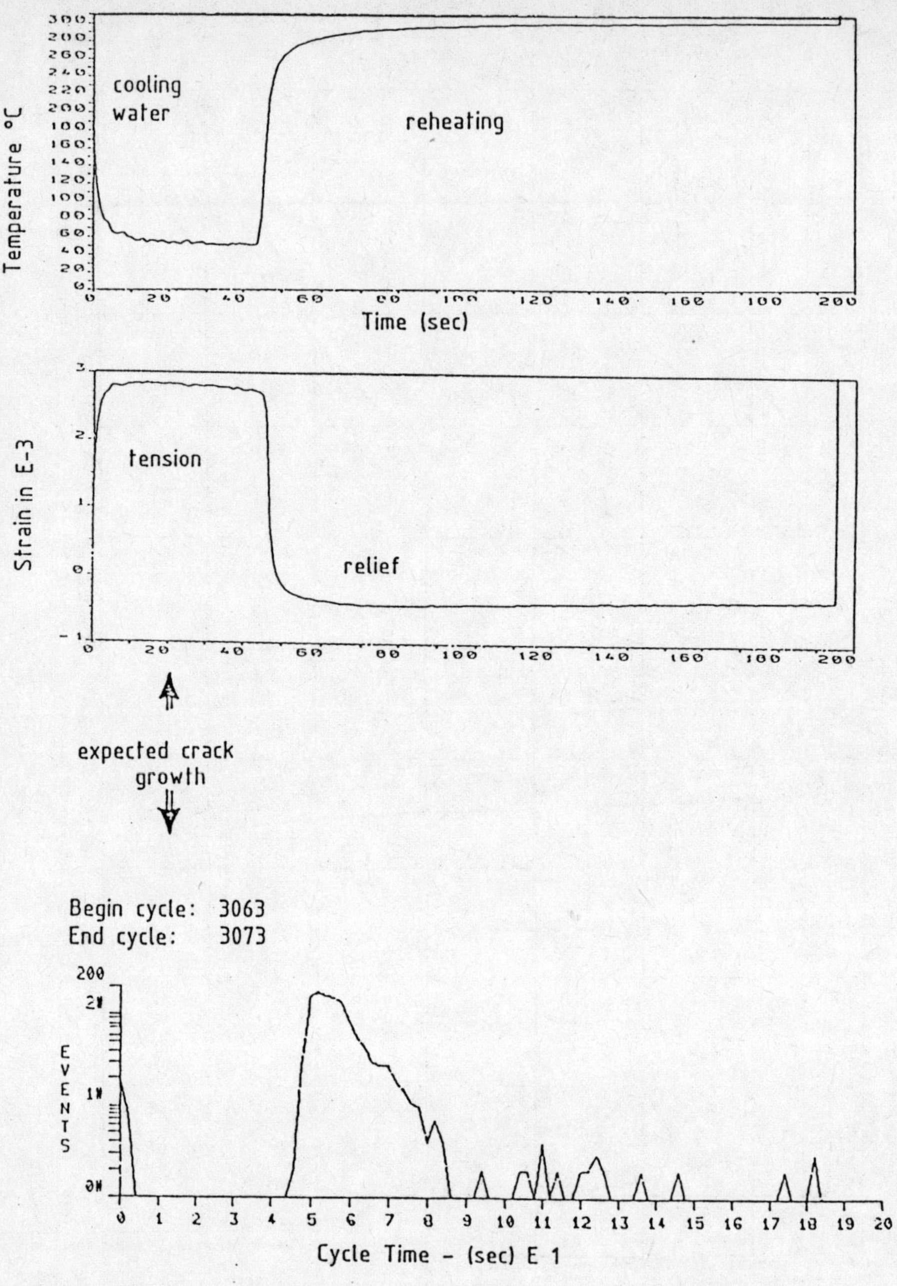

Time distribution of acoustic emission events within the cycles

Fig.14

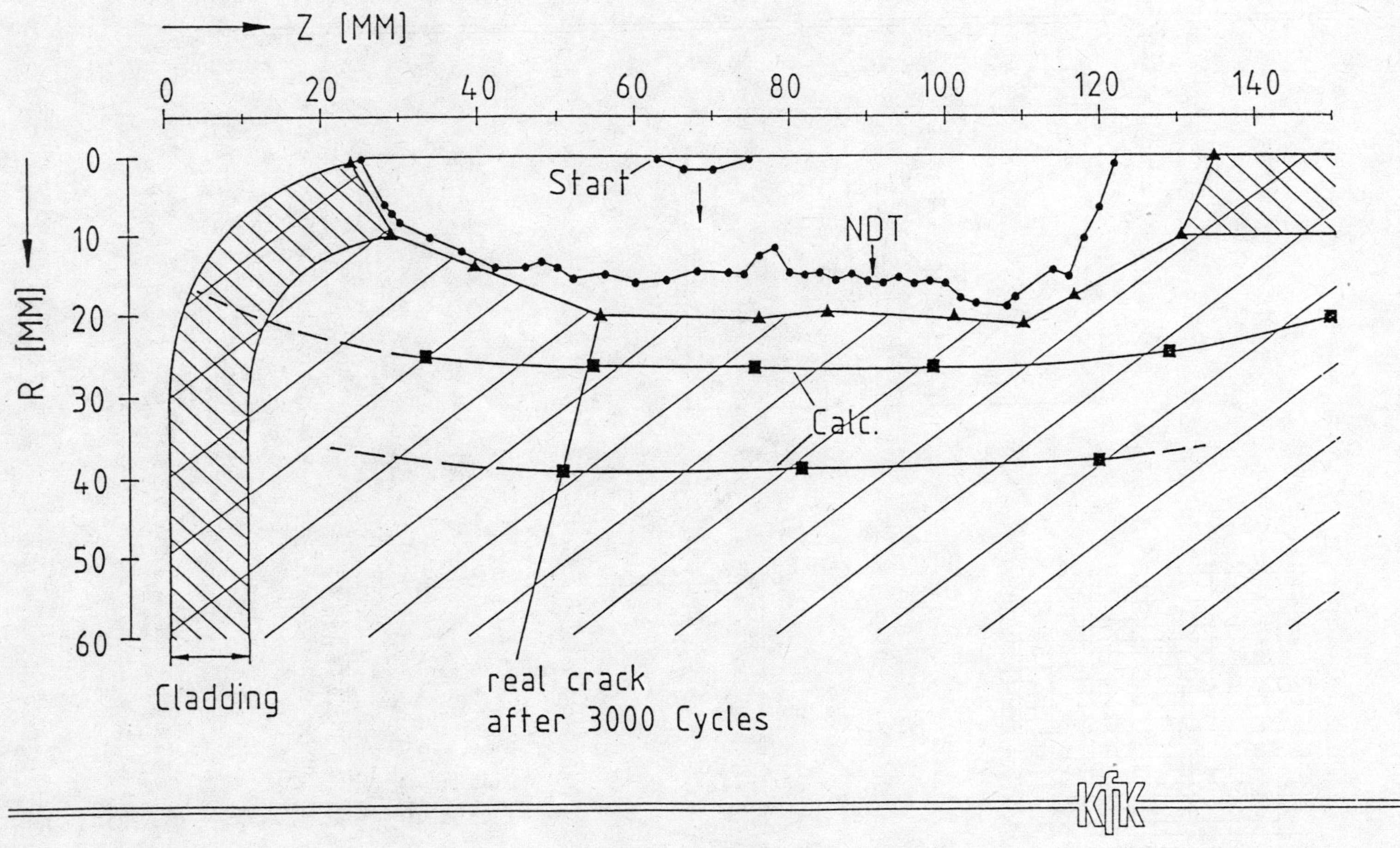

Thermal Shock Crack on RPV-Nozzle

Fig.15

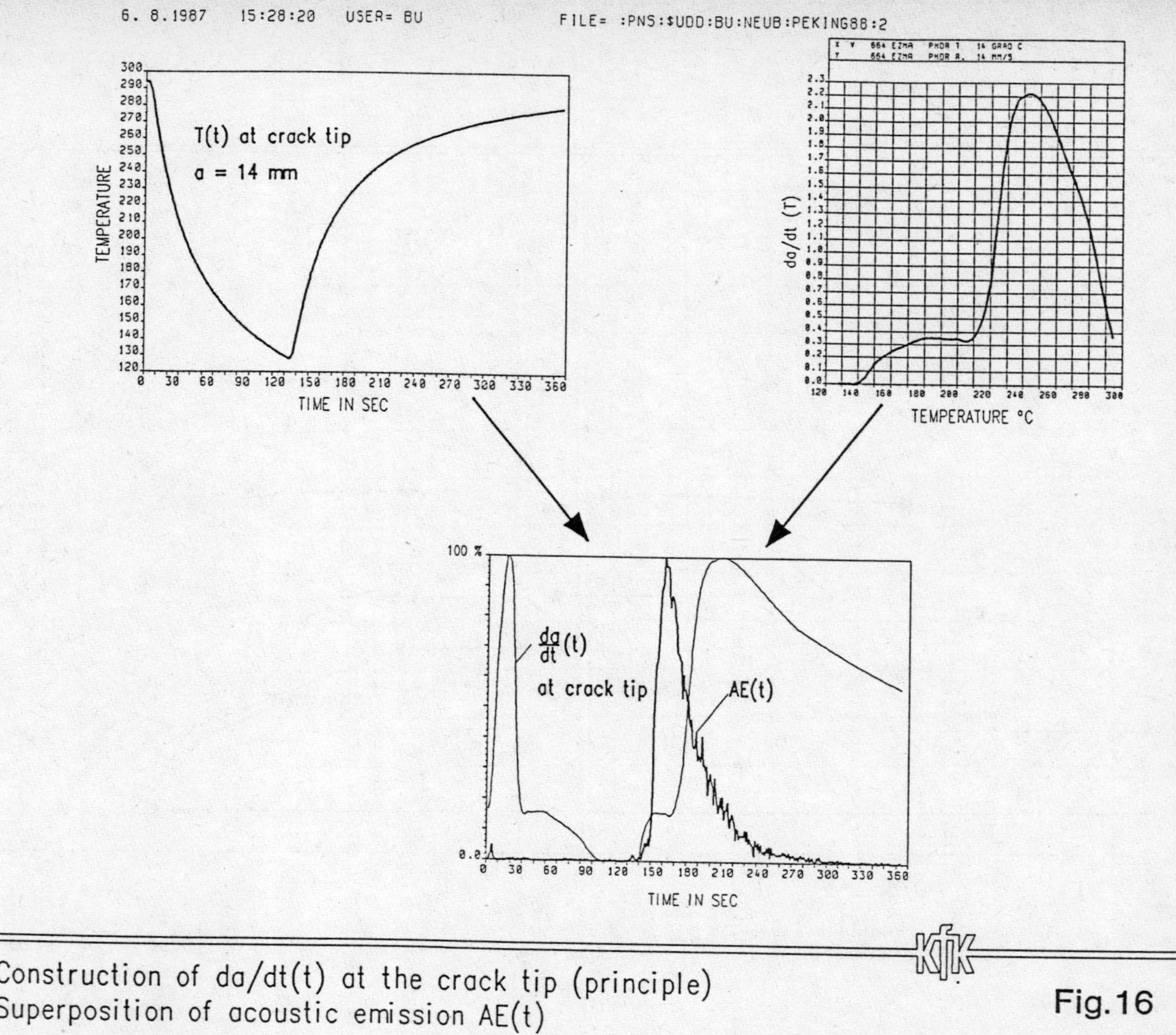

Construction of da/dt(t) at the crack tip (principle)
Superposition of acoustic emission AE(t)

Fig.16

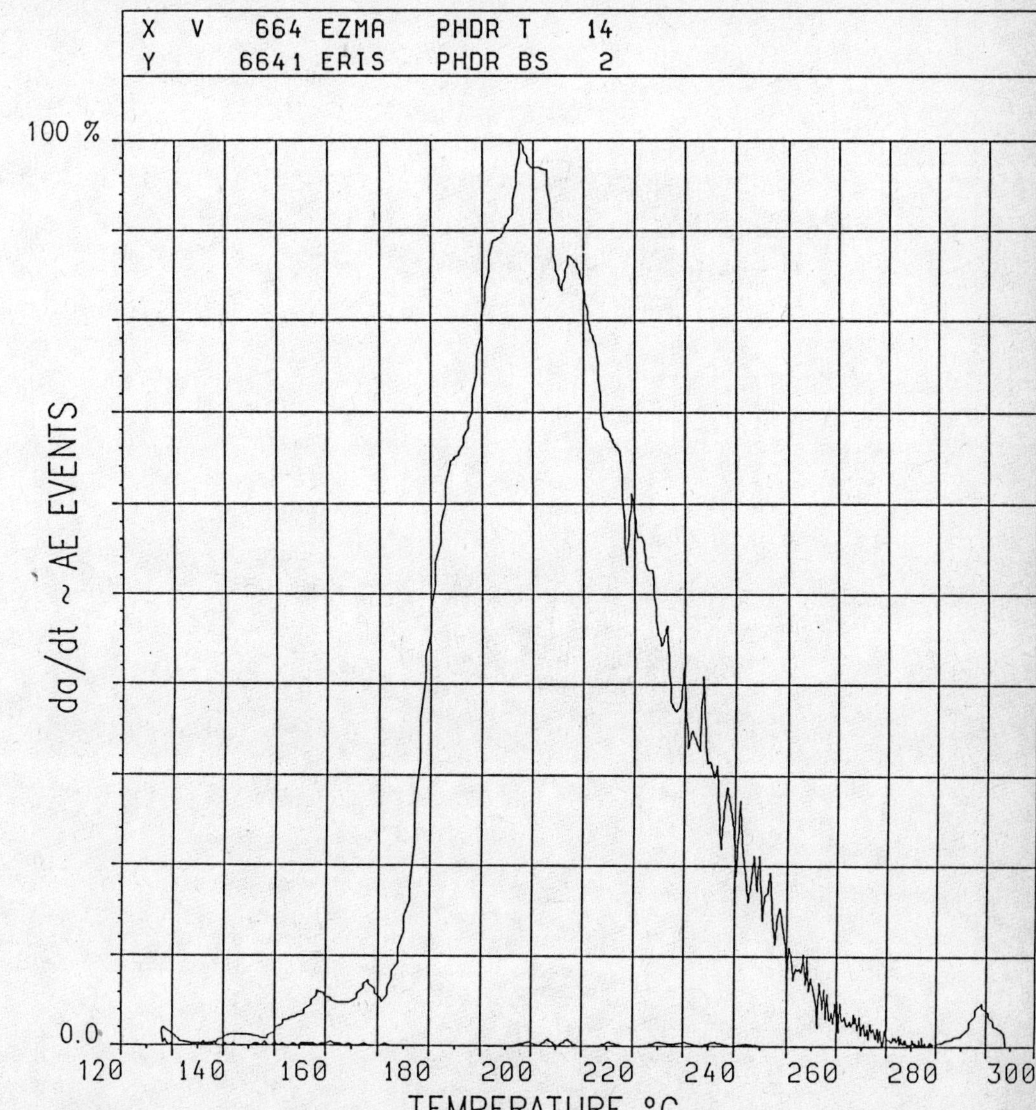

Assumption: real da/dt(t) proportioinal to AE(t)
Construction: da/dt(T) from T(t) and AE(t)

Fig.16a

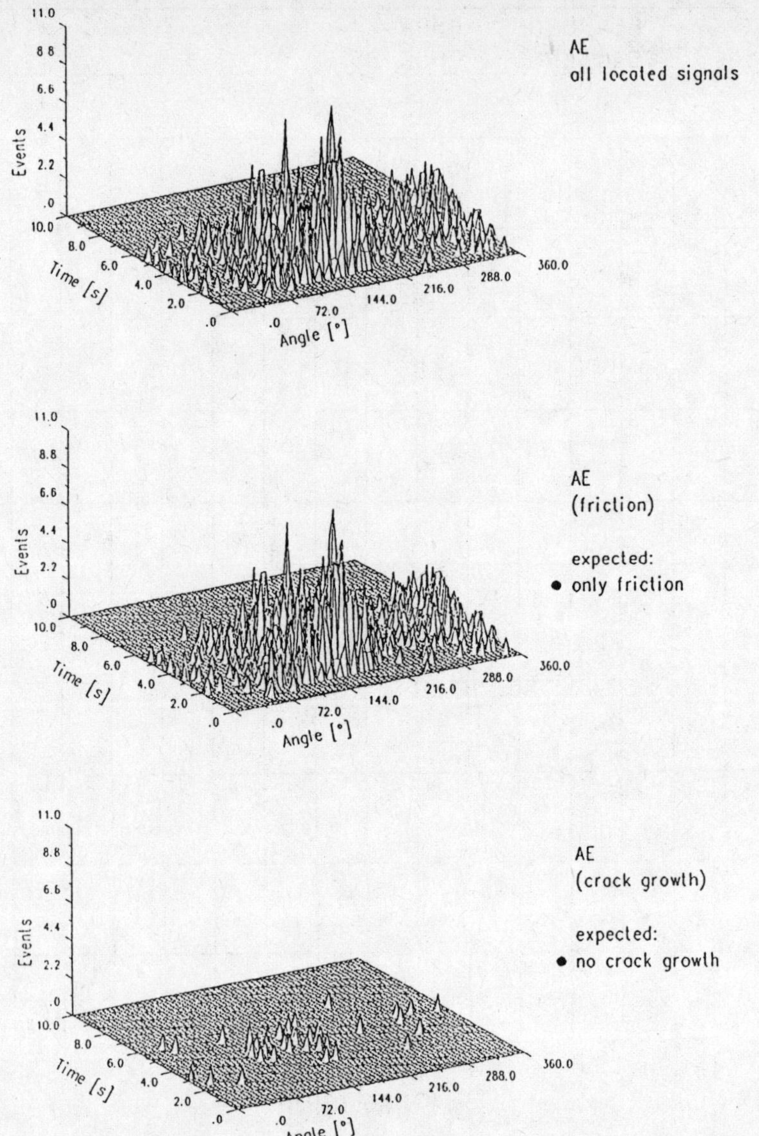

Result: In agreement with fracture mechanics

Acoustic emission in the first 10 seconds
after start of cooling

Fig. 17

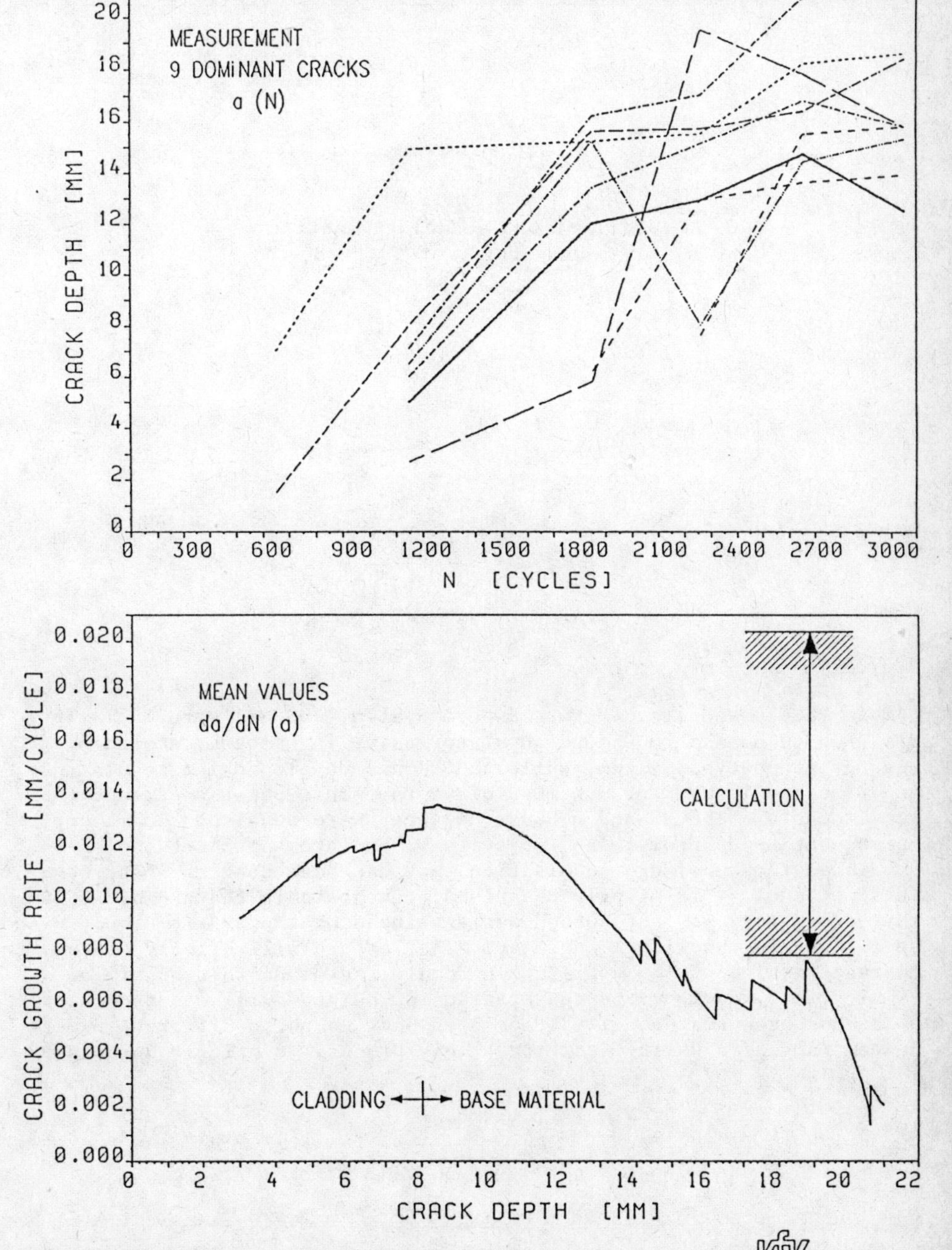

Processing of non-destructive test results
Comparison with calculated results

Fig.18

<u>ON THE USE OF EXPERT SYSTEMS TO HANDLE CORROSION KNOWLEDGE</u>

Dr. Andrew Basden

Information Technology Institute,
University of Salford,
Salford,
U.K.

<u>ABSTRACT</u>

SCCES and its derivate Expert System, AUSCOR, were intially
designed to handle knowledge about corrosion in austenitic steels. SCCES
was the proptotype system, while AUSCOR was developed for regular use,
and remained in use for a number of years. This paper describes the
requirements of SCCES and AUSCOR, how they were developed and their
subsequent use. From these two systems, and others, a methodology
developed for knowledge acquisition that has been used to great benefit.
This methodology is briefly described. One notable feature of AUSCOR and
the methodology is that AUSCOR was developed by the expert alone, without
an intervening knowledge engineer. The lecture will briefly discuss the
characteristics of the application domain that made this possible.
Finally, the knowledge representation method used will be described.
SCCES has been successfully translated between three different Expert
System languages during its life. This process is briefly discussed.

INTRODUCTION

Expert Systems are now well-advanced in their application in engineering. This paper describes two such systems designed for the area of corrosion prediction, of which one has been in regular use for some years now. It concentrates on the technology of Expert Systems rather than corrosion science, and discusses lessons learnt from these and other Expert Systems which are of general application.

First the paper describes the two systems, SCCES and AUSCOR, paying particular attention to how they were developed. From these systems, and others, a methodology for knowledge acquisition emerged, which is breifly set out. While SCCES was developed by the author, acting as knowledge engineer for the corrosion experts, AUSCOR was developed by one of the experts himself, Dr. John Hines at ICI plc. The development of an Expert System directly, without an intervening knowledge engineer, is of great interest, and we discuss the factors that made this possible. Finally SCCES and AUSCOR, have been encoded in three languages, and the paper discusses the problems of portability of Expert System knowledge bases.

A REVIEW OF TWO EXPERT SYSTEMS

The SCCES Expert System

Requirements

SCCES was the first Expert System developed in ICI plc, and was seen as a means of testing the (then) emerging Expert System technology. While most Expert System groups at the time were trying out the technology on artificial problems and developing knowledge representation techniques, we adopted a different approach: to accept the software then available and investigate how well it could be used in a real application. All we knew at the time was that Expert Systems were supposed to be good for handling uncertainty and could explain their reasoning. We therefore looked for a collaborator within the company who had a problem that may be amenable to Expert Systems technology.

Fortunately, Dr. Hines had such a problem - corrosion. He had been using computer-based mathematical models to predict corrosion modes (Edeleanu and Hines, 1983), and, though he had had some success, there were certain limitations (Basden and Hines, 1986).

This model, Change Conds, was based on solving simultaneous equations to describe separate anodic and cathodic polarization curves, with other equations to adjust the coefficients of the main ones to account for changes in operating conditions. While it was recognised that some of the latter equations may be arbitrary, it was argued that as many as possible should have a sound theoretical basis. This would minimise difficulties with interpolation and extrapolation. Most of the information used was from electrochemical experiments, but it was also found possible to incorporate two other types of information: data from other branches of science and information from experience in the field.

Change Conds, after some development, gave reasonable predictions over a wide range of conditions for two alloys and two corrosion modes. But it was found difficult to handle uncertain information and difficult to incorporate uncertain knowledge; there is much uncertainty in the parameters in real-life chemical plant. Moreover, more experiments were needed than had been expected to obtain the parameters for the equations.

So a computer technology that could handle uncertainty appeared attractive, and two corrosion experts, John Hines and Peter Moreland embarked on an experimental project with the author as knowledge engineer to build SCCES. The method employed was for the author to interview the experts and elicit their knowledge, encapsulate it in the computer using the software we had available, and then test it against the experts' expectations and real life cases. Eventually SCCES developed into AUSCOR, and it was found that this did not supplement Change Conds, but rather complemented it; within its limitations Change Conds was particularly good for detailed analysis.

Two decisions had to be made early on which had a profound effect on the shape of SCCES:

1. The domain of expertise. Stress Corrosion Cracking (SCC) was chosen, on the grounds that it is unpredictable and therefore less amenable to mathematical treatment and that it was potentially a high pay-off area.

2. The task that SCCES should perform. At that time, most Expert Systems had been aimed at diagnoses of faults or diseases from symptoms (such as MYCIN: Shortliffe, 1976); should SCCES be built to diagnose causes of SCC? The problem with this was that occurrence of SCC was rare, owing to generous engineering designed to prevent it, so it would be difficult to find sufficient cases to test SCCES, and its future benefits would be limited. The other option was that SCCES should advise on design, or design modifications, to reduce the engineering safety factors needed, but the available software did not seem capable of this type of task. In the end we opted to build a model of the SCC process, which could be used in both diagnosis and design.

The structure and operation of the SCCES

The SCCES was originally developed using an early Expert Systems package called AL/X. But, though this very easily handled uncertainty by means of a Bayesian mechanism, it was soon found to be limited, since it could not handle numeric information. It was then translated to other, more versatile software called SAGE, and finally, via AUSCOR, to Savoir (ISI, 1984).

The SCCES holds corrosion knowledge and operates by asking the user a sequence of questions about a given piece of plant and its conditions, and then declaring the likelihood of SCC occurring. It is thus a form of simulation model. The basic form of SCCES was that of the

inference net. This is a network of items and relationships, in which
the items represent meaningful statements, such as 'A good protective
scale will form', or numeric values, such as 'The pH in the bulk of the
fluid'. The former items possessed an attribute which give the
probability of the item's statement being true in this plant and set of
conditions. Functions are available for transferring information between
the two types (eg. from a pH to the probability that we can consider the
fluid to be 'strongly acidic') and thus SCCES performed both qualitative
and quantitative reasoning. These items are linked by inference
relationships, which specify from which other items an item's value may
be deduced or calculated. Part of the inference net is shown in Fig. 1.
The SCCES operated by means of backward chaining, since this was what was
available in SAGE.

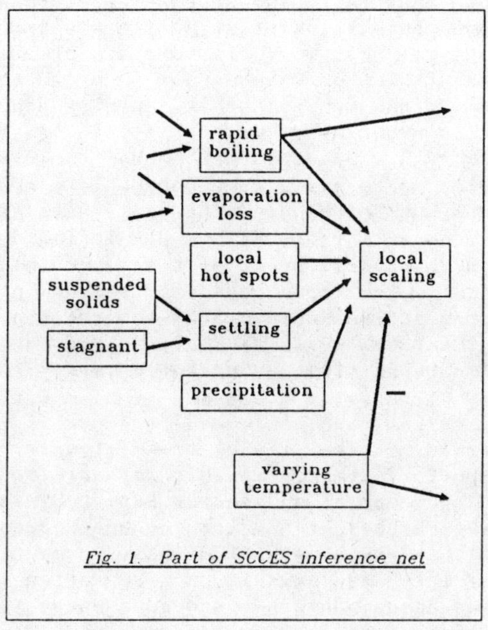

Fig. 1. Part of SCCES inference net

Uncertainty in the knowledge is expressed by means of
probabilistic links, and uncertainty in information supplied by the user
is expressed by the probabilities of the question items. The method of
combination is by Bayesian accumulation of evidence and other
probabilistic or fuzzy operators. While there is much academic argument
about the suitability of these, we found them to be very adequate in
practice.

It is because the 'program' of an Expert System comprises
meaningful items and relationships that it is relatively easy to get it
to 'explain' its reasoning. Most Expert System packages offer low-level
explanation facilities, which trace rules fired or links between one item
and another, but the more useful explanations derive merely from
displaying the current values of items; since they represent meaningful
concepts, simply displaying their value is meaningful. This allows the

user to search the knowledge base and find what are the main factors that contribute to a high risk of SCC, or which might turn a low into a high risk. SCCES and AUSCOR allow what-iffing to aid the user in this process.

Developing the SCCES

Little was known at the time about how to build such inference nets, and we developed a method that has stood the test of time, and is described later. But in summary the process was as follows. Inference nets have 'goal' items - those on the right hand side that are the results. Having decided that SCCES should be a prediction model, the main goal item must be the risk of SCC occurring. Initially the author asked the experts questions like, "What information do you use to predict the risk of SCC occurring?" and obtained a few high-level parameters. But it soon became clear that a more sophisticated approach was needed. One problem was that we obtained a mix of answer types from this, such as 'pH of the fluid' on the one hand and 'whether crack initiation is possible' on the other; one is causally dependent on the other to some degree. So eventually, we took all these different types of item and laid them out in a network that showed the causal relations among them. This enabled us to place some on the left hand side, and these were the questions put to the user in the final system. Others, which represented the probability of various physical and chemical mechanisms operating, such as scaling or pitting, were placed in the middle, while yet others, which represented macro-level processes like crack initiation and propagation, were placed towards the right hand side and fed directly into the final goal. Then links between them were assigned, and, with some repositioning, we ended up with a useful diagram.

This diagram could very readily be translated into an inference net and thence into Expert System code. This emphasis on causality was found to be crucial in a number of subsequent Expert Systems projects, and is discussed below. Further, the inference net representation was found to be very useful; it was a natural means of communicating with the experts, while being a close diagrammatic representation of the knowledge base. Eventually it was accommodated on around a dozen A4 sheets of paper, which made indexing easier.

Some Bayesian weightings were required, which were obtained by asking the experts questions of the form, "Given the situation in which the SCCES is to be used, in how many cases out of a hundred would such-and-such be true?" and then performing some simple arithmetic. After a very short time this became second-nature to the experts, and they were able to assign weightings directly.

Once the inference net had been implemented in code, to produce the SCCES, it would be run by the experts for validation. The validation was in terms of the experts' expectations, based on a number of test cases they were aware of. Any problems encountered were either due to wrong weightings on the evidence or, on a few occasions, some missing or wrong knowledge. There were two forms of the latter case. The first was a wrong structuring of knowledge. If an item had more than four or five inputs, this suggested that there ought to be intermediate items, and the experts were questioned as to what these might be. The second concerned missing knowledge, and the wrong results helped to identify what was needed; in fact the process of building this knowledge base helped to high-light certain gaps in the experts' knowledge and steps were then taken to fill those gaps.

Results

The SCCES took nine months to build, and gave reasonable predictive results that matched the experts' expectations over a wide range of conditions for a handful of alloys. It dealt only with crack initiation, the knowledge concerning propagation being considered to be relatively simple to implement. The whole exercise demonstrated beyond doubt that Expert Systems technology was able to handle complex predictive knowledge, which included a large measure of uncertainty both in the knowledge and in the parameters that describe the conditions of the plant. The SCCES was not yet a truly usable status, though it was occasionally consulted as a back-up for the experts.

It was then put on the shelf for a year or so, while another Expert System was built. This one, SYSLAG, was diagnostic in nature and gave advice on insulation systems to employ for pipe lagging. It was developed almost entirely by the experts themselves (John Hines and Robert Terrell, a chemist), without recourse to a knowledge engineer, using the methods learned during the building of SCCES. This was a notable achievement, since the AI community at the time were talking about dispensing with the knowledge engineer as a future goal. With this arrangement the expert had more direct control over the knowledge base. SYSLAG was developed right through to a usable state and went into service in 1985. Over half the development effort went into the man-machine interface and usability features.

After this, the SCCES was developed into AUSCOR (below).

In addition to the conclusion that Expert Systems technology had some part to play in generating prediction and other advice systems, the SCCES project also highlighted a number of facts about the roles that Expert Systems could play. While most are expected to fulfil a consultancy role, where they act as a mini-consultants, giving advice and explanations, other roles were encountered. One was the Knowledge Refinement role (Basden, 1983) where they can be used, not just to give expert advice, but also to refine human knowledge, and arose from the experience of finding gaps in the experts' knowledge that could be filled. Some of the required knowledge was obtained by means of experiments. Another was a Communication Role, which allowed two or more experts that had different perspectives to see, in fairly precise terms, where those differences lay. The methods we adopted, of seeking the underlying causal understanding of the domain, allowed us to overcome the oft-cited problem of disagreement among experts. It provided a principled way of bringing together the differing specialities of different experts. These findings are discussed below.

The AUSCOR Expert System

AUSCOR was developed from SCCES by John Hines, with some assistance from Rob Terrell, using the experience described above. The main developments were to extend the system to cope with more corrosion modes (hence the change of name) and more alloys, and to add knowledge about crack propagation. But, whereas the purpose of SCCES was to investigate Expert Systems technology and its capabilities, that of AUSCOR was to make expertise available to engineers throughout the company, so the development also included the addition of usability features. As with SYSLAG, no knowledge engineer was needed for most of the work, though the author was called upon occasionally to provide general consultancy.

Like SCCES, AUSCOR is a type of causal model, but was extended to cover 28 alloys. It can handle a wide range of liquid environments, including strong acids and alkalis, waters, and more concentrated neutral solutions, in the temperature range 0 to 300 deg. C. It provides probabilities that the alloy will be passive and that general corrosion (active or transpassive), localised corrosion (pitting, crevice corrosion or corrosion under deposits), and SCC will occur. It also provides a qualitative estimate of the likely deterioration rate in terms of the probability that the rate will be acceptable, and an assessment of suitability of the alloy for the duty which takes account of all these detailed considerations.

Various levels of detail are available by way of result screens, optional explanations and a Help system. One set of displays indicates the sensitivity of the main conclusions to changes in the more important aspects of the duty, and the user may examine the effect of changing selected features on a 'what-if' basis, including the alloy, pH, concentration of halides or oxidising and reducing agents, or the evidence relating to temperature. When making each such change, all the other evidence remains unchanged, but the effect is intelligently propagated by the Savoir software so that if a previously unimportant aspect becomes important, AUSCOR takes this into account automatically and the user may be asked for further relevant information. Such assessments may be repeated using the sensitivity measures as a guide, until the user has an adequate picture. The Help system goes far beyond that normally provided by computer software, in holding a considerable amount of knowledge about corrosion and alloys etc. in addition to the normal 'how to do it' help.

Testing and validation was performed in a similar way to that described above for SCCES. The validation process had a progressive nature, in which three stages can be discerned:

a)   AUSCOR was tried on 'standard' cases, of progressively increasing scope, and the weights were tuned accordingly. As some of the weights had been first guesses, this may not be considered real testing.

b)   AUSCOR was then tried by other users on less standard cases

where the answer was known. These included the recent experience
of individuals and a wider range of standard cases.

c) It was then used for real problems on a tentative basis by
users who were sufficiently experienced not to be seriously
misled.

The main difference between AUSCOR and SCCES was the use of the
what-iffing facility. Isocorrosion charts, derived from experience and
laboratory tests, hold contours of constant corrosion rate against
concentration and temperature; by altering these parameters AUSCOR's
predictions could be compared with these charts over a range of
conditions and its weights could be modified accordingly. The forward-
chaining process of Savoir was sufficiently fast to make this a realistic
process, even though the total knowledge base contained around 3000
rules.

But, as with other projects with which the author has been
involved, it is a mistake to separate validation from knowledge
acquisition. The act of validating provides useful information that can
lead to further acquisition or tuning. Essentially, things like the
isocorrosion charts hold knowledge of their own, and the cyclical what-
iffing is the process whereby this knowledge is gradually transferred to
the Expert System. Moreover, different parts of the system might be in
different stages at different times.

AUSCOR was brought into use in 1985 and its usage was monitored.
It was used by a number of materials engineers and others. Occasionally
it was used for demonstration purposes by a wide range of people. The
main problem was that it was hosted on a central VAX computer, rather
than on a PC, and the interface was textual rather than windows, and this
has limited its growth. Plans have been made to put it on a PC.

SOME GENERAL ISSUES ILLUSTRATED BY SCCES AND AUSCOR

The experience with SCCES and AUSCOR raises a number of general
issues, with relevance to many other Expert System projects.

The application of Expert Systems technology

Especially in the early days, many Expert Systems projects were
seen as Artificial Intelligence research, sometimes with an objective of
obtaining a PhD for somebody. While research into knowledge
representation etc. is needed, many projects have suffered because the
same research attitude has continued, though now it takes a different
form: "Our firm wants to get into Expert Systems; we will try to build
one, to see if our knowledge can be represented." While evaluation of
the technology is valid, it is generally no longer necessary to expend
major effort on evaluating the knowledge representation techniques.

Though this may be a controversial statement, I believe that
knowledge representation is no longer the main issue with this type of
Expert System; the main issue is the objectives and roles. Or, to put it

another way, we should not be concerned about the technology, but about the client for whom the Expert System is being built. The UK Alvey Community Club report (Alvey, 1987) contain examples of projects where expensive software and hardware were unnecessarily used or where time was unnecessarily wasted on knowledge representation investigations, simply because these issues were not properly analysed at an early stage. While many Expert Systems packages based on Production Rules may indeed be limited in knowledge representation terms, the author's experience (discussed later) has been that those based on inference nets are not so limited, and more natural. Almost any predictive Expert System can be built with them, so long as it does not require much of either wide-ranging (common-sense) or sensory knowledge. This is not just an experiential statement; there are philosophical undertones, but this paper will not discuss them.

However, the field of Expert Systems is now reaching a measure of maturity, as evidenced by the shift in research activity away from demonstrating that they can be built to the development of methodologies for building them.

The most common structure assumed for ES projects is one of three main stages. After an initial stage, which is not defined too closely but which includes such things as setting objectives, there come three stages:

    a)   knowledge acquisition,
    b)   knowledge representation,
    c)   validation.

There are variants on these. Knowledge acquisition is sometimes split into two: such as domain conceptualization and knowledge elicitation (Motta and Eisenstadt, 1986). Parsaye and Chignell (1988) add a fourth stage for maintenance. But these are the three pillars of current-day ES methodology.

With the growing interest in methodologies, there has been a welcome and growing recognition that Expert Systems (ESs) are not so totally separate from conventional computing as had at first been assumed, and there are attempts to bring the experience of software engineering to bear on the development of ESs. Some have argued that the more structured approach of software engineering, of phases of specification, design, implementation and testing can and should be applied in Expert Systems in a modified form. The most common version is to see knowledge acquisition as just a glorified form of specification and design wrapped together, knowledge representation as a glorified form of implementation, and validation as nearly the same as testing. Balzer et. al. (1983) take a slightly different line, arguing that the pattern of incremental development that grew out of AI can be seen in conventional terms as incremental development of a specification of the final program, after which implementation can proceed by conventional means, and thus gain the advantage of long-tried project management techniques.

Partridge and Wilks (1986) have argued, on the other hand, that the methods of conventional software engineering cannot be successfully applied to ESs, for fundamental reasons. They contrast what they call SAT (Specify and Test) methodologies, which is the assumed starting point for conventional software engineering, with the RUDE (Run-Understand-Debug-Edit as a cyclical process) methodology that has grown up in AI. SAT requires that a formal specification be drawn up, but there are some problems for which a formal specification is an impossibility, and thus a RUDE methodology is needed, in which understanding of the application is built into the implementation process. In particular, there are many problems today to which computers are being applied, but for which there is no stopping rule or there is no single true solution. An example of this is when an ES is used in the Communication or Knowledge Refinement roles (Basden, 1983). But RUDE does not offer a full methodology. It says nothing, for instance, about embedding a system in use in its organisational environment, nor about usability, nor about maintenance.

For all their differences, these three methods have one thing in common. They are all technology centred. They all concentrate, more or less, on the activities that the programmer carries out with the technology, such as Edit, Implement, Represent Knowledge, Test, Debug, Validate. This technology centredness leads to problems:

a) The whole process is obscure to the client. S/he cannot understand the activities involved, since they are related to the technology and described using the jargon. Thus the client has too often been forced into a position of waiting for the 'deliverable' and hoping for the best, and the benefits of direct participation by the client are usually sacrificed.

b) The 'deliverable' too often does not meet the client's real needs. Legion are the stories of "It may meet the spec, but it's not what I want". Real needs tend to emerge over a longer period of time than is usually allocated to drawing up specifications, and also require better mutual understanding than can be achieved in the time. The RUDE method and the traditional AI method should in theory overcome this, since they allow the system to be developed without a specification in an evolutionary manner, yet in practice they are too unstructured to guide this evolution.

c) The milestones of the project are either missing or artificially constructed to suit the technology rather than the client. RUDE offers none. Conventional techniques offer specification, design and implementation milestones, which are of little intrinsic interest to the client (unless s/he has learned the 'realities' of computer projects!).

What is needed, and what grew out of the experience with SCCES, AUSCOR and others, is a client-centred methodology for building Expert Systems. It attempts to talk in terms that may be more meaningful to the client. It allows for the gradual emergence of a realisation of what is really needed in the system, and has mechanisms to encourage the client and users to express this; the methodology is tolerant of the possibility that the client's needs may only be clarified as they see the ES in operation. Yet it has milestones, but these relate more to what the client may see than to the needs of the technology.

Long (1986) complained that "there seems to be a singular lack of human factors work with a serious software engineering base and vice versa." This methodology can be seen as an attempt to bring the two together. It tries to integrate usability into system development in a more deliberate way than do other methodologies, because usability is something that touches the client and users very closely.

The methodology involves seven stages in the building of an Expert System (plus maintenance). In some projects, especially where a simple system is involved, some of the stages may not be needed or may have already been completed. In others it may be necessary to bring forward some parts of the later stages. But it is useful to retain the conceptual distinction between the stages presented here, since they serve to name and emphasise certain important aspects of an ES project.

The seven stages are:

1. Start
2. Skeleton system
3. Demonstration system
4. Working system
5. Usable system
6. Saleable system
7. System embedded in use

As can be seen, the stages are in terms of deliverables rather than activities and describe an evolutionary process, and this helps to make the process more meaningful to the client. The purpose of the skeleton system is to allow the knowledge engineer and experts to developing an understanding of each others' expertise and agree an initial functional specification. The typical skeleton system would ask a few questions, give a few results, etc. but would contain little domain knowledge. The demonstration system does contain knowledge, and should give reasonable results in a narrow range of cases. It is demonstrated to the client and can be used as a project decision point. The Working system, developed from the Demonstration system, gives adequately accurate results over an adequately wide range of conditions, and is the stage at which the main acquisition of domain knowledge can be said to be finished. The SCCES was somewhere between a Demonstration and Working system.

Many early Expert Systems stopped at the Working system stage, where the system gives correct results. But a Working system may be clumsy in use. For real-life use, usability features must be added, to generate a Usable system, which can bring true business benefit. While a Working system satisfies the experts, the Usable system satisfies the end-users. But it may still be a bit rough round the edges and thus only be used by those that are sympathetic to it. The Saleable system is a Usable system that has been polished up, with attention given to ergonomics, wording of screens, documentation, etc. It can be 'sold' to a wider range of users. Finally, the Expert System has to be embedded in use, and attention needs to be given to training of users, not only on which buttons to press but also on how to make effective use of it. AUSCOR was somewhere between a Usable and Saleable system, with some embedding in use.

The stages usually overlap, especially in that some usability features are added during stages 3 and 4, but their objectives are kept separate. The methodology is discussed in more detail in Basden (1989).

The importance of roles, users and benefits

Identification of roles and users is vital at an early stage since it helps to set the context for the rest of the project and the knowledge acquisition phases. Much ES literature speaks about 'the user', sometimes in the plural. In fact, there are three types of 'user':

- the end-user, who actually operates the ES,
- the expert, who feels some of the effects of the use of the ES, and
- the client, who provides the resources for building the ES.

Each type of user can expect a different set of benefits and/or constraints. Sometimes one person will be two types of user. The knowledge engineer should attempt to ensure that all users feel benefits from the ES, to ensure acceptability. The benefits (and constraints) that can be felt relate to the roles in which the ES is used. There are eight main roles, described in more detail in Basden (1983):

- Consultancy
- Checklist
- Program
- Monitoring
- Communication
- Training
- Knowledge Refinement
- Demonstration.

The Consultancy, Communication and Knowledge Refinement roles have been mentioned above. The roles are orthogonal to what Stefik et. al. (1982) call expert tasks; roles are of more concern to the client than are tasks. It is important to realise that a given ES may have several roles.

The importance of roles is that they help to identify both what kind of knowledge will be needed, and also what benefits and constraints may be expected when using the ES. Fig. 2 shows how roles, users and benefits are related in order to assess the probability of success of an ES. Each role leads to a number of benefits and constraints. But not all of these will be relevant to the objectives of the ES, and we arrive at a list of benefits and limitations we can expect. Considering these in relation to acceptability and the ability to encapsulate knowledge provides a means of identifying the critical success factors for the ES project.

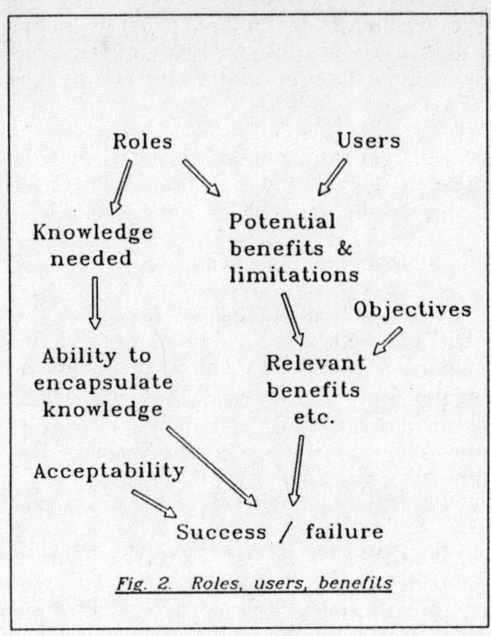

Fig. 2. Roles, users, benefits

For instance, in the Consultancy Role an ES may have the following benefits: For the end-user, speed, reliability, accessibility and understandability, for the expert, a saving of time on simpler cases and for the client, fewer errors, a spreading of expertise, some standardisation, and a data-gathering exercise. Likewise each type of user may be subject to certain limitations in the use of the ES; for instance, the client may have to provide formalised mechanisms for obtaining the data the ES needs and for distributing its results, or the end-user may have to learn keyboard skills. In other roles other benefits assume greater importance; to the user these include reliability in the Checklist Role, the ability to standardise and exercise knowledge in the Communication and Knowledge Refinement Roles, etc. We will not discuss them in detail here.

Thus we see the crucial position that role analysis has in Stage 1. Notice that many methodologies have some, but not all, of this mechanism. Many speak of objectives and of the ability to encapsulate knowledge. Some consider acceptability. But few employ the type of role analysis suggested above, yet it is just this type of analysis that the client can often cope with, since it is performed using his/her language and concepts.

Knowledge acquisition, knowledge representation and validation activities

Conventional Expert Systems methodologies see Knowledge acquisition, knowledge representation and validation as three separate stages, but in this methodology they are not mentioned explicitly. Instead they are seen as activities that are carried out in three of the stages in a cyclical form. In each stage there is an indefinite number of cycles of these activities, with more knowledge acquisition at the beginning and more validation towards the end. It is in this way that we combine the advantages of Partridge and Wilks' RUDE methodology with some measure of project control.

These three activities are needed, as might be expected, when producing the Demonstration and Working systems. But they are also needed to produce the Usable system. Usability is too often seen as 'mere programming' and 'addition of frills' - something simple that may be added after all the 'real' work has been done. But true usability goes beyond ergonomics and can be quite a complex matter; it is often reported that around half the project effort goes into such things. So, rather than treat this stage as a programming exercise, there are advantages to treating it as a knowledge engineering exercise, similar in activity to the two previous stages. There will be cycle(s) of knowledge acquisition, implementation and validation, but the difference is that the knowledge to be sought now is from end-users rather than domain experts. It concerns how they wish to use the system and what facilities they need in order to procure real business benefit. In the case of AUSCOR, for instance it was found that, while some of the usability features, especially what-if facilities, could be added during earlier stages, the best way of obtaining input information and providing the results took time to develop, and this could only be done after the system had been brought to Working status. It is often only at this stage that the real characteristics of the users can be determined; who they are, what they know and what they need. It is at this stage that a full realization of the expected benefits and limitations is sought.

Organising an Expert Systems project in this fashion does, I believe, start to address the complaint of Long (1986) that system development has not been integrated with usability.

<u>Knowledge acquisition</u>

From our experience with SCCES, SYSLAG and AUSCOR, we concluded that knowledge acquisition was not the bottleneck that most people seemed to find it. This conclusion has retained its validity during several other Expert System projects undertaken by the author (Jones and Crates, 1985, Brandon, et. al., 1988), and has been developed into a methodology described by Attarwala and Basden (1985). Why this is so is now discussed.

The bottleneck refers to the apparent inability of experts to express their knowledge in ways that can be used by knowledge engineers to build Expert Systems. I now believe that the single most important reason that no such bottleneck was encountered in these systems was that we sought the causality or, more generally, an understanding of the domain, rather than merely the problem-solving heuristics. This has proved significant. Most other methodologies either assume that the knowledge engineer should seek problem-solving heuristics, presumably in order to avoid the over-simplifications of 'book-knowledge', or do not specify what kind of knowledge should be sought.

As demonstrated in Attarwala and Basden (1985) it is the seeking

of heuristics that leads to the well-known difficulties encountered when
building Expert Systems:

- fragility of the system,
- poor explanations
- difficulty in understanding the knowledge base,
- difficulty in handling disagreement among experts.

These all arise because problem-solving heuristics, as with other aspects
of experience, are subjective rather than objective. Understanding, on
the other hand, is relatively objective, and not dependent on the
individual experience or preferences or context. It encompasses what
many see as 'Deep Knowledge', but goes beyond this. As shown in Fig. 3,
experience can be seen as a partial instantiation of understanding by
means of context-dependent problem-solving. This latter includes
information about the working context, the user of the domain knowledge
(eg. the expert) and the problem-solving strategies employed.

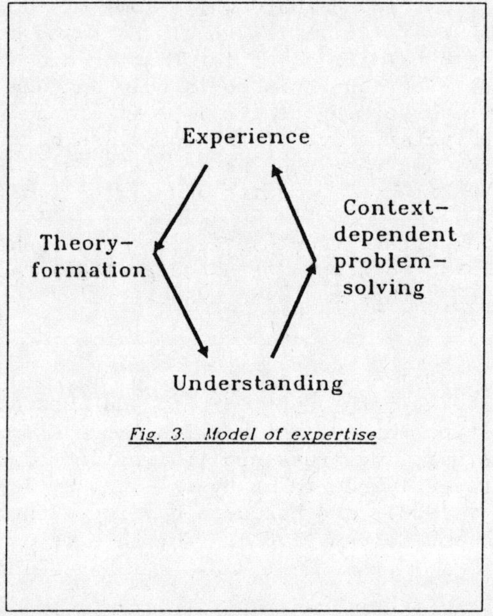

Fig. 3. Model of expertise

In knowledge acquisition the aim should be to uncover the
underlying understanding and, at the same time, make explicit the
context-dependent problem-solving. A methodology has been developed that
enables the knowledge engineer or expert to do this. Very briefly, it is
as follows. First, a number of heuristics are obtained, by various
methods discussed in the literature on knowledge acquisition. Then, for
each, the questions "Why?", "What else?" and "When not?" are asked.
Asking "Why?" uncovers understanding, and thus is similar to Silvestro's
(1988) methodology, which is centred on explanation. Asking "What else?"
and "When not?" deals with the contextual aspects, and offers an easy way
for the expert to enhance the completeness of the knowledge base. It is
this that increases robustness and reduces the fragility of the Expert

System. The ELSIE system (Brandon, et. al., 1988), which provides guidance on the early planning of office developments, was built using this methodology and has been used successfully on buildings other than offices.

The problem of disagreement among experts is also overcome by seeking understanding. If experts disagree when they agree about the understanding the usual reason is that their experience relates to different circumstances and they are generalising too far. Not only does this mean the disagreement is likely to be fairly easy to sort out, but it also gives a way of refining and adding to knowledge. Both are much more difficult if heuristics are used for the knowledge base without a causal basis. (Thus we might surmise that the Knowledge Refinement role cannot be fulfilled without understanding, but this conjecture has not been tested.)

Note that nothing has been said about what Motta and Eisenstadt (1986) call domain conceptualization, the identifying of the classes of objects etc. relevant to the system. They see this as a separate stage, prior to detailed knowledge acquisition, and originally designed the KEATS software accordingly. Now, though, they are attempting to bring the two stages together because they have found there is no such sharp chronological demarkation. In the methodology presented here, it is very much seen as part of the overall knowledge acquisition process, and it has been found that asking the "Why?" question automatically leads to some domain conceptualization.

Both SCCES and AUSCOR are predictive systems, with the direction of causality being the same as the direction of inference. Is the methodology appropriate also for diagnostic systems, where the directions are opposite? Since only a few diagnostic knowledge bases have been built using this methodology, no definite answer can be given to this question, but it appears to be appropriate, though requiring some modification to the detailed method. One method is to build first a causal model and then reverse the arrows in the diagram. Another is to do this on a local basis, asking the three questions for each diagnostic heuristic and reversing the arrows immediately.

Knowledge acquisition without a knowledge engineer

As mentioned above, AUSCOR and SYSLAG were constructed by the domain experts with very little aid from a knowledge engineer. They were given a short course on using the Expert Systems software, and then built their own systems. This was a very interesting development, in view of the fact that such a state of affairs was at the time considered a long-term goal of the AI community. How did it happen? Can only certain experts do this? Or is it that their domain is particularly suited to the task? This is what we discuss in this section, based on a talk given by Hines (1985).

First, it seems to be part of a natural sequence that has been followed in areas like simulation. In the early days we wrote our own programs, then we hired programmers to write them for us, now we are again doing it ourselves, using packages. Savoir etc. can be seen as such a package for knowledge engineering.

There seemed to be five things that made it possible for a materials expert to do his own knowledge engineering in these cases:

a)   Engineers are often used to a pictorial representation, and the inference net can therefore be very easily understood and employed as a technique.

b)   The domain of knowledge has many parts that are susceptible to accurate, mathematical definition, and engineers find it very natural to express such knowledge as expressions in a computer language.

c)   Where it exists, the theory works. So the understanding sought by the methodology is relatively public knowledge, and obtaining it is therefore not onerous.

d)   But materials science is difficult to treat in a completely mathematical way as it has areas which are as yet only vaguely understood and it is markedly multi-dimensional. Savoir was able to accept whatever degree of understanding there was - precise mathematical expressions, qualitative relationships or mere propositional statements - and combine them into an overall knowledge base.

e)   Owing to the inherent modularity of the inference net formalism, knowledge can be added incrementally.

It should also be remembered that one of the experts involved had had close involvement in the construction of the SCCES, and so the methodology for eliciting knowledge was not new to him, though the details of the software were.

Similar results have been found in the Counsellor Expert System, which advises farmers on application of fungicides on winter wheat crops (Jones and Crates, 1985). Biological areas of knowledge like fungal growth are similar to materials science, in that much of the causality is scientifically known, even though it may be less mathematical.

Six types of knowledge are required in the process of building an Expert System:

a)   of the domain,
b)   of the software,
c)   of the usage of the final Expert System,
d)   project management skills,
e)   the political skills of a project champion, and
f)   of how to promote the final system.

When considering the activity of knowledge acquisition by the expert our main concern is with the first two. These can be seen as a spectrum, shown in Fig 4. If A and B are large compared with C, and C can be provided by consultancy, then a domain expert can perform most of his/her own knowledge acquisition. Usually, however, it is useful to have at least two people involved in the project, in order to bounce ideas around, but they need not be from different disciplines. As far as (c) to (f) above are concerned, whether other personnel are needed beside the domain expert will depend on how many of those skills the expert has.

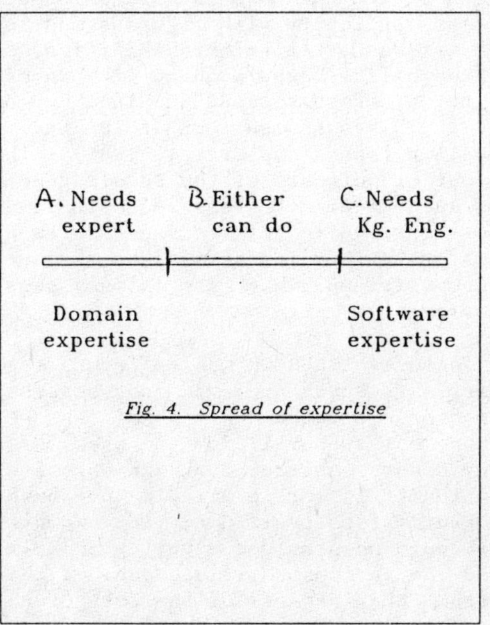

Fig. 4. Spread of expertise

Knowledge representation and Expert Systems software

It is the features of Expert Systems technology, especially as represented by the final software used, Savoir (ISI, 1984) that made it possible to develop AUSCOR within a reasonable time, and by the experts themselves. Extension from SCCES to AUSCOR into many alloys and modes was made feasible by the fact that the items of an inference net can represent meaningful states of the duty, rather than just being arbitrary variables in a computer language. The ability of Savoir to link to a database in which alloy attributes were stored, to process numeric as well as probabilistic information, and to allow expert-defined functions to be written were also important. Regarding the latter, a curve function was written so as to represent eg. polarization curves, which had the property that even re-entrant curves could be handled. This was made possible by the fact that Savoir represents numeric information not as single numbers but as a High-Low pairs.

The feature of intelligent propagation of changed information was provided automatically by the forward-chaining activity of the software operating on structure of the inference net, and that of seeking relevant information in a context-sensitive manner was provided by the backward-chaining operation. While some Expert Systems software have either one or the other, it is usually important that both are available in the software.

A significant feature of materials science and corrosion in particular is that many properties have inherent scatter. The knowledge is often most sensibly expressed as a curve or surface with a scatter band. This fact, together with the need to balance opposing tendencies, are the primary source of difficulty with rigorous models like Change Conds. But as the scatter bands reflect probabilities, such data closely fit the concept of a probability-based package like Savoir. An example of the use of this, pitting, was discussed in Hines and Basden (1986) and Basden and Hines (1986). This was also found to be true in a completely different area - Counsellor (Jones and Crates, 1985). This seems a strong argument in favour of software of the Savoir type for these types of systems; Production Rules seem much less suitable, even if they use certainty factors. Moreover, while inference net software has a strong representation of items, relationships and values, Production Rule software is based on propositions, which are in many senses a more cumbersome form of knowledge.

One usability feature of AUSCOR was to reduce the amount of data needed from the end-user. This was achieved in two ways. In one, information was obtained from a database if conditions were such that it was appropriate to do so, but approval from the user for the value was sought if there was any doubt. Another took the opposite line: allow the user to supply information if it were available, but have it calculated from other information automatically if not. For example, the concentration of acids could be provided either precisely or by means of approximate ranges, and the various pH values could either be provided by the user, calculated from other aspects of the duty or obtained as typical values from a database. For such complex treatment of default values the Expert Systems software had to provide facilities for testing the state of the knowledge base. In other similar projects, such as ELSIE, which advises quantity surveyors during early stages of construction projects and is described by Brandon et. al. (1988), the software also had to provide a measure of non-monotonic reasoning. Savoir was sufficient for all these means of obtaining information.

In general - and this has been backed up by experience in other Expert Systems projects - we can say that it is the structure of the inference net that made it possible to encapsulate complex knowledge, and it was the features of forward and backward chaining and the ability to perform default and non-monotonic reasoning that made it possible to produce a usable program.

In many Expert Systems projects, choice of software is given high priority. But in the author's view, software selection is not so critical as many believe, as long as one has software with a certain amount of flexibility. More important than the language in which knowledge is expressed is the deep structure of the knowledge, in terms of items, relationships and the values of attributes. Inference net software can handle these very aspects of deep structure, though perhaps in a limited way. The most desirable additions would be classes, to allow generalised rules, transition networks, to ease the setting up of Help systems, and certain more sophisticated methods of displaying and obtaining data, to ease the production of usable displays. Egeria (ISI, 1988) has some of these features. 60% of AUSCOR comprised code to handle the Help system, the control knowledge and the displays.

As mentioned earlier SCCES and AUSCOR together have been held on three separate pieces of Expert Systems software throughout their life. First, SCCES was built with AL/X, then moved to SAGE (SPL, 1983). It was in SAGE that AUSCOR was initially developed from SCCES, and it was then translated as a more complex system into Savoir before the main usability features were added. Contrary to expectation, translation from one to the other was relatively painless. It comprised first a number of global edits to change most of the syntax; for instance, in addition to translating keywords, multi-line text in SAGE was enclosed by double quotes and each line had to be terminated with double quotes, while in Savoir it is enclosed in single quotes and each line is terminated without quotes. There were some tricky translations of syntax, but these were few. This was followed by manually changing some parts of the knowledge base, especially the control structure, to take advantage to new facilities. For instance, SAGE had only backward chaining and procedural statements, while Savoir had forward and backward chaining and its control was by demons. This however did not take long.

The major reason for the ease of translation was probably that all three packages were based on inference nets, and so there was little need for semantic translation in addition to syntactic.

CONCLUSION

SCCES and AUSCOR were two Expert Systems constructed to model the corrosion processes in austenitic steels, with SCCES being a prototype system from which AUSCOR was developed. AUSCOR is in real-life use.

A number of lessons were learnt from building them, and other Expert Systems, and some of these have been discussed in this paper:

- A life-cycle methodology for Expert Systems has been suggested, which consists of seven client-centred stages, and in which consideration of roles, users and benefits plays a major part.

- A methodology for knowledge acquisition has been developed which is based on seeking the underlying understanding (causality) and context-dependent problem-solving behind expert heuristics. This methodology reduces some of the problems normally found in building Expert Systems.

- AUSCOR was built without a knowledge engineer. The conditions which allow this to happen have been discussed, and relate largely to characteristics of the domain of expertise.

- The knowledge representation used was inference nets. This allowed natural expression of domain expertise, together with relative portability between different software packages.

ACKOWLEDGEMENTS

I would like to thank Dr. John Hines for very many helpful additions to and comments on this paper. He usefully revitalised my memory of events long past.

REFERENCES

Alvey (1987), Report on the Grand Meeting of Community Clubs, 6th February 1987. Alvey Directorate, London.

Attarwala, F.T., Basden. A. (1985) 'A methodology for building Expert Systems', R&D Management.

Balzer, Cheatham, Green (1983) 'Software technology in the 1990's - using a new paradigm', IEEE Computer, Nov 1983.

Basden A. (1983) 'On the application of Expert Systems', IJMMS, $\underline{19}$, p.461.

Basden A. (1989) 'A client-centred methodology for building Expert Systems', presented at HCI-89, Nottingham, UK.

Basden A., Hines J.G., (1986), 'Implications of relation between information and knowledge in use of computers to handle corrosion knowledge', British Corrosion Journal, $\underline{21}$ (3), pp. 157..162.

Brandon P.S., Basden A., Hamilton I., Stockley J. (1988) 'Expert Systems - the strategic planning of construction projects', The Royal Institution of Chartered Surveyors, London.

Edeleanu C., Hines J.G., (1983), British Corrosion Journal, $\underline{18}$ (1), pp. 6..9.

Hines J G, (1985), Talk given at the ICI Expert Systems Users Group.

Hines J.G., Basden A., (1986), 'Experience with the use of computers to handle corrosion knowledge', British Corrosion Journal, $\underline{21}$, (3), 151-156.

ISI (1984) Savoir, Expert System software, Intelligent Systems International, Redhill, Surrey, UK.

ISI (1988) Egeria, Expert System software, Intelligent Systems International, Redhill, Surrey, UK.

Jones M.J., Crates D.T. (1985) 'Expert Systems and Videotext: an application in the marketing of agrochemicals' in 'Research and Development in Expert Systems - proceedings of the Fourth Technical

Conference of the British Computer Society Specialist Group on Expert Systems' ed. Bramer M.A., Cambridge University Press.

Long J. (1986) 'People and computers: designing for usability', in Harrison M.D., Monk A.F. (eds), 'People and computers: designing for usability', Cambridge University Press.

Motta E., Eisenstadt M. (1986) 'KEATS: the knowledge engineer's assistant', Technical Report No. 20, Human Cognitive Research Laboratory, Open University, UK.

Parsaye K., Chignell M. (1988) 'Expert Systems for experts', Wiley.

Partridge D., Wilks Y. (1986) 'Does AI have a methodology different from software engineering?', Report No. MCCS-86-53, New Mexico State University.

Shortliffe E H, (1976), 'Computer-based medical consultation: MYCIN', Elsevier, New York.

Silvestro K. (1988) 'Using explanations for knowledge-base acquisition', IJMMS, $\underline{29}$(2), p.159.

Systems Programmers Ltd., (1983), SAGE, Expert Systems Software.

Stefik M., Aikins J., Balzer R., Benott J, Birnbaum L, Hayes-Roth F, Sacerdoti E. (1982) 'The organisation of Expert Systems, a tutorial', Artificial Intelligence, $\underline{18}$(2), p.135.

A.I.-BASED CORROSION RISK ANALYSIS IN OIL REFINERY OPERATION

Marc J.S. Vancoille, Walter F.L. Bogaerts
Faculty of Engineering - Dept. MTM
Katholieke Universiteit Leuven
W. de Croylaan 2, B-3030 Leuven, Belgium

Felix Perdieus
Esso Refinery Antwerp
Polderdijkweg, B-2030 Antwerp, Belgium

Abstract

The assessment of corrosion and related risk behavior in oil refinery operation is both based on practical experience and basic understanding of the prevailing corrosion mechanisms. It was thought necessary to automate to a larger extent the process of corrosion risk analysis. This automated process would enable corrosion and material experts to give a faster and more uniform evaluation of corrosion risks. The usefulness of artificial intelligence techniques in this domain was investigated. Some preliminary studies have been made and prototypes built to demonstrate what can be done with state of the art technology. Right from the conception of the project, much attention was given to produce a useful working product.

Keywords: artificial intelligence, expert system, corrosion prevention, materials selection, risk analysis, oil refinery, PRIME, active interface, explanation

1 Introduction

1.1 The CEC and the ESPRIT programme

The potential and limitations of Artificial Intelligence (A.I.) techniques are poorly known to most company managers. Two major attitudes prevail. The advocates of A.I.-techniques try to put these to work as much as possible, whereas the mostly sceptical managers and experts are reluctant to engage into A.I. as they are sceptical about the return and performance of these techniques. Most major U.S. companies are actively involved in A.I. research and applications.

Europe is still lagging behind, despite ambitious plans such as the CEC-sponsored *Esprit programme.*

The Commission of the European Communities has concluded that, in order to make the most of the opportunities offered by information technology, a strategic action was required to meet the challenges ko this growing world market. It is foreseen that in world trade, Information Technology (IT) and electronic equipment will overtake the automobile sector in the 1990's. With worldwide R&D spending on IT rising from $35 billion in 1986 to some $90 billion in 1990, IT will remain one of the dominant sources if technological advance until the end of the century. The blueprint of the strategic action developed by the commission of the CEC has been established in the 1980's with the European Strategic Programme for Research and Development in Information Technology (ESPRIT).

ESPRIT is a supra-national industrial cooperation programme which aims to promote IT industry in the European Community (EC). For the first phase, an overall effort of more than $1500 million was foreseen, with the Community's research budget providing 50 percent and the participant providing the remaining half. The second phase started in early 1988 and provides even more resources, with an additional budget that is more than doubled and a foreseen research effort of about 30,000 manyears.

1.2 A.I. put to work.

Although some larger companies are actively involved in this kind of research, it is hard to say whether they have achieved much. It is well known that starting something in the field of A.I. requires a considerable amount of time, resources and manpower before something useful comes out. In Europe, A.I. seems to be still very much a research topic with few industrial applications.

The first reason why most managers in Europe still hesitate to adopt this technology is probably its difficultly quantifyable benefits in terms of competitive advantages. Most people in the field however, agree that they might benefit from the introduction of A.I.-techniques as an expert at hand or a second opinion.

The second reason is that there are still few people explicitly trained in the use of A.I.. So if a company wants to start something in the field of A.I., they either have to hire new people or train them themselves, which of course takes a lot of time and money. A third alternative is to seek some cooperation with universities. Such cooperation may provide a valuable test bed for their theoretical work.

Thirdly, the cooperation between industry and university is not always an easy one when it comes to exchanging information. It is obvious that in order to make a useful expert system, a lot of knowledge has to be incorporated into the system. Most of this knowledge however is proprietary and it is not always easy to convince people involved that it is essentially this kind of information that will make the system a useful and valuable one.

The present joint University-Industry project was set up as a feasibility study to determine how the company could benefit from this technology. Most of the development work has been done in the M.T.M.-A.I. research lab at Leuven University and is based on information provided by the company, but also based of generally available literature. The combination of both knowledge sources assures that the information contained in the system is at the same time general enough to be applied to any oil refinery, but on the other hand contains enough details to give answers to very particular problems in the company involved.

Different aspects were worked out in detail and will be discussed hereafter. A dedicated corrosion risk analysis module based on the in-house developed expert system environment PRIME has been developed [Vancoille 1988]. PRIME, which stands for **P**rocess **I**ndustries' **M**aterials **E**xpert, is a domain specific expert system environment which is primarily aimed at providing assistance for the selection of materials for various process equipment parts. It has been developed in the framework of the ESPRIT-project P857/990.

Knowledge acquisition for the present project was carried out primarily from written documents. Clarifications were provided by human experts when needed. This knowledge was then put into the appropriate format (e.g. knowledge bases, rules, database tables, numerical procedures) and linked with the control system in PRIME. The responses of the system were verified by extensive trials in cooperation with the company's experts.

The following parts will describe some of the modules that have been realized and will give an overview of how the system behaves.

2 Problem Specification

2.1 Oil Refinery Practices.

Economy and safety are two of the major concerns in oil refinery operation. Corrosion and corrosion prevention play a major role in both of them. If left uncontrolled, this corrosion can cause costly equipment replacement, loss of revenue from system shut down, hazardous leaks causing major safety issues, and environmental contamination. Cleanup of the latter is extremely expensive and time consuming. For example, one major refinery in Los Angeles County has estimated $85,000,000 for the cleanup during the next two decades. Fire or explosion hazards as a result of (corrosion initiated) leaks may even be more worrying. The combined presence of high temperatures combined with highly inflammable liquids and gasses is a constant safety concern during plant operation.

There is however a major dilemma: how to work as safe as possible without major expenses? From a materials point of view, a plant can be made extremely safe with lots of corrosion resistant materials. They are however prohibitively expensive. Moreover, operating conditions may vary rapidly without the possibility to change the equipment. From an economic point of view, all materials have to be as cheap as possible. Carbon steel will be preferentially used whenever possible. This necessitates the need for a consistent and redundant risk analysis system and a consciously carried out inspection strategy.

The major problem with this approach however is that it is not always easy to assess possible corrosion damage due to changing operating conditions. Also aspects such as creep and fatigue have to be take into consideration. It is typical for oil refineries to process the cheapest crudes they can buy. Unfortunately, these may often be the most corrosive (e.g. sulphur content, naphtenic acid corrosion, high salt content that yields hydrogen chloride in the atmospheric distillation unit, ...). A typical solution to these problems is to make a blend of different crudes in order to keep chemical composition and physical parameters in the optimal range. All this has major implications on the corrosion behavior. Not only is the corrosion behavior a rather complex one, it is also very difficult to predict.

In oil refinery operation, it is customary to use a lot of approximations to determine the rate and extent of the corrosion damage. Most of these practices are based on experimental data and experience. Although much effort has been put in structuring the information to make it easily accessible, it is still a major task to find all relevant information, if it can be found at all. This can be a problem especially in a situation where a quick solution has to be found. In such instances, experts rely very much on their own experience. It should be no surprise that sometimes the better solutions are overlooked.

2.2 User Demands for a "computer-based" corrosion assistant.

Two distinct user demands prevail. On the one hand, corrosion experts expect a computer-based assistant to act as a *Corrosion Simulator* thus allowing them to analyze different **what if** scenarios when changing operating conditions: what if crude composition, temperature, pressure, addition of inhibitors, ... change; what kind of corrosion problems can be expected; what will be the corrosion rate; which corrosion phenomena can be expected; how dangerous are they; ... ? This demand has been implemented in an expert system module called *PETROCRUDE* (paragraph 3.2.1).

A second demand is that the expert system should be able to act as an intelligent *information bank* allowing them to quickly assess problems such as corrosion processes and materials selection for each unit within the plant. The experts acknowledge themselves, that although they have all information at hand in their library, it is almost impossible to gather and grasp it all at the time it is needed. This has been implemented in a *Materials Selection* Expert System Module (paragraph 3.2.2).

This kind of flexible problem solving requires a quick response. The specialists felt they should be able to interact with the system, change operating conditions and get a response almost in real time. In this way they would be able to quickly get a good idea of how the system evolves when operating conditions change and what to expect. A detailed analysis could then be performed. This kind of approach has the advantage that different scenarios can be looked at without having to analyze them all in depth.

Both modules should of course be able to interact. It would be helpful if the *PETROCRUDE*-module would be able to suggest some preventive or corrective measures (e.g. choice of material, coating, inhibitor selection, ...) in combination with the *Materials Selection Module*.

The next paragraphs will show what the possibilities of expert system technology are in addressing these kind of problems.

3 Implementation

This project has been based on the work that has previously been done at the departments of Chemical Engineering and Metallurgy and Materials Engineering of the University of Leuven in the framework of the Esprit project P857/990. The very modular structure of the expert system PRIME enabled us to build a first prototype rather quickly.

3.1 Knowledge Acquisition.

The most important part of the whole project was of course the knowledge acquisition and implementation. Although some A.I.-experts claim that interviewing domain experts is the most appropriate way to do this, it is our experience that this is often not the case. It is very difficult for the knowledge engineer to ask the right questions at the time of the interview; it is equally difficult for the domain expert to give a detailed, complete and structured overview of his knowledge in a particular field at the time of the interviewing. It is usually easier to start from some written material. If a company has a written, concise overview of their design or corrosion control practices, this can be used as general as overview of what specific (corrosion) problems might occur in certain operation.

This general text can then be supplemented by other literature, reports, experimental data and rules of thumb. Of course, the final material has to be reviewed by the domain experts. It is our experience that this way of working is appropriate. Certainly when one has to deal with different experts; it already resolves part of the problem of conflicting knowledge. It is clear that conflicting knowledge should either not be entered into the final system or further detailed in order to resolve the contradictions. We won't elaborate on how this problem could be resolved. However, starting from a general text everyone can agree upon, already alleviates part of the problem.

3.2 Knowledge Representation.

The expert system has been implemented by means of the expert system building tool KEE [1] and the A.I.-programming language Lisp. The objects of the application domain are represented

---

[1] KEE is a registered trademark of IntelliCorp, Inc., Mountain View, CA, USA

as *units*. These units can be hierarchically combined in *knowledge bases* which are a collection of units. The knowledge base provides a hierarchical framework (or *frames*) for organizing the objects, thus creating a convenient interface for creating a model of the application domain. The objects contain or point to most of the declarative or procedural knowledge. The attributes or characteristics of units are called *slots*.

In the present project, major problem classes that are addressed by the system (and that are represented as knowledge bases) are corrosion types (e.g. name, characteristics, ...), material properties and description, equipment description (e.g. what are the different parts of a -generic-pump, what is the relation between those parts, ...) and properties of all kinds of chemicals that are used in the chemical process industry. An example of such a knowledge base is given in Figure 1 that displays part of the knowledge base on corrosion types. Another example is the knowledge base that represents different crudes (Figure 2).

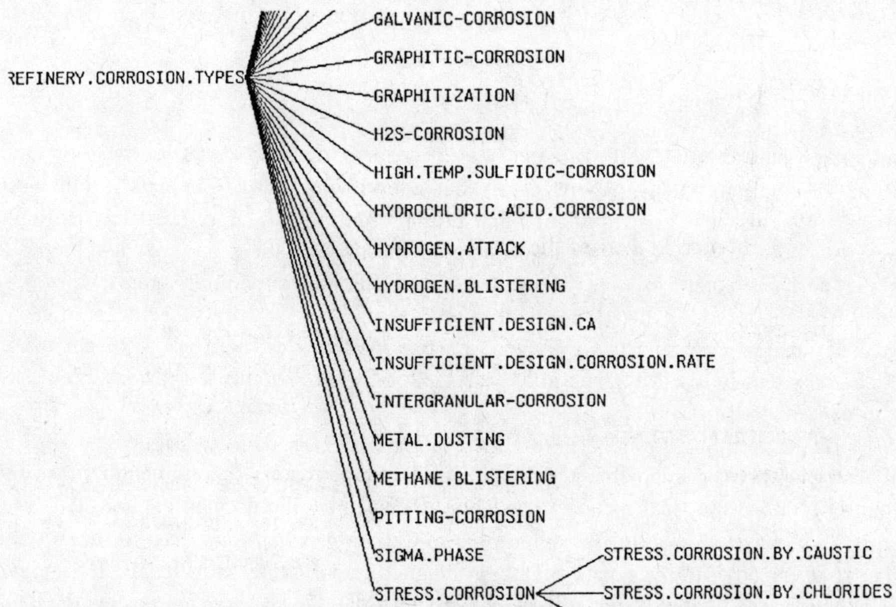

Figure 1: Part of the corrosion type knowledge base.

The refinery plant is also represented as a knowledge base. Figure 4, for instance, depicts the atmospheric distillation unit. The graphical representation of the plant is coupled with an underlaying knowledge base. (Figure 3) that represents the different parts of the unit together with its properties. The way of representing a unit in a hierarchical combination of different parts has the advantage that various properties which are common to the whole part or unit, can be defined at a higher level; they are *inherited down* to the specific units. For instance, all pumps belong to a generic superclass "pumps" in the knowledge base "equipment". Since most

366

pumps have similar properties, they only have to be specified once at the highest level. This kind of knowledge representation has been used throughout the whole project.

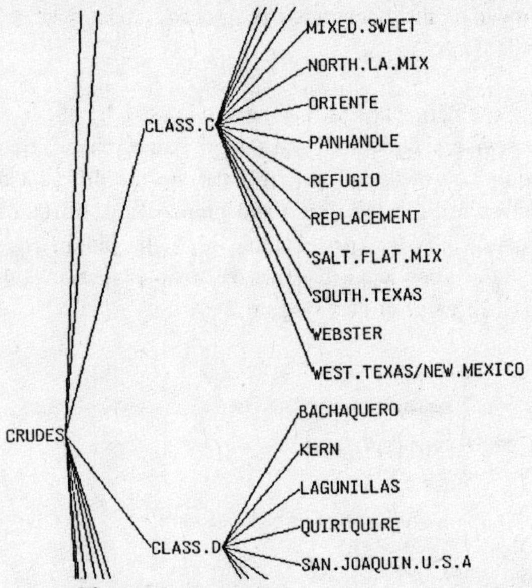

Figure 2: Part of the crude knowledge base.

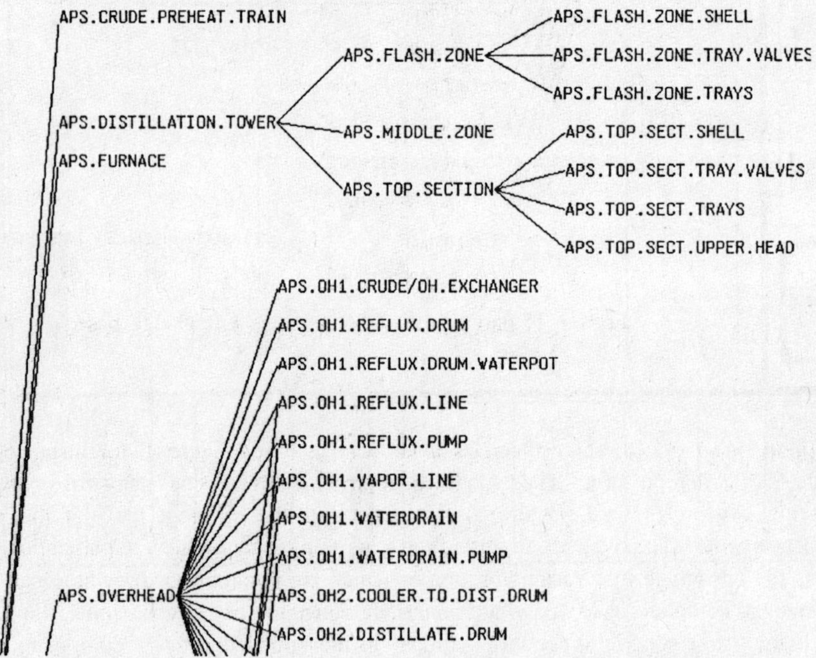

Figure 3: Knowledge base representation of the atmospheric distillation unit.

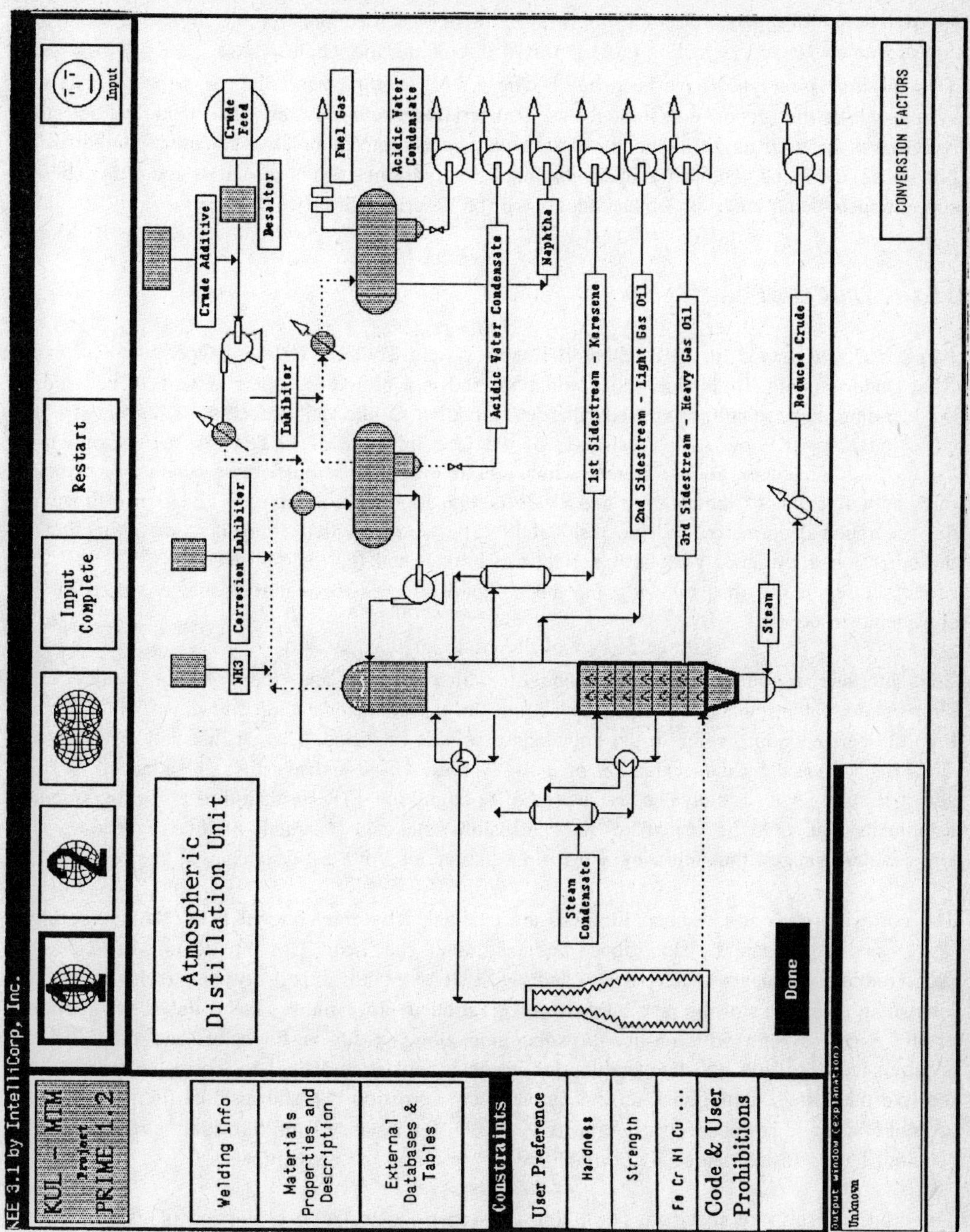

Figure 4: The atmospheric distillation unit.

(Copyright © by M.J.S. Vancoille and W.F.L. Bogaerts)

3.3 Prototype Design.

Two different prototypes have been build. One prototype implements the real time risk analysis system whilst the second one implements a materials selection system. The materials selection system can be used as an aid in the design of new equipment or as a guideline for corrosion preventive measures. The latter does not require a real-time response and is therefore able to support an in-depth analysis. Both modules will be described briefly.

3.3.1 PETROCRUDE.

PETROCRUDE stands for **P**RIME-based **E**xpert system **T**ool for **R**efinery **O**peration **C**orrosion **R**isk **U**nderstanding, **D**eduction and **E**valuation, and is a prototype expert system, to be used in the blending of petroleum crudes and the determination of allowable process parameters (from a corrosion point of view). An illustration of the user interface of PETROCRUDE is shown in Figure 5. It combines all parameters which are relevant for a quick assessment of corrosion rates as a function of temperature and crude composition. Furthermore, it gives an estimate of the accumulated corrosion and the residual lifetime of a specified component as a function of the corrosion allowance. Various materials can be selected (e.g. carbon steel, 5% Cr steel and Monel in Figure 5), thus allowing for the comparison between materials that are used or are considered to be used.

The interface is built with *active images* which change and monitor slot values. The temperature, for instance, can be set by sliding the thermometer bar up and down. At the same time the corresponding value in the knowledge bases is modified. This change will also produce side-effects caused by the evaluation of *active values*. These active values are attached to slots and fire whenever a slotvalue is accessed or modified. These data-directed programming techniques are used to calculate the corrosion rate and residual lifetime whenever the temperature changes, thus allowing a real time interaction with the system.

The corrosion rates and residual lifetimes are calculated by conventional FORTRAN programs, based on experimental data, thus demonstrating the seamless integration of classical programming languages such as C and FORTRAN with expert system technology. An interesting side effect of the possibility to integrate other programming languages is the coupling of the expert system with on-line sensors, thus allowing for on-line corrosion control. The accumulated corrosion and temperature for instance can be measured by corrosion probes and fed into the system, thus giving an indication of the corrosion rate and residual lifetime. If there are more accurate corrosion measurements available, for instance from ultrasonic inspections, the accumulated corrosion can easily be modified by means of the active images.

The crude can also be specified, permitting the experts to make an evaluation of its corrosivity (Figure 6). The crude can be specified either by name or by composition. Also different crudes (a blend) can be specified at once. The properties of this mixture are derived mostly by weighted averages of the individual components.

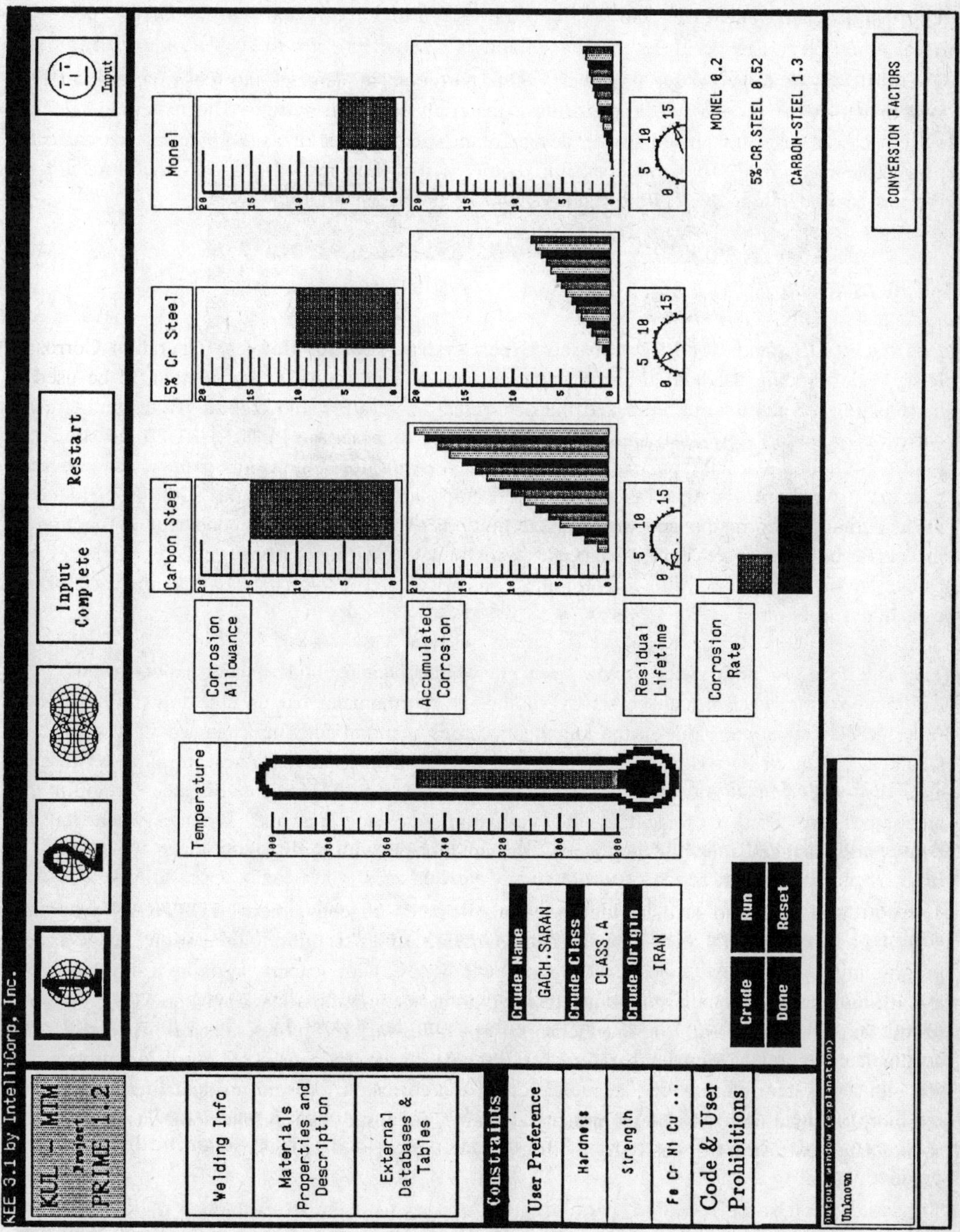

Figure 5: Prototype interface to the PETROCRUDE module.

(Copyright © by M.J.S. Vancoille and W.F.L. Bogaerts)

As an option, the system can also be set to do a more thorough evaluation of possible corrosion risks. After every change in temperature (or other parameters), it will scan its knowledge bases and determine what specific corrosion problems can be expected. This procedure however takes somewhat more time.

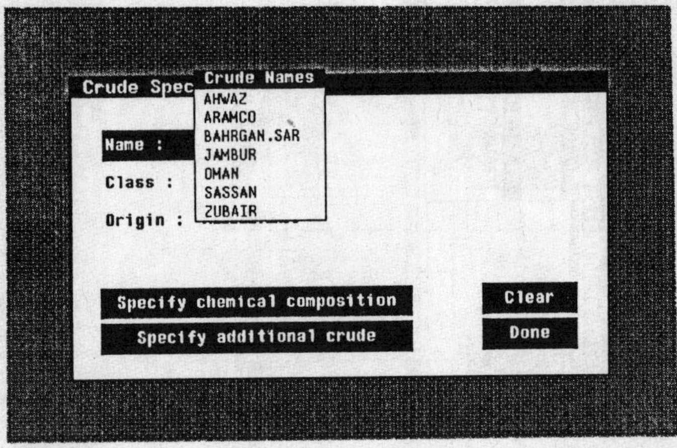

Figure 6: Crude specification input.

(Copyright © by M.J.S. Vancoille and W.F.L. Bogaerts)

3.3.2 Materials Selection Module.

Appropriate material selection is one of the principal preventive measures that can be taken to limit corrosion damage. Much of the required knowledge is rather easily accessible. It would be a shame not to use this valuable information. Moreover, it could be easily implemented taking advantage of most of the work that previously has been done on the PRIME expert system.

A consultation with the PRIME oil refinery module is an **interactive process**, where the user, in effect, can select and fill in all the information he has at hand by means of different corrosion engineering worksheets with icons, pop-up menus etc.

On-line user help facilities, and a graphics-based interface provide a user friendly environment that transcends conventional query based interfaces and yields dramatically enhanced productivity.

A session with the system might then be as follows. The user indicates in the **first Corrosion Engineering Worksheet** (Figure 7) that the industry is 'Petroleum Industry - Atmospheric Distillation Unit', the environment is 'oil' and the equipment is for instance the flash zone of the atmospheric distillation unit (Figure 4). It has to be noted that the interface will adapt to the

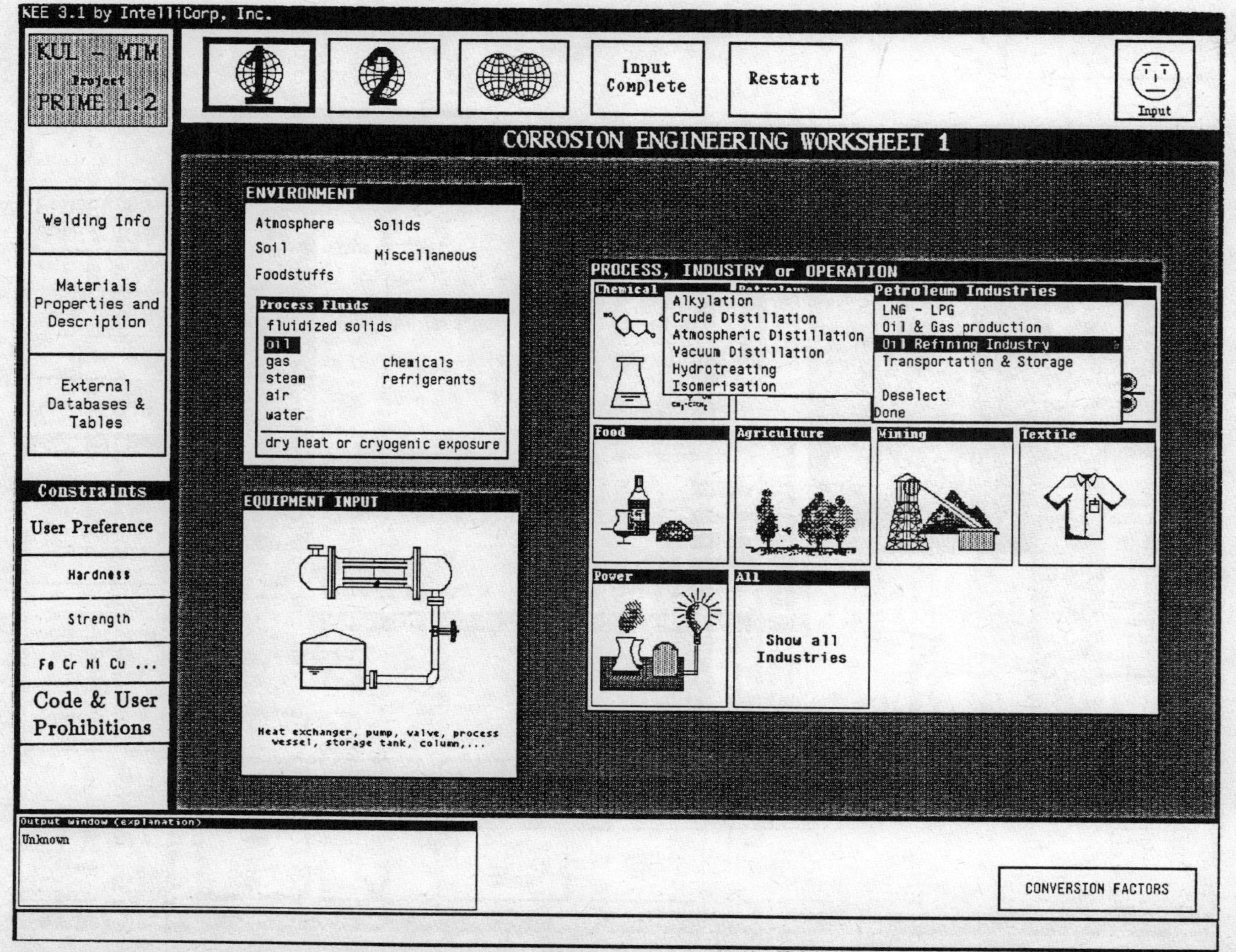

Figure 7: First Corrosion Engineering Worksheet.

(Copyright © by W.F.L. Bogaerts and M.J.S. Vancoille)

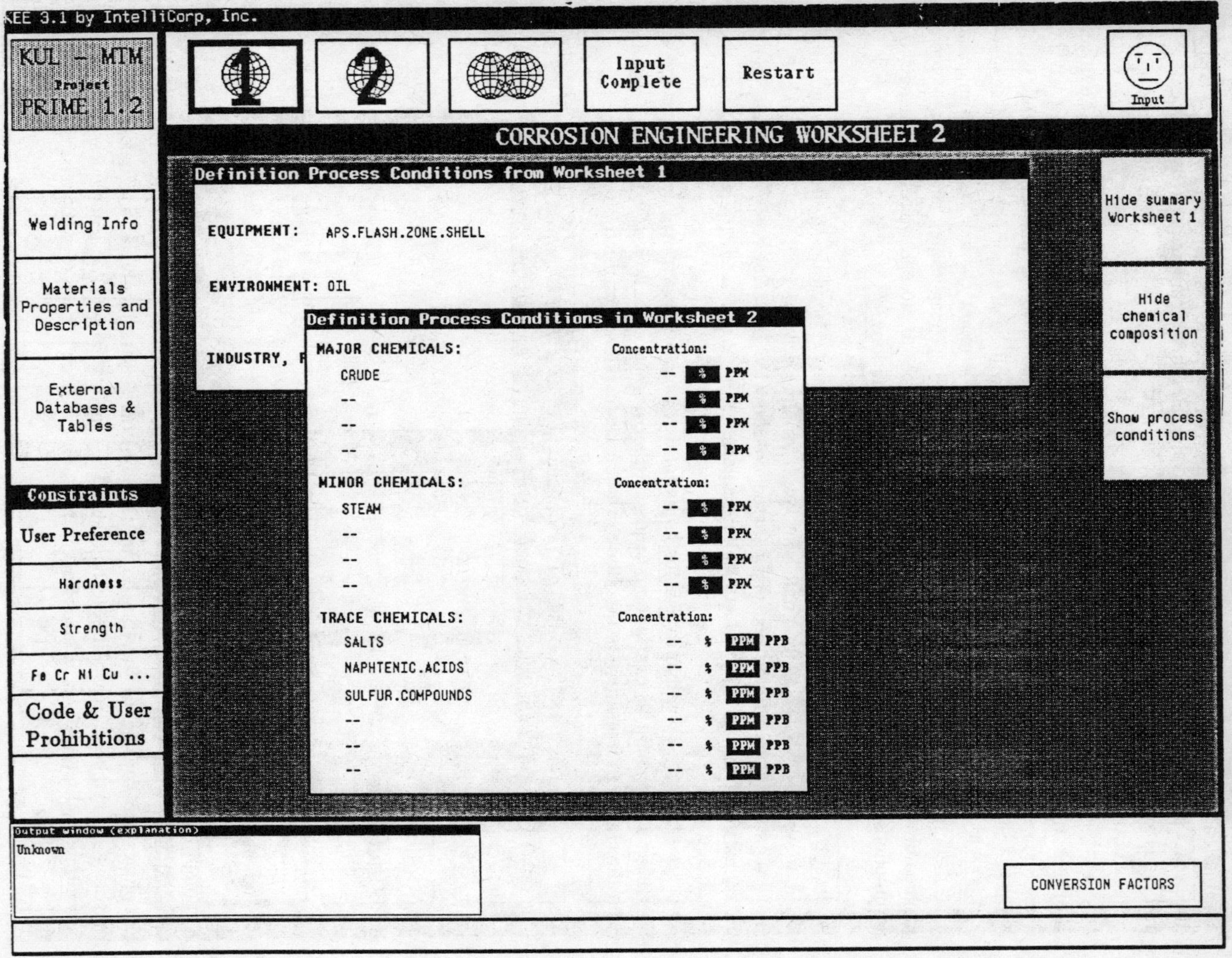

Figure 8: Second Corrosion Engineering Worksheet.

(Copyright © by W.F.L. Bogaerts and M.J.S. Vancoille)

information already given (*Active Interface*). The graphical representation of the atmospheric distillation unit will only pop up if the user specifies both parameters (environment is oil, process is atmospheric distillation unit).

On specifying 'Input Complete', the system starts reasoning. A **second Corrosion Engineering Worksheet** (Figure 8) appears which has already some values filled in. The system has made some assumptions regarding chemicals that are most likely to be present. For the example described above (flash zone), it is obvious that 'crude' will be a major chemical. Moreover, the system has also made the assumption that steam will be a minor chemical (steam is injected to heat the crude) that certain salts are trace chemicals (e.g. magnesium and sodium chloride, naphtenic acids, sulphur compounds). The user is free to accept, reject or change these assumptions at any time, or to fill in any known concentrations.

Eventually, the system will come up with some **questions**. For instance, assume that the user failed to specify what kind of crude was used. Another window will pop up (Figure 6) that enables him to specify the crude class (based on corrosion characteristics), the crude name or the origin or a blend of different crudes (cfr. supra).

At every stage, it is possible to ask for **explanations** why the system has deduced certain facts by means of the built-in explanation facility. This explanation facility called *HyperExplain* has been developed in house and is described in more detail in [Arents 1989].

Aside from the lists of the materials, one can also look at the properties of the different materials (e.g. corrosion rate, corrosion risks, chemical composition, mechanical properties, welding information, price, vendors, ...) by means of the *knowledge browser* which is incorporated in the *HyperExplain* module.

4 Conclusion

The recent acceptance of experts systems as a legitimate technology and the subsequent growth in commercial applications are just the beginning of a radical change in the way computers are viewed and used.

Although the project has only recently come out of its prototyping phase, it has already clearly demonstrated what can be done with state of the art hard- and software. Expert system techniques are capable of handling complex problems that can only be adequately dealt with bymeans of symbolic reasoning techniques. The application of A.I.-techniques to corrosion control and risk prevention in oil refinery operation has been chosen to demonstrate this clearly.

Furthermore, the use of powerful expert system building tools, combined with a well designed, user friendly graphics oriented interface can contribute to the more widespread use of expert system technology. The prototype has been conceived in such a way that the user does not require any precognition of A.I.-techniques and programming languages.

Acknowledgement

We would like to express our gratitude towards Mr. Somers, Arents (K.U.Leuven), Bartels, Demeyer and Desmedt (Esso Refinery Antwerp) for their valuable contributions and support in the realization of this project.

References

[Arents 1989]   Arents, H.C., Bogaerts, W.F.L. (1989). Use of Hypertext for Corrosion Analysis, *this volume*.

[Vancoille 1988]   Vancoille, M.J.S., Bogaerts, W.F.L., Rijckaert, M.J. (1988). PRIME, The European Project on Expert Systems for Materials Selection, *Proc. Corrosion'88*, ed. NACE, Houston, Texas, USA

USE OF HYPERTEXT FOR CORROSION ANALYSIS

Hans C. Arents, Walter F.L. Bogaerts
Faculty of Engineering - Dept. MTM
Katholieke Universiteit Leuven
W. de Croylaan 2, B-3030 Leuven, Belgium

ABSTRACT

Hypertext forms a novel approach to the problem of handling large amounts of heterogeneous information. Its underlying concepts appear to be well suited to the computer implementation of diagnostic guides, and should also enable us to devise more adequate explanation and help facilities for use in diagnostic expert systems. In this paper we will describe research on applying hypertext techniques to corrosion analysis, and highlight the advantages compared to more conventional techniques.

KEYWORDS: hypertext, corrosion analysis, expert systems, explanation/help facilities

1 Hypertext: a new way of handling information

Although the original concept of hypertext dates back for more than 40 years (Bush, 1945), only the past few years true hypertext systems have become commercially available. A hypertext system is essentially a complex database management system that allows one to connect information entities using associative links into a network of interrelated information (Fig. 1). These different information entities, usually consisting of screen or window-size workspaces, can contain text, graphics, images and audio or video data. Their networked interconnection gives the user a quick, intuitive access to the information through a mouse and menu-based interface.

1.1 Using hypertext

In a hypertext system, information is divided into discrete units of information, called *nodes*, that contain a single concept or idea. The information that is contained in such a node is usually displayed on the computer screen, either in a window or taking up the full screen. *Links* are used to connect these nodes into a network of interrelated information by defining the node's relationship to other nodes in the database. Such a link bears some resemblance to a footnote or reference in an ordinary text in the sense that it directly refers to a piece of related information. The links themselves are embedded in the text or the graphics displayed on the screen, and can be accessed and traversed by simply clicking with the mouse on a *link point*. A link point is a single character, token or icon that symbolizes a link in a hyperdocument. In addition to these link points there are also *buttons* that can trigger the display of additional information, perform a query to the database or activate an external program (Fig. 2).

Navigating through the database is done by simply following links to the various nodes or by activating buttons to find the information that is being sought after. This allows the user to search through the information by applying one of five distinct search strategies: *scanning* (covering a large area without detail), *browsing* (following links until a goal is achieved), *searching* (seeking an explicit piece of information), *exploring* (looking around at the extent of the information), *wandering* (purposeless and unstructured jumping around) (Canter *et al.*, 1985). In large databases however, the user runs a real risk of getting disoriented during his search, so most hypertext systems provide a *graphical browser*. A graphical browser is an utility that displays a structural diagram of the network of nodes, a global map of the network. With this map the user can orient himself, or zoom in directly to an area he is interested in by selecting a node or set of nodes with a mouse. Some hypertext systems also provide *paths* (default routes through the database) or *viewing filters* (filters that suppress additional levels of detail).

1.2 The potential of hypertext

The most urgent problem confronting today's post-industrial corporate environment is sheer *informational overload*, resulting in a serious cognitive overload. As the relentlessly expanding information technology keeps generating more and more data, no company can keep up with the daily flood of information that it has to process and assimilate. It is here that hypertext systems can play a crucial role as the core of future *knowledge support tools*, helping a company to apply its collective knowledge more efficiently (Drexler, 1987). Providing effective and lasting support for personnel working with knowledge will become a crucial asset in a company's competitive strategy (for a number of practical hypertext-based solutions see Morrison, 1989).

The two most important characteristics of hypertext systems that make them the ideal core of future knowledge support tools for personnel in the engineering industry are:

- *intuitive data handling*: flexible handling of interrelated collections of information through the possibility of establishing meaningful links between multi-source information
- *significant data compression*: hiding irrelevant and superfluous detail through the possibility of user-driven query and instant retrieval of only the appropriate information

These qualities make hypertext systems destined to become the core of user-friendly analysis and documentation systems. We will illustrate this by discussing the application of hypertext techniques to a corrosion guide and to the explanation/help facility of a corrosion expert system.

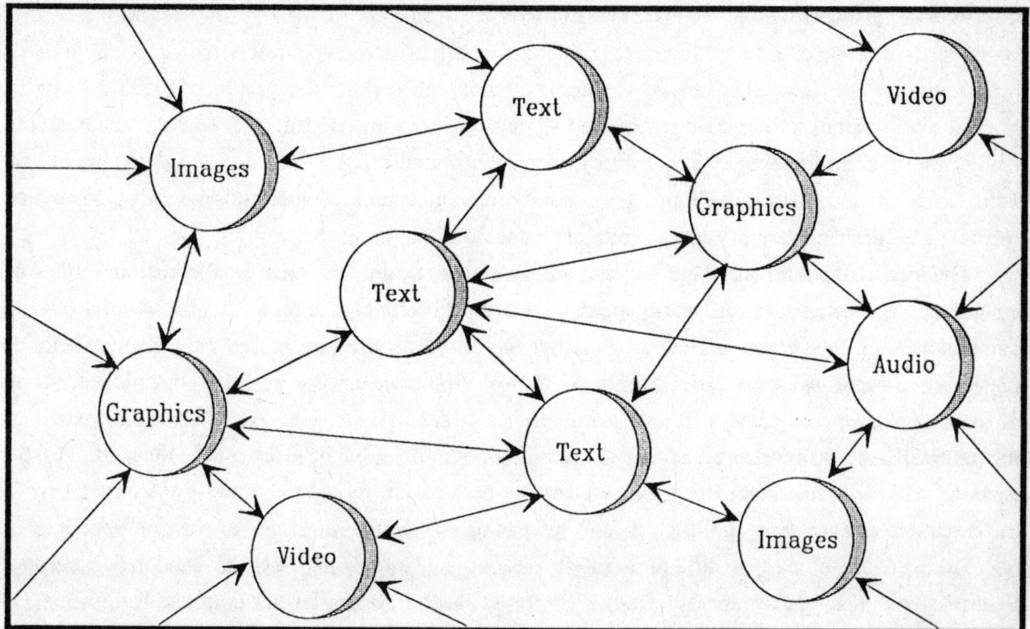

Figure 1. A hypertext information network.

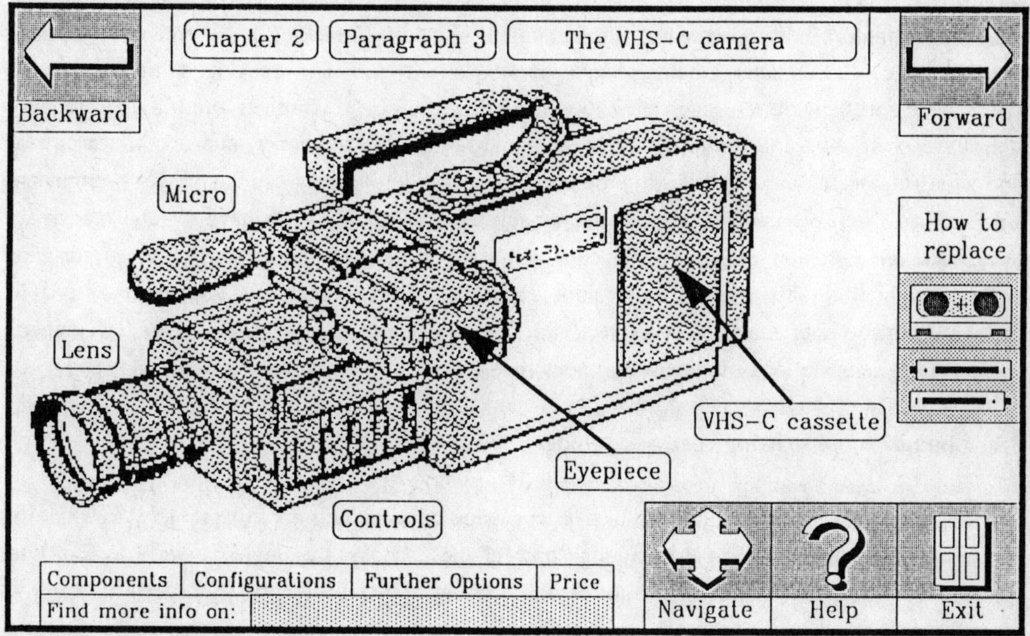

Figure 2. An example of a hypertext screen.

2 Combining hypertext and corrosion guides

Corrosion analysis has always been an important issue in the structural safety assessment of many industrial components. The various *corrosion phenomena,* together with other phenomena such as *creep* and *fatigue,* belong to the most frequent causes of materials degradation, which substantially limit a plant's safety, availability and lifetime.

The optimal strategies for structural safety enhancement, as well as for residual lifetime assessment and extension, are partly passive and partly active in nature. In the *passive phase,* components are inspected periodically and/or non-periodically, using several testing methods (destructive material investigation, replicas, X-ray, ultrasonic and other NDE techniques, strain measurements, etc.). These methods allow one to assess, as precisely as possible, the state of the material, the risk of any defects occurring, or the component's remanent lifespan. In the ensuing *active phase,* protective and/or corrective actions are thought out and implemented, with the intention of extending the lifespan and increasing the structural safety of the component.

During both phases of structural safety assessment, *corrosion analysis* is a major burden. Corrosion diagnosis is a daunting task, even for skilled and experienced corrosion engineers, as corrosion phenomena are extremely complicated and determined by many factors. For example, (microscopic) cracks or voids detected during inspection may not only result from creep, creep fatigue, stress relief cracking, high cycle or high strain fatigue processes, but may equally well have been produced by a variety of environment sensitive 'cracking' phenomena (e.g. intergranular attack (IGA), intergranular or transgranular stress corrosion cracking, knife line attack, hydrogen embrittlement, hydrogen penetration and blistering, ...). Furthermore, fatigue cracking may have been corrosion initiated, corrosion propagated, etc. Moreover, due to the scarcity of corrosion engineers, most corrosion problems occurring in industry are handled by engineers who lack an extensive knowledge of corrosion science. These engineers have to rely intensively on guides on corrosion diagnosis and case study reports, which are neither easy to use nor widely available. This results in a poor corrosion (risk) assessment and, very often, in completely inadequate remedial or corrective actions. Hypertext systems can form the core of computer implementations of corrosion guides that address these important shortcomings.

2.1 Limitations of existing corrosion guides

Most existing guides and handbooks on corrosion analysis and control problems can hardly be called 'easy to use'. When consulting some of these books, the novice as well as the more experienced corrosion engineer has to plough through piles of theory in order to find the specific practical information he is looking for. This problem is further aggravated by the fact that the corrosion experience needed for corrosion analysis and control is not centralized, but is widely scattered throughout many scientific journals, publications and handbooks.

For these reasons, many companies are forced to adopt a policy of compiling an in-house corrosion data handbook built of corrosion case studies and corresponding remedial procedures relevant to their particular type of production processes and equipment. This however has the disadvantage that those data are not available or useful outside the company, since they consist of proprietary or company-specific information. And the company itself is also faced with the problem that any change in the plant situation can make the data of the company's handbook useless, as most corrosion data are specific for one single situation and retain their predictive or corrective value only within minute deviations of the original situation.

2.2 Advantages of hypertext-based corrosion guides

To remedy the deficiencies of the conventional diagnostic guide, we have to create a new type of corrosion 'handbook' that can be used by the corrosion expert as well as the novice, a corrosion guide that links together in an intuitive manner theory with practical information. This can best be realized using a hypertext system, which combines together the fundamentals of corrosion theory, photographs of corrosion phenomena, corrosion case studies and information on specific corrosion prevention and correction methods.

Such a hypertext-based corrosion guide will form an interactive computer-based system that can be used as a tutoring tool as well as an on-the-spot trouble-shooting help. It will serve as the ideal source-book as it can be read in many different ways according to the user's degree of interest or level of expertise, resulting in a less steep learning curve. But more importantly, it will also allow the (would-be) corrosion engineer to perform a quick assessment of a particular corrosion problem by simply comparing the corrosion phenomenon under investigation with high resolution graphics pictures of corrosion damage case studies from the past. At the same time, he will have immediate access to the corresponding preventive and corrective measures that are recommended by more experienced corrosion engineers or by scientific researchers.

2.3 A session with a hypertext-based corrosion guide

In the framework of the ESPRIT-project *KWICK*, the R&D unit of our department is now actively involved in the development of what has been called the *Active Book*™ *on Corrosion*. This *ABC* will be a hypertext-based implementation of the renowned Elsevier *Corrosion Atlas* (During, 1988), supplemented by a comprehensive system of background information and expert knowledge. It is being prototyped in a collaborative effort with Elsevier Science Publishers.

The *ABC* hypertext system should offer the user computer-guided step by step procedures for identifying the origin of and the mechanism responsible for material degradation or failure. It will display case-studies from an atlas of pictures on corrosion and related damage, will contain information on theory and practice of recommended preventive and corrective measures, and will also include an adequate glossary and library facility (annotated bibliographic database).

The *ABC* will present to the user a hierarchical, multi-layered view of the available corrosion information, allowing him to consult only those parts of the different documents he is really interested in. In addition to the text, high resolution graphics, including a comprehensive set of macro- and micrographs of material damages, will constitute the graphical 'backbone' of the system. These micrographs can be viewed at different degrees of magnification, highlighting the more important microscopic features, and will contain link points to information contained in the same document or in other related documents of the corrosion guide. Easy access to pop-up menus and additional clarifying notes will allow the user to perform a detailed analysis.

In its final form, the *ABC* will contain illustrations of all predominant processes which may result in serious future damage to or eventual complete mechanical failure of a material. These processes include mechanisms initiated and sustained by environmental attack ('corrosion' in sensu stricto), as well as mechanisms which have characteristics and also effects similar to those resulting from environmental attack. Whenever possible, illustrations will be provided of the working surface from which the damage has initiated, illustrations of voids, cracks or fracture surfaces at high and at low magnifications, and a metallographic section through the defect or through a typical metallographic feature associated with the damage.

For each distinct failure mechanism with its commonly associated damage phenomena, the *Active Book*™ *on Corrosion* will provide an immediate and computer-guided access to:
- the crucial conditions leading to the appearance of the damage
- the environmental circumstances contributing to the further development of the damage
- the alternative mechanisms which might be confused with the responsible mechanism
- the ways in which mechanisms of similar appearance can be discerned and discarded
- the recommended analytical procedures
- the recommended protective and/or corrective actions

Whenever it is not possible or advisable to present to the user all (essential) information on a single screen, additional information will be hidden behind *expansion buttons*. Clicking on these expansion buttons will display more detailed information on a given point of interest.

The following paragraphs briefly describe some of the typical actions the non-experienced user can perform during a session with the prototype *Active Book*™ *on Corrosion*. After starting the *ABC* the user can enter into the hypertext 'book' through different access points (Fig. 3):
- through CORROSION CUBE, taking him to an innovative 3-dimensional table of contents that gives a visual perspective on the connections between the book's contents
- through CORROSION ATLAS CASE STUDIES, taking him to a collection of case studies that were taken from the Elsevier *Corrosion Atlas* and other bibliographical sources
- through PREVENTIVE MEASURES, taking him directly to protective and/or corrective measures
- through LIBRARY, taking him to a library of information on corrosion theory and practice
- through GLOSSARY, taking him to a glossary of the terms and concepts used in the book

Clicking on any of the icons brings the user to the corresponding subcomponent of the system.

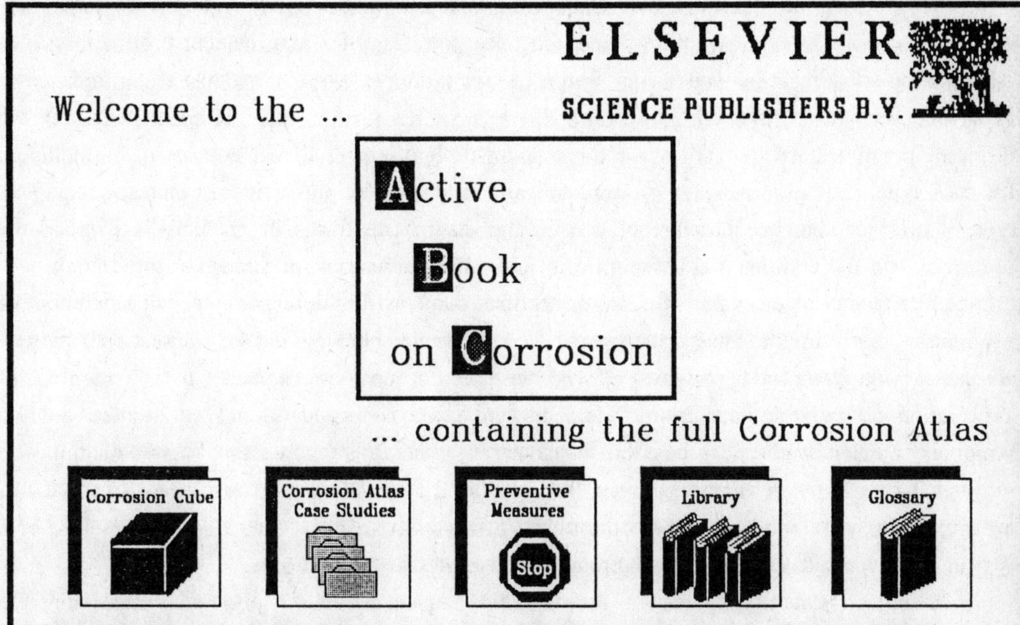

Figure 3. The opening screen of the *Active Book on Corrosion*.

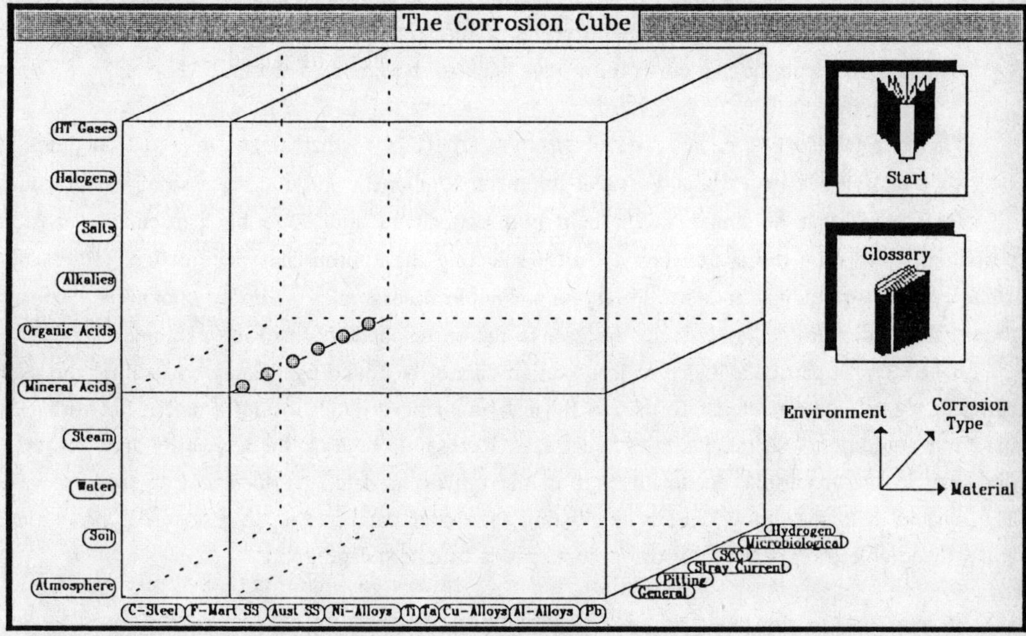

Figure 4. The Corrosion Cube table of contents of the *Active Book on Corrosion*.

After clicking on the Corrosion Cube icon, the user is offered a 'cube of contents' with entry points for MATERIAL, ENVIRONMENT, and CORROSION TYPE (Fig. 4). By selecting an appropriate combination of a material and an environment, for example FERRITIC-MARTENSITIC STEELS and MINERAL ACIDS, the user gets a display of an ordinary 1-dimensional (linear) table of contents, listing the different possible corrosion types for this combination of material and environment (including for each type: descriptions, ways of identification, mechanisms, preventive measures, etc.). This type of table of contents initially only shows the main items, but will gradually expand as the user clicks on the sections and subsections of the table he wants to consider in more detail.

Selecting one of the table of contents sections displays the corresponding (sub)document or information node in the hypertext system. For example, clicking on the subsection PREVENTIVE MEASURES AGAINST UNDERGROUND CORROSION allows the user to jump immediately to a screenfull of explanation on cathodic protection. Here he can easily access additional information on two important types of cathodic protection: by SACRIFICIAL ANODES or by IMPRESSED CURRENT. Clicking on SACRIFICIAL ANODES lets him choose between having a look at an indepth DESCRIPTION, an ILLUSTRATION or a VIEW OF A FIELD INSTALLATION. Clicking on VIEW OF A FIELD INSTALLATION gives him a scanned picture of a typical setup for cathodic protection through the use of sacrificial anodes.

Clicking on ILLUSTRATION gives the user a quick impression of the principles underlying the technique of cathodic protection by sacrificial anodes (Fig. 5). In this illustration he can point at ANODE or SACRIFICIAL ANODES for a definition in the glossary, or go straight to the GLOSSARY itself. At any moment he can get back to the beginning of the book by clicking on START, and restart the entire consultation, or continue another line of thought. He can also return through CONTENTS to the table of contents he started from, or JUMP back to the previous screen.

The user is also offered a powerful SEARCH FACILITY (Fig. 6) that allows him to find quickly, by only specifying a material and/or an equipment component and/or a corrosion phenomenon, all case studies that he thinks might help him understand and solve his particular corrosion problem. Activating the search facility, after selecting the appropriate specifications, filters out all relevant case studies from the library of available case studies. Simply clicking on one of these case studies immediately takes the user to the corresponding corrosion example.

If however a particular piece of information cannot be found by the search facility, the user can still try to locate related information by just browsing through the information network. In the future, additional search facilities will make it possible to look for keywords and free text, not only in the hypertext documents but in the corrosion data, graphics and figures as well. The system will also allow access to existing relational databases on e.g. material properties, and will be able to export its corrosion conclusions to external programs.

The *ABC* is being implemented for IBM® and IBM®-compatible PCs, using a dedicated and enhanced version of a commercially available hypertext tool for PCs, and is now being designed with a future release on CD-ROM in mind.

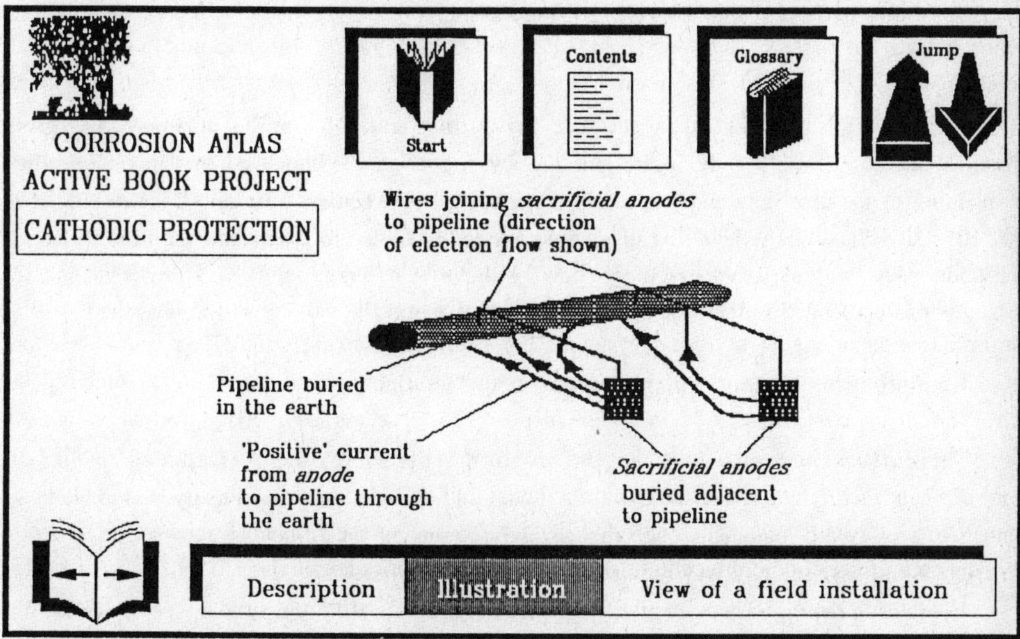

Figure 5. A screen illustrating the principles behind cathodic protection.

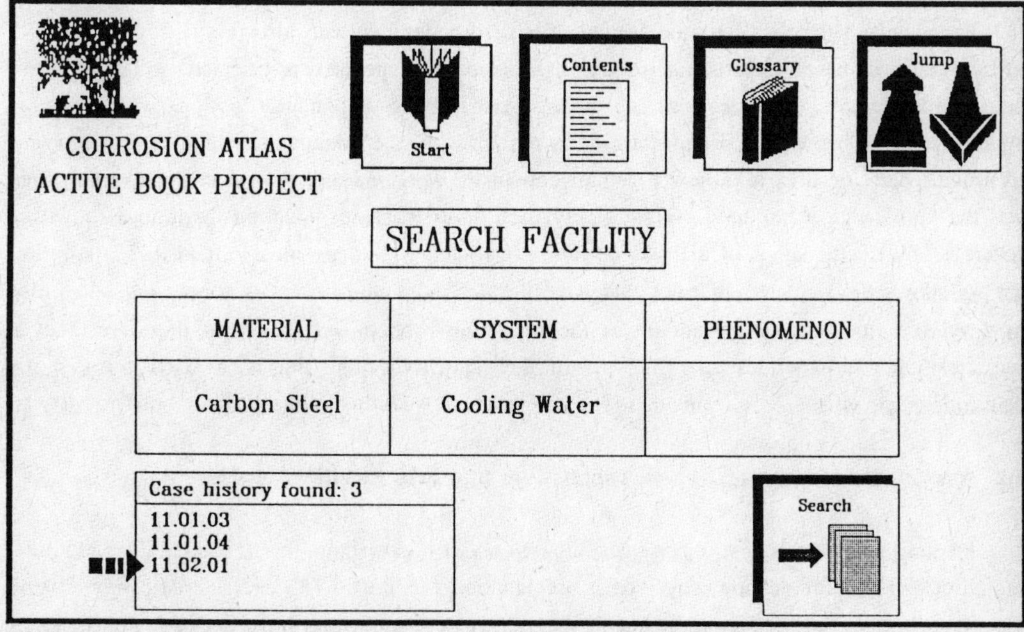

Figure 6. The search facility of the *Active Book on Corrosion*.

3 Combining hypertext and corrosion expert systems

An expert system without an adequate explanation and help facility only allows a given problem to be solved <u>for</u> the user, and not <u>by</u> the user. Without an explanation facility, explaining to the user how and why the system arrived at a certain conclusion, the user fails to see the reasoning strategy behind a given solution and is forced to accept that solution by an act of faith. And without a help facility, allowing the user to understand the knowledge and the assumptions embedded in the system, the user cannot assess the validity of the answer.

3.1 Limitations of existing explanation and help facilities

The *explanation facility* in an expert system describes the system's reasoning to the user and the *help facility* elucidates the system's model and knowledge base. Ideally, the explanation and help facility are intended to accomplish one or more of the following complex tasks in an intelligent and user-friendly way (adapted from Rolston, 1988):
1) assisting the knowledge engineer in building and validating the system
2) informing the user about the why and how of a given result
3) clarification of terms, concepts and knowledge used by the system
4) increasing the user's confidence in (and acceptance of) the system
5) enhancing the user's personal level of expertise (interactive tutoring)

Present-day rule-based expert systems try to achieve this type of user-oriented explanation and help facility in essentially one way: by offering two primary explanation options *why* and *how*. When the user asks *why* a certain question is put forward, the system shows the rule it is trying to prove, when the user asks *how* a certain conclusion was reached, the system shows the rules and the knowledge it has used. This widely used approach to explanation facilities is however severely limited: the scope of the *why* option is restricted to the current rule, and the *how* option simply shows the sequence of rules that have the desired statement as its conclusion. In order to develop a complete understanding of the reasoning process the user must therefore enter a series of *how* and *why* questions. The system then shows the rules that it has used in a <u>system-determined sequence</u>. As a result this type of explanation facility offers little or no flexibility.

3.2 Advantages of hypertext-based explanation and help facilities

Integration of hypertext techniques allows one to overcome the fundamental flexibility limitations of the simple how/why-type of explanation facilities. The characteristic property of hypertext systems is precisely the capability to follow non-sequential links between given pieces of information. Adopting this approach when developing an explanation and help facility for the PRIME expert system (see Vancoille *et al.*, 1988) resulted in a window and mouse-based explanation environment, called *HyperExplain,* which allows the user to retrace a proof path in a non-linear <u>user-determined sequence</u>. Incorporated into this explanation environment is a rule parser that transforms compiled representations of rules into a more 'human-readable' format. This new type of explanation environment helps the knowledge engineer when checking the validity of the knowledge and the reasoning built into the system. But more importantly, it also allows the non-expert user to explore only the parts of the proof path he thinks are important, which enables him to gradually increase and refine his knowledge of the domain and the system itself. This results in a system-supported tutoring process which nevertheless leaves absolute freedom to the user, thereby increasing his acceptance of and confidence in the system.

3.3 A session with a hypertext-based explanation facility

To fully understand the possibilities of a hypertext approach to an explanation facility, we will describe a typical session with our hypertext-based explanation facility *HyperExplain*, a module developed independently from the work done on PRIME (for more detailed information on the PRIME expert system see Bogaerts and Vancoille, 1989).

After the PRIME system has completed its forward chaining reasoning process, it presents a shortlist of recommended materials for corrosion prevention. The user then has the possibility of clicking on one or more of the recommended materials to ask for an explanation (Fig. 7). Selecting this explanation option shows the rule(s) that led to this recommendation, and the user is now capable of following the entire reasoning path in a backward direction, back to the facts the system has used (either built-in facts or facts provided by the user during the consultation). The actual user-directed replay of the reasoning process proceeds in the following manner:
  - the system shows the final rule that led to the advice on material selection
  - the user clicks on one or more of the rule's premises to be explained, if he fails to understand how the system proved this premise, or if he wants to explore in more detail the reasoning subpath that was used in proving this particular premise
  - the system shows the rule that it has used to prove the premise selected by the user, and highlights the corresponding conclusion in the rule (Fig. 8)

The rules that are being traced back are each shown in a separate *rule-window*, positioned on the screen and sized interactively by the user, and are displayed in a format that is easy to read and understand by an inexperienced user. This approach allows the user to walk through the entire reasoning path, by letting him explore different subpaths, return to a previous rule, close rule-windows he is no longer interested in, or descend until he reaches the level of the facts that were asserted by himself, or built into the system. He can now see for himself how the system arrived at certain conclusions, what kind of built-in facts or data supplied by the user it used, or where it went wrong due to a lack of built-in knowledge. As a consequence, the user is able to make a clear distinction between <u>system-dependent reasoning</u> and <u>user-dependent reasoning</u> that took place because of the user's own assertions and initial assumptions.

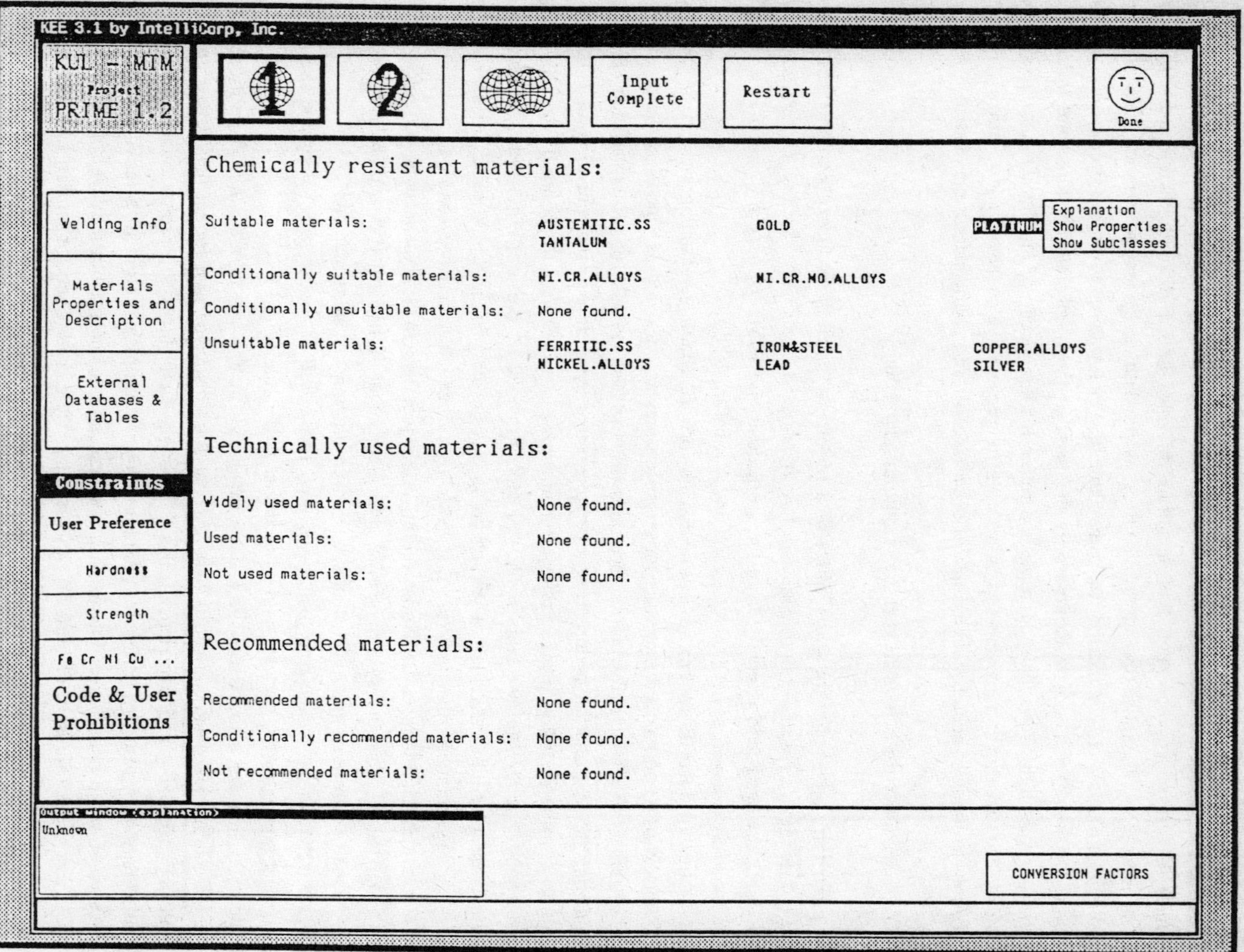

Figure 7. The hypertext-based explanation facility *HyperExplain*.

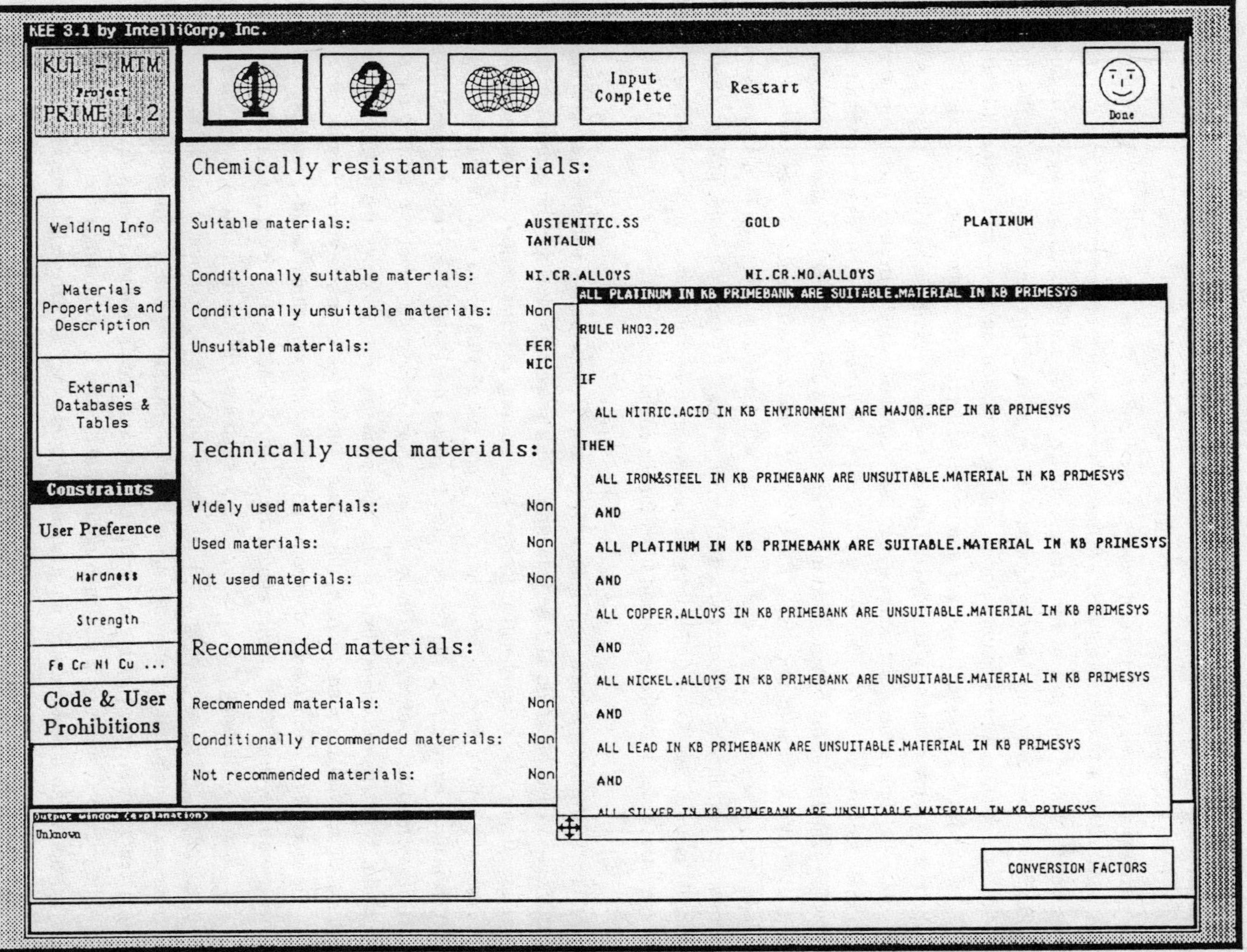

Figure 8. A replay of the reasoning process with *HyperExplain*.

4 Conclusions

The development work done sofar on the *Active Book*™ *on Corrosion* clearly demonstrates the potential of a hypertext approach to the computerization of 'corrosion guides'. Hypertext's inherent flexibility and ease of use, the possibility to link and address heterogeneous forms of information, the ability to hide unnecessary information details until they are really needed, make it the ideal tool for implementing a guide to corrosion analysis, prevention and control.

A number of important interface and conceptual problems inherent to the design of large hypertext systems remain to be solved (for an overview see Halasz, 1988). Issues such as handling the link structure, managing the data complexity and effective user navigation still demand extensive research. The *Active Book*™ *on Corrosion* will serve as an important testbed for the evaluation of the hypertext design methodologies that are only now slowly emerging.

The final version of the *Active Book*™ *on Corrosion* will be released as a commercially available CD-ROM product at the end of 1990. It will serve as a prototype product to assess the commercial viability of hypertext-based reference works. As the CD-ROM technology slowly but steadily matures, we can undoubtedly look forward to an ever increasing number of a new breed of hypertext-based knowledge support tools for the engineering industry.

The PRIME explanation and help environment *HyperExplain* confirms the superiority of hypertext-based explanation and help facilities with respect to flexibility and user-friendliness. The use of this type of explanation facility as a tutoring tool has serious weaknesses however: the user can explore subpaths of the reasoning process of little or no real tutorial value, he can fail to make a clear distinction between the deep and the shallow knowledge embedded in the PRIME system, etc. Addressing some of these shortcomings will involve extending the tutorial capabilities of *HyperExplain,* by e.g. indicating in the rule-windows the degree of tutorial interest of a given reasoning subpath, the kind of knowledge used in a given premise, etc.

In order for *HyperExplain* to become an intelligent explanation facility, assisting the user when working with PRIME, future enhancements should include the following functionalities:
- a *glossary*, explaining the terms and the concepts used in the formulation of the rules
- a *knowledge-browser*, giving a visual representation of the contents of and relations in
  the knowledge bank(s), allowing the user to understand the system's knowledge
- a *reasoning-browser*, giving a visual representation of the entire proof path, allowing
  the user to zoom in on a particular point of interest in the reasoning process

By extending the object-oriented programming methodology to hypertext systems, in the sense that each hypertext node is considered as an object that responds to different requests in different ways, it should become possible to adopt a more knowledge-oriented view of hypertext (as advocated by Weyer, 1988). In this way the explanation facility can be made to adapt itself to the user's preferences, and can become directed by his previous actions and present goals.

Acknowledgements

We would like to thank ir. P. Van Wiechen, dr. A. de Ruiter and J. Blecker (E.S.P.), and C. Dekker (M.I.D.), for their stimulating and enlightening discussions on the implementation aspects of the *ABC*. We would also like to thank dr. K. Agema (K.U.Leuven) for his comments on the corrosion aspects of this paper, and ir. M. Vancoille (K.U.Leuven) for his continuous programming support during the development of *HyperExplain*.

References

Bogaerts, W.F.L. and Vancoille, M.J.S. (1989).   Practical building of expert systems for materials selection and corrosion problems.   In *Expert Systems in Structural Safety Assessment* (ed. MPA Stuttgart), Springer-Verlag, Heidelberg, in press.

Bush, V. (1945). As we may think. *Atlantic Monthly*, **176**, 101-108.

Canter, D., Rivers, R. and Storrs, G. (1985).   Characterizing user navigation through complex data structures. *Behaviour and Information Technology*, **4** (2), 93-102.

Drexler, K.E. (1987).   *Hypertext publishing and the evolution of knowledge*.   Stanford University Press, Stanford, **7**.

During, E.D. (1988). *Corrosion Atlas. A Collection of Illustrated Case Histories*, *Vol. 1 and 2*. Elsevier Science Publishers B.V., Amsterdam, Oxford, New York, Tokyo, 242 pp.

Halasz, F.G. (1988).   Reflections on NoteCards: Seven issues for the next generation of hypermedia systems. *Communications of the ACM*, **31** (7), 836-852.

Morrison, A.W. (1989).   Hypertext and expert systems - Experiences and prospects. In *Proc. Fifth International Expert Systems Conference* (ed. Learned Information (Europe) Ltd), Bocardo Press Ltd, Didcot, Oxon, pp. 1-10.

Rolston, D.W. (1988).   Chapter 7. Explanation facilities. In *Principles of Artificial Intelligence and Expert Systems Development*. McGraw-Hill Book Company, New York, pp. 113-136.

Vancoille, M.J.S., Bogaerts, W.F.L. and Rijckaert, M.J. (1988).   PRIME, The European project on expert systems for materials selection. *Proc. Corrosion '88* (ed. NACE), NACE Press, Houston, in press.

Weyer, S.A. (1988).   As we may learn.   In *Interactive Multimedia: Visions of Multimedia for Developers, Educators, and Information Providers* (eds Ambron, S. and Hooper, K.), Microsoft Press, pp. 87-103.

For a more general introduction to hypertext see:
Fiderio, J. (1988). A grand vision. *BYTE*, **13** (10), 237-244.

REASONING ABOUT DATA: THE COMBINATION OF MULTIPLE AND NON HOMOGENEOUS SOURCES OF KNOWLEDGE

Andrea Servida

Commission of the European Communities, Joint Research Centre,
Ispra Establishment, 21020 ISPRA (VA), Italy
and
Petten Establishment, 1755 ZG PETTEN, The Netherlands

Abstract

In structural engineering critical is the combination of multiple and uncertain sources of knowledge. Commonly, these sources are any set of data, expert opinions, physical models or outcomes from instruments or techniques monitoring a particular event. Whilst the structure of the composition problem can differ from one case to another, we argue that a general pattern is preserved. Thus, the investigation and characterization of this pattern can help to highlight the nature of the dependencies between the different sources. The paper aims at presenting a rigorous and general approach to managing the imperfection of data. The proposed analysis is performed in the frame of the theory of fuzzy uncertainty measures. In fact, the expressiveness of these measures makes them suitable to represent a wide range of situations. In the paper we present two examples of application of the above analysis to real problems. The perspective of the analysis towards a knowledge engineering approach is emphasized.

KEYWORDS

data fusion, fuzzy uncertainty measures, theory of evidence, expert opinions.

INTRODUCTION

The growing interest in the analysis of engineering systems and the development of innovative computer technologies enhanced the capability of tackling the variety of information imperfection. Moreover, the semantics of uncertainty and imprecision has been thoroughly investigated [17], [6]. Uncertainty usually refers to the truth of data in the sense of conformity to a specified reality. On the other hand, a piece of information is imprecise if its content is not well defined, either in relation to a scale of comparison (eg.: imprecision of a measurement), or in the sense that its contour is vague or fuzzy.

Having made this distinction, many authors focused on studying the multifacet aspects which can be highlighted in the nature of uncertainty and imprecision. Meanwhile, in the theory and practical developments of expert systems, the management of uncertainty and imprecision has been often left to **ad hoc** schemes for representing and combining rules. With this practice sound theory and clear semantics were often disregarded. However, the contribution of the expert systems theory to the subject is undeniable.

In the following only uncertainty is addressed. The analysis of engineering systems is mainly concerned with the numerical characterization of uncertainty. Hence, disregarding the investigation of symbolic modelling, some numerical techniques are presented. For an extensive presentation of the state of the art in approaching uncertainty, reference can be made to several authors [2], [5], [12], [6]. Fuzzy uncertainty measures (FUM) and evidence theory (ET) are two of the most important and complete frameworks for dealing with uncertainty, numerically. The formal presentation of these theories, which is outside of the interests of the papers, is briefly reviewed and attention is paid to some examples of data fusion problems.

The paper is divided in three parts. In Chapter 1 are the fundamentals of FUM and ET which are relevant to the following applications. The aim of the chapter is at defining a comprehensive framework were the semantics of the various techniques for modelling uncertainty is thoroughly explained. Examples of application data fusion are presented in Chapters 2 and 3. References to the papers containing the detailed treatment of the examples is also provided.

1. DEALING WITH UNCERTAINTY IN A CONSISTENT NUMERICAL FRAMEWORK

This chapter aims at explaining the basic ideas which are behind the uncertainty modelling techniques used in the case studies detailed in Chapter 2 and 3. In this respect, the fuzzy uncertainty measures and the theory of evidence are introduced. It is highlighted the possibility of defining proper composition structures as a tool to enhance both the representation capabilities as well as the characterization and the understanding of the interdependencies among the being compound sources of knowledge.

1.1 FUZZY UNCERTAINTY MEASURES

In the Seventies the concept of fuzzy uncertainty measure (FUM) [16] was introduced in order to define a comprehensive framework for dealing with subjective and epistemic uncertainty. In this regard, it can be argued that probability theory and Bayes' theorem have extensively and successfully been used for managing these types of uncertainty. However, two are mainly the motivations that prompted Sugeno to define the large class of fuzzy measures. On the one hand, is the rigidity of combination structure in probability theory, which relies on a set of axioms which do not completely adhere to the semantics of epistemic uncertainty. On the other hand, are the analysis and study of uncertainty semantics.

For sake of simplicity, let X be a finite universe, a fuzzy measure (or Sugeno measure) is defined as a set function $g(\cdot)$ mapping from a σ-algebra ß on X to $[0,1]$, such that

i) $g(\Phi) = 0$

ii) $g(X) = 1$

iii) $\forall\, A, B \in ß$ if $A \subseteq B$ then $g(A) \le g(B)$.

Thus, a fuzzy measure is a monotonic measure, not decreasing in the sense of inclusion, which differs from the classical probability measure by relaxing the additivity property. The measure $g(A)$ can be interpreted as a measure of the subjective state of knowledge about the occurrence of A.

The above axioms define a very large class of measures, which can be particularized in different ways. Particularly, the class of decomposable form has been proposed, where the uncertainty measure for union of events (likewise, for intersection) can be obtained directly by their respective measures of uncertainty [7]. In this respect, the following axiom has been introduced:

iv) $\forall\, A, B \in ß$, if $A \cap B = \Phi$ then $g(A \cup B) = g(A) * g(B)$

where $*$ is some operator under which $[0,1]$ is closed. This means that, whenever A and B are disjoint, the measure of the union event is given by a simple function of the two original measures. The algebraic structure of ß induces compatibility constraints on $*$ which lead to the definition of $*$ as a triangular conorm [7]. The class of triangular conorms can be represented using parametrised families [3], [7]. Here are the main triangular conorms

$$\max(a,b) \le a+b-a{\cdot}b \le \min(1,a+b) \le T_w^{*}(a,b) = \begin{cases} a & \text{if } b=0 \\ b & \text{if } a=0 \\ 1 & \text{otherwise.} \end{cases}$$

Each choice for the operator * in axiom iv) characterizes a particular measure of uncertainty. For instance, FUM includes, as a special case, the probability measure. According to the axiom above, one can say that each uncertainty measure and the semantics of uncertainty it represents, are fully depicted by its composition pattern.

Possibility measure $\Pi(\cdot)$, obtained replacing * by $\max(\cdot,\cdot)$, is investigated in some details as it will be used for the analysis in the next chapters. For this measure the axiom iv) becomes

$$\forall \ A, B \ \varepsilon \ \ss: A \cap B = \Phi$$

$$\Pi(A \cup B) = \max(\Pi(A), \Pi(B)). \qquad (1)$$

Consequently, the possibility measure is the lower bound among the measures belonging to its class, and its composition behaviour is the most cautious one. Moreover, the semantics of the measure is completely clarified by the following relations:

$$\forall \ A \ \varepsilon \ \ss$$

$$\max(\Pi(A),\Pi(\overline{A})) = 1$$

and since $\Pi(X) = 1$,

$$\max_{x_i \varepsilon X}(\Pi(x_i)) = 1.$$

These relations mean that at least one of two contradictory events is always possible, and that among the singletons at least one is completely possible. Finally, it can be pointed out that the normalization condition does not allow one to infer anything about the negation of an event that is completely possible, in fact

$\forall\ A\ \epsilon\ \text{\ss}$

$$\text{if } \Pi(A) < 1 \quad ====> \quad \Pi(\overline{A}) = 1$$

$$\text{if } \Pi(A) = 1 \quad ====> \quad \Pi(\overline{A}) = ?.$$

$\Pi(A)=\Pi(\overline{A})=1$ signifies a situation of complete ignorance, where every event is possible. In this respect, the semantics of possibility measure exactly corresponds to the one usually ascribed to the common concept of "possible".

The representation of the state of knowledge concerning an event A by means of g(A) is incomplete. In fact, g(A) is not exhaustive in the sense that it does not explicitly quantify the uncertainty of the contrary event not A. This information is provided by the dual fuzzy uncertainty measure which may be defined replacing the axiom iv) by

v) $\forall\ A,B\ \epsilon\ \text{\ss}$ and if $A \cup B = \mathbf{X}$

$$g'(A \cap B) = g'(A) * g'(B).$$

The measure $g'(\cdot)$ derived from the axioms i), ii), iii) and vi) is a decomposable measure which is related to the previous $g(\cdot)$ by the equality

$\forall\ A\ \epsilon\ \text{\ss}$

$$g'(A) = 1 - g(\overline{A})$$

and

$$g(A) = 1 - g'(\overline{A}).$$

These duality relations clarify the meaning of the link between dual fuzzy measures. The operator in v) belongs to the class of triangular norms $S(\cdot,\cdot)$. We refer the interested reader to [7] where the formal characterization of the different measures and their semantics are thoroughly investigated. Here are the inequalities which hold among the main triangular T-norms:

$$\min(a,b) \geq ab \geq \max(0,a+b-1) \geq T_w(a,b) = \begin{cases} a & \text{if } b=1 \\ b & \text{if } a=1 \\ 0 & \text{otherwise} \end{cases}$$

To the scope of the paper, it is worth emphasizing that the description of uncertainty by means of fuzzy measures usually requires the definition of two numbers: $g(\cdot)$ and $g'(\cdot)$. For any event A, $g(A)$ is the direct measure of the uncertainty of A and $g'(A)$ provides an indirect quantification of the uncertainty of A, as it results from the uncertainty analysis of all the events external to A. In case of possibility measure the dual measure is the necessity, defined by the T-norm $\min(\cdot,\cdot)$, $a*b = \min(a,b)$. That is

$\forall$ A, B ε ß, and $A \cup B = X$

$$N(A \cap B) = \min(N(A),N(B)) \qquad (2)$$

and the duality relations are

$\forall$ A ε ß

$$\Pi(A) = 1 - N(\overline{A}) \qquad (3)$$

$$N(A) = 1 - \Pi(\overline{A}). \qquad (4)$$

The significance of the measure coincides with the common concept of necessity, as degree of impossibility of the contrary event. This correspondence is further clarified by the relations

$\forall$ A ε ß

$$\text{if} \quad N(A) > 0 \quad \text{then} \quad N(\overline{A}) = 0 \quad ===> \quad \Pi(A) = 1$$

and

$$\text{if} \quad \Pi(A) < 1 \quad \text{then} \quad \Pi(\overline{A}) = 1 \quad ===> \quad N(A) = 0$$

that is to say, every necessary event is always completely possible. Conversely, if an event is not completely possible then it cannot be necessary at all. From the definition it is easy to verify that

$\forall$ A ε ß

$$\Pi(A) \geq N(A).$$

In this respect, it can be said that the uncertainty of A is bounded between the lower limit $N(A)$ and the upper limit $\Pi(A)$.

1.2. THE THEORY OF EVIDENCE

Alternatively to FUM, theory of evidence (ET) provides another formalism for modelling uncertainty. The general principles of the theory are found elsewhere [15], in the following fundamentals will be described while drawing the attention to those aspects which make ET a suitable framework for representing the epistemological uncertainty.

Let X be the frame of discernment, that is the universe of the possible values of a variable, as stated by the analyst; $P(X)$ the set of all the subsets of X. Then a belief function $Bel(\cdot)$ is a set function mapping from $P(X)$ into $[0,1]$, and the axioms [15] define $Bel(\cdot)$ as a monotonic measure of order infinity in the sense of Choquet's capacity theory [4]. Banon demonstrated that a $Bel(\cdot)$ is a FUM [6]. Shafer demonstrated that a belief function can be uniquely defined through a basic probability assignment (bpa) $m:P(X) --> [0,1]$ such that

1) $m(\Phi) = 0$

2) $\sum_{A \varepsilon P(X)} m(A) = 1$

where $\forall A \varepsilon P(X)$

$$Bel(A) = \sum_{B \subseteq A} m(B).$$

Any subset A of $P(X)$ such that $m(A)>0$ is called focal element. The union of all the focal elements of a belief function is called the core.

The quantity "$m(A)$ measures the belief that one commits exactly to A, not the total belief that one commits to A" [15]. In other words, $m(A)$ represents the exact value of our confidence that the truth is in A, as it results from our state of knowledge about the event. $Bel(A)$ is the inferred value that the truth is in A starting from the basic knowledge assessment. While $m(a)$ is the basic information which, because of the structure of our knowledge, cannot be atomized any more. $Bel(A)$ states the total degree of uncertainty about A as it is supported by our state of knowledge (bpa).

Being $Bel(\cdot)$ a FUM and according to the what explained in the previous section, a dual measure $Pl(\cdot):P(X) --> [0,1]$ can be defined as $PL(A)=1-Bel(\bar{A})$. This function is named plausibility and can be obtained from the bpa as

$$PL(A) = \sum_{A \cap B \neq \Phi} m(B).$$

The properties of measures Bel(·) and Pl(·) depend on the structure of the basic probability assignments. The suitability of ET to represent epistemological uncertainty is thoroughly investigated in [10], [15], to which we refer the interested readers. However, it can be reminded argue that this type of uncertainty has to do with

i) lack of specificity, that is the focal elements are compound events and endowed with granularity;

ii) the impossibility of defining correctly the sole admissible and exhaustive frame of discernment;

iii) the temporal and spatial limits of the knowledge;

iv) the selection of measures to quantify the uncertainty of the events would reflect the judgment of the analyst and subjective preference.

Moreover, it is found that the theory of evidence can easily deal with situations where

- the information gathered, or represented, are not mutually exclusive;

- the structure of the information is "informative", such as

. consonant structure (internal coherence and consistency among the events);

. conflicting structure (disjoint events);

. simple support function (homogeneous evidence);

. separable support function (resulting from the merging distinct homogeneous evidences);

- the ignorance differs from disbelief because of the distinction between lack of belief and disbelief.

The uncertainty modelling envisaged by ET is consistent with the one proposed by FUM. For instance, it can be underlined that possibility and necessity measures, which were introduced as particular FUM, correspond to belief structures defined by consonant basic probability assessments. In this respect, the structure of bpa is quite important for understanding the correct applicability of the measures.

1.3 COMBINATION OF UNCERTAINTY AND COMPOSITION STRUCTURE

Epistemic uncertainty is intimately coupled to the lack of precise knowledge about the conformity between a piece of information and the real system to which the information refer to. Expert system technology provides some effective modelling tools to represent in a coherent framework all the different types of uncertainty which can be identified. The consistency of the frame is justified by soundness of the formal theory involved and by the analysis and the characterization of the nature of uncertainty that is entailed.

In the previous sections two general approach for dealing numerically with uncertainty were presented. At representation level the theories are quite satisfactory, but at propagation level they are not. In fact, apart the particular problem of combining distributions which belong to the same universe of discourse and that is performed according to the definition axioms of the measure, critical is case of composition of distributions defined on different universes. As a general pattern we can highlight, apart from possible normalization factors, the following general structures

$$(m_1 * m_2)(A) = \bigvee_{\substack{ij \\ A_i*B_j=A}} (m_1(A_i) \wedge m_2(B_j)) \qquad (5)$$

and dually

$$(m_1 * m_2)(A) = \bigwedge_{\substack{ij \\ A_i*B_j=A}} (m_1(A_i) \vee m_2(B_j)) \qquad (6)$$

where m_1 and m_2 are distributions or assessments, according to the adopted framework; $\vee$ and $\wedge$ are generalized union and intersection operators; $*$ can be any set theoretic operation. The particularization of the structure is done by assigning the operators and operations. It is clear that the choice is made according to the needs of the composition problem to be modelled and complying with the features of the distributions or assessments involved. For instance, in ET the combination of two independent belief structures, respectively defined by m_1 and m_2, is given by a bpa m obtained by applying Dempster rule as follows

$$m(A) = 1/K \cdot (m_1 \cap m_2)(A)$$

where

$$(m_1 \cap m_2)(A) = \sum_{\substack{ij \\ A_i \cap B_j = A}} (m_1(A_i) \cdot m_2(B_j)) \tag{7}$$

which is apart from a normalization factor a particularization of (5). Similarly, bearing in mind the equivalence between fuzzy set and possibility distribution [19] one can recognize in the (5) and (6) the pattern of the fuzzy composition. However, abstracting from the particular case, the point is how to characterize the single composition rule. That is to say, how should the meaning be ascribed to the various operators and operation involved.

As envisaged in previous papers [10] the composition should be defined at two different levels: the Relation level and the Projection level. The Relation level of a composition is the informative merging step in which the uncertainty measures or distributions, attached to the combining pieces of knowledge, are combined. The operation deals with granules whose universes of discourse are generally different. In this respect, the choice of the operation and the operator to be used should preserve the structure and properties of the composition situation being modelled. On the other hand, the Projection level provides the internal informative operation to be used in the universe defined by the Relation. The Projection step can be envisaged as necessary for collecting all the information that, while resulting by the aggregation of separate pieces of information, do address the same entity. Therefore, the Projection involves quantities belonging to the same universe, and its formal properties will correspond to the identified structure of the composition situation.

For sake of simplicity only the semantics of the operators is addressed, referring the interested reader to the available literature for a complete treatment of the matter [10]. The Relation level is the step where the various links and interactions highlighted among the different addends can be represented. The choice of a particular operator should be carried out bearing in mind both the type of interactions, among the sources of knowledge that have provided the addends, and the composition properties that those interactions require. The formal frame to describe $\vee$ and $\wedge$ is that of triangular norms and conorms. These classes of operators are dual one, that is a duality relation is defined which allows to compute, by the knowledge of an operator belonging to a class, the correspondent operators to the other class. In this respect, we will explain in detail the significance ascribable to triangular conorms, referring the interested reader to specialized literature. To better understand this point, let us start from the operators instead that from the interactions. Let us consider the generalized intersection which is modelled by the class of triangular norms. Dual argumentations hold for the generalized union operators. In section 1.1 is a list of the main t-norms. The interpretation of the composition scenarios modelled by the different operators is obtained looking at their properties. Referring the interested reader to the detailed investigations [10], one may have this very general classification

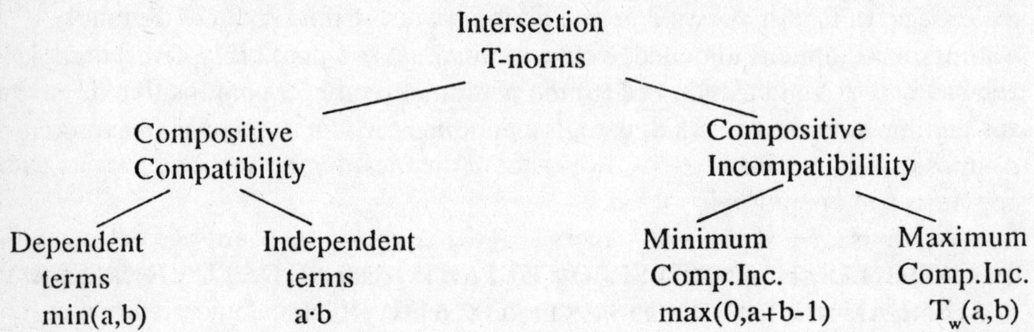

where compatibility and incompatibility refer to the fact that in the first case, given that both the two terms are greater than zero, the intersection would be meaningful (greater than zero). On the other hand, in the second case some compositive compatibility relations hold

i) $T_w(a,b) > 0$ iff $\max(a,b) = 1$

ii) $\max(0,a+b-1) > 0$ iff $a+b > 1$.

The second remark is related to the operators semantics. The min operator models the composition of terms completely dependent, because the idempotency of composition may be assumed to be the peculiar property, being valid in the limiting case of absolute and complete dependence. Independent terms are compound by means of the product operator, whose semantics is clear from the probability heritage. The semantics of $\max(0,a+b-1)$ and $T_w(a,b)$ is related to the compositive compatibility relations. In fact, we can say that the "$\max(\cdot,\cdot)$" operator is well suited for minimum compositive incompatibility because it defines the intersection combination for all those events whose overall uncertainty measure is high ($a+b>1$). On the other hand, the $T_w(\cdot,\cdot)$ operator is well suited in the case of maximum compositive incompatibility, as it defines the intersection combination only in the case that at least one term is completely possible. In choosing the $\wedge$ operator at relation level one should bear in mind the above classification, and the highlighted operator properties. However, starting from this frame it would be easier to classify the real composition, too. Dual argumentation hold for the choice of operators at Projection level.

The particularization of (5) and (6) is always the result of a decision process whose inference path can not be represented exhaustively. We can think of this process in terms of cognitive process, that is, a reasoning action about the knowledge. When dealing with uncertainty, the processes of this type are quite common, because the main source of uncertainty is epistemic one, that is, related to the cognitive approach to real systems. In this respect, expert system technology provides one with some very basic tools for coping with and reasoning about knowledge. In fact, it allows one to represent not only the knowledge about a problem, structured as knowledge base (domain knowledge) and inference engine (how to use the knowledge), but at the same time we are provided with a frame that can captures the metaknowledge, extremely important for the optimum exploitation of the knowledge itself.

2. POSSIBILISTIC ANALYSIS OF EXPERIMENTAL DATA YIELDED BY MANUAL ULTRASONIC TESTING CAMPAIGNS

It this section we shall see how FUM can be applied to the analysis of experimental data. First, the main features of the experimental setup are presented and it is explained why it was necessary to carry out the analysis in the frame of fuzzy measures. Second, the data and methodology used for the analysis are described. Finally, the results and conclusions are presented. The complete analysis is in [14] to which the interested reader is referred.

2.1. THE DATA ANALYSIS EXERCISE

The experiment, which yielded the data, aimed at characterizing the dispersion connected with the application of ASME procedure [1] which should be enforced when performing manual ultrasonic testing. The problem is that of the analysis of data on errors of location and sizing of cracks (artificially implemented in a weld) as result of manual ultrasonic testing campaigns. The complete explanation of both the experiment setup and the evaluation of the outcomes is reported in [14] to which the interested reader is referred. The testing had been performed by four different teams (T1,T2,T3,T4), on the same welding sample, in the same conditions, at the same place, using the same equipment and according to the same ASME procedure. The results of the destructive examination of the sample, performed by a steering committee, were provided to all the teams. On that basis, histograms of absolute errors in locating and sizing the cracks, along the longitudinal axis X and across weld axis Z, have been calculated. These histograms were the data available to the analyst.

To understand why fuzzy measures have been used to elaborate these data, one must look at the nature of the conditions that had defined the experiment. Manual ultrasonic testing, as stated in the ASME standard, is extremely affected by the contribution of the operator. The main source of scatter, which biases the results of these testing campaigns, is the human factor. This refers to a whole series of actions and interpretation steps which in practice the operator carries out subjectively. Closer investigation of the question shows additional difficulties having an epistemic nature [13]. Naturally, due to the equipment and the recommended procedure, the data is also affected by "pseudo recurrent" error. In fact, this type of imperfection can not be defined as truly stochastic, because it includes not only measurement errors, but also technique errors arising from the equipment and the common core of the procedure, which all the operators share in performing the testing campaign. Given these premises, it becomes clear why a pure statistical analysis could not have been carried out. The boundary conditions which define the data are not homogeneous among the teams. Thus, the epistemic nature of the imperfection seems to justify the analysis in the frame of fuzzy uncertainty measures.

2.2. POSSIBILISTIC CHARACTERIZATION OF THE ORIGINAL DATA

As mentioned above, each team has calculated the absolute frequency histograms for defect detection and the errors committed in dimensioning the parameters which had been selected to characterize the events of defect location L and defect sizing S.

We shall use absolute frequencies of occurrence to calculate necessity and possibility measures. Let τ be the generic parameter and $X = \{x_1,...,x_n\}$ the discrete space of values on which τ is defined, where x_j is the mid-point of the corresponding interval $[x_j-\delta x_j, x_j+\delta x_j]$. The experimental data, regarding the generic parameter τ, can be rearranged so to obtain the normalized frequency distribution $p(x_j)$ of x_j, with

$$\sum_{j=1}^{n} p(x_j) = 1.$$

Following Shafer's characterization of necessity and possibility measures, Dubois and Prade [8] have introduced a necessity measure defined as the amount by which the normalized frequency of an event exceeds the maximum normalized frequency of any of the contrary events, that is

$$\forall x_j \in X$$

$$N(x_j) = Max(p(x_j) - p(\overline{x_j}),0) \qquad (8)$$

where

$$p(\bar{x_j}) = \underset{\substack{x_i \varepsilon X \\ x_i \neq x_j}}{\text{Max}}\ p(x_i).$$

When a compound event is considered a similar relation is obtained.

According to the duality relations (3) and (4) a possibility measure $\Pi(\cdot)$ for x_i can be inferred as

$$\forall x_i \varepsilon X$$

$$\Pi(x_i) = \sum_{j=1}^{n} \min(p(x_i),p(x_j)) \qquad (9)$$

and the possibility of a compound event A is immediate from the axioms. It is worth emphasizing that the measures above are coherent to the axioms of definition and they also satisfy the inequalities

$$\forall A \varepsilon X$$

$$N(A) \leq P(A) \leq \Pi(A) \qquad (10)$$

which state that, whenever a probability measure for an event A is defined, it must belong to the interval $[N(A),\Pi(A)]$. In the problem addressed $P(A)$ cannot be defined, and the meaning of the above relation should be understood by thinking of $P(A)$ in terms of relative frequency of occurrence of A, $p(A)$. Thus, the ordinal information expressed by $\Pi(A)$ and $N(A)$ would encompass the original data $p(A)$ if the construction of the measure will satisfy the inequalities (10).

2.3. INITIAL UNCERTAINTY MEASURES

In the safety assessment of welded structures, such as pressure vessels, critical are location and size of cracks across the welds. Thus, focus ought to be on the interpretation of manual ultrasonic testing data concerning the dimensions along the axis Z and, for sake of brevity, only on the frequency of occurrence of errors of sizing (ESZ)is reported. In Table I are the normalized frequencies of occurrence of ESZ for the teams T1, T2, T3 and T4.

Team	ESZ (mm)							
	-40	-20	0	20	40	60	100	120
T1	.06	.26	.26	.16	.26			
T2	.05	.25	.30	.25	.15			
T3		.20	.20	.15	.05	.10		.30
T4	.05	.21	.16	.16	.21	.16	.05	

Table I. Normalized frequency distributions of errors (in mm) in sizing cracks along Z axis (ESZ), for teams T1, T2, T3 and T4.

For each team, using formula (9) of Section 2.2, necessity and possibility measures have been calculated. However, we will focus only on possibility measures because, as explained elsewhere [11], the necessity measures are not operationally significant, unless one is interested in the uncertainty values of compound events. The main concern of this analysis is the uncertainty distribution of the atomic events, therefore we will keep track only of possibility measures. Table II summarizes the possibility distributions of ESZ, calculated for all teams.

Team	ESZ (mm)							
	-40	-20	0	20	40	60	100	120
T1	.30	1.0	1.0	.70	1.0			
T2	.25	.95	1.0	.95	.65			
T3		.90	.90	.75	.30	.55		1.0
T4	.35	1.0	.90	.90	1.0	.90	.35	

Table II. Possibility distributions of errors (in mm) in sizing cracks along Z axis (ESZ) for all teams.

The meaning of the above distributions will be clarified further on. However, we must point out that their shapes retain, in their ordinal nature, all the descriptive properties of data consistent with the semantics of possible.

2.4. CHARACTERIZATION OF THE POSSIBILITY MEASURES AND FINAL RESULTS

It is clear that both the frequency distributions and the corresponding uncertainty measures are conditional upon the detection event D. An upper bound for the effective distribution can be obtained through a simple conjunction operation. For instance, the effective possibility distribution of error ESZ, $\Pi^*_{ESZ}(\cdot)$, results from the intersection of the conditional possibility $\Pi_{ESZ}(\cdot|D)$ and $\Pi(D)$

$$\Pi^*_{ESZ}(\cdot) = \Pi_{ESZ}(\cdot|D) \wedge \Pi(D)$$

where $\wedge$ is any triangular norm. Whenever, as in our case, non interactivity holds, it is

$$\Pi^*_{ESZ}(\cdot) = \min(\Pi_{ESZ}(\cdot|D), \Pi(D)). \tag{11}$$

Given these premises, the uncertainty measures of detection event D and of its negation D~ have been calculated for each team, as shown in Table III

Team	p(D)	p($\overline{D}$)	N(D)	N($\overline{D}$)	Π(D)	Π($\overline{D}$)
T1	.70	.30	.40	0.0	1.0	.60
T2	.74	.26	.48	0.0	1.0	.52
T3	.74	.26	.48	0.0	1.0	.52
T4	.70	.30	.40	0.0	1.0	.60

Table III. Uncertainty measures $N(\cdot)$ and $\Pi(\cdot)$ for the events D and $\overline{D}$, obtained from the corresponding relative frequency distribution $p(\cdot)$.

For all the teams $\Pi(D)=1$. Thus, from (11) and from a similar relation for $\Pi^*_{ESZ}(\cdot)$, it follows that each absolute error distribution is always equivalent, in values, to the conditional one.

In the following we shall be looking at the main features of both the above initial possibility distributions, and the possibility measures obtained by applying the intersection and union operators to them.

2.4.1. INITIAL POSSIBILITY DISTRIBUTIONS

The possibility distributions for ESZ retain the ordinal properties of the initial data. The meaning that can be ascribed to the distributions is related to the semantics of possibility. The greater the value of possibility of an event, the more likely, or plausible, is its occurrence. This result follows from the ordinal ranking established on the universe of the events by the distribution itself. The quantification of uncertainty is, in this respect, not absolute, rather ordinal one. Nevertheless, such information is, in this case, much more meaningful than any ''pseudo-probability'' measure, as it preserves, in its definition axioms, the imprecision paradigm characterizing the experiment itself.

Looking now at the tables II, we can argue that T1 and T2 are the only teams capable of correct sizing. However, only T2 can be reliable in sizing, as for T1 possibility measure equal to 1 is also ascribed to error classes identified by -20 and 40 mm.

2.4.2. POSSIBILITY MEASURES FOR THE INTERSECTION

The possibility distributions for ESZ can be merged to obtain the intersection distribution. This tells us about the common unavoidable error which affects every ultrasonic measurement. This scatter may be interpreted as being caused by what is shared among the teams during the measurement campaign: the technique, as defined in Section 2.1.

In general the intersection of several different sources of information can be defined by the formula
$$\forall\, x_i \in X$$

$$\Pi_{INT}(x_i) = \Pi_1(x_i) * ... * \Pi_n(x_i) \tag{12}$$

where the operator * belongs to the class of triangular norms. The choice among the available operators is carried out by looking at both the semantics of the operators [11], and the relations between the sources. Tables IV reports the results of (12), having replaced * with the main T- norms. The tables show the computed and normalized values of the intersection possibility distributions.

	ESZ (mm)			
T-norm	-20	0	20	40
min(a,b)	.90	.90	.70	.30
Normal	1.0	1.0	.78	.33
a.b	.855	.81	.4489	.195
Normal	1.0	.9474	.525	.2281
max(0,a+b-1)	.85	.80	.30	0.0
Normal	1.0	.9412	.3529	0.0
$T_w(a,b)$	0.0	0.0	0.0	0.0
Normal	=	=	=	=

Table IV. Possibility distributions for the intersection event, Π_{INT}, regarding the error of sizing ESZ (in mm). Each row of the table gives the results of (12) obtained by replacing * with the corresponding first column T-norm. The results are shown in both the calculated and in the normalized upon the supremum forms.

The table above displays how the intersection conditions, applied to the existing compound distributions, become stronger going from the min to the T_w operator. The meaning of this behaviour is related to the semantics ascribed to operators with regard to the compositional pattern being modelled by them as reported in section 1.3. Whenever the various Π_i in (12) are not interactive, the relation holds

$$\forall\, x_i \,\varepsilon\, X$$

$$\Pi_{INT}(x_i) = \min_{j=1,..,n} \Pi_j(x_i). \qquad (13)$$

In our problem the Π_j are assumed to be non-interactive, and (13) produces the results shown in Fig. 1 where the normalized distribution is also presented. The intersection possibility distribution represents ordinal information (which can therefore be renormalized) about the scatter caused by the available testing technique.

As discussed in Section 2.1, this type of error is pseudo- recurrent; once it has been identified, it can be treated as pseudo stochastic. Applying reverse computation to formula (9), we can get, from Π_{INT} in its normal form, a probabilistic characterization of the technique errors, Figs. 2. This result could have not been achieved by statistical analysis, as the nature of the data does not allow one to differentiate between the various sources of error that affect the data. However, by following a completely possibilistic approach, it has been possible to separate the effects of the inmost recurrent and, at the extreme, stochastic source of dispersion.

2.4.3. POSSIBILITY MEASURES OF THE UNION

Applying dual argument to the possibility distributions of errors, we can evaluate the union possibility distributions for ESZ:

$\forall \, x_i \, \varepsilon \, X$

$$\Pi_{UNI}(x_i) = \Pi_1(x_i) + ... + \Pi_n(x_i) \qquad (14)$$

where the operator + belongs to the class of triangular conorms. As in the case of intersection, the operator suitable, to be used instead of +, is chosen by looking at both the semantics of the operators and the relations between the sources. However, the meaning to be ascribed to the union operation is that of the sum of all the different sources of imperfection that would affect any manual ultrasonic testing campaign. In fact, these distributions can be viewed as providing an overall assessment of type of errors arising from all those aspects which characterize a measurement: the technique and the human factor. Table V shows the results obtained from (14) having replaced + with the main T-conorms. In the tables we have indicated with C1, C2, C3 and C4, the T-conorms max(a,b), a+b-ab, min(a+b,1) and T_w^*, respectively.

The selection of the operator in (13) should reflect the properties highlighted in the compositional structure. For instance, assuming that the initial distributions are "equivalent" to four different pieces of information, but belonging to the same universe of discourse, we should choose the max operator. In fact, this is the operator which we find in the definition axioms of possibility measure and which operates on distributions of the same universe. Fig. 3 shows the possibilistic information about the dispersion of the measurement testing obtained through the max operator.

T-con.	ESZ (mm)							
	-40	-20	0	20	40	60	100	120
C1	.35	1.0	1.0	.95	1.0	.90	.35	1.0
C2	.6588	1.0	1.0	.9996	1.0	.9550	.35	1.0
C3	.90	1.0	1.0	1.0	1.0	1.0	.35	1.0
C4	1.0	1.0	1.0	1.0	1.0	1.0	.35	1.0

Table V. Possibility distributions Π_{UNI}, for the union event regarding the error of sizing ESZ (in mm). Each row of the table gives the results of (13) obtained by replacing + with the corresponding first column T- conorm.

3. BELIEF STRUCTURE FOR EXPERT POOLING

The aim of this chapter is to describe an application of ET to expert judgments pooling. Particularly, experts are asked for percentiles of the seismic capacity of a certain class of structural components. Then, the expert judgments are merged to estimate the failure fractions of these components following a (horizontal) seismic ground acceleration. The suitability of ET to deal with the issue and the proposed methodology are investigated.

3.1. THE CASE STUDY

A group of experts (five experts) provides the estimation on 10th and 90th percentiles of the fragility curve. The goal is at building the fragility curve which results from merging the provided estimates. The estimates and the generic description of the corresponding knowledge background (or expertise) are found in Table VI. It is assumed that three factors mainly contribute to qualify the knowledge background: data (D), methods (M) and thoughts (T). Data stay for the availability of information concerning experience on failure and testing of comparable components at high acceleration. Methods refers to reliable analytical model and computer codes. Lastly, thoughts concerns the skill and practice the expert has in the field.

The problem can be dealt with by relying upon Bayes' rule. However, this approach does not grasp the complexity of the problem, as it introduces some very restricting hypotheses. The limits of the Bayesian methodology in facing the expert opinions pooling problem have been thoroughly investigated in a previous paper [10] to which the interested reader is referred.

```
------------------------------------------------------------------------
Experts        Opinions on percentiles (L)          Knowledge
                 10%           90%                   backgroud
-------------|---------------|--------------|------------------------------
   E1          1.0           2.0                     D, M, T
   E2          1.4           2.25                    D, M
   E3          1.5           3.0                     D, M, T
   E4          1.5           3.0                     T
   E5          2.5           4.0                     D, M, T
------------------------------------------------------------------------
```

Table VI. Experts opinions on 10% and 90% percentiles. Values of L are in units of Safe Shutdown earthquake. Knowledge Background may be characterized in terms of data (D), analytical models (M) and thoughts (T).

Conversely, ET is suitable to handle this type of problems as it provides one with a representation framework which may easily preserve consistency in the analysis. In fact, both the flexible structure of the assessment (bpa) as well as the availability of a number of composition structures enlarge the scope of this modelling technique. It is worth highlighting that, looking at the single expert as a separate body of knowledge, the judgments pooling problem may be considered exactly equivalent to that of combining different sources of knowledge. Thus, abstracting from the named application, the procedure and the pattern adopted for approaching this problem would be of general application.

3.2 THE METHODOLOGY

The definition of 10th and 90th percentiles makes the representation of the estimates provided by each expert straightforward. In fact, the estimates can be interpreted as focal events of a belief function concerning the failure event of the structural component. Thus, let $\vartheta = [0, 1]$ be the frame of discernment; $E = \{E_i\}$ with i=1,...,5 be the group of experts; and let

λ_{i1} be the estimate by expert E_i concerning the upper limit of interval $[0, \lambda_{i1}] \subseteq \vartheta$, having failure fraction at 10%;

λ_{i2} be the estimate by expert E_i concerning the upper limit of interval $[0,\lambda] \subseteq \vartheta$, having failure fraction at 90%;

Consequently, to each expert E_i one can associate a bpa $m_i = \{m_{i1}, m_{i2}, m_{i\vartheta}\}$ where

$$m_{i1} = m_i(0,\lambda_{i1}) = 0.10, \quad m_{i2} = m_i(0,\lambda_{i2}) = 0.90,$$

$$m_{i\vartheta} = m_i(0, L) = 0.10.$$

The above assignment defines a consonant belief structure in which m can be interpreted as the measure of how likely is, according to the expert, the occurrence of the event outside its frame. The structure of bpa is direct consequence of both the definition of percentile as well as the subjective nature of the estimates. Once the bpas are built for all the experts, the combination of the various assessments is performed by pointing out the relationships between the different bodies of evidence (as was argued in section 1.3.). It is worth underlining that the composition structure, being adopted for merging the original data, is defined on the basis of a clear investigation of the nature of dependencies and links existing between the sources of knowledge. Thus, the basic pattern of section 1.3 is completely specified by the demanded choice of operators and operations.

3.3. THE RESULTS

The flexibility of the approach is shown by presenting some limiting cases: i) equivalent and independent experts, ii) non-equivalent and independent experts and iii) equivalent and dependent experts.

3.3.1. EQUIVALENT AND INDEPENDENT EXPERTS

In this case Dempster's rule can be applied and the computation is straightforward. The results are presented in figure 4 and 5. The distribution $N(0,\lambda)$ represents the necessity of the failure in $[0,\lambda]$, for which the possibility is always 1. Conversely, the possibility distribution of the opposite event $\Pi(\lambda,L)$ is found in Fig. 5. One may also interpret this distribution in terms of occurrence of the single event λ, so a correspondence is established between the two above measures on one hand and cumulative probability function and the probability values, on the other.

3.3.2. NON-EQUIVALENT AND INDEPENDENT EXPERTS

The case of non equivalent experts is a puzzling one. In the contest of ET the combination of non equivalent bodies of knowledge has been dealt with by introducing a discounting factor [15]. In the following it is argued that instead of having a factor which may hardly be justified, the decision maker has for each expert to provide an assessment (bpa) which takes into account whether part and/or the whole assessment is supposed to be "oversized" and/or "undersized". Then, the bpa of the decision maker can be combined with the original piece of data by straightforward application of Dempster's rule. The possible structure of the decision maker bpa is derived from the nature of the problem and by complying some consistency rules [9].

3.3.3. EQUIVALENT AND DEPENDENT EXPERTS

Whenever belief structures are dependent, Dempster's rule cannot be applied any more in its original form. As it was presented in Chapter 1, it is argued that a generalized composition structure should be used. To obtain the proper composition structure one has to define in rule (5) both the suitable union and intersection operators as well as the set theoretic operation to be applied. The particularization of the structure is done by pointing out which are the target characteristic that should be preserved in the composition. For instance, whenever idempotency is considered the target and the bpa are consonant, the rule (5) becomes

$$(m_1 \cap m_2)(A) = \mathop{Max}_{ij} (min(m_1(A_i), m_2(B_j))$$
$$A_i \cap B_j = A$$

where * is the intersection, $\vee$ and $\wedge$ are the max and min operators. Figs. 6 and 7 show the results of the example in case of dependent operators.

CONCLUSION

The general problem of combining different sources of data has been dealt with in the framework of fuzzy uncertainty measures and evidence theory. The suitability of the approach has been tested against two real example of data analysis. The perspectives of Expert system as natural environment for facing this type of problem was highlighted. Moreover, the particularization of the representation model has been pointed out as knowledge engineering problem.

ACKNOWLEDGMENTS

The author is indebted to Prof. S. Garribba (CESNEF, Politecnico di Milano), Prof. G. Volta and Dr. A.C. Lucia (JRC, Ispra) for the fruitful discussion regarding the above analysis. Special thanks go to Dr. S. Crutzen (JRC, Ispra), who provided the experimental data of chapter 2, and to Dr. J. Bressers (JRC, Petten) for the support during the preparation of the paper.

REFERENCES

[1]ASME Code, Section XI, 1974.

[2]Bonissone, P., Decker, K. S. (1986). 'Selecting uncertainty calculi and granularity: an experiment in trading-off precision and complexity'. In Kanal and Lemmer (Eds.) Uncertainty in Artificial Intelligence, Elsevier (1986).

[3]Bonissone, P. (1987). 'Summarizing and propagating of uncertain information with triangular norms'.Int. J. Approx. Reasoning, **1**, no. 1, pp. 71-101.

[4]Choquet, G. (1953). 'Theory of capacities'. Ann. Institut Fourier, **5**, 131-295.

[5]Cohen, P. R., Grinberg, M. R. (1983). ' Theory of Heuristics Reasoning about Uncertainty'. The AI magazine, **4**, 2, 17-23.

[6]Dubois, D., Prade, H. (1980). Fuzzy Sets and Systems: Theory and Application. Academic Press, New York.

[7]Dubois, D., Prade, H. (1982). 'A class of fuzzy measures based on triangular norms. A general framework for the combination of uncertain information'. Int. J. General Systems, **8**, 43-61.

[8]Dubois, D., Prade, H. (1983). 'Unfair coins and necessity measures: Towards a possibilistic interpretation of histograms'. Fuzzy Sets and Syst., **10**, 15-20.

[9]Garribba, S., Servida, A. (1988). 'Evidence aggregation in expert judgments'. In Bouchon, Saitta and Yager (Eds.): <u>Uncertainty and Intelligent Systems</u>, Springer-Verlag (1988).

[10]Garribba, S., Servida, A., Volta, G. (1989). 'Generalized structures for the composition of evidence'. <u>Proceedings of the 3rd IFSA World Congress</u>, Seattle, WA (USA), August 6-11, 1989.

[11]Garribba, S., Lucia, A.C., Servida, A., Volta, G. (1988). 'Fuzzy measures of uncertainty for evaluating non-destructive crack inspection'. <u>Structural Safety</u>.

[12]Lesmo, L., Saitta, S., Torasso, P. (1985). 'Evidence combination in Expert Systems'. <u>Int. J. of Man-Machine Studies</u>, **22**, 307-326.

[13]Servida, A. (1985). <u>Applications of fuzzy logics to non destructive testing: a model for the interpretation of ultrasonic measurements</u>. Doctoral thesis, School of engineering, Politecnico di Milano, Milano, (in italian).

[14]Servida, A. (1989). 'Fuzzy uncertainty measures: a tutorial approach'. <u>Proceeding of the Ispra Seminar on "Structural Reliability"</u>, May 11-15, 1987, Reidel Publishing, 1989.

[15]Shafer, G. (1976). <u>A Mathematical Theory of Evidence</u>. Princeton University Press, Princeton, NJ.

[16]Sugeno, M., (1974). 'Theory of fuzzy integrals and its applications. Ph.D. thesis. Tokio Institute of Technology, Japan.

[17]Zadeh, L. A. (1965). 'Fuzzy sets'. <u>Information and Control</u>, **8**, 338-353.

[18]Zadeh, L. A. (1978). 'Fuzzy sets as a basis for a theory of possibility'. <u>Fuzzy Sets and Systems</u>, **1**, 3-28.

[19]Zadeh, L. A. (1979). 'Fuzzy sets and information granularity'.In Gupta, Ragade, Yager (Eds.), <u>Advances in Fuzzy Set Theory and Applications</u>, 3-18, North Holland Publishing Co., NY, 1979.

POSSIBILITY DISTRIBUTION
Intersection Event

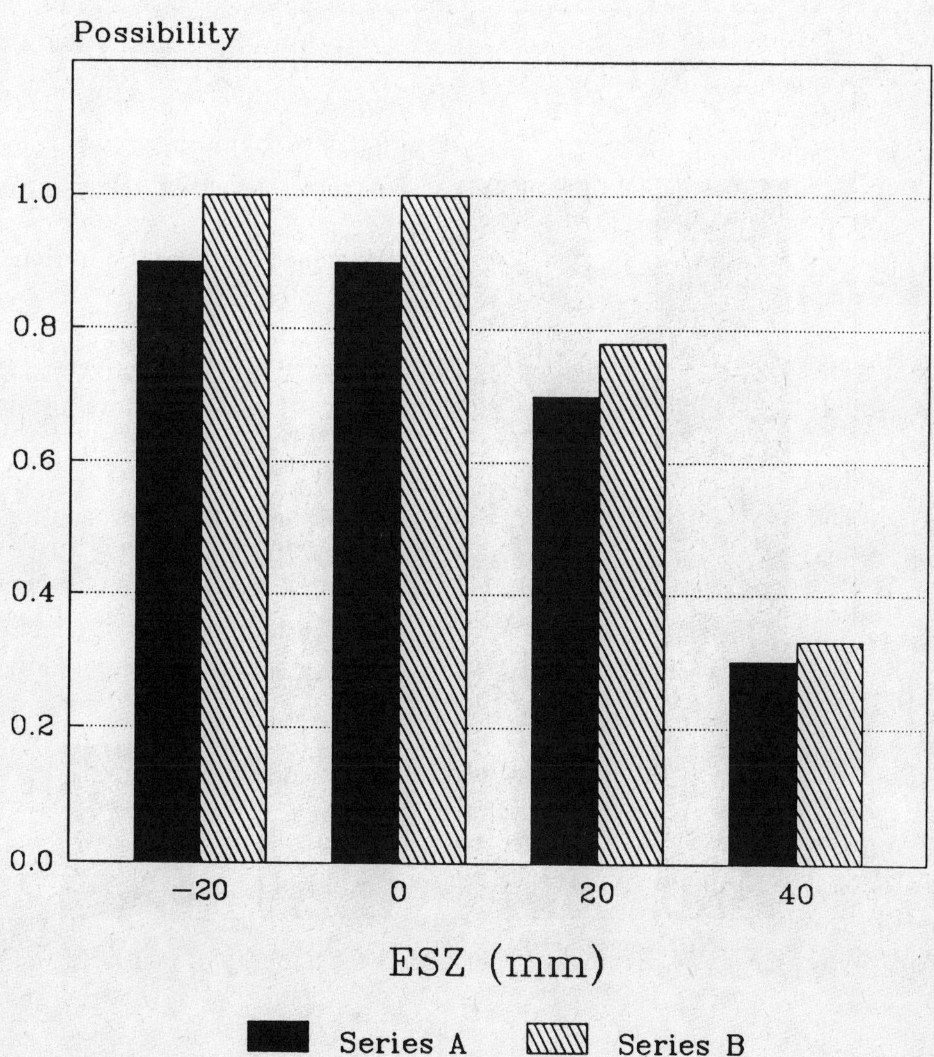

Series A = Computed Distribution
Series B = Normalized Distribution

Fig. **1**. Computed and normalized (upon the supremum) possibility distributions for the intersection event regarding the error of sizing ESZ (in mm).

PROBABILITY DISTRIBUTION
Technique Error

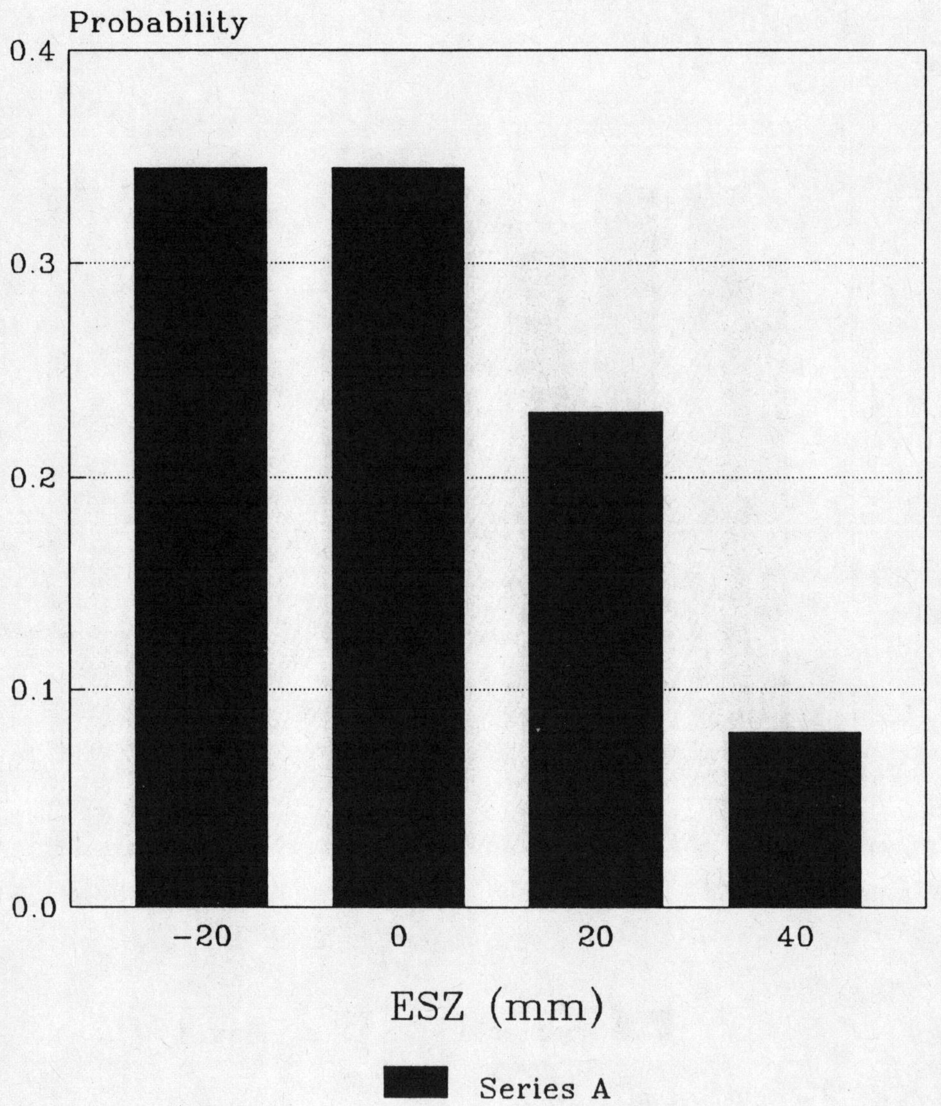

Fig. 2. Probability distribution for the technique error regarding the error of sizing ESZ (in mm).

POSSIBILITY DISTRIBUTION
Union Event

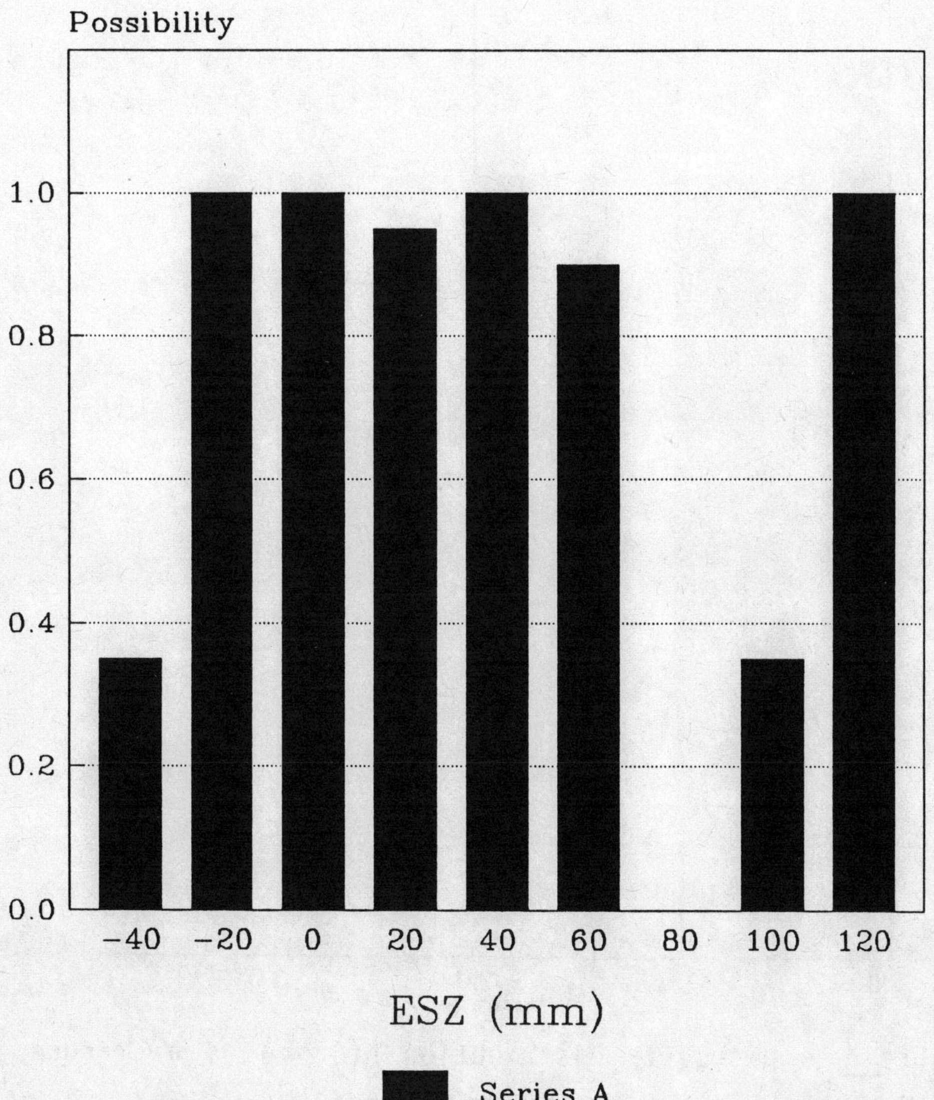

Fig. 3. Possibility distribution for the union event regarding the error of sizing ESZ (in mm).

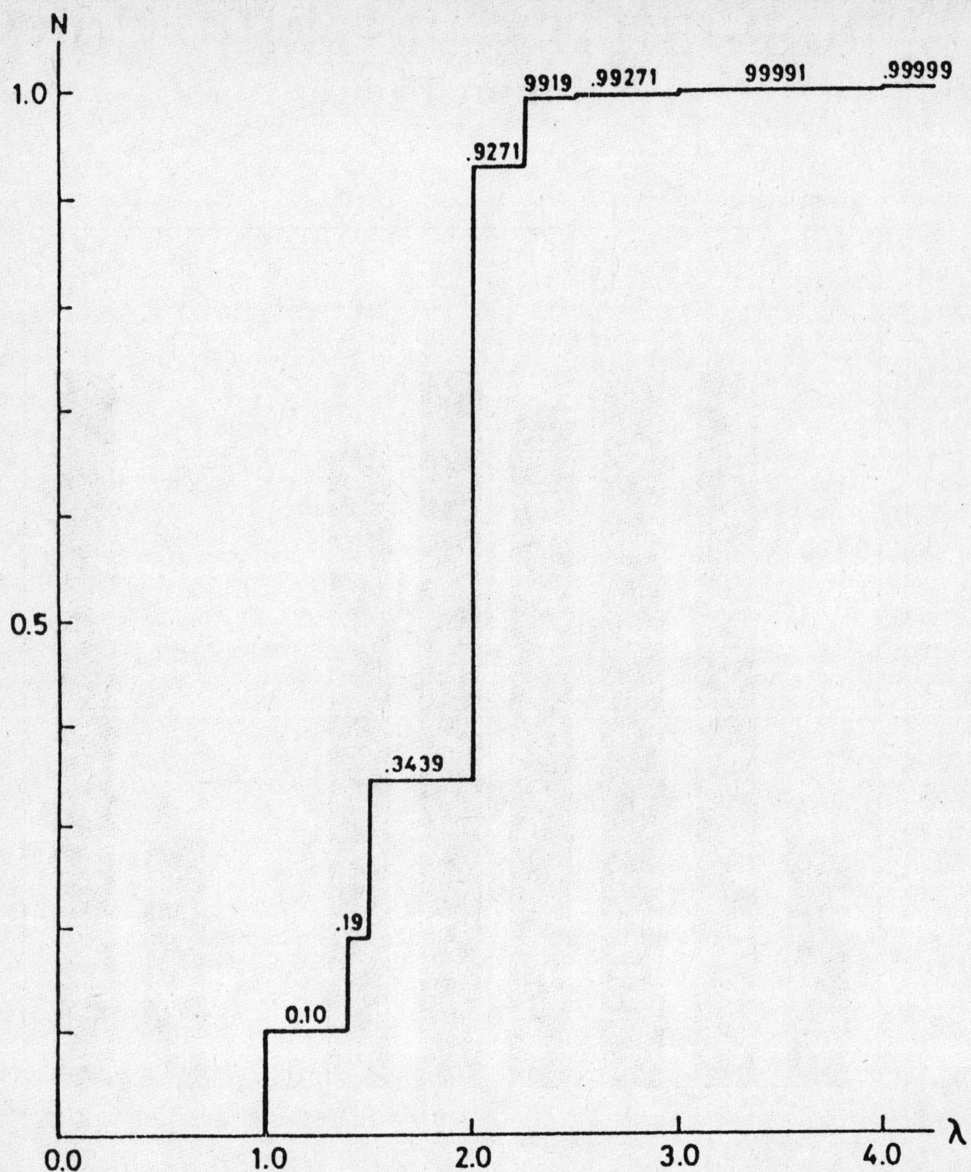

Figure 4 - Necessity distribution N(0;λ) as a measure for the occurrence of the failure event in [0,λ]. Case of composition of opinions from equivalent and independent experts.

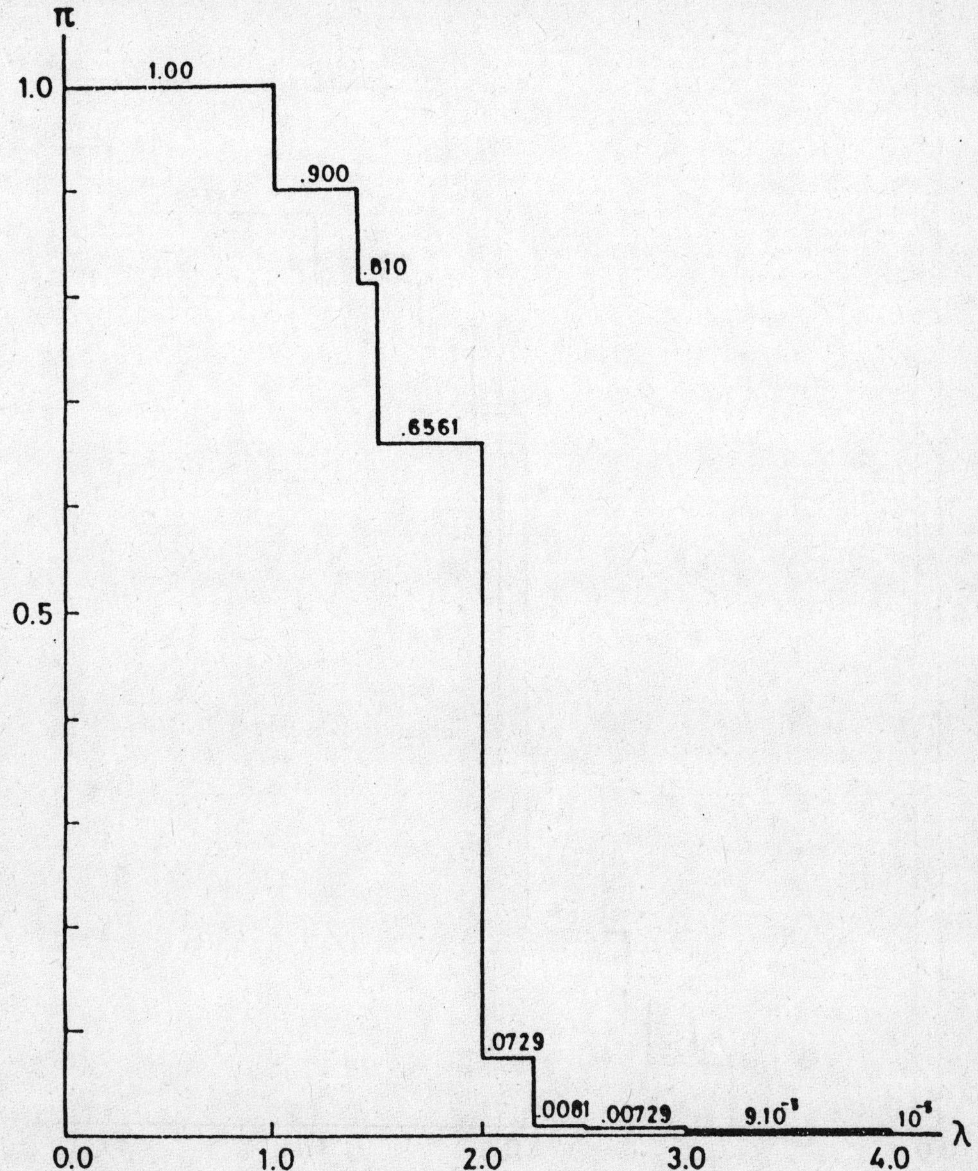

Figure 5 - Possibility distribution $\pi (\lambda ; L)$ as a measure for the occurrene of the failure event in $[\lambda, L]$. Case of composition of opinions from equivalent and independent experts.

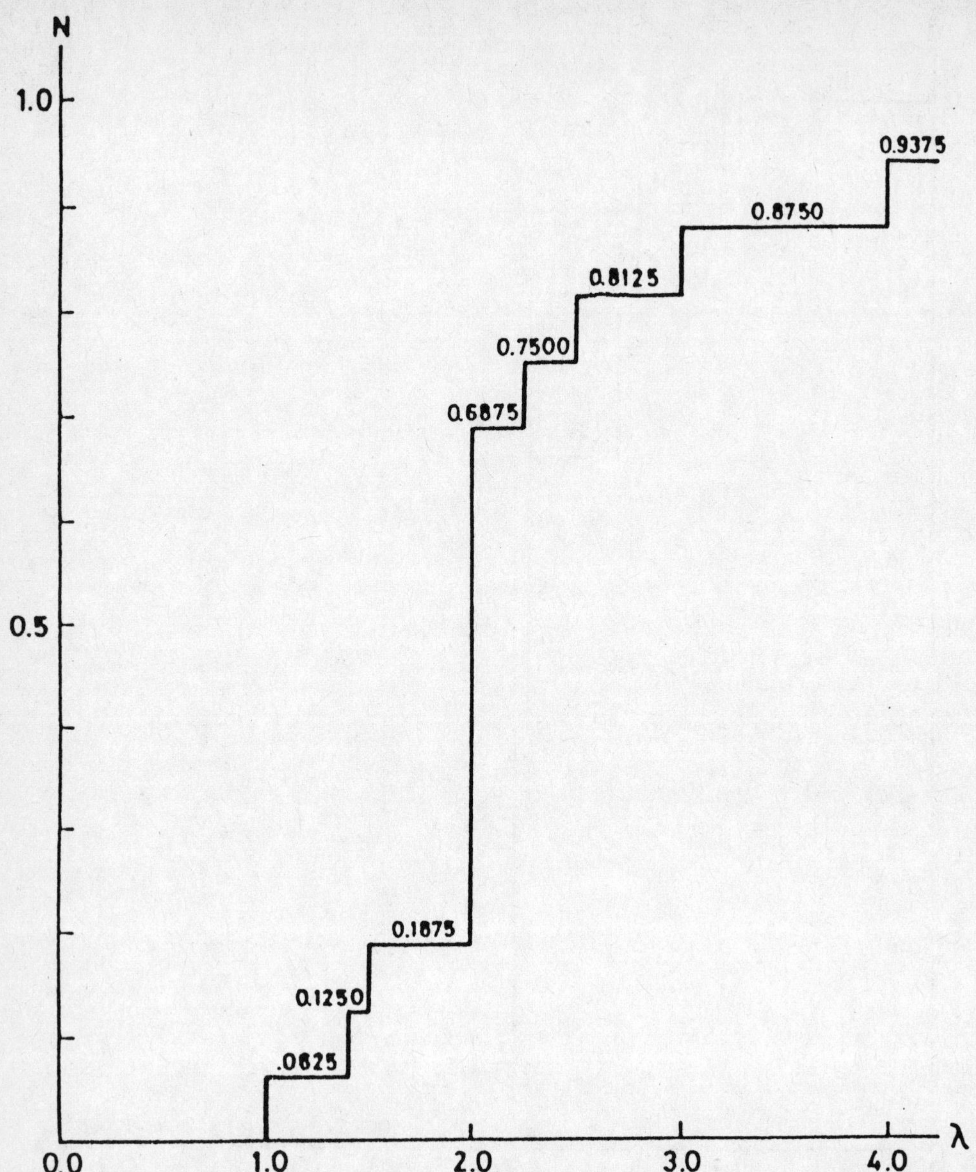

<u>Figure 6</u> - Necessity distribution N(0; λ) as a measure for the occurrence of the failure event in $[0,\lambda]$. Composition of opinions for the case of "percfect" dependence among equivalent experts.

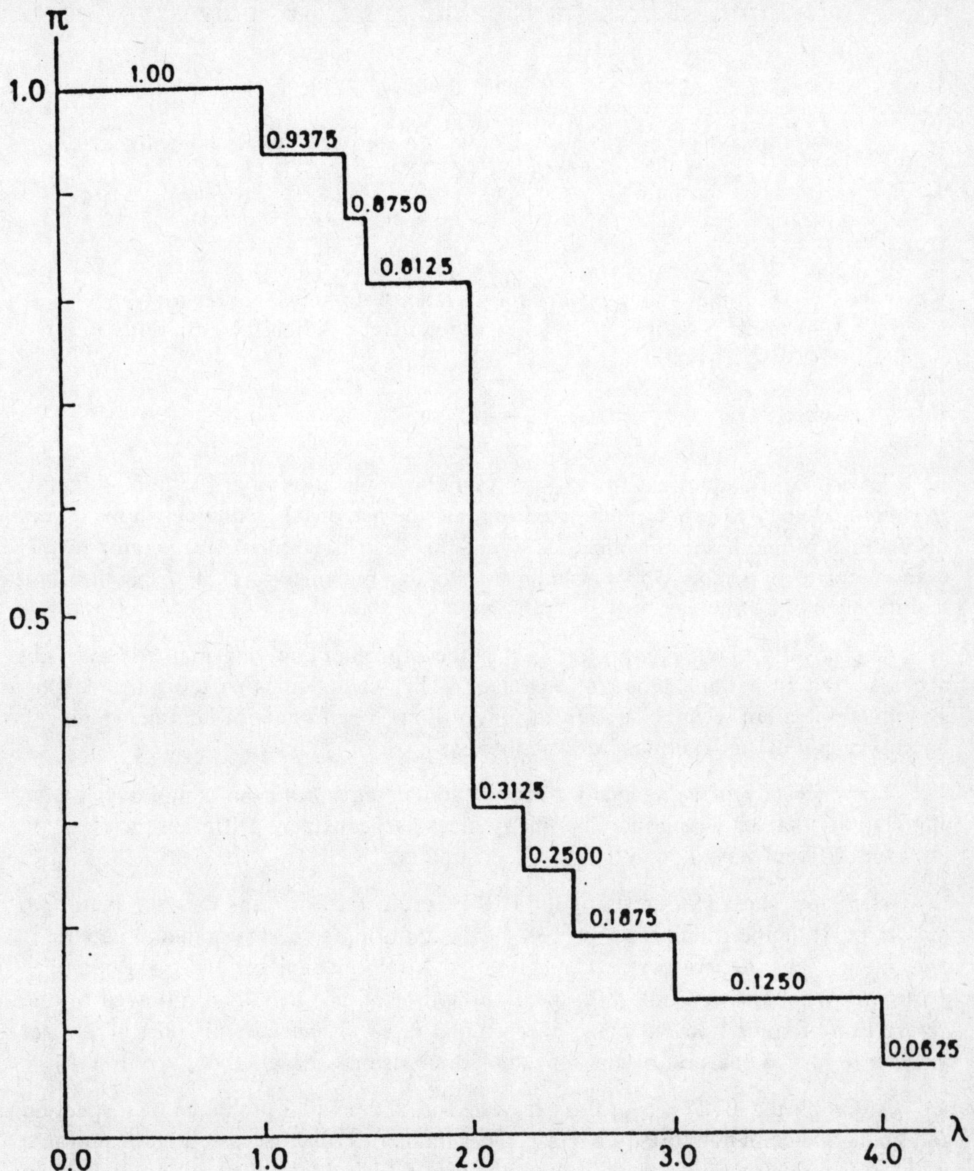

Figure 7 - Possibility distribution $\pi(\lambda; L)$ as a measure for the occurrence of the failure event in $[\lambda, L]$. Composition of opinions for the case of "perfect" dependence among equivalent experts.

Inference Engine Building & A Case Study

Vee H. Khong

Commission of the European Communities, Joint Research Centre
Ispra Establishment, I-21020 Ispra, (VA), Italy

Keywords: Artificial intelligence, expert systems, domains, bureaucracy,
multi-paradigm, inspection, maintenance, reliability, reliability index

1.0   Introduction

The aim of this paper is to describe a unique application of artificial intelligence (AI) methodology to structural engineering, i.e. the use of AI techniques in managing the various problem solving methods available to an operator. An operator in this case refers to a person who is charged with the responsibility of inspection and maintenance of a structure.

The novelty about this project is that these problem solving methods may not necessary be in a homogeneous language. The challenge is to design an expert system architecture which bears the attributes of an "open system", i.e. one which can be linked up with tools of many different sorts.

The necessity of using many different tools stems from the complexity of the application domains where there are many factors to consider. Different methods are used for different sets of conditions and constraints.

The paper starts by describing the BRITE project; its aim and its work plan. An important AI theme which is adopted in the design of the expert system architecture is presented. The expert system is called RAMINO which stands for Reliability Assessment for Maintenance & INspection Optimisation. It is then followed by the description of some relevant experience gained in a very successful ESPRIT project in order to form a basis for discussions and comparison.

Although the BRITE project is targetted to address the problems of inspection and maintenance in both offshore and nuclear industries, the discussion here is limited to the offshore structures. Some methodologies which this project are committed to investigate and which are relevant to the expert system are briefly discussed. This is followed by the discussion on the role of artificial intelligence in this project.

Acknowledgement: The author would like to thank Mr. A C Lucia & all the partners in the project for their assistance and comments in the preparation of this article. The views expressed here are purely those of the author's and not necessarily those of the BRITE P2124 and the KRITIC consortium.

The general architecture of RAMINO will be described and each important module will be expounded.

2.0 The BRITE Project 2124   -

Enhancement of inspection & maintenance of industrial structures using reliability based methods and expert systems

BRITE (Basic Research in Industrial Technology for Europe) is a framework programme partially supported by the European Commission. The aim of P2124 (BRITE Project 2124) is to improve the effectiveness of inspection and maintenance of industrial structures and to reduce costs. These industrial structure include offshore oil platforms and high pressure vessels in a power generating plant.

The aim of P2124 can be elaborated in the following way:
  I. to determine the processes by which structures may eventually fail and to predict the progress of degradation during service life
  II. to ensure a given reliability level at the lowest cost.

To fulfil criteria (I) & (II) above, the following work plan was drawn up
  1. defining the format of the data base for the project and collecting the relevant data such as material properties and mock-up test results
  2. researching into stochastic modelling of the time dependent process at element as well as system level
  3. researching into the optimisation of inspection and maintenance strategies of structures
  4. specifying and designing an architecture of an expert system, which formulates and utilises the knowledge accumulated in the three previous work packages

The collaborators of this projects include Elf Aquitaine, Framatome, Siemens (the applications domains), Technical University of Munich, Synthesis srl, Polytechnic of Milan and the CEC Joint Research Centre.

In order to put the novelty of this project into a proper context, a brief discussion is given here on some of the applications.

3.0 AI Application in Structural Engineering - A Brief Review

There are many applications of expert systems in structural engineering which are concentrated in the design phase. Systems such as ALL-RISE [13] and HI-RISE [9] aim at assisting the user in the designing of a building. They include informations such as legal requirements and constraints imposed by certain configurations.

HI-RISE is implemented on PSRL, a rule based language which rests on top of a frame based representation structure called SRL [17]. The knowledge in HI-RISE is divided into 2 general levels, the process knowledge and the structural knowledge level. The process knowledge deals with the expert's approach to solving the design problem. Such knowledge includes the procedures and the logical sequence of considering various factors as well as the order of which the evaluation should be carried out. The structural knowledge concerns with the structural descriptions, inter-relationships and classification of components. It also includes heuristics that are specific to individual components. Such kind of knowledge is generally encoded in the frame based language which, by the nature of its design, lends itself very well in representing relations such as *part-of*, *alternative-to*, etc. It is important to note that the two levels of knowledge (process and structural) are respectively equivalent to the general concept of <u>meta-level</u> knowledge and <u>object-level</u> knowledge.

There is another part of the work of Maher et al [9] called SPEX which is more interesting to the P2124 project because it bears some of the attributes which the P2124 project is trying to achieve. SPEX is an expert system aimed to aid the use of analysis using finite element method (FEM). It is independent of HI-RISE and is intended to to <u>intelligent pre- and post- processing</u> for an external FEM program. SPEX uses a blackboard inference architecture. In brief, SPEX examines a problem at hand and through using heuristic approach, classifies the problem. It then tries to match the identified problem with a FEM programme. The heuristics in SPEX use parameters such as *component-type*, *material-properties*, *component properties*, the *governing standard*, and the *limitation of component behaviour*. A similar strategy used in HI-RISE is also adopted in SPEX, i.e. driving towards a solution space by iteratively selecting the feasible constraints to satisfy.

In summary, SPEX's approach cannot be used directly in the P2124 project because the latter has a more diverse types of external programmes and not just algorithmic programmes on FEM. Secondly, SPEX and HI-RISE are targetted to address the design phase of an engineering task where the problem is more formalised. On the contrary, the problem of inspection and maintenance is not. This means that the task of implementing a knowledge base for inspection and maintenance requires some extra effort in problem formulation.

The works which is closest in characteristics to this BRITE project are SPERIL (Structural PERIL) [18] and SRA (Seismic Risk Advisor) [3, 16]. Both systems are targetted to evaluate the conditions and consequently the risk factor of structures (buildings) after having been subjected to perturbations such as earthquake. They are similar to this project because they gather information from many sources and analyse it for the decision making process in the expert system.

The earlier generation of SPERIL was implemented on C and has forward chaining capability. It takes into account of structural behaviour of a building such as damping, stiffness, creep and buckling. Based on such knowledge, it can deduce the damage state of the building by considering input data such as accelerometers readings and visual observation. It uses fuzzy logic in handling uncertainty. The later generation of SPERIL was implemented in a dialect of Prolog and has forward and backward chaining. It has more or less the same functionality but uses Dempster-Shafer algorithm in handling uncertainty.

The unique case of this BRITE project is that the application area of offshore structures is relative new in as far as expert systems are concerned. In addition, this project looks at a wider spectrum of analytical methods and the integration of a these methods with RAMINO whose architectural design is to serve this very purpose.

To begin designing RAMINO's architecture, some philosophical arguments are examined in order to get some hints on the feasibility of the ideas. The next section looks into the idea of multiple paradigm approach, its history and its relevance in the BRITE project.

4.0   The Theme of This Expert System

Minsky [11] said that human intelligence is a repertoire of many problem solving skills. These skills have been accumulated through the process of evolution and trial and error. If we are to build a machine which is able to solve problems, it stands to reasons that one problem solving method is not sufficient; we need multiple problem solving methods. What is more important is that we need to research into the organisation of these problem solving methods when they are put to work together, i.e. we need to devise some kind of *bureaucracy*.

The necessity of using more than one problem solving method was realised before, albeit in a different form. For example when LOOPS [14] was launched in the mid-70s, it was heralding an era of a programming environment with multiple paradigms, i.e. object oriented, access oriented and rule oriented approach.

The concept of bureaucracy in LOOPS, although not explicitly pronounced, nevertheless existed. By using a well thought out scheme, these different paradigms are integrated and able to coexist and function harmoniously. The well thought out scheme in LOOPS refers to the use of the object oriented language which forms the basis upon which other approaches can be built.

Such was the beginning of the multi-paradigm knowledge representation structure. In AI, there are many methodologies, both specific and general. To give a few examples, they include production rules, goal directed rules, multiple world reasoning, truth maintenance system and analogical reasoning.

Multi-paradigm approach is not restricted to the area of AI. In structural engineering, there are ways to estimate the mechanical characteristics of a component. An example would be the choice a simple 'beam' method or a comprehensive 'finite element analysis'. The final choice depends on the current circumstances and constraints.

Apart from the philosophy of architectural design, there is another important aspect in the practice of applied AI, that is the organisation of the work and the flow of information between workers: AI programmers on one hand and domain experts on the other. In the next section, a brief description of a completed project is given to illustrate the point.

5.0   The Experience in The KRITIC Project

KRITIC was an ESPRIT project with 32 man-years over 3 years. There are 4 partners in all: two industrial partners which forms the application domains (telecommunication maintenance and power systems control), 1 software company and 1 academic institution. The objective of the project was to study the knowledge representation and inference techniques for industrial control in these two domains.

The important decision made at the outset of the project was that it was to design and implement inference and knowledge representation tools instead of buying commercial ones. The decision was influenced largely by the uncertainty in the returns obtainable, and also by the high initial capital cost. This was coupled by the fact that, at the time, the domains were not sufficient understood (as far as AI was concerned) to allow the choosing of a right shell.

The personnel in the project were divided into 2 groups; one group called the *kernel* was charged with the duty of designing tools based on the understanding on the needs in the applications, and the other group, called the *domains* who test the prototypes implemented by the kernel in their respective application areas and provide feedback for improvement. "Boundary-crossing" by personnel took place at official and unofficial level. This was the contributing factor to the success of the project because it improved mutual understanding of the problems on either side. Furthermore, it managed to pool together valuable design ideas wherever existed in the partnership.

In retrospect, the decision to implement in-house tools and finally the environment was well justified. Due to the differences in the two application domains, it was necessary to tailor the inference engine (called MIKIC [8]), the knowledge representation language (called AVALON), the truth maintenance system and other tools to suit the common needs.

Before returning to the main stream discussion of the paper, it is necessary to recap the points. Firstly, for an expert system project to be successful, it is very important for the "informaticians" (the expert system builders, the knowledge engineers, etc) to know and to study the subjects of the application domains. It is equally important for the domain workers to find out about the work in the expert system building process. Secondly, the output of the KRITIC project was a collection of AI tools tailored to address the problems in the domains. These tools need further refinement and packaging if they are to be used by people other than the developers themselves.

For the BRITE project, it is not possible to used the same policy adopted in the KRITIC project, i.e. to build in-house tools, because the emphasis in the BRITE project is on the application of existing AI tools rather than building new ones. In addition, the expert system built in the BRITE project will have to be distributed to the partners and this calls for the need for a set of well packaged AI tools.

The next section describes the problems of offshore structure which forms part of the BRITE project target application domains.

6.0   The BRITE Offshore Structure Problems

6.1   Background Information

In practice, the maintenance department of Elf looks after platforms which are located in many different parts of the world. They have to provide a round-the-clock service to cover any emergencies. In case of an accident, they advise the affiliates on the safety of the platform and consequently whether it is necessary to evacuate the platform. In less urgent cases, they have to ensure that the platforms are checked in a <u>timely</u> fashion so that their safety levels are within the acceptable limit.

The word <u>timely</u> means that the inspection of a platform is done at the point when the benefit is maximum. It can be interpreted as the time beyond which the safety of the structure would become precarious. Conversely, the time before which inspection is unnecessary.

At present the time-interval between two successive inspections is fixed at some nominal value. For example, a platform which is "important" should be inspected every 3 years and a platform of little importance can be inspected every 9 years. The importance is currently judged by the level of petroleum output and the degree of disruption if the platform went out of action. This method fails to incorporate up-to-date knowledge. In the service life of a platform, there are events which could lead to the need for more frequent inspections in the future. On the other hand, over pessimistic assumptions in the early estimations may have to be corrected as the structure is "better understood."

The cost of inspection is high, both in terms of direct cost and the disruption in production. A wrong diagnosis can have costly consequences too. It is the aim of this project to improve the efficacy in this area. From the early rounds of knowledge elicitation, areas which can potentially be improved were identified. The intention was to identify the way which the expert system can be most beneficial.

One application area for RAMINO is to alleviate the problem of communication. It was realised that between observations (through inspection) and repair, there is sometimes a necessity for mathematical analysis. The inspection and the mathematic analyses are done by two different departments: the maintenance and the engineering department respectively. It was found that there is a delay introduced by the communication time of the two departments. This is undesirable because the process of a structural diagnosis would invariably be slower if the maintenance expert has to refer to the engineering department each time he has the need of a mathematical estimation, even though it is a simple case.

The first objective of RAMINO in this area is to make available some of the engineering analytical tools (programmes) to the maintenance experts. This is not to say that the function of an engineering department is eliminated as a result. The idea is to provide the maintenance expert with analytical tools so as to allow him to make a rough estimation of, say, the mechanical properties of the structure quickly and to allow him to continue with the work. If rigorous analysis is still needed, he can then refer to the engineering department.

Another objective of RAMINO comes from within, that is to oversee and control the operations carried out by various analytical programmes. These two objectives will be used as the guidelines to defined the general architecture of RAMINO.

Apart from the expert system, this BRITE project also looks into new ways to estimate the reliability of a platform. These include uncertainty modelling, effect of human errors, time-variant system reliability and so on. Here, descriptions of some of those aspects which are relevant to RAMINO are given.

6.2 Collection of Tools in The BRITE Project

In this section, a frame of reference is described in order to facilitate subsequent discussions. This is the Reliability Index $\beta_i(t)$ vs. Time. Figure 1 shows two such curves and the various important levels. The lower of the two curves in figure 1 describes the degradation of the system based on information available at the time. One can use this curve to plan future inspection and maintenance.

In principle, the situation can change. For example, the discovery of new cracks may lower the curve and upset the already planned programme of inspection and maintenance. Conversely, after an inspection, one may find that the degradation is less severe than originally anticipated (through mathematical calculations) and the curve may be "pushed" up, thus resulting the curve $\beta_{i+1}(t)$. This can also upset the inspection programme.

Each time some new data arrive, the $\beta_i(t)$ curve is modified using some updating techniques, the estimation of crack growth is re-calculated using the appropriate modelling techniques, and the effect of human error and uncertainty is re-assessed. These different techniques are some of the topics studied in this project in addition to the expert system. However the results will be coupled with RAMINO and some brief descriptions on the purpose of these techniques are given here.

6.3 Some Important Analytical Modules

The following gives a brief summary of some of the research work being carried out in this BRITE project. The purpose of this section is to complement the forthcoming discussions on the expert system architecture.

In Section 2, a 4-part work plan of this project are given. The collection of data will produce a wealth of data in a well defined format. This will facilitate the work in the second part, i.e. research in stochastic modelling. The research works described in this section are all from this areas. The third section deals with inspection strategies which does not play such an important part at this stage of the expert system development. The forth concerns the RAMINO design.

Four pieces of on-going research work in the area of stochastic modelling are briefly described in the following sub-paragraphs.

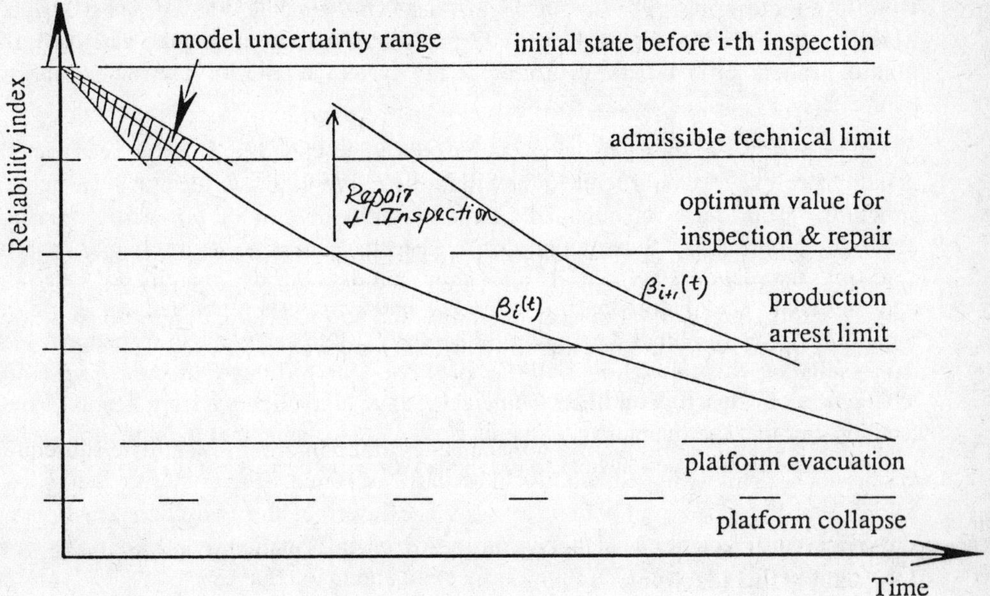

Figure 1 : Reliability Index (β) vs. Time

Uncertainty modelling Curve $\beta_i(t)$ in figure 1 has a small shaded region at the left hand side. This indicates the possible variation of the reliability index (for a given time-point) due to uncertainty. This task studies how uncertainty in the data collected, such as the mock-up test data, material properties, etc, can influence the decision making process. The output of this topic is more likely to be incorporated in other modules, either by "hard-wiring" or through the deliberation of RAMINO at consultation time.

Time variant system reliability This task looks into the stochastic modelling of fatigue crack nucleation, crack growth and environmental assisted fatigue crack growth. It is important because it calculates the damage accumulation and produces prediction models which are needed in the estimation of the degradation behaviour of the structure. This task requires statistical data which have been prepared previously and produces output that can be used in updating the "believed state of the system" and in "planning for the optimal times for inspection and maintenance".

Input-output structural relationship When it is not possible to use an analytical approach to estimate the system reliability, a response surface methodology is used. This maps out the response surface of the propagation of a crack for the required range. This work is done "off-line". That is to say that as far as RAMINO is concerned, the result are the data in the data base and can be retrieved when necessary.

Updating techniques The output of this task complements the two just described. Referring to curve ß of figure 1, this task provides a way to update the data base, the time dependent probabilistic measures of the system reliability, and other relevant parameters.

The purpose of these modules can be briefly explained in a simple scenario. A maintenance expert would refer to the reliability curve to get an idea of the state of a structure. This curve is obtained partly through the use of *time-variant system reliability* programme. Some input data for the time-variant system reliability can be read from the *response surface*. If *uncertainty* on data has not been incorporated, he may run some calculations on the reliability index to get an adjusted value. If the reliability curve, and other data, point to the necessity to carry out an inspection, then the results of this operation will be processed and filtered by some *updating techniques*. In this process there could be many interactive cycles, questions, data feeding, etc, that the maintenance operator has to carry out. It is the intention of this project to create an expert system to aid the person who carries out this task.

7.0   RAMINO -   The expert system for Reliability Assessment for Maintenance & INspection Optimization

RAMINO has two important functions. Firstly it is intended to bridge the communication gap between the maintenance & inspection team and the engineering group who does the mathematical analyses. Secondly, it serves as a linchpin for all the programming modules by playing the role of co-ordinating the individual operations carried out by those modules.

7.1   General Architecture

Issues concerning the architecture The intrinsic constraint imposed by the project is that there are computer programmes which are generated independently and in

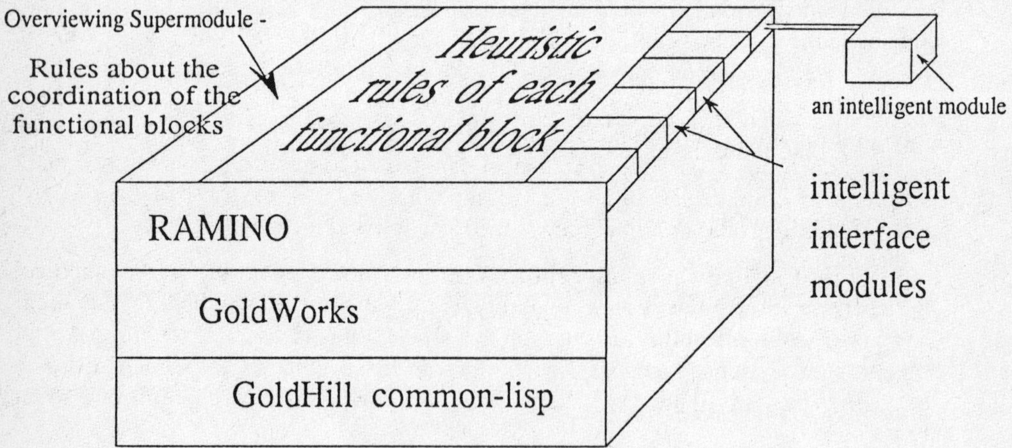

Figure 2 : A schematic view of the RAMINO architecture

different languages. The design of RAMINO must be such that these modules operate in their own environments and can be called up when necessary. It must also be able to accommodate new programmes in the future without major modification in the existing code. Conversely, a removal or modification of a currently attached module should not result major changes in the RAMINO programme. The architecture should allow for the evolution, i.e. improvements, of the external module. If the input-output format of an external modules is modified, it should not be necessary to change the other modules in order to adapt to the new standard. In other words, interaction with the outside should be abstracted and packaged into independent units.

Description of the general architecture The general architecture of RAMINO consists of 3 levels. Each level carries knowledge of a certain abstraction. The lowest of the 3 levels is made up of *Intelligent Interface Modules (IIM)*. An IIM is analogous to the input-output port of a computer where external devices are connected. The IIM is essential in allowing RAMINO to communicate with external programmes. The second level is made up of knowledge modules on specific topics in structural engineering. The middle layer can be labelled as the *object-rules* layer and it is intended to hold rules that are closely related to the domain. The top-most level is called the *Overviewing Supermodule*. It contains meta-knowledge [2] which serves to control the sequencing of the knowledge modules and sometimes the external programmes. The following section explains the functionality of each of the 3 levels.

The Intelligent interface modules (IIM) Imagine an operator who has a variety of computer tools available to him. Given a problem, he has to select one programme which is the most "appropriate". The concept of appropriateness involves an iterative process of screening. The list of all programmes are successively "short-listed" until one programme which can satisfy all the constraints is found. In each iteration, an additional constraint, e.g. available time, cost, accuracy, etc, is introduced and there should be heuristics about the way to order the constraints. It is a problem of constraint satisfaction [7] which involves the design of the representation scheme for the knowledge.

The IIMs are to provide information on the characteristics of the external programmes when it is necessary for RAMINO to communicate with them. It is envisaged that there is exactly one IIM per external module. The information contained in an IIM is typically the function of the programme, e.g.

- the accuracy achievable by the programme
- the assumptions used by the programme
- the parameters required by the programme
- the range of applicability
- the input/output format
- the preparation time required
- an indication of the cpu time required

The middle layer - object-rules While the IIMs in the bottom level take care of the details of individual external programmes, the content in this layer of the architecture is composed of knowledge of higher abstraction. It is expected that this layer contains knowledge about

- the engineering interpretation of an output
- the inter-consistency between results generated by different external programmes
- the way to select, from a set of external programmes, one which is the most suitable
- the inter-relationship between various subject areas, e.g. whether it is necessary to incorporate uncertainty on some data, such as material properties
- how to schedule and re-schedule the inspection and maintenance programme when the reliability index has be updated
- the purpose of each of the subject included in the project such as
    * gross errors   * uncertainty modelling
    * data bases   * forecast of behaviour
    * loading & load effects   * response surface
    * inspection techniques   * planning of inspection

As it is done in the IIM (bottom) level, the knowledge at this level is also bundled into modules which are independent of one another. This is similar to the use of knowledge sources in the blackboard architecture [5] or other schemes such as MIKIC [8,15].

These modules exchange information through a working memory which is only accessible to themselves. The direct invocation of one module by another is limited. The major part of it will be done through an "intermediary" by first posting a request. The format of this request includes information which is necessary for the intermediary to decide which request should be granted. This format is still to be defined and only possible when more knowledge elicitation has been done. The intermediary is a common concept and is usually called meta-knowledge. This meta-knowledge will be place in the top-most level of the RAMINO architecture.

The Overviewing Supermodule This level contains meta-knowledge as describe earlier. One can view some of the knowledge here as being further removed from the actual domain of application and it tends to be of common-sense. For example, if there are 2 modules in the object-rules layer competing for resources, then it is logical that the one which is most urgent should be allowed first. This kind of heuristics can sometimes be domain independent.

The knowledge here is not always domain independent. For example, in trying to establish whether a defect is one of the following

- service induced crack
- fabrication crack
- slag-inclusion

If experience says that it is more fruitful to test for Slag-inclusion first before anything else, then the knowledge here is domain dependent.

7.2 Philosophical Compliance

The general architecture has been designed according to the philosophy and requirements described earlier in Section 4 and 6.1, i.e.

- to act as a manager of the bureaucracy of the different problem solving methods
- to provide an "open" architecture which facilitates the evolution of external programmes as well as the addition or removal of external programmes
- to enhance communication between the maintenance group and the engineering group

The target system of this project will consists of many different problem solving methods, such as those described earlier, i.e. time variant system reliability, input-output relationship, up-dating techniques, etc. This complies to the design philosophy of a multi-paradigm expert system environment such as LOOPS and what Minsky [11] has claimed. The smooth integration of these different problem solving methods, i.e. the bureaucracy, is built into the RAMINO architecture:

* the IIMs at the bottom-most level provide the gateways to the external modules and contains declarations of the purposes and utilities of these modules

* the knowledge modules in the object-rules layer can either be self-contained problem solving methods or acting as the "governor" of an external module. The role of the governor is to control the execution of the external modules and maintain some kind of logical consistency in the operation

* the Overviewing Supermodule has the role of controlling the whole process of problem solving

The IIMs in the target system serve to fulfil the requirement of an "open" structure. For example, when the I/O format of an external programme is modified, only the corresponding IIM need to be up-dated. If the functionality of an external programme is modified, only the corresponding knowledge module in the object-rules layer needs to be up-dated. If necessary, only a small area of the Overviewing Supermodule need to be modified.

By way of implementing the two features described above, the aim of bridging communication between the maintenance and engineering people is achieved through a side-effect. As it stands now, RAMINO serves the purpose of co-ordinating engineering analyses. To extend its capability so that it also serves the maintenance people, additional knowledge modules or external programmes can be implemented. These additional knowledge modules can be the interface to the users and the experiential knowledge from the maintenance experts.

7.3 Progress Description

At present, the project has reached a half-way stage. Presently, a small prototype has been constructed. The other work which has been carried out in the RAMINO development process includes the following.

1. the selection of hardware & software
2. the process of knowledge elicitation
3. a small prototype

Apart from these three, the general architecture has been defined and has been described earlier.

Selection of hardware & software It was decided at the beginning of the project that for the ease of software exchange between partners, personal computers should be used. Furthermore, since the industrial partners are already using IBM personal computers and for the advantage of an European wide support, the IBM PS/2 system has be chosen.

In the case of software, there were two possibilities. Firstly, we can implement our own inference and knowledge representation tools. Secondly, we can use a commercial shell. The advantages of the first choice are likely to be the same as those described in the ESPRIT experience (Section 5), i.e. one can tailor the tools to the specific needs of the project. However there was a stringent constraint on the availability of man-power. Therefore it was decided that a commercial expert system shell should be used.

The next problem was the selection of a correct shell for the project. The constraints presented to the selection process were

- the software must run on the IBM personal computers

- the software must have a reasonably good knowledge representation and inference capability and provide a user-friendly and intelligent environment

- the software must have the capability to link up with external computer programmes, running also on the same machine, in particular "dBASE III plus" which has been defined as the standard for the data storage in this project.

There are many expert system shells available in the market. From them, 7 most promising ones were selected for rigorous examination. These seven were
- KEE (Knowledge Engineering Environment)
- BESB (Basic Expert System Builder)
- KES (Knowledge Engineering System)
- OPS5
- GoldWorks
- Nexpert Object

Inference	**Knowledge Representation**	**Development Environment**
Rules - forward, backward, mixed	Frames	Readability
Pattern matching	Procedural attachment	(rules & data)
Conflict resolution	Object oriented	Rule editor
Priority assignment	Inheritance	Data/Knowledge base editor
Nexted conditions	Multiple inheritance	On-line help facility
Agenda for rule firing	Definable relations	Break points
Numerical functions	Access oriented	Explanation
Partitioning	Multiple world representation	Graphic interface
Meta-control	**Support**	Graphic toolkit
Uncertainty handling	Documentation	
Truth maintenance	Courses, On-line support	

Figure 3: Criteria used in selecting a shell

In order to make an objective evaluation, a set of features was drawn up. The fields were cross-compared among all the shells and a table was used to facilitate the decision making process. The result of the study points towards the selection of GoldWorks. Figure 3 shows the set of evaluation criteria used and brief explanations are given on the features.

Inference Most expert system shells have a rule based inference system in one form or another. Few systems come in as just a knowledge representation language, such as only the frame based language. The *forward* and *backward* chaining are the two most common rule inference modes but few systems provide *mixed* mode chaining. *Mixed* mode chaining allows a more free expression of heuristics from the experts. *Pattern matching & Conflict resolution* form part of the recognise-act cycle in inference in a production system. The pattern matcher has a direct influence on the flexibility of the rule language. The more powerful is the pattern matcher, the more expressive is the rule language. Conflict resolution is important in that its behaviour is directly related to the overall behaviour of the system. *Priority assignment* allows the user to impose a predefined order of rule firing. This is useful only in very specific cases. Apart from the power enabled by the characteristics of the pattern-matcher, the *Nested conditions* capability in the rule language allows a combination of complex conditions to be expressed succinctly in one rule. The available of an *Agenda* gives the possibility of 'peeking' into the recognise-act cycle of an inference process. This facilitates better knowledge base development. If access to the agenda is allowed, the control of the inference process can also be improved, in the sense that the behaviour of the inference process can be tailored according to the needs. *Numerical functions* are important especially in engineering applications. Expert systems shells written on C or Pascal languages usually have little problems in providing a good library of numerical functions. Shells which are built on Lisp depend largely on the available mathematical library of the dialect of Lisp used. Sometimes, the rule language is constructed in such a way that some of the mathematical functions are not directly accessible. *Partitioning* is important because it contributes directly to the good organisation of the knowledge base and consequently providing a handle to the control of inference. *Meta-control* in a shell provides a layer of language on top of the ordinary rule language. The rules in this layer co-ordinate the execution of the ordinary object-rules. This however, is not commonly provided in shells. What one should look for in this aspect is whether the architecture in the shell facilitates or hinders the eventual implementation of a meta-knowledge level. *Uncertainty handling* usually means that some numerical values can be used to represent belief. Bayesian method is used in many shells. It is difficult to find shells which allows one to implement a specific kind of uncertainty handling without resorting to very involved code writing. The *Truth maintenance* capability is one which keeps track of the consistency of the relationships between data which is not handled properly by the simple if-then rule system. In a rule based system, assertions are made by the detection of certain conditions. When a truth maintenance system is available, the assertions made will be retracted when these conditions are no longer valid.

Knowledge Representation The heading Knowledge Representation is slightly inaccurate especially in view of the fact that data and rules are both knowledge about the world. Another name for it could be Data Representation. The basic object-attribute-value triplet is widely used in representing data. By grouping all data relevant to an object in a unit and by providing a extended structure for describing the details, we arrive at the *Frame* concept [10]. The basic frame structure should provide a rudimentary *Inheritance* capability and the set of commands to manipulate the descriptions in the frames. In *Procedural attachment*, the extended description structure sometimes allow some Lisp code to be cached and invoked when necessary, thus improve the clarity of the data base and effecting some kind of detail hiding. The *Object oriented* structure can be viewed as a frame based system with an added power of processing information by message passing. The philosophy of object oriented programming and their advantages can be found in a variety of literature [4,12]. Some systems provide *Multiple inheritance* capabilities. For example, Zetalisp in Symbolics machines allows one to define the inherited value of a slot. *Definable relations* gives the user a extra handle on the inheritance behaviour. In addition, it allows easy definition of orthogonal relationships, such as *A is part-of B*, as in the case of HI-RISE (see Section 3). *Access oriented* is sometimes called daemons and it is a piece of procedural code which is 'hooked' onto a piece of data so that when the data is modified or deleted, that procedural code is executed. The difference between access oriented programming and rule forward chaining is that the latter runs the risk of being triggered unexpectedly, especially in a large rule base. *Multiple world* comes in large shells such as Knowledge Craft, ART and KEE. It allows the spawning of hypothetical world-contexts in its inference process. This facility allows the exploration of various assumptions or hypotheses in less deterministic situations.

Development Environment & Support The items here are self explanatory and will not be expanded here.

The process of knowledge elicitation An iterative approach has been used in drawing knowledge from the experts. At the beginning, a survey was made on the tools currently used in Elf Aquitaine to manage inspection and maintenance. There were two. The details of these two systems were studied in order to understand the nature and the scale of the problem. At this stage, we realised some of the current problems in maintenance of offshore platforms such as the inflexible inspection interval, the responsibility of the engineering department, the subtleties in choosing an analytical tool for a job and some of the principle concepts in this field. As a result, we identified some of the requirements in RAMINO. For example, to overcome the statical inspection interval, RAMINO must have the heuristics to relate it with the appropriate estimation techniques. However, a detailed method for achieving this is still to be worked out.

The second round of discussion was with the maintenance experts. As a result, we learned that the efficiency of inspection can be improved if the maintenance experts have direct access to some simple analytical facilities. Presently the maintenance experts rely heavily on the engineering department on mathematical calculation, even simple ones. There is an inevitable time lag in the communication process. We also learned how they use the Reliability Index ß as a reference to plan future inspections.

After the two rounds of talks, the general architecture was defined. To further define the structure and the contents of each of the IIMs and the knowledge modules in the object-rules layer, work began to elicitate knowledge from the appropriate experts. For many external programmes, the experts are the collaborators who are developing new techniques described earlier in this paper. (These new techniques will come as the external programmes connected to RAMINO via the IIMs.) As a first step, a questionnaire was designed (see appendix) and a list of operators was drawn up. (An operator is equivalent to an external programme.) The implementor of a programme is made responsible to fill in a questionnaire for his or her work. Currently, we are still in the process of collecting them.

The small prototype Implementation of small prototypes began as soon as some answers to the questionnaire came in. At present we have implemented some knowledge modules in the object-rules layer of RAMINO which deals with the identification of 3 kinds of defects: slag inclusion, fabrication crack or service induced crack. Also we have emulated the monitoring of the transition of incubation period to propagation stage of a defect.

The structure of GoldWorks has made the implementation up to now easy. GoldWorks supports forward chaining and backward chaining. Forward rules in GoldWorks can be packaged into rule-sets. One purpose of doing this is to emulate the effect of goal-directed reasoning but in a "contained" search space. Also, by partitioning the rules, the knowledge base becomes more organised and easy to maintain. Each rule-set has a triggering-pattern which, when matched by an external query-pattern, triggers the rule-set. The aforementioned knowledge was implemented as individual rule-sets in GoldWorks. There is only a small amount of heuristics available in the Overviewing Supermodule. The meta-knowledge dictates that one should check for slag-inclusion before going onto other possibilities of defects.

8.0 Conclusion & Future Work

The application of expert system techniques in RAMINO to structural engineering is different from most systems in the same faculty. Systems such as SPERIL and SRA use AI to improve the decision making process by finding ways to handle uncertainty and to deal with the combining functions within the expert system itself. RAMINO on the other hand, leaves the detailed mathematical calculations to the external programmes and concentrates on the abstract knowledge on the functionality and the logical contents of the input and output of each external programme. RAMINO is not an expert system designed to tackle a narrow aspect of structural engineering problem solving but the management many of such a process.

This paper has presented the general goal of the BRITE 2124 project and of the way adopted in building RAMINO. The principle design philosophy of the expert system and a brief description of the architecture hitherto defined has been given. LOOPS was used as a case to argue for the use of multiple methods and bureaucracy.

Currently, the effort is focussed on the collection and compilation of engineering knowledge (the knowledge of input data preparation and checking and output data interpretation) of the external programmes. This will serve the purpose to finalise the definition of all the necessary IIMs. At the same time, some of the knowledge in the object-rules layer will be defined.

Later on, attention will be re-focussed on the maintenance experts for their judgmental knowledge on inspection and planning of maintenance. Knowledge such as the frequency of certain problems on structures located in a certain geographic areas and the suitability of a certain inspection techniques will be gathered at this stage. This will enrich the knowledge in the object-rules layer as well as the Overviewing Supermodule.

References

1. Adeli H. (Ed), *Expert systems in construction and structural engineering,* Chapman & Hall 1988
2. Davis R., Content Reference: Reasoning about rules, in *AI 15 (1980)* pp233-239
3. Dong W-M, Shah H.C. & Wong F.S., Condensation of the knowledge base in expert systems with applications to seismic risk evaluation, in Adeli H. (Ed) *Expert systems in construction and structural engineering*, Chapman & Hall 1988.
4. Fikes et al 85 Fikes R. & Kehler T., The role of frame based representation in reasoning, in *Communication of the ACM*, September, 1985
5. Hayes-Roth F. & Lesser V.R., Focus of attention in Hearsay-II system, in *Proc. IJCAI 5*, 1977
6. Ibbs C.W. & De La Garza J., Knowledge engineering for a construction scheduling analysis system, in Adeli H. (Ed) *Expert systems in construction and structural engineering*, Chapman & Hall 1988.
7. Khong, V.H., Bigham, J. & Mamdani, E.H., CLARA - A constraint language for aiding room allocation, in *Proc. of 2nd IFSA Conference*, Tokyo, July 1987.
8. Khong, V.H., *MIKIC Manual*, ESPRIT Project 387 Technical Memo 10, Department of Electrical & Electronic Engineering, Queen Mary College, (Univ. of London), London E1 4NS.
9. Maher M.L., Fenves S.J. & Garrett J.H., Expert systems for structural design, in Adeli H. (Ed) *Expert systems in construction and structural engineering*, Chapman & Hall 1988.
10. Minsky M., A framework for representing knowledge, in P H Winston (ed) *The psychology of computer vision, New York,* McGrawHill.
11. Minsky M., Private communication in New developments in artificial intelligence: mental functions and memory process, held at FAST, Milan, 4 May, 1989.
12. Robson D., Object oriented software systems, in *BYTE Vol.6, No.8*, August, 1981, pp74-86
13. Sriram D., ALL-RISE: A case study in constraint-ba27-6-89sed design, in Artificial Intelligence in Engineering, 1987, Vol.2, No.4.
14. Stefik M., Bobrow D., Mittal S. & Conway L., Knowledge programming in LOOPS, in *The AI Magazine*, Fall 1983, pp 3-13.
15. Williamson G., Khong, V.H., King, S. et al, Using a KBS in telecommunications, in *Proc of ESPRIT Technical Week III* , Brussels, Sep 1986.
16. Wong F.S., Dong W-M & Blanks M., Coupling of symbolic & numerical computations on a microcomputer, in *Artificial Intelligence in Engineering*, 1988, Vol.3, No.1.

17. Wright J.M. & Fox M.S., *SRL/1.5 User Manual*, Robotics Institute, Carnegie-Mellon University, Pittsburgh, Pa., December 1983.
18. Zhang X.J. & Yao, J.T.P., Expert systems for condition evaluation of existing structures, in Adeli H. (Ed) *Expert systems in construction and structural engineering*, Chapman & Hall 1988.

Appendix - (Questionnaire form for knowledge acquisition)

Expert System RAMINO

Identification of the attributes of an operation

1. Operator name____

2. Argument names

   a. Module names____

   b. Data bases____

   c. Input stream____

   d. others____

3. Input definition

   a. Input variables & parameters (and their location)___________

   b. Suggestions on default values, ranges, etc_________________

   c. others____

4. Input acceptability criteria

   a. Criteria of admissibility___________________________________

   b. Criteria of consistency_____________________________________

   c. Criteria of completeness____________________________________

   d. others____

5. Output acceptability criteria

   a. Physical admissibility______________________________________

   b. Consistency (range, logicality)_____________________________

   c. Engineering interpretation of numerical output_____________

   d. others____

6. Algorithm / inferential process
   a. Level of sophistication______________________________________

   b. Assumptions & applicability limits______________________________

   c. Documentation of the model____________________________________

   d. Software____

   e. Sample case____

   f. Computational cost (man power and cpu time)_____________________

   g. others____

7. Uncertainty modelling
   a. Sensitivity analysis____

   b. others____

8. Interaction with other systems' operations
   a. Operations which should have been executed before (pre-requisites) –> warning –>
      errors____

   b. Operations which could be executed after –> suggestions__________

   c. Admissible interactions with databases
      –> access for reading from____________________________________

      –> access for writing to_______________________________________

   d. others____

USE OF INQUIRY FOR ACQUISITION OF ENGINEERING KNOWLEDGE FOR EXPERT SYSTEMS APPLIED IN THE STRUCTURAL SAFETY ASSESSMENT

A. Jovanovic
Staatliche Materialprüfungsanstalt (*MPA*)
Universität Stuttgart
Federal Republic of Germany

Abstract

Acquisition of engineering knowledge is a major problem in development of expert systems, in particular those in the field of structural engineering. In practical applications, the generic theoretical concepts must be usually modified and/or combined with the ad hoc developed methods, allowing to resolve the application specific problems. This has been confirmed by the *MPA* experience with and inquiry involving over 40 structural engineering experts. The inquiry has been used as a way of elicitation of engineering knowledge necessary for the development of the expert systems in the domain of structural analysis, at the first place for the expert system prototype for leak–before–break analysis. The results of the inquiry and the experiences gathered during their elaboration are shown in the paper.

1. Introduction

The co–operative expert systems, helping in data search, diagnostics, decisional and similar problems, are emerging in the current engineering practice. However, application of such systems has been, so far, much more spread and efficient in the fields like medicine, financial management, or geological searches, where users deal mainly with the qualitative data. Accordingly to the experience which has been gathered so far at the Staatliche Materialprüfungsanstalt (*MPA*), in order to make the expert systems comparably useful in the field of structural engineering, the following research is needed:

a) adapt the existing knowledge engineering (KE) methods, so that they can be used in interaction with other elements of the usual engineering practice, such as codes, standards, procedures of the numerical analysis, testing, etc.

b) provide the "background data" necessary for a successful application of the knowledge engineering (KE) methods, at the first place of the methods regarding the interfacing between the user and the expert system (make the domain expert, the user and the expert system *speak the same language*).

In other words, to make an information, like the one that "the difference between the yield and the ultimate stress values, by the high strength steels, is usually small", usable for an expert system and to allow its coupling with the numerical analysis, it is necessary both to adapt the KE methods and to provide data on the total of the information contained in the qualitative terms used in the expert knowledge communication (e.g. in the term "high strength steel").

This idea has been practically realized at *MPA* in form of a pilot–inquiry, performed in late 1988 and early 1989 at *MPA* (involving its employees and the guests of the *14th MPA* Seminar) and in several other occasions which followed. Its particular goals have been to obtain data necessary for further development of the expert system prototype for the Leak–before–Break analysis (Jovanovic, 1989; Jovanovic and co–workers, 1989b) and to gather new experiences in elicitation of the engineering knowledge. The expected/obtained results of the inquiry should allow to use and to treat more systematically the experience (knowledge) formulated in heuristic rules and in heuristic knowledge in general (Jovanovic and co–workers, 1989a), which is involved in structural analysis (e.g. "if the energy stored in the pipe is high, a break is more probable than a leak"). In particular, the data are important for combination of contardicting rules in expert systems. The paper shows the main features and the results of the inquiry and introduces some new approaches to the elaboration of these results.

2. Knowledge Elicitation in Engineering – General Problems

It has been recognized and confirmed that acquiring knowledge from a human expert (engineer) represents a major problem when building a knowledge based, i.e. expert systems (Boose, Bradshow, 1987). The task to acquire engineering knowledge is a very complex one, including in most of the cases, at least the following aspects:

- elicitation of distinctions,
- decomposition of the main domain problem
- combining uncertain information
- specification and/or choice of the most appropriate elicitation/acquisition techniques
- integration of different data/information types
- use of multiple sources of knowledge
- use of different sources of knowledge
- organization and formulation of elicitated knowledge (as a preparation for the the knowledge representation)
- feed–back between the knowledge and the domain engineer
- verification of compiled/formatted knowledge

In reality, the knowledge acquisition process rarely ends abruptly, it often continues in parallel with the formulation and practical realization of the knowledge base, and the application of the expert system. In such cases, the *refining of the knowledge base* has also a series of single steps (Eshelman and co–workers, 1987), involving elicitation and contacts with the domain experts. The first step is usually *differentiating of knowledge*. It includes differentiating between the anticipatory and circumstantial knowledge, qualifying of knowledge and adjusting the support values. The following steps are usually represented by the *covering* and *combining* of knowledge. However, in engineering applications, these general concepts are often to be modified and/or combined with ad hoc solutions (e.g. Cung and Ng, 1988, or the approach applied in this work).

3. The MPA Inquiry

3.1. Practical Goals and Form of the Inquiry

Main practical goal of the inquiry has been to check the understanding of the linguistic terms normally used in formulation of the heuristic rules applicable in the engineering analysis of structural safety. A list of 8 such terms has been chosen for the inquiry:

- Charpy V–notch impact energy (upper shelf)
- the "Reference NDT–Temperature", according to ASME code,
- yield stress
- temperature
- crack depth
- energy stored in piping containing three types media (Gas, Steam, Liquid).

To each of the terms has been assigned a list of linguistic estimators, usually used to describe the term (e.g. "very deep crack", "extremely low temperature", etc.). Each of this composed expressions had to be related to a range, i.e. to an interval of numerical values, as shown in figure 1.

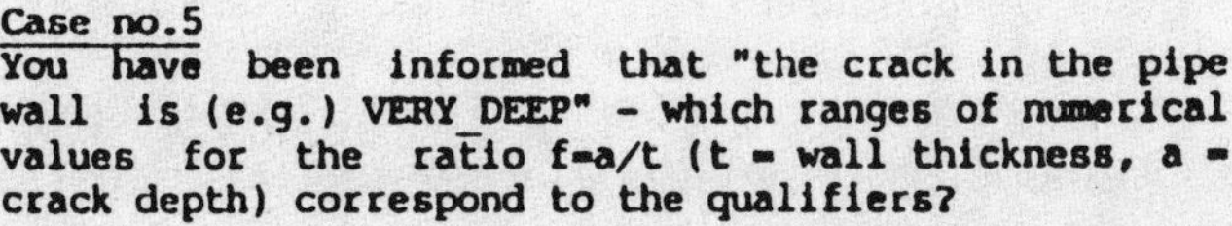

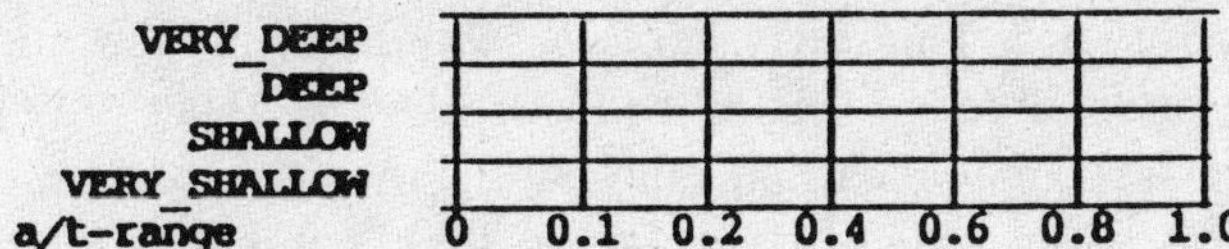

Fig.1 – Sample question from the inquiry

In addition, the subjective probability estimators, such as "very low chance" or "certain", had to be related to the subjective expectance of realization of an event coupled with these estimators, expressed as subjective probability, in percents. A list of 11 such estimators had to be evaluated in the inquiry, both in English and in German.

Format of the inquiry corresponded to the imposed goals and to the methodology adopted for use of the results. The methodology based on fuzzy sets (Zimmermann, 1988; Dubois, Prade, 1986) has been chosen as a tool for coupling between the numerical and symbolic analysis, i.e. it has been chosen for the transition between the numerical and linguistic variables used in/by the expert system. A practical goal for each question has thus been to obtain a reliable formulation of the membership function, i.e. the corresponding fuzzy set/number (figure 2), for each of the analyzed cases.

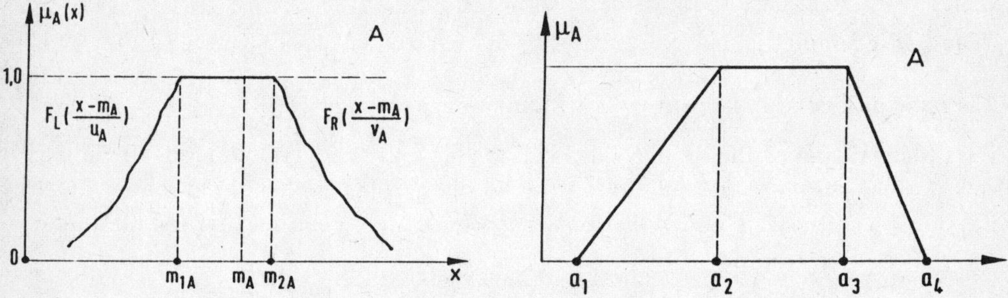

Fig.2 – Generic (left) and trapezoidal (right) fuzzy nubers (TrFN's) with a flat

3.2. Results

As shown in Table 1, quite a number engineers involved in structural safety analysis has taken part in the inquiry. Even the generic inquiries of the type, not limited to the engineering or any other specific group of people, had not had many more participants (e.g. one of the most widely used inquiries for the problems of subjective probability assessment, the one of Beyth–Marom (1982), had 63 participants).

The general statistic of the inquiry is given in Tables 1 to 12. The participants have been related to structural integrity and assessment problems from various subdomains: material science, stress analysis, fracture mechanics analysis, nondestructive examination, expert systems, etc. This facts explains why some of the questions have had a lower answering rate (table 2). Engineers from industry (over 50% of all participants) have been particularly cautios when completing the inquiry forms, avoiding to give answers on the questions which where not in their competence range. The questions involving extremes of the subjective probability estimators remained also with a relatively low number of given answers, which indicates that many of the participants might had some difficulties in understanding and/or accepting the concepts of subjective probability examined in the Question Nr. 7 of the inquiry.

Institution	Number of Participants
Total	41
14th MPA-Seminar	24
Femcad-Conference	2
Mail	15
Answers in English	18
Answers in German	23
Research Institutes	8
Industry	24
MPA employees	8
other	1

Table 1 – Overall statistics of the inquiry

Question	Number of Answers
Nr.1	28
Nr.2	26
Nr.3	28
Nr.4	30
Nr.5	29
Nr.6a	27
Nr.6b	27
Nr.6c	27
Nr.7 (english)	18
Nr.7 (german)	20

Table 2 – Statistics of answers in the inquiry – all questions

Question Nr.1: Charpy V-notch impact energy (uppershelf)-$A_{v-u.s}$						
Estimator	number of assigned ranges					
EXTREMELY HIGH				3	13	22
HIGHG			4	18	14	2
AVERAGE		11	22	14	2	
LOW	5	20	5			
EXTREMELY LOW	24	5				
$A_{v-u.s}$-range (J)	< 50	50-100	100-150	150-200	200-250	> 250

Table 3 – Statistics of the inquiry: question Nr. 1

Question Nr.2: Reference NDT-temperature (RT_{NDT})					
Estimator	number of assigned ranges				
HIGH			4	16	25
AVERAGE		11	19	19	1
LOW	3	22		8	
EXTREMELY LOW	24	4			
RT_{NDT}-range ($^{\circ}C$)	< -50	-50-0	0-50	50-100	> 100

Table 4 – Statistics of the inquiry: question Nr. 2

Question Nr.3: Yield Stress (σ_y)						
Estimator	number of assigned ranges					
VERY HIGH				1	9	28
HIGH			1	11	22	4
AVERAGE		3	17	19	4	
LOW	3	19	17	1		
EXTREMELY LOW	24	9				
σ_y-range (MPa)	< 150	150-200	200-300	300-500	500-800	> 800

Table 5 – Statistics of the inquiry: question Nr. 3

The material science, and the stress analysis and design engineers have been prevailing among the participants. About 50% of the participants which completed the inquiry forms in English have had English as their first language. Lack of answers or inappropriate completing in question Nr.7 is somewhat higher among the German speaking

Question Nr.4: Temperature (T)						
Estimator	number of assigned ranges					
EXTREMELY HIGH				1	5	29
HIGH			2	15	27	8
AVERAGE			19	16	2	
LOW	2	22	15			
EXTREMELY LOW	26	2				
T-range (°C)	< -50	-50-0	0-250	250-400	400-600	> 600

Table 6 – Statistics of the inquiry: question Nr. 4

Question Nr.5: ratio Crack Depth/Wall Thickness (a/t)						
Estimator	number of assigned ranges					
VERY DEEP			1	8	20	28
DEEP		2	14	20	14	1
SHALLOW	9	21	8	2		
VERY SHALLOW	27	1				
a/t-range	< 0.1	0.1-0.2	0.2-0.4	0.4-0.6	0.6-0.8	0.8-1.

Table 7 – Statistics of the inquiry: question Nr. 5

Question Nr.6: Energy stored in the Pressure Vessel (medium: liquid)						
Estimator	number of assigned ranges					
HIGH		1	4	7	11	18
LOW	2	9	8	7		
NEGLIGIBLE	20	7	5			
p-range (bar)	< 1	1-10	10-50	50-100	100-150	> 150

Table 8 – Statistics of the inquiry: question Nr. 6a

participants, probably reflecting some kind of their reluctance towards the application of uncertainty based concepts in engineering.

Question Nr.6: Energy stored in the Pressure Vessel (medium: steam)						
Estimator	number of assigned ranges					
HIGH		2	13	16	17	19
LOW	3	14	8	1		
NEGLGIBLE	18	2				
p-range (bar)	< 1	1-10	10-50	50-100	100-150	> 150

Table 9 – Statistics of the inquiry: question Nr. 6b

Question Nr.6: Energy stored in the Pressure Vessel (medium: gas)						
Estimator	number of assigned ranges					
HIGH			9	13	16	18
LOW		3	15	8	3	1
NEGLIGIBLE	16	2				
p-range (bar)	< 1	1-10	10-50	50-100	100-150	> 150

Table 10 – Statistics of the inquiry: question Nr. 6c

German Version		English Version	
Estimator	number of answers	Estimator	number of answers
UNWAHRSCHEINLICH	16	UNLIKELY	20
EVENTUELL	18	LOW CHANCE	20
VIELLEICHT	17	PERHAPS	19
KANN SEIN	17	MAY BE	18
MÖGLICH	17	POSSIBLY	18
WAHRSCHEINLICH	18	PROBABLY	19
SEHR WAHRSCHEINLICH	18	HIGH CHANCE	19
ZIEMLICH SICHER	17	VERY HIGH CHANCE	19
BEINAHE	15	VERY LOW CHANCE	21
OFT	17	OFTEN	19
SELTEN	16	SELDOM	18

Table 11 – Statistics for question Nr.7: 11 estimators

German Version		English Version	
Estimator (extreme values)	number of answers	**Estimator** (extreme values)	number of answers
UNWAHRSCHEINLICH	12	UNLIKELY	16
EVENTUELL	14	LOW CHANCE	16
VIELLEICHT	14	PERHAPS	16
KANN SEIN	15	MAY BE	16
MÖGLICH	13	POSSIBLY	16
WAHRSCHEINLICH	13	PROBABLY	16
SEHR WAHRSCHEINLICH	14	HIGH CHANCE	16
ZIEMLICH SICHER	12	VERY HIGH CHANCE	16
BEINAHE	11	VERY LOW CHANCE	17
OFT	15	OFTEN	16
SELTEN	13	SELDOM	18

Table 12 – Statistic for question Nr.7: 11 estimators – extremes

4. Elaboration of Data

4.1. Possibility of Transition from Probability to Possibility Measures

Main issue in the elaboration of results has been the transformation of the obtained probability measures (directly obtainable from the statistics presented in Chapter 3) into a possibility measure (i.e. possibility measures). In other words, it was necessary to relate the possibility of events to their frequency of occurrence. In the open literature, there is a lack of a commonly agreed and by everyone accepted approach to the problem. Main reason for this is the fact that it is not possible to define a universal, single solution for reduction of the richer structure of the probability (p resp. P) concept (having the property of additivity), to the terms of possibility (π resp. Π), i.e. when "replacing" additivity with comparability. The practical solutions are dependent on the consistency requirements which one wants/has to apply in the particular case. In any case however, some mutual consistency requirements must be satisfied (Dubois, Prade, 1986). The most relevant and important one is probably the intuitive verbal requirement "the more necessary an event (A), the more probable, and the more probable, the more possible", leading to the following inequality (Dubois, Prade, 1980):

$$\forall A \qquad N(A) = 1 - \Pi(\neg A) \leqslant P(A) \leqslant \Pi(A)$$

where $\neg A$ denotes "*non–A*" and N is the necessity measure. Any method claiming to derive possibility measures from the possibilistic data must satisfy the above given condition in order to be intuitively accepted.

4.2. Practical Transformation

As shown by Dubois and Prade (1986), a simple pointwise transformation through a linear pointwise transformation of the probability assignment $p = (p_1 \ldots p_n)$, based on the statistical data regarding the universe (Ω) given as $\Omega = (w_1 \ldots w_n)$, to the possibility assignment $\pi = (\pi_1 \ldots \pi_n)$, via

$$\forall i \quad \pi_i = k p_i.$$

does not hold for non-elementary events, and hence, cannot be used in the case of this inquiry. Therefore, a non-pointwise transformation has been applied.

The transformation starts from the definition of the necessity (figure 3) given by

$$N(A) = \sum_{w_i \in A} \max(p_i - p_A, 0)$$

where $p_A = \max\{ p_i \,|\, w_i \notin A \}$. The possibility measure is calculated from the necessity by duality:

$$\pi_i = 1 - N(\Omega - \{ w_i \}),$$

i.e.

$$\pi_i = \sum_{k=1}^{n} p_k - \sum_{j \neq i} \max(p_j - p_i, 0).$$

Furthermore, the preceding equation defines a bijective mapping between the probability and the possibility assignment sets

$$\left\{ p = (p_1 \cdots p_n) \,|\, \sum p_i = 1 \right\}$$

$$\left\{ \pi = (\pi_1 \cdots \pi_n) \,|\, \max_i \pi_i = 1 \right\}$$

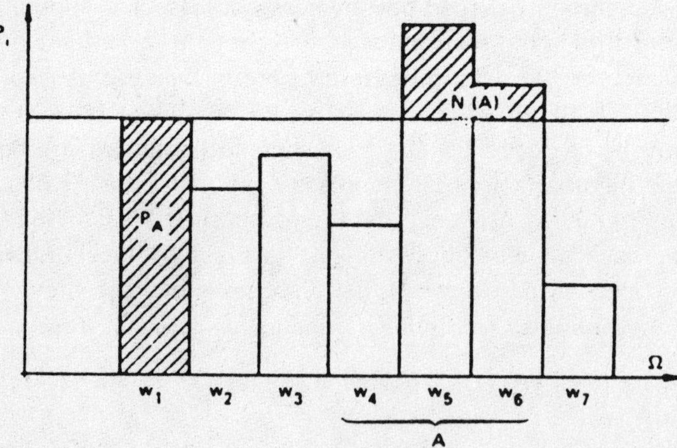

Fig.3 – Definition of the necessity measure from the statistical data (Dubois and Prade, 1986)

and brings the methodology in congruence with the axioms of the Shafer's theory (1976) and the Bayesian on. The mapping namely allows also to define the probability assignment if the only possibility assignment is known (Dubois, and Prade, 1986). This has however been of a modest interest for this inquiry.

4.3. Nested Intervals

In order to be able to obtain a possibility measure from the statistical data (considered here as a sort of random experiments) one must (theoretically) impose the requirement that the set of obtained intervals be nested. As can be shown by the results of this inquiry, this requirement is definitely too strong, nevertheless its application leads to some interesting results. A more realistic but weaker requirement would be the one that all the intervals somehow overlap, and this requirement has been fully satisfied in our case. However, the theoretical advantage of working with nested intervals is very important, because it allows a direct definition of the corresponding fuzzy set μ, which is in this case equal to the basic probability assignment m (as defined by Shafer, 1986), to the plausibility Pl and to the possibility, namely

$$\mu_F(x) = Pl(\{x\}) = \sum_{x \in I_i} m(I_i) = \Pi(\{x\})$$

The nested intervals, in other words, allow to reach the same results as those obtained by the direct derivation of $\mu(x)$ from the statistical data, considering each of he intervals as a focal element. This way of the direct derivation of a membership function (see Zimmermann, 1988) from the statistical data is well known (Mac Vicar–Whelan, 1979).

A sample derivation of the nested intervals in the inquiry is shown in figure 4.

4.4. Practical Derivation of Fuzzy Sets and TrFN's

In the practical definition of fuzzy sets corresponding to the initial linguistic variables (expressions) the method denoted above has been followed, introducing however, several alternatives in the elaboration. These alternatives and modifications reflect the wish to encompass the following factors:

- The questionaire allowed necessarily some questions to be understood in a "possibilistic way", i.e. the initial, statistical probability measure already contained some of the possibilistic elements.

- The usual definitions of the level–1 and level–0 of the possibility measure, where the level–1 corresponds to the most probable interval in the statistics and the level–0 to interval limits, was assumed to be unrealistic.
  This, as well as the wish to follow, by the parameters of the defined TrFN the integral of the possibility distribution, lead to a simple numerical algorithm for calculation of the TrFN parameters. The intuitive meaning of this concern is

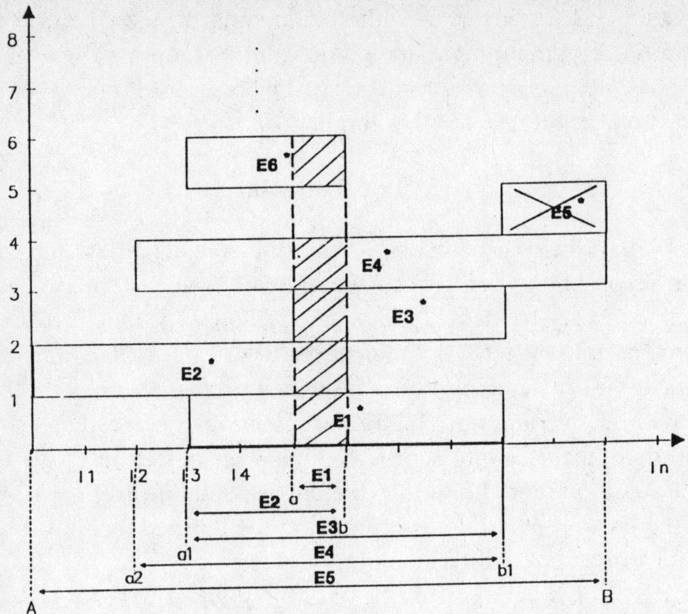

Fig.4 – Practical derivation of the nested intervals in the inquiry

shown in figure 5, which shows how a significantly different possibility assignments can lead to the same TrFN.

The algorithm mentioned in above makes corrections of the derived membership function taking into the above mentioned concern as mentioned in figure 6. Figures 7 and 8 show how significant this correction can be in some cases, in the given case for the estimator "may be" in the inquiry. The final results, a summary of which is given in the appendix, show however, that this difference is not as big in other cases.

- Not every of the 42 participants has had the possibility to discuss the questionaire with *MPA*. Furthermore, both the interviewers and the majority of the interviewees, had a very limited preceding experience with such inquiries. Therefore, some missunderstanding cannot be excluded. It is very probable that some of these factors gave some hardly understandable extremes in the inquiry. Accordingly to the usual practice in the classic statistics, these extremes have been neglected.

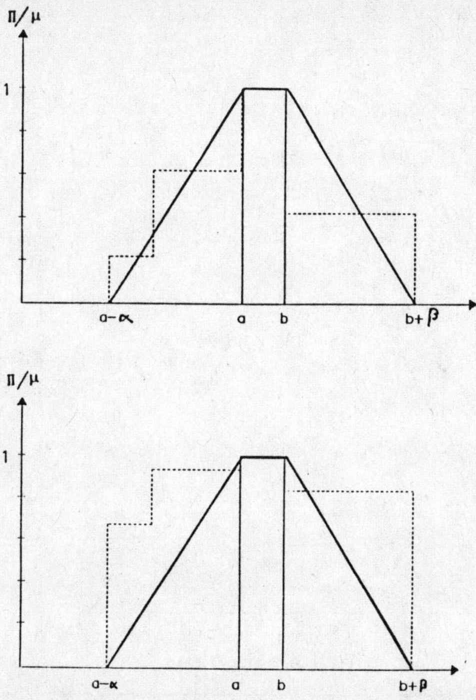

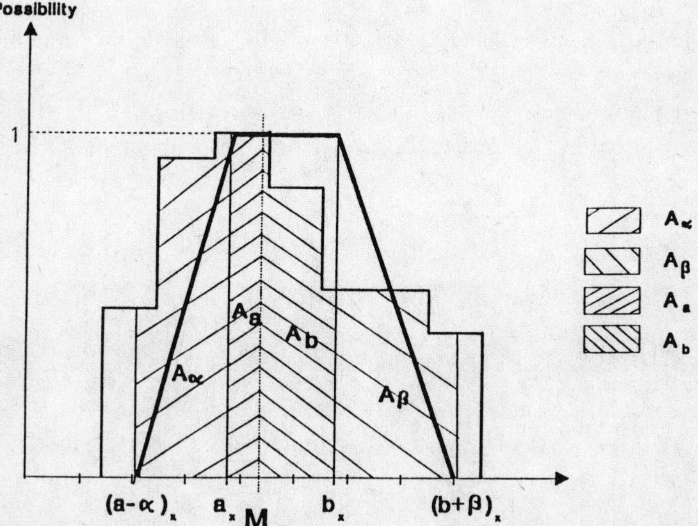

Fig.5 – Two different possibility assignments leading to the same TrFN

Fig.6 – Definition of the TrFN on the basis of the possibility assignment
eliminating extrems and taking into account shape of its distribution

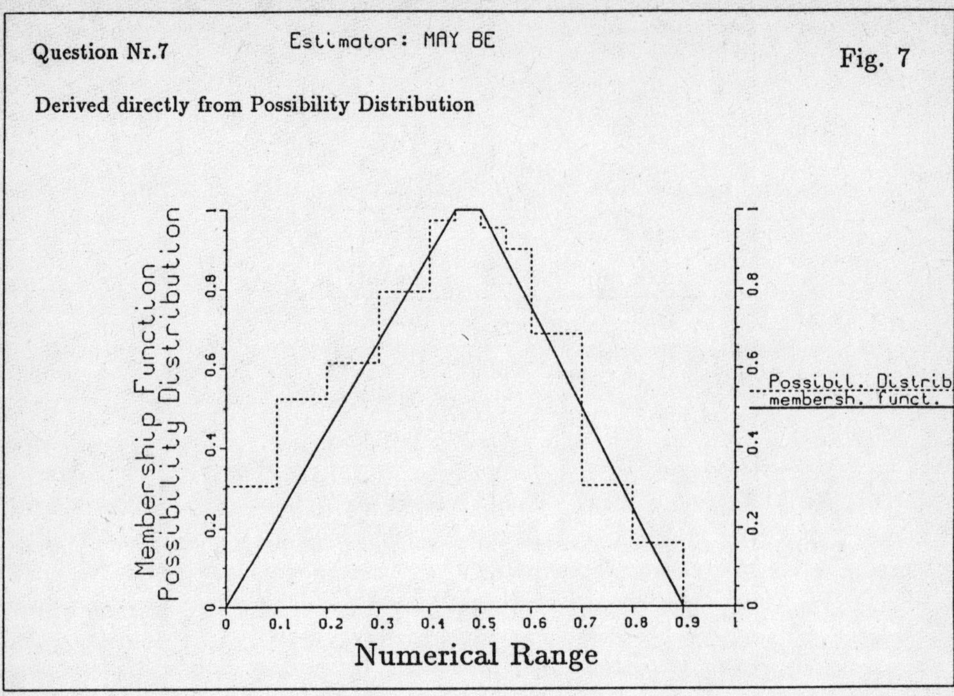

Question Nr.7
Estimator: MAY BE
Fig. 7
Derived directly from Possibility Distribution
Membership Function
Possibility Distribution
1
0.8
0.6
0.4
0.2
0
0   0.1   0.2   0.3   0.4   0.5   0.6   0.7   0.8   0.9   1
Numerical Range
Possibil. Distrib
membersh. funct.

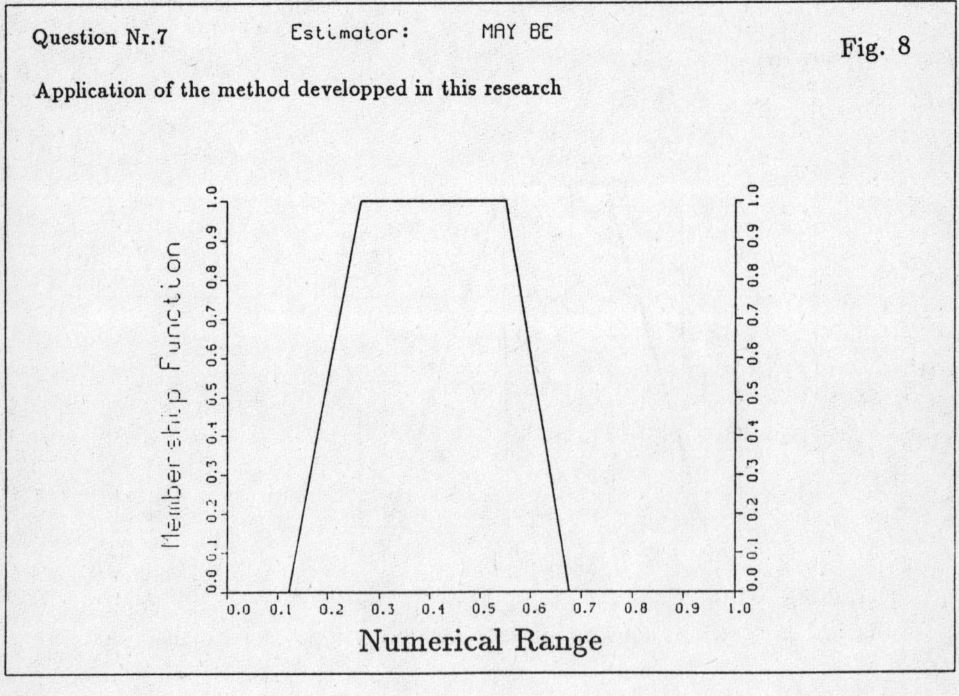

Question Nr.7
Estimator: MAY BE
Fig. 8
Application of the method developped in this research
Membership Function
1.0 0.9 0.8 0.7 0.6 0.5 0.4 0.3 0.2 0.1 0.0
0.0   0.1   0.2   0.3   0.4   0.5   0.6   0.7   0.8   0.9   1.0
Numerical Range

5. Results

The preliminary elaboration of results (Jovanovic and Hassler, 1989) shows that the inquiry enabled to collect a significant amount of new information. As for the difference with the preceding research in the domain of the linguistic terms for the subjective probability definition (Beyth–Marom, 1982), it has been possible to extend this work in two directions:

- Extremes of the terms have been analyzed separately (figures 9 and 10), allowing to the inquirees to distinguish better between the "normal cases" and the "hypothetical ones" (this was actually a question raised very often in the discussions with the interviewees).

- It has been possible to check the transferability of the possibility assignments for the estimators by the language translation, in this case, from English to German. The answer is essentially positive, while most of the understanding of the underlieing subjective probability, attached to the estimators, showed small changes when translated from English to German and vice versa. The example shown in figures 11 and 12 (the estimator "seldom") confirms this statement. More difficulties have been found with estimators where a direct translation is not possible, like in the given example. So for instance, the presumably similar English "low chance" and the German "eventuell" cannot be used as equivalents from the point of view of underlieing subjective probability assignment. This has been of a particular importance because the degree of acceptance of the original or the derived English terminology in the field of structural safety assessment is far more lower than the corresponding degree of acceptance in the field of knowledge engineering, where the English terminology is quite widespread in Germany too.

The results directly regarding the application domain have also been very consistent. . Terms required to be evaluated by the domain experts (e.g. "very deep crack", "deep crack", "shallow crack" and "very shallow crack" – figures 13 to 16) have been brought into a form directly usable in/by expert systems, i.e. in the rules containing these terms in their premises (e.g. "*when* the crack is very deep, *then* the failure through leak becomes more probable").

This part of the results of the inquiry is planned to be further discussed and commented by the domain experts in order to obtain a reliable assessment of their dependency with other factors of the leak–before–break analysis. This should significantly improve current capabilities of the *MPA* expert system for the leak–before–break analysis (Jovanovic and co–workers, 1989b) and represent a form of introduction of the deep knowledge (Jovanovic and co–workers, 1989a) in a compiled form (the interdependencies and the underlieing phenomenology will be formatted as new rules).

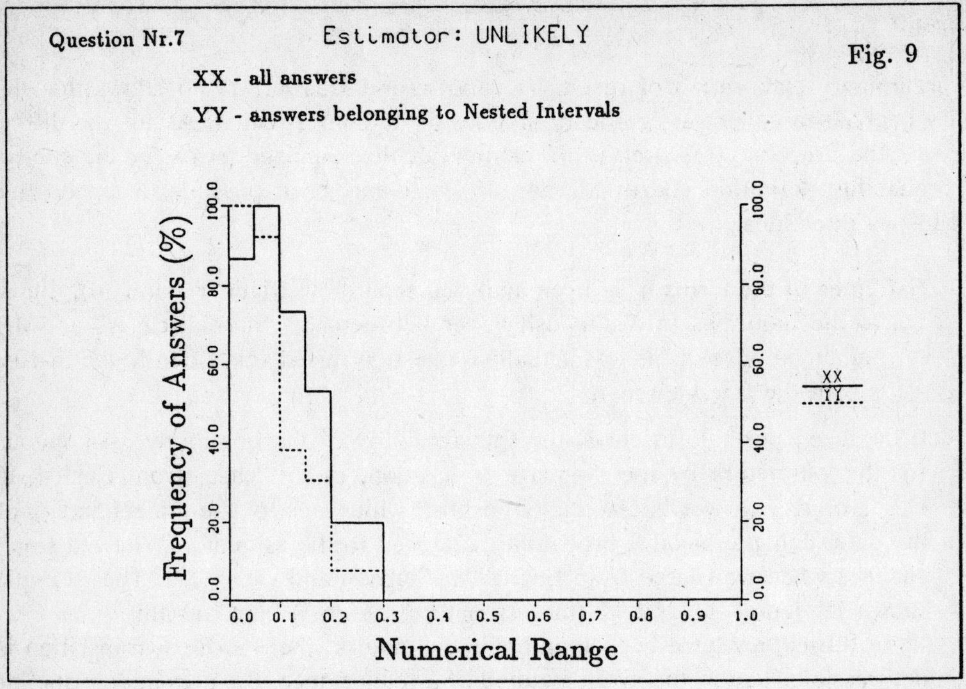

Question Nr.7
Estimator: UNLIKELY
Fig. 9
XX - all answers
YY - answers belonging to Nested Intervals
Frequency of Answers (%)
Numerical Range
XX
YY

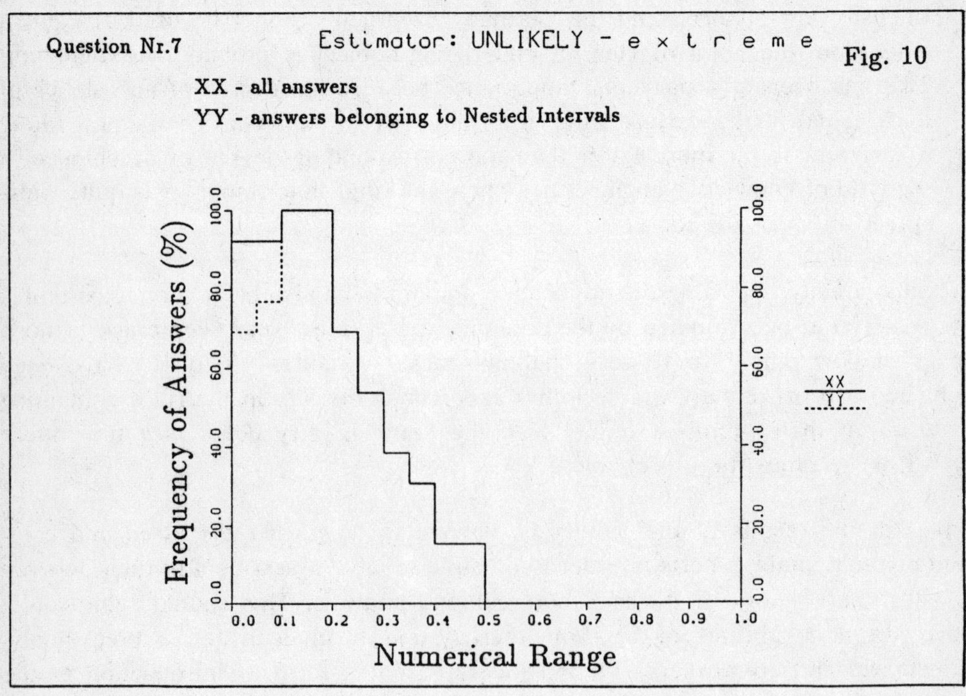

Question Nr.7
Estimator: UNLIKELY - e x t r e m e
Fig. 10
XX - all answers
YY - answers belonging to Nested Intervals
Frequency of Answers (%)
Numerical Range
XX
YY

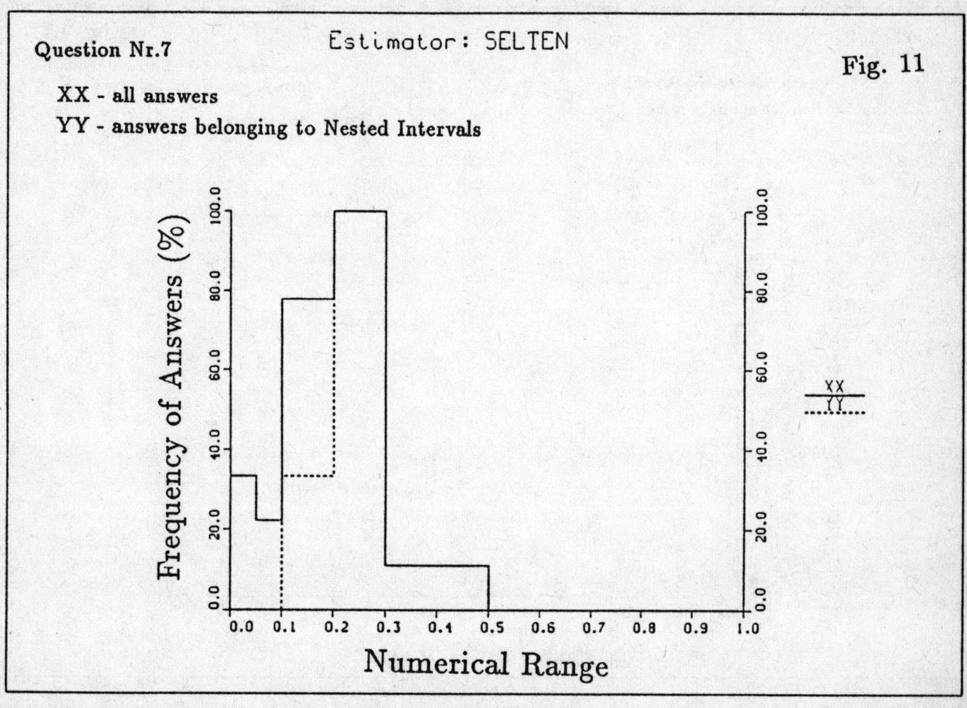

Question Nr.7
Estimator: SELTEN
Fig. 11
XX - all answers
YY - answers belonging to Nested Intervals
Frequency of Answers (%)
Numerical Range
XX
YY

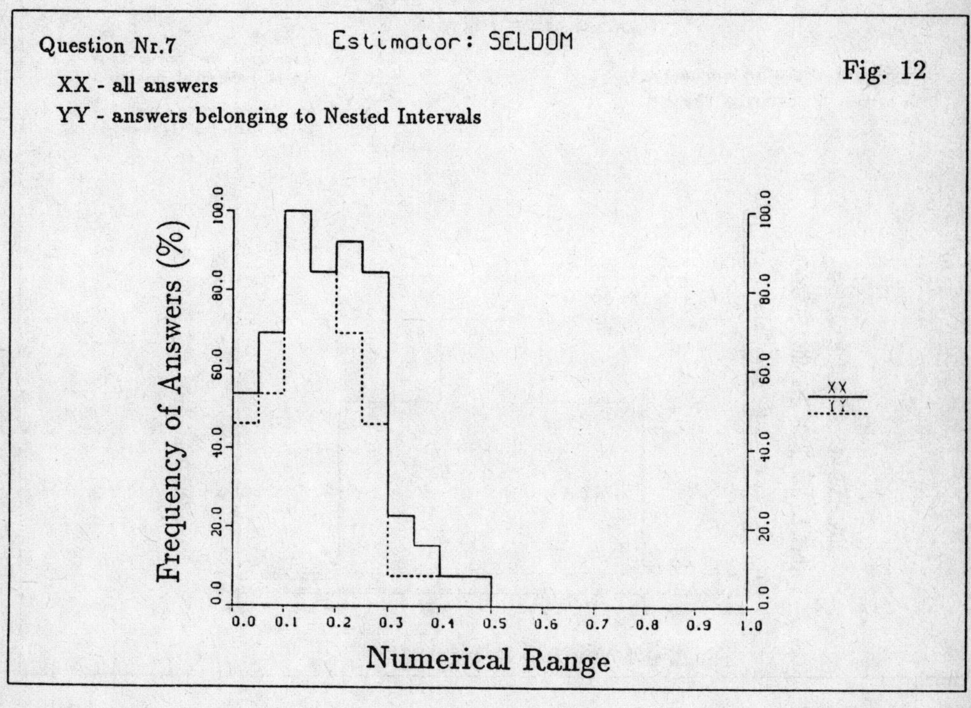

Question Nr.7
Estimator: SELDOM
Fig. 12
XX - all answers
YY - answers belonging to Nested Intervals
Frequency of Answers (%)
Numerical Range
XX
YY

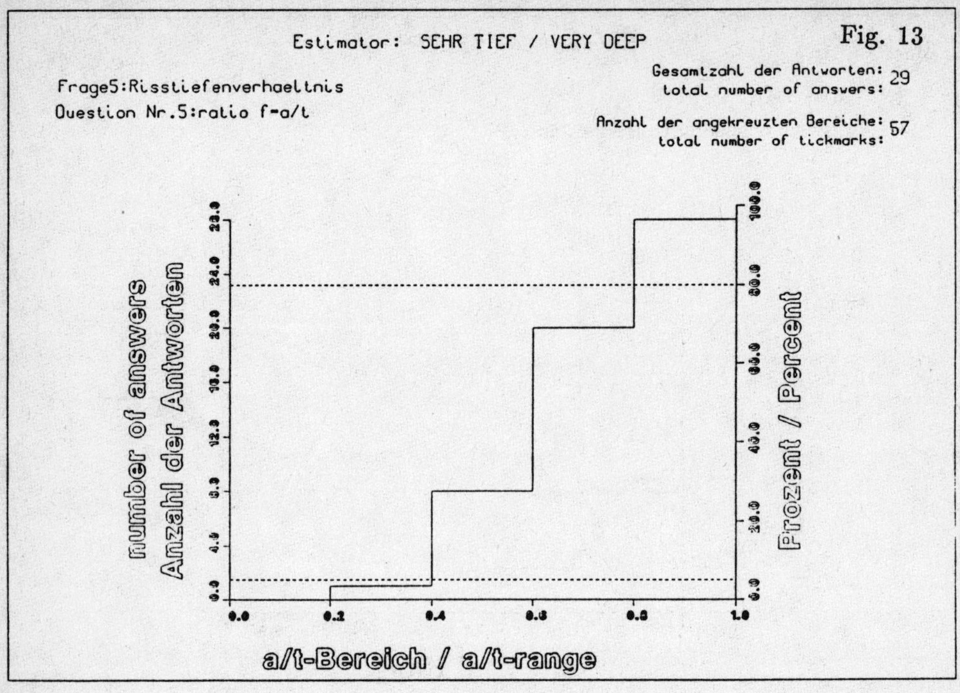

Estimator: SEHR TIEF / VERY DEEP
Fig. 13
Frage5:Risstiefenverhaeltnis
Question Nr.5:ratio f-a/t
Gesamtzahl der Antworten: 29
total number of answers:
Anzahl der angekreuzten Bereiche: 57
total number of tickmarks:
number of answers
Anzahl der Antworten
Prozent / Percent
a/t-Bereich / a/t-range

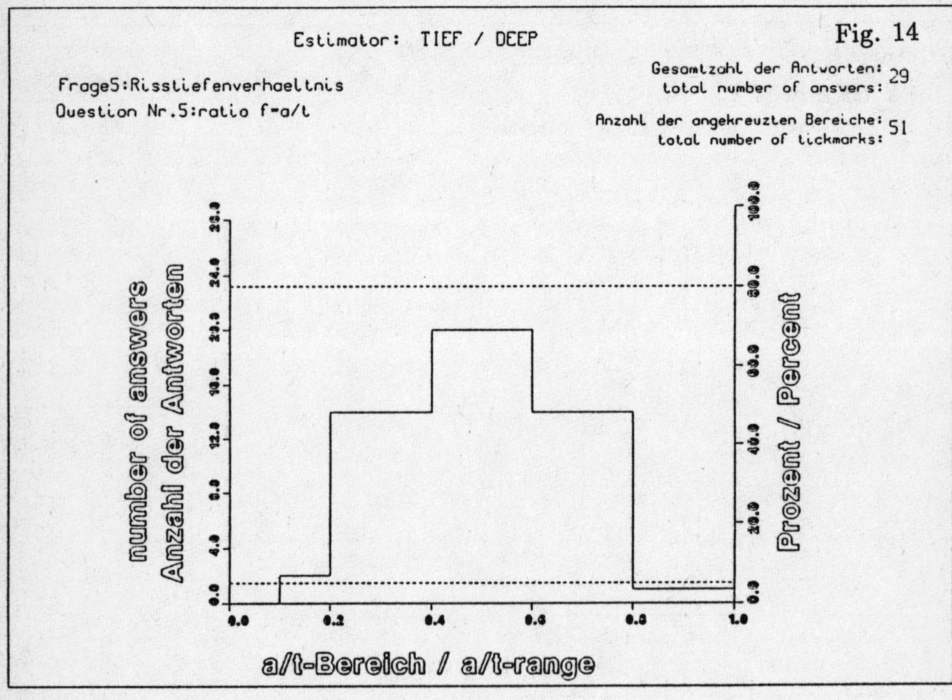

Estimator: TIEF / DEEP
Fig. 14
Frage5:Risstiefenverhaeltnis
Question Nr.5:ratio f-a/t
Gesamtzahl der Antworten: 29
total number of answers:
Anzahl der angekreuzten Bereiche: 51
total number of tickmarks:
number of answers
Anzahl der Antworten
Prozent / Percent
a/t-Bereich / a/t-range

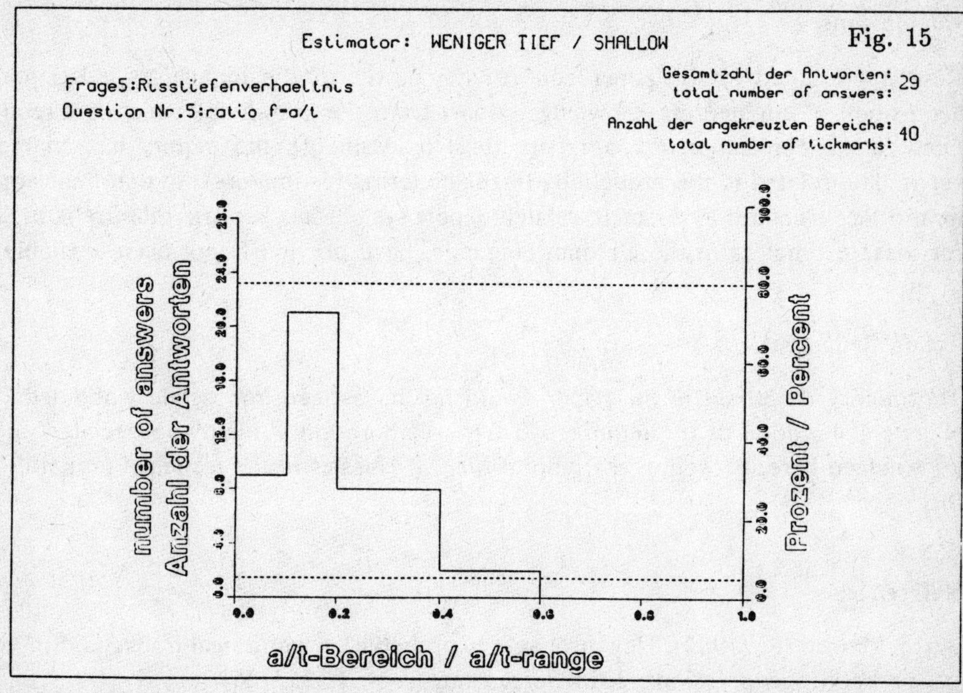

Estimator: WENIGER TIEF / SHALLOW
Fig. 15
Frage5:Risstiefenverhaeltnis
Question Nr.5:ratio f-a/t
Gesamtzahl der Antworten: 29
total number of answers:
Anzahl der angekreuzten Bereiche: 40
total number of tickmarks:
number of answers
Anzahl der Antworten
Prozent / Percent
a/t-Bereich / a/t-range

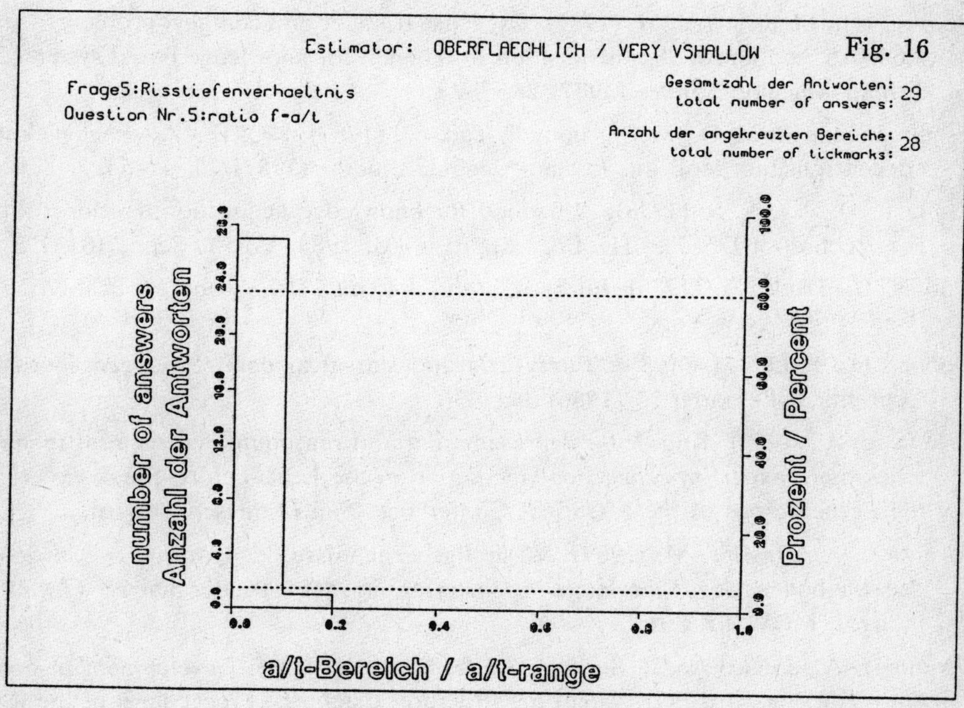

Estimator: OBERFLAECHLICH / VERY VSHALLOW
Fig. 16
Frage5:Risstiefenverhaeltnis
Question Nr.5:ratio f-a/t
Gesamtzahl der Antworten: 29
total number of answers:
Anzahl der angekreuzten Bereiche: 28
total number of tickmarks:
number of answers
Anzahl der Antworten
Prozent / Percent
a/t-Bereich / a/t-range

Conclusions

Results presented in the paper confirm the power of the inquiry as a KE tool for acquisition of engineering knowledge. Apart from the knowledge which is directly related to the application domain (structural assessment), the inquiry has enabled to verify data related to the probability linguistic terms (estimators), so that their application in the structural assessment related expert system has become more reliable. Data for these estimators in the German language, have previously not been available.

Acknowledgments

The inquiry described in the paper would not have been impossible without the precious collaboration of the inquirees. Their collaboration is highly appreciated and acknowledged here, as well as the help of Mr. M. Hassler in the technical preparation of this paper.

References:

Beyth–Marom, R. (1982): How probable is probable? A numerical translation of verbal probability expressions. J. of Forecasting, Vol. 1, 257–269

Boose, J. H., Bradshaw, J. M. (1987): Expertise transfer and complex problems – using AQINAS as a knowledge–acquisition workbench for knowledge based systems. Int. J. Man–Machine Studies (1987) 26, 3–28

Eshelman, L., Ehret, D., McDermott, J., Tan, M. (1987): MOLE – A tenacious knowledge–acquisition tool. Int. J. Man–Machine Studies (1987) 26, 41–54

Cung, L. D., Ng, T. S. (1988): A method for knowledge acquisition developed in the construction of DESPLATE. Eng. Appli. of AI, 1988, Vol. 1, Sept., 181–193

Dubois, D., Prade, H. (1980): Fuzzy sets and systems: Theory and applications. Academic Press, New York

Dubois, D., Prade, H. (1986): Fuzzy sets and statistical data. European Journal of Operational Research 25 (1986) 345–356

Jovanovic, A. (1989): Knowledge representation and reasoning in structural reliability assessment expert systems: Implementation in the Leak–before–Break expert system, Proceedings of the ICOSSAR Conference, San Francisco, August

Jovanovic, A., Hassler, M. (1989): Vorläufige Ergebnisse der Umfrage – Anwendung der Methoden des Knowledge Engineering in der Strukturanalyse (Zwischenbericht). MPA Stuttgart

Jovanovic, A., Lucia, A. C., Servida, A., Volta, G. (1989a): Development of a deep knowledge based expert system for structural diagnosis, Proceedings of the ICOSSAR Conference, San Francisco, August

Jovanovic, A., Kobes, E., Sauter, A., Sturm, D. (1989b): An expert system in the domain of fracture mechanics, Proc. of the SMiRT Conference, Los Angeles

Mac Vicar-Whelan, P.J. (1979): Fuzzy logic: An alternative approach. Proceedings of of the 9th Symp. on Multiple-Valued Logic, Bath

Shafer, G. (1976): A mathematical theory of evidence, Princeton University Press, New York

Zimmermann, H. J. (1988): Fuzzy set theory and its application (3rd printing). Kluwer Academic Publishers Group, Boston-Dordrecht-Lancaster

A p p e n d i x

The appendix contains statistics (figures A1 to A10) and elaboration of data regarding the estimators from question Nr. 7 (figures A11 to A20). The data regard 10 estimators for which a comparison with the Reference Membership Function (Beyth-Marom, 1982) was possible.

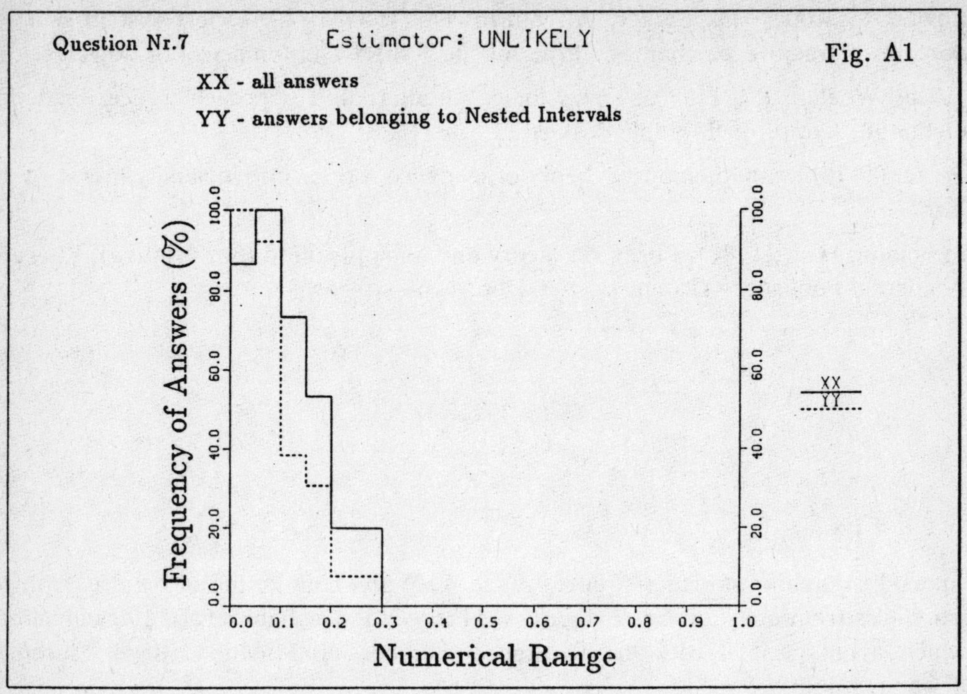

Question Nr.7
Estimator: UNLIKELY
Fig. A1
XX - all answers
YY - answers belonging to Nested Intervals
Frequency of Answers (%)
Numerical Range
XX
YY

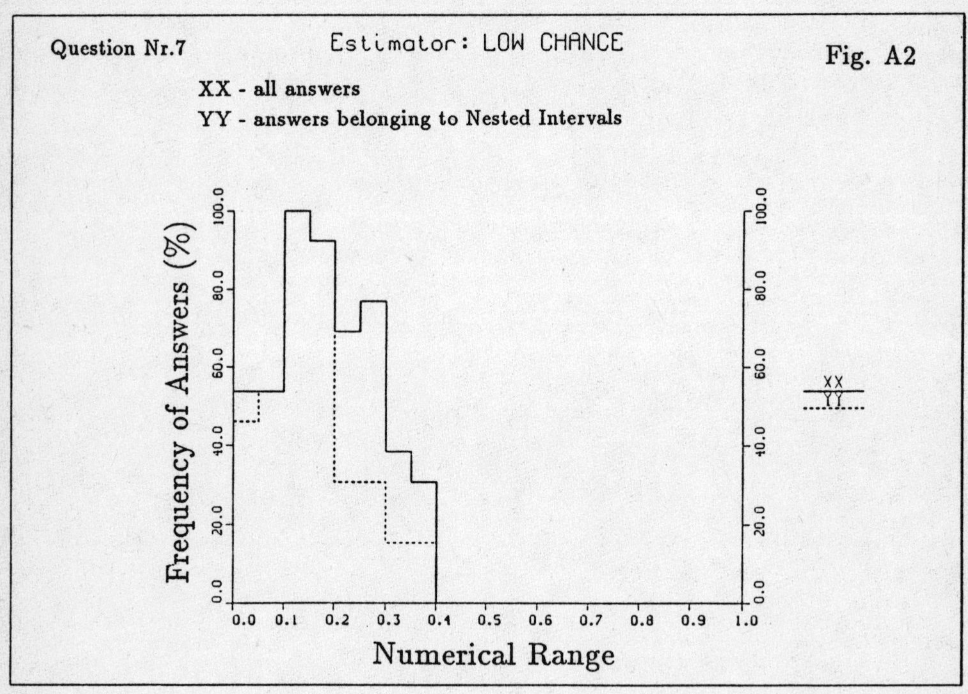

Question Nr.7
Estimator: LOW CHANCE
Fig. A2
XX - all answers
YY - answers belonging to Nested Intervals
Frequency of Answers (%)
Numerical Range
XX
YY

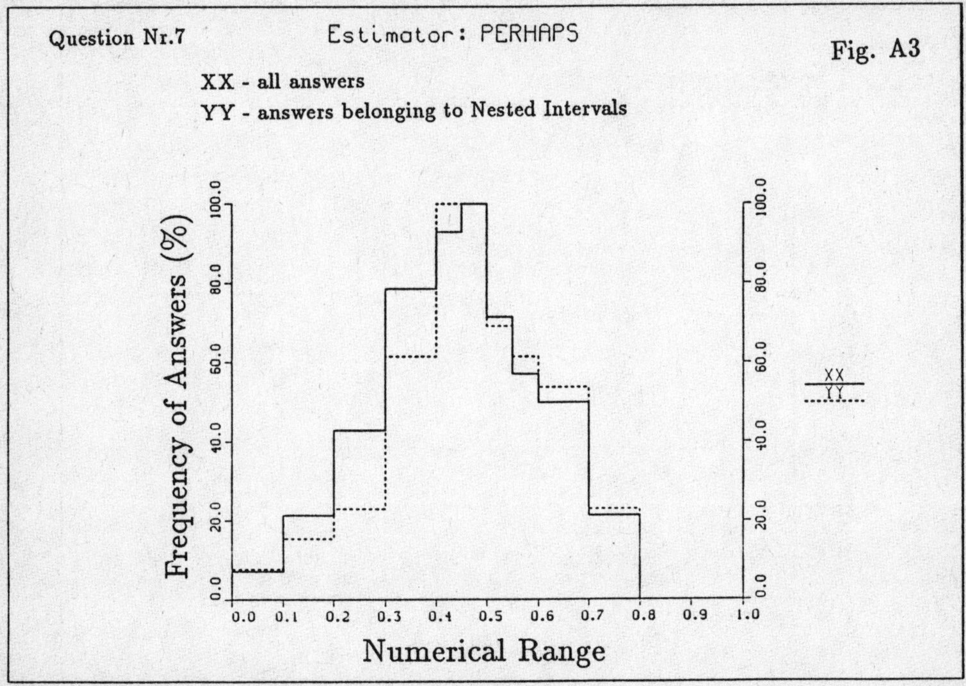

Question Nr.7
Estimator: PERHAPS
Fig. A3
XX - all answers
YY - answers belonging to Nested Intervals
Frequency of Answers (%)
Numerical Range
XX
YY

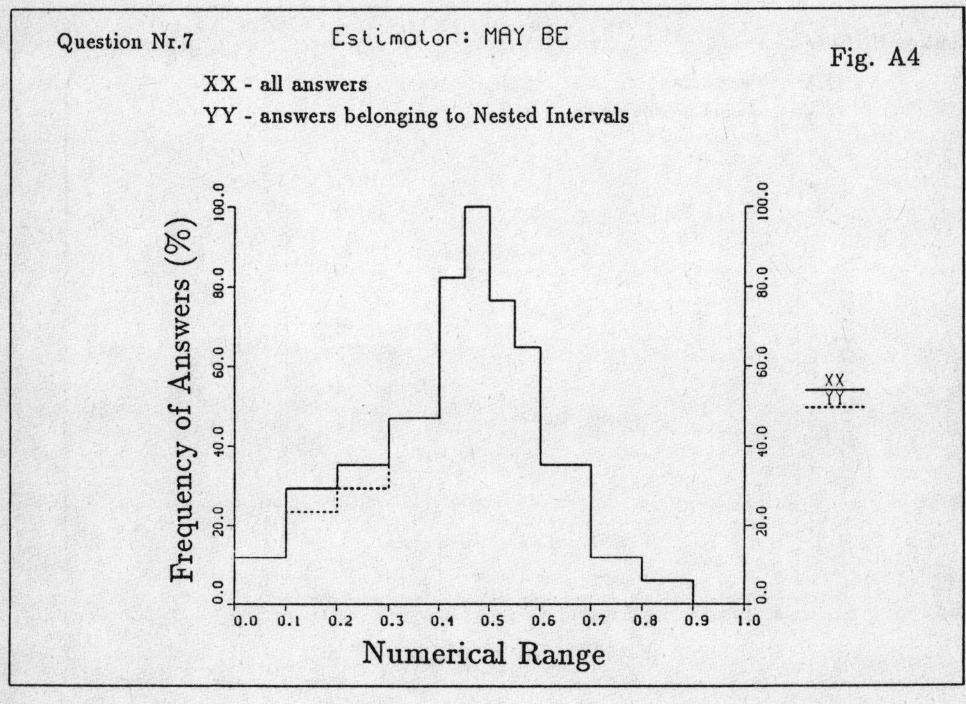

Question Nr.7
Estimator: MAY BE
Fig. A4
XX - all answers
YY - answers belonging to Nested Intervals
Frequency of Answers (%)
Numerical Range
XX
YY

464

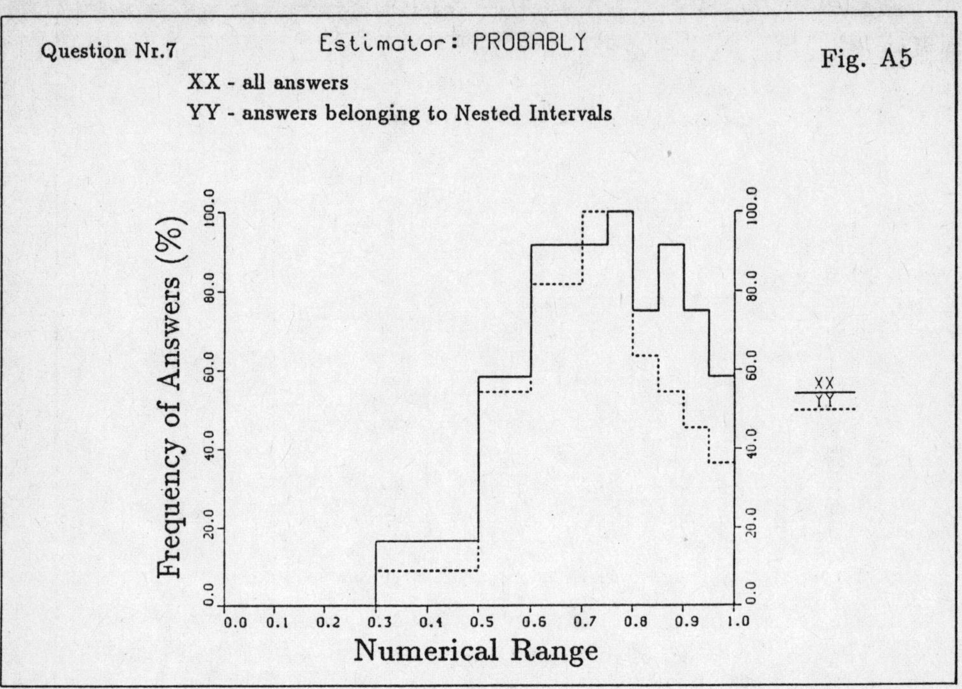

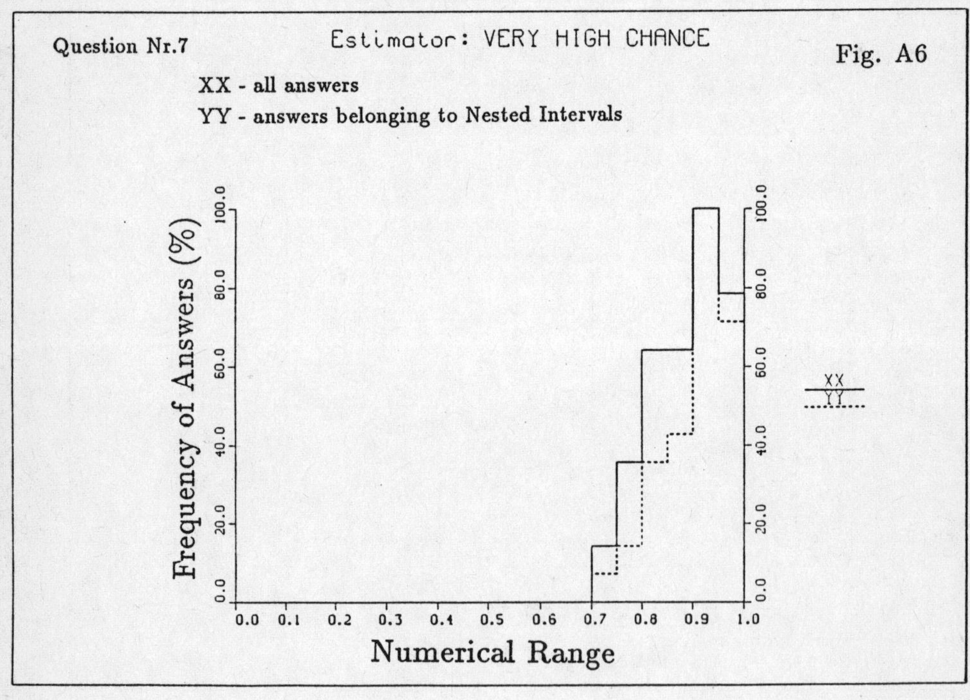

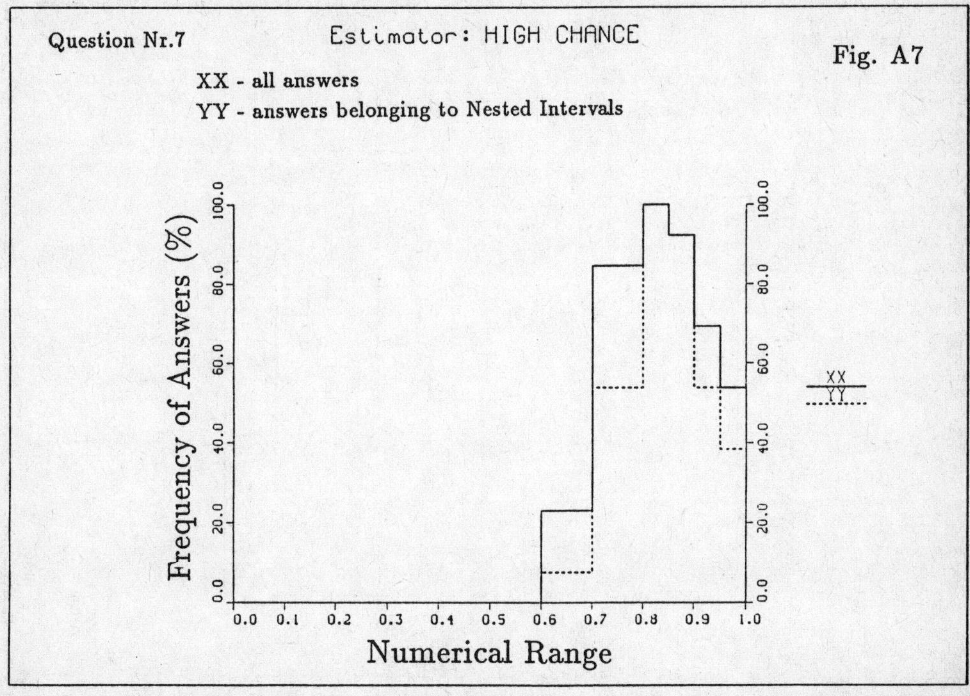

Question Nr.7
Estimator: HIGH CHANCE
Fig. A7
XX - all answers
YY - answers belonging to Nested Intervals
Frequency of Answers (%)
Numerical Range
XX
YY

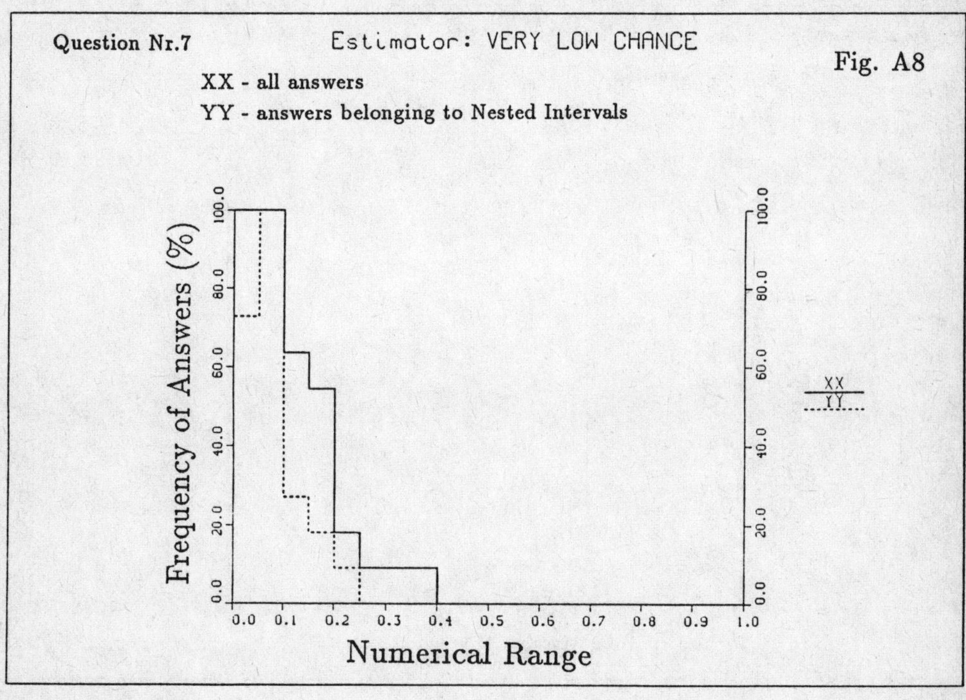

Question Nr.7
Estimator: VERY LOW CHANCE
Fig. A8
XX - all answers
YY - answers belonging to Nested Intervals
Frequency of Answers (%)
Numerical Range
XX
YY

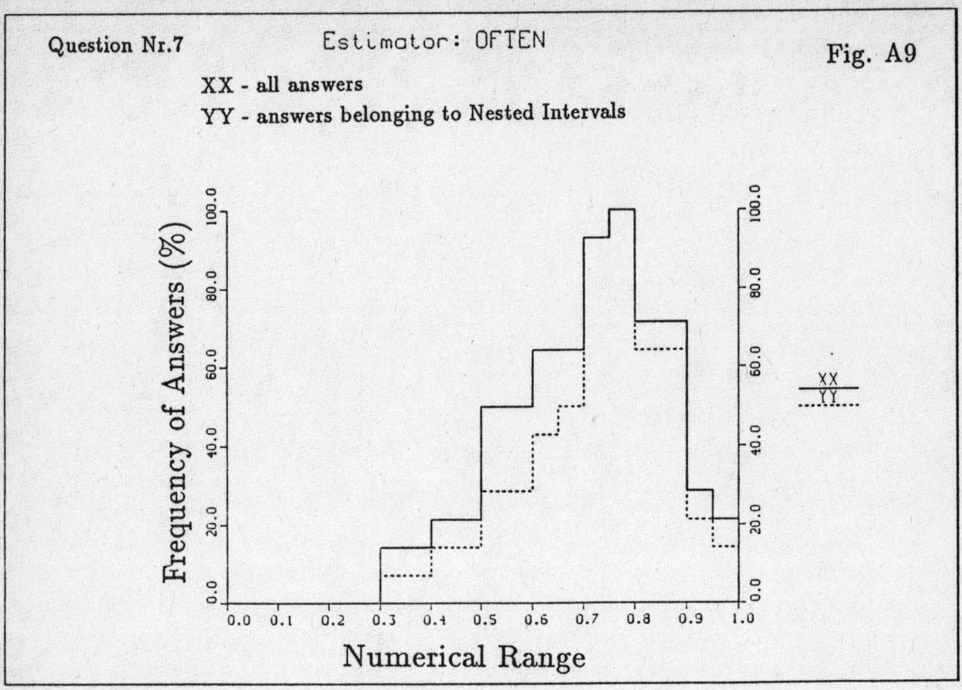

Question Nr.7
Estimator: OFTEN
Fig. A9
XX - all answers
YY - answers belonging to Nested Intervals
Frequency of Answers (%)
100.0
80.0
60.0
40.0
20.0
0.0
0.0 0.1 0.2 0.3 0.4 0.5 0.6 0.7 0.8 0.9 1.0
Numerical Range
XX
YY

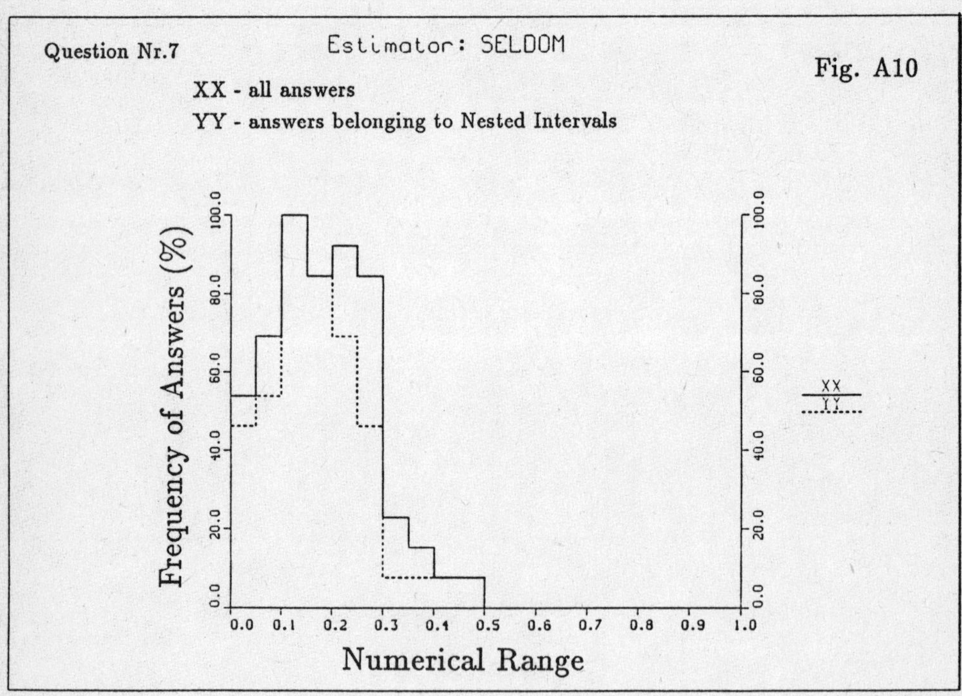

Question Nr.7
Estimator: SELDOM
Fig. A10
XX - all answers
YY - answers belonging to Nested Intervals
Frequency of Answers (%)
100.0
80.0
60.0
40.0
20.0
0.0
0.0 0.1 0.2 0.3 0.4 0.5 0.6 0.7 0.8 0.9 1.0
Numerical Range
XX
YY

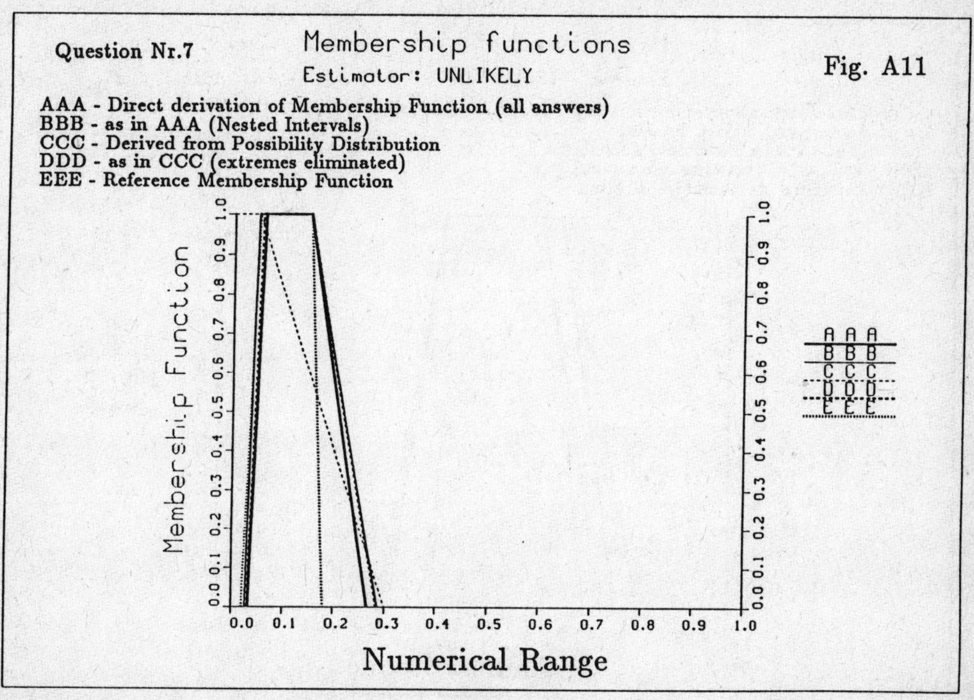

Question Nr.7
Membership functions
Estimator: UNLIKELY
Fig. A11
AAA - Direct derivation of Membership Function (all answers)
BBB - as in AAA (Nested Intervals)
CCC - Derived from Possibility Distribution
DDD - as in CCC (extremes eliminated)
EEE - Reference Membership Function
Membership Function
Numerical Range
AAA
BBB
CCC
DDD
EEE

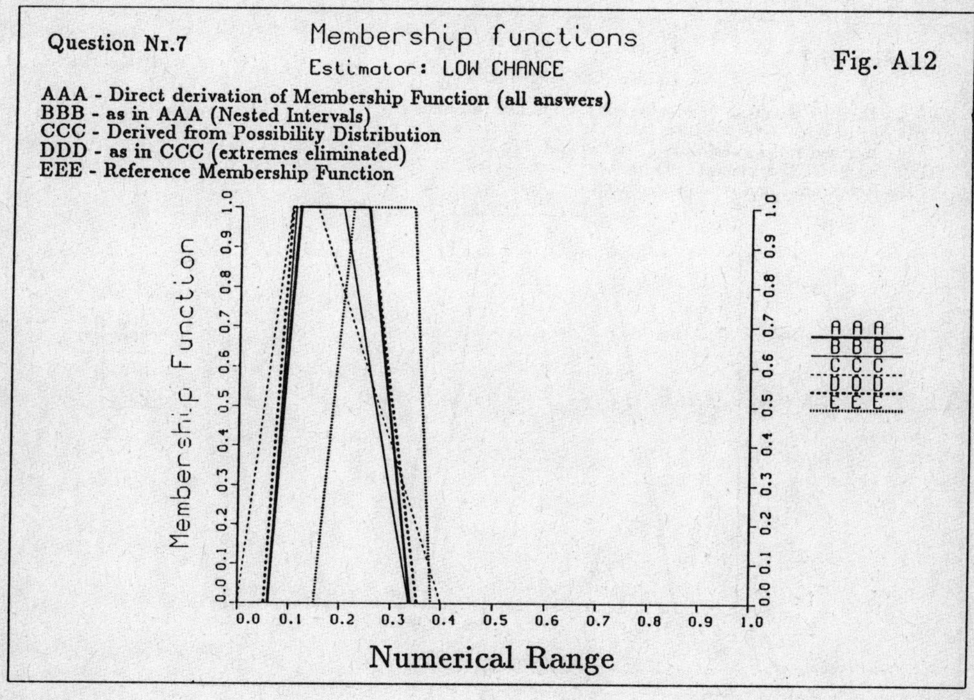

Question Nr.7
Membership functions
Estimator: LOW CHANCE
Fig. A12
AAA - Direct derivation of Membership Function (all answers)
BBB - as in AAA (Nested Intervals)
CCC - Derived from Possibility Distribution
DDD - as in CCC (extremes eliminated)
EEE - Reference Membership Function
Membership Function
Numerical Range
AAA
BBB
CCC
DDD
EEE

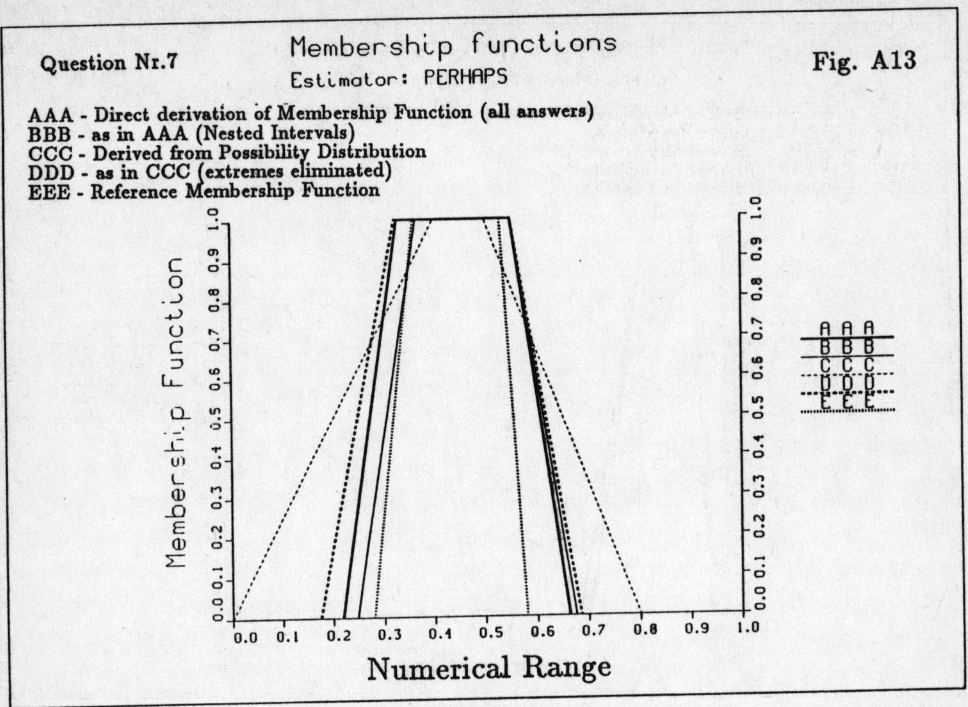

Question Nr.7
Membership functions
Estimator: PERHAPS
Fig. A13
AAA - Direct derivation of Membership Function (all answers)
BBB - as in AAA (Nested Intervals)
CCC - Derived from Possibility Distribution
DDD - as in CCC (extremes eliminated)
EEE - Reference Membership Function
Membership Function
Numerical Range
A A A
B B B
C C C
D D D
E E E

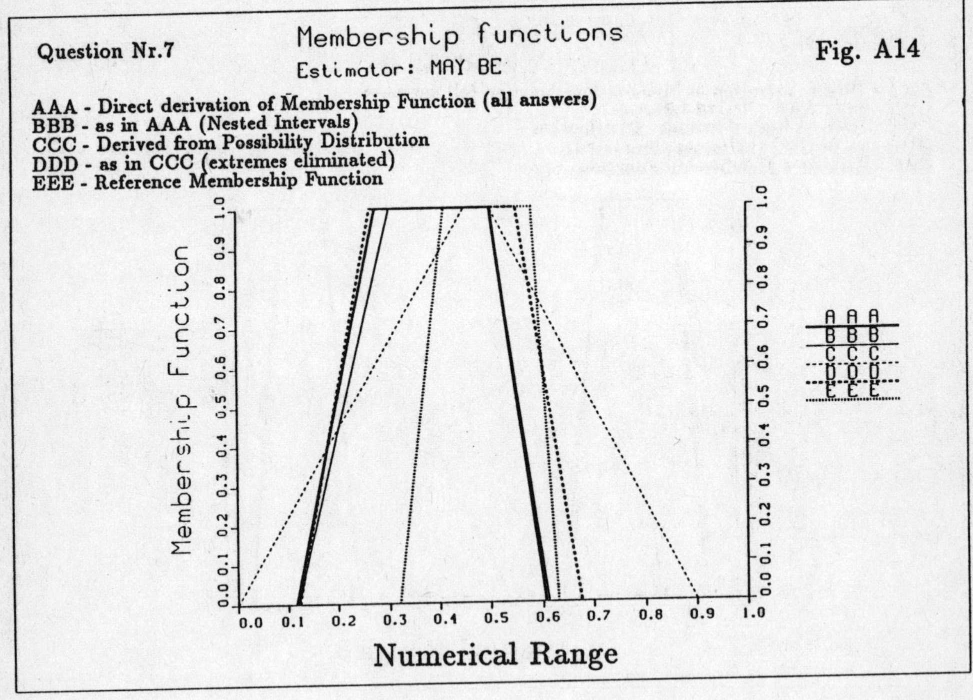

Question Nr.7
Membership functions
Estimator: MAY BE
Fig. A14
AAA - Direct derivation of Membership Function (all answers)
BBB - as in AAA (Nested Intervals)
CCC - Derived from Possibility Distribution
DDD - as in CCC (extremes eliminated)
EEE - Reference Membership Function
Membership Function
Numerical Range
A A A
B B B
C C C
D D D
E E E

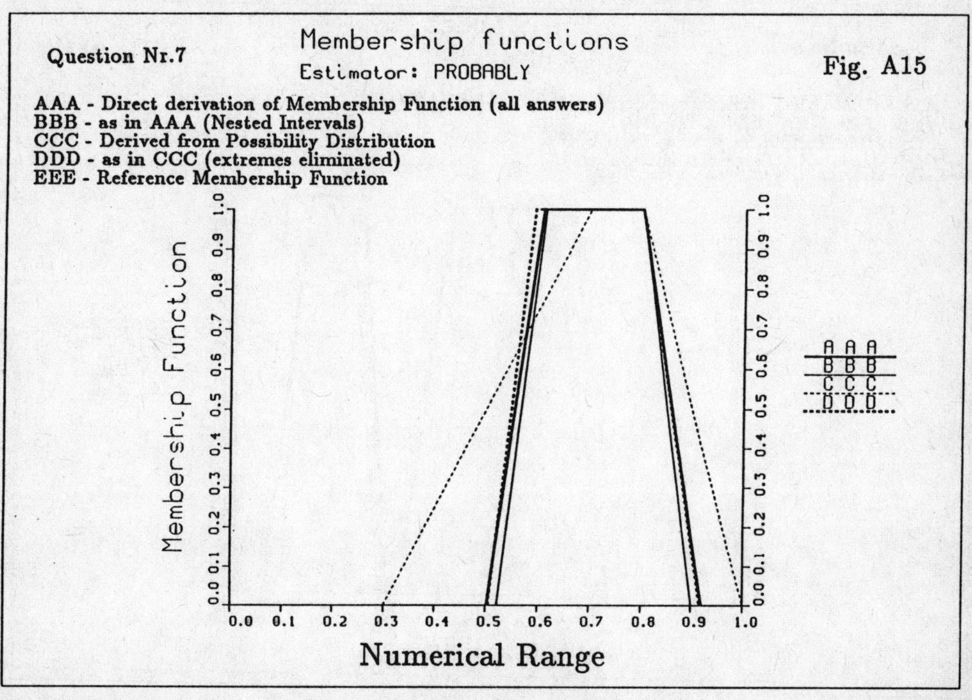

Question Nr.7
Membership functions
Estimator: PROBABLY
Fig. A15
AAA - Direct derivation of Membership Function (all answers)
BBB - as in AAA (Nested Intervals)
CCC - Derived from Possibility Distribution
DDD - as in CCC (extremes eliminated)
EEE - Reference Membership Function
Membership Function
Numerical Range
AAA
BBB
CCC
DDD

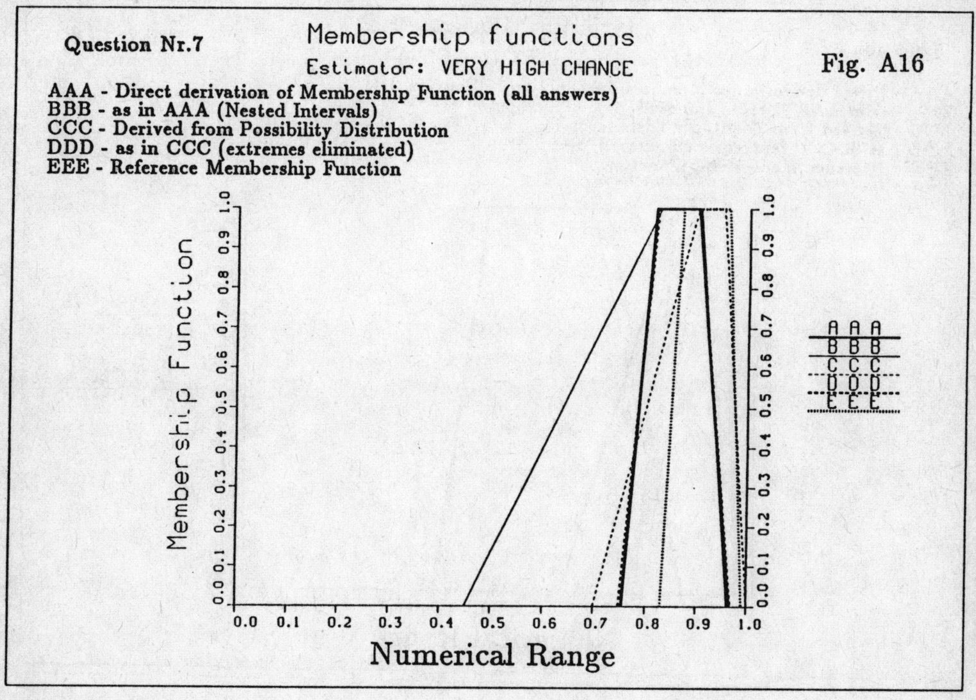

Question Nr.7
Membership functions
Estimator: VERY HIGH CHANCE
Fig. A16
AAA - Direct derivation of Membership Function (all answers)
BBB - as in AAA (Nested Intervals)
CCC - Derived from Possibility Distribution
DDD - as in CCC (extremes eliminated)
EEE - Reference Membership Function
Membership Function
Numerical Range
AAA
BBB
CCC
DDD
EEE

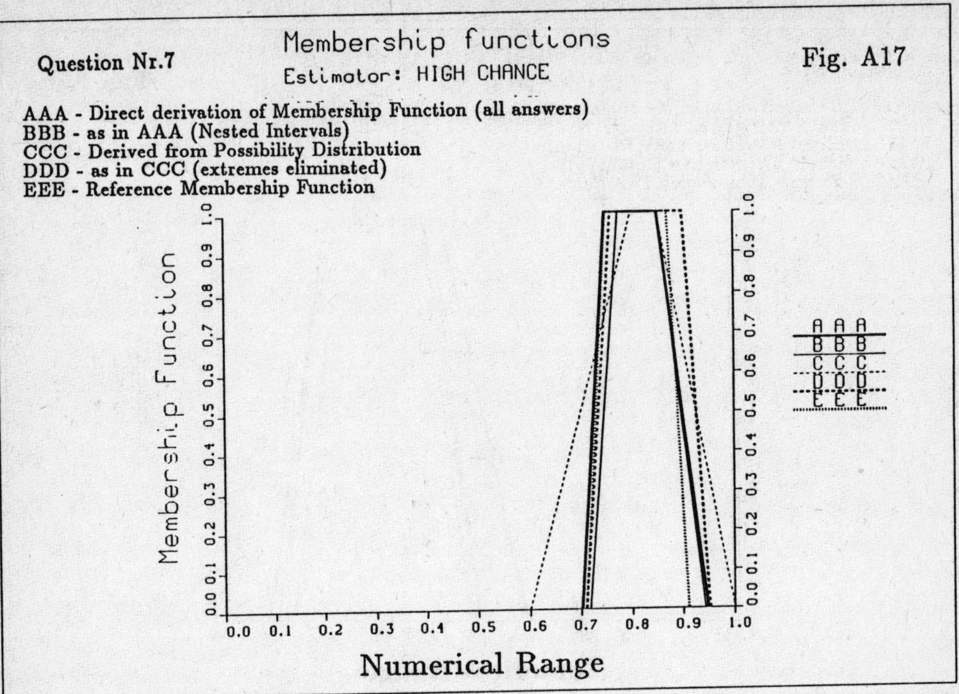

Question Nr.7
Membership functions
Estimator: HIGH CHANCE
Fig. A17
AAA - Direct derivation of Membership Function (all answers)
BBB - as in AAA (Nested Intervals)
CCC - Derived from Possibility Distribution
DDD - as in CCC (extremes eliminated)
EEE - Reference Membership Function
Membership Function
Numerical Range
A A A
B B B
C C C
D D D
E E E

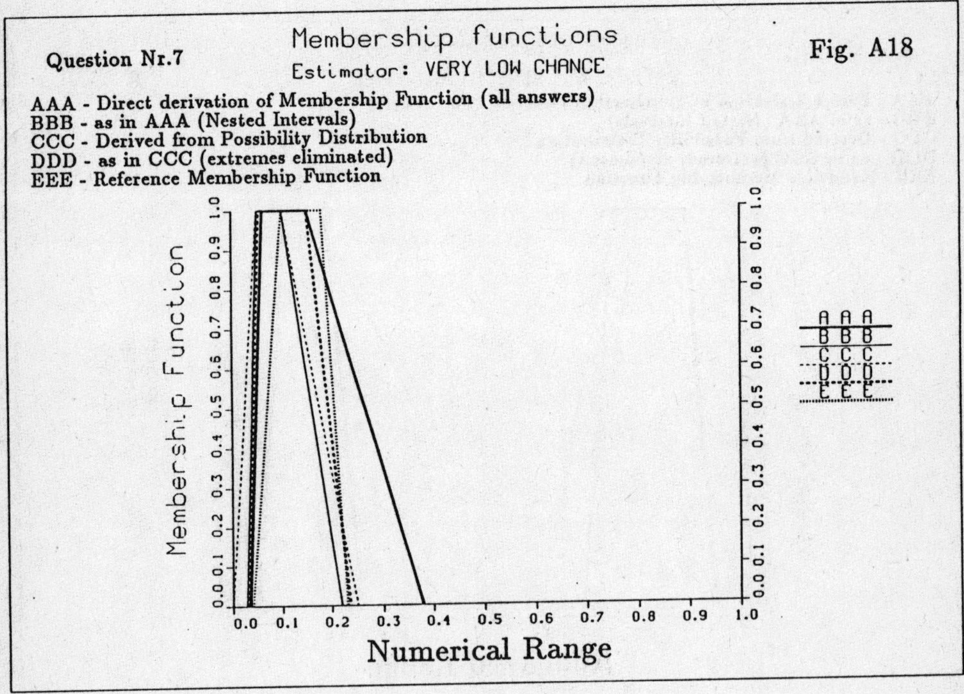

Question Nr.7
Membership functions
Estimator: VERY LOW CHANCE
Fig. A18
AAA - Direct derivation of Membership Function (all answers)
BBB - as in AAA (Nested Intervals)
CCC - Derived from Possibility Distribution
DDD - as in CCC (extremes eliminated)
EEE - Reference Membership Function
Membership Function
Numerical Range
A A A
B B B
C C C
D D D
E E E

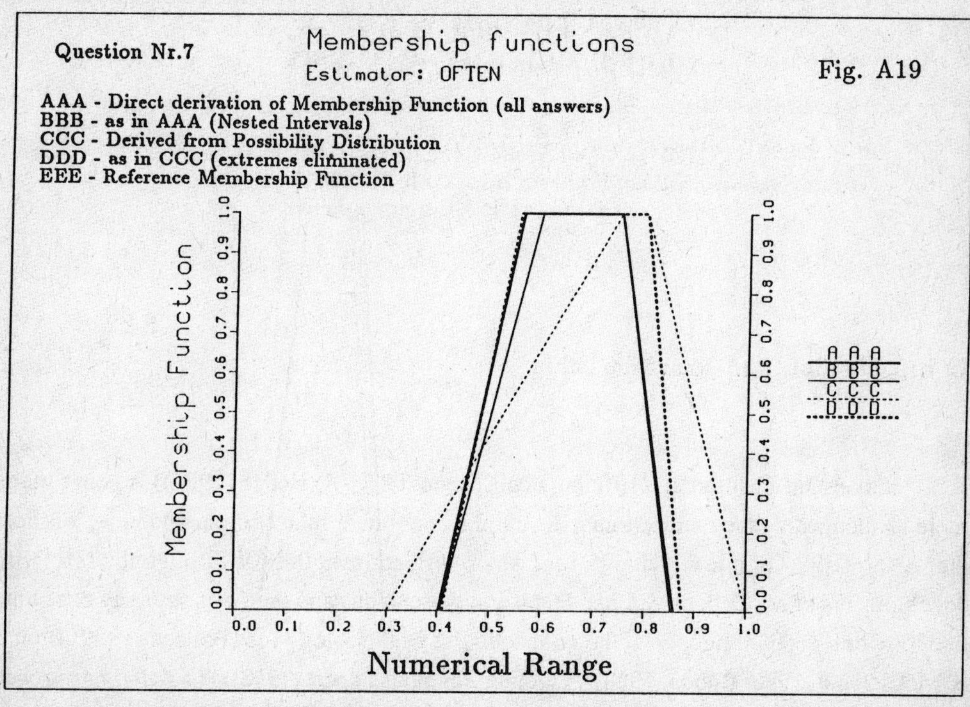

Question Nr.7
Membership functions
Estimator: OFTEN
Fig. A19
AAA - Direct derivation of Membership Function (all answers)
BBB - as in AAA (Nested Intervals)
CCC - Derived from Possibility Distribution
DDD - as in CCC (extremes eliminated)
EEE - Reference Membership Function
Membership Function
0.0 0.1 0.2 0.3 0.4 0.5 0.6 0.7 0.8 0.9 1.0
Numerical Range
A A A
B B B
C C C
D D D

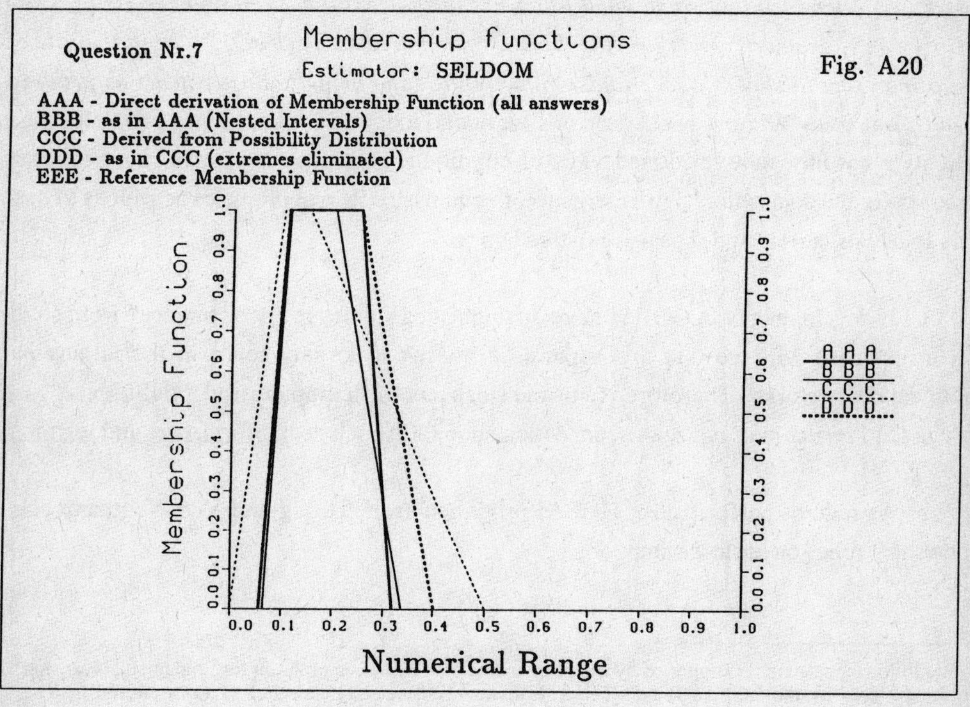

Question Nr.7
Membership functions
Estimator: SELDOM
Fig. A20
AAA - Direct derivation of Membership Function (all answers)
BBB - as in AAA (Nested Intervals)
CCC - Derived from Possibility Distribution
DDD - as in CCC (extremes eliminated)
EEE - Reference Membership Function
Membership Function
0.0 0.1 0.2 0.3 0.4 0.5 0.6 0.7 0.8 0.9 1.0
Numerical Range
A A A
B B B
C C C
D D D

CONNECTIONISM vs GOFAI:
A BRIEF CRITICAL ANALYSIS

Hugues Bersini[1]
IRIDIA - Université Libre de Bruxelles
50, Av. Franklin Roosevelt CP 194/6
1050 Bruxelles- Belgique

1. Introduction and basic concepts

Good-Old-Fashioned-Artificial-Intelligence (GOFAI (Boden, 1988)) is being more and more challenged by the connectionist developments. Rather than being brand new, it is actually the "resurrection" of a research program which thrived from the 40s through the 60s (with the developments of McCulloch & Pitts, Hebb and Rosenblatt) and then was severely retrenched in the 70s (for a detailed history of the connectionist methodology see (Pollack, 1989; Rumelhart & McClelland, 1986; Boden, 1988)). Quoting Seymour Papert (1988): *"this bloody retrenching work was attempted by two staunch followers of the "Artificial Intelligence sister", Marvin Minsky and Seymour Papert (in their famous perceptron book (1969)), cast in the role of the hunstman sent to slay "Connectionist Snow White" and bring back her heart as proof of the deed"*. But Snow White was not dead and we attend today an explosive infatuation for this once old style but now new fashioned view of cognition. This paper will briefly try to grasp the reasons for the connectionist re-emergence, to emphasize its real promises as well as to examine cautiously its current and most speculative hopes.

If developments in GOFAI have distinguished studies in "performance" from studies in "learning", we will see that this separation doesn't make any sense in the largest part of connectionist works. Therefore, two main aspects of "computerised intelligence" will be discussed in order to organize the connectionism/GOFAI debate: <u>performance</u> and <u>learning</u>.

As regards performance, GOFAI originates from Turing and Von Neumann classical views and relies on basic assumptions:

---

[1] The following research is supported by the Belgian National incentive-program for fundamental research in AI initiated by the Belgian State - Prime Minister's Office - Science Policy Programming. The scientific responsibility is assumed by the author.

1) the separation in treatment and description between the software and the hardware level, and consequently the "processing autonomy" of the software level.

2) the physical separation of the central processor and the memories. The processor calls, in an intermittent way, for some memory items, processes them, transfers them back in memory, and iterates again for other necessary memory items.

3) memory is generally not content-addressable

4) reasoning is organized in a linear and "transparent" way. Every elements of thinking - facts, inferential rules and reasoning mechanisms - are visible, accessible and easily modifiable.

For all those aspects, connectionism forces a conceptual shift. Following a <u>reductionist line</u> (because roughly inspired by what is known about the architecture of the brain), <u>it mixes intimately software and hardware. It rejects the existence of an autonomous knowledge or symbolic level</u> (Newell, 1982) <u>indifferent to the question of what could be the supporting architecture</u> (Fodor & Pylyshyn, 1988). <u>It works without a central processor.</u> Indeed, <u>memories and processors are equally spread out in the neuronal network. Memory is content addressable. Search and extraction from memory are inherently parallel and implicit to a large extent.</u>

Connectionism draws some inspirations from neurosciences. A set of units (sorry for irritating neuroscientists in using the terms "neuron" and "synapse") are connected among themselves by synaptic links. These links have different intensities (weight) and can be either excitatory (positive) or inhibitory (negative). The whole set of neurons functions simultaneously. Each neuron has an individual trivial responsability. It receives activations from its connected neighbours. These activations, being modulated by the weights of the respective connections, are combined. Then the neuron, having a certain finite threshold, "decides" yes or no to transmit the activation. The most elementary mathematical formulation of this activation transfer is:

$$a_i (t + dt) = f (\Sigma_j w_{ij} a_j (t))$$

a_i = activation of neuron i

w_{ij} = synaptic weight between neuron i and j

f = a monotonous function generally not linear (see fig.1).

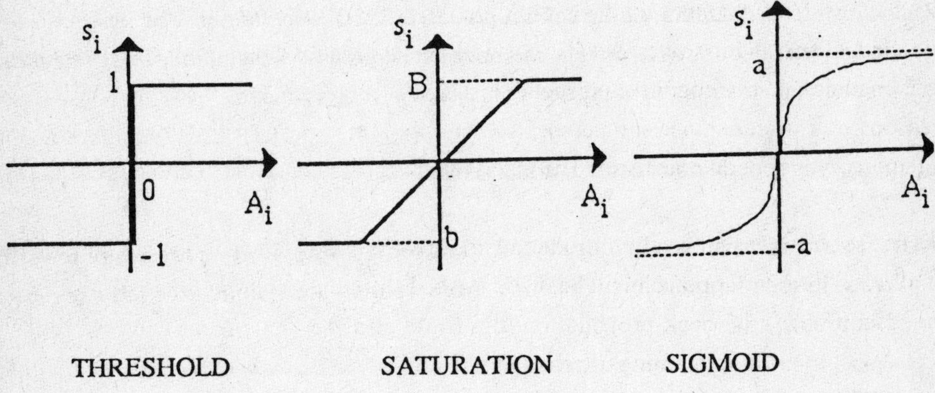

fig. 1 Three possible functions for the activation transfer

As regards learning, the connectionist views (essentially in the distributed line, but that point will be deepened later) give up with a "symbolic, declarative learning" where facts or inferences are explicitly added in response to the needs. Indeed, connectionist learning must be envisaged in a "behaviorist-biological" perspective and its basic formulation traces back to Hebb's writing (1949). Memory is stored in connections. Based on input/output examples of behaviour, the synapses modify themselves in order to gradually reproduce these examples. The modification principle is immediate and quoting Hebb himself : "*When a cell A is near enough to excite a cell B and repeatedly or persistently takes part in firing it, some growth process or metabolic change takes place in one or both cells such that A's efficiency, as one of the cells firing B, is increased*". Here again, the most direct mathematical formulation for this synaptic learning could be:

$$w_{ij}(t+dt) = w_{ij}(t) + g(a_i(t), a_j(t)) \quad \text{with } a_i(t), a_j(t) \text{ being imposed by the example}$$

We will discuss the different expressions of "g" in the next chapter.

A substantial number of researchers puts forward aspects of neurosciences, neuropsychology and cognitive sciences for justifying their current "connectionist passion" (McClelland & Rumelhart, 1986; Grossberg, 1988; Sejnowski et al., 1988; Dreyfus, 1988; Churchland, 1986). AI and cognitive sciences have always demonstrated difficult if not conflictual relationships, made of long periods of love and frequent infidelities. AI has always drawn and will draw again important inspirations from cognitive sciences (semantic network, schema-based representation, uncertain and imprecise reasoning formalisms, naïve physic etc.) but diverges from them when it exploits those techniques in a "confined engineering perspective". <u>An increasing part of AI studies intelligence "per se" without worrying about the "cognitive realism" of the mechanisms grounding the programs</u>. Aeronautic engineers are not ornithologists. Thus, if for some reasons that will be surveyed, connectionism appears more and more unavoidable in a cognitive perspective, a capital point concerns the same relevancy of these reasons in a strict engineering perspective, privileging "<u>transparency</u>", "<u>optimality</u>" and "<u>controllability</u>" to "cognitive realism". The rest of the paper will be dedicated to this point.

The first chapter will briefly introduce the different basic connectionist methodologies. First of all, the "localist" approach will be distinguished from the more "distributed" one. In the distributed approach, the back-propagation algorithm and the Boltzmann machine will be sketched. As an interesting learning alternative, the unsupervised competitive learning will be mentioned. The second chapter will critically examine the contributions of connectionism to cognitive sciences both in performance and learning, to, finally, re-appreciate these contributions in a more engineering vein.

2. The different connectionist approaches

2.1 The localist approach

The localist exploitation of connectionist ideas (whose main advocates are Feldman (1988), Fahlman (1979), Shastri (1988)) does not totally give up with GOFAI methodological philosophy. Knowledge representation is structured and the connectionist network becomes isomorphic to the semantic network it implements. In its extremist version, <u>a neuron can represent either a concept or a conceptual relation</u>. Inferences take place by means of spreading activation through the network. A concept calls another one if activation circulates from the first to the second one. Uncertainty aspects are inherent in the degree of neuronal activation and in the synaptic strengths. <u>In this approach, STM (Short Term Memory) is the restricted activated part of the network, LTM (Long Term Memory) is the large desactivated part of the network</u>. Fig.2 is an example of localist connectionism applications (Shastri, 1988). This figure depicts the interaction between a fragment of an agent's restaurant routine and a part of his memory network. In this routine fragment, the task of deciding on a wine results in a query to the memory network about the taste of food, and the decision is made on the basis of the answer returned by the memory network.

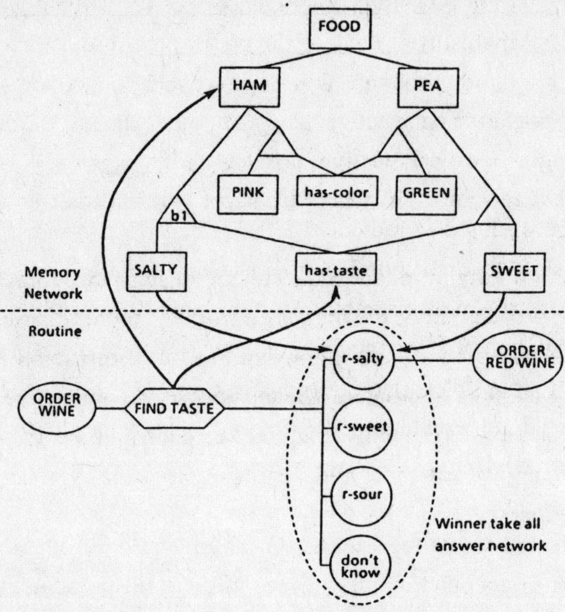

fig.2 An example of localist connectionism (from (Shastri, 1988))

Basic differences with GOFAI are:

- memory is content-addressable.

- parallelism: different inferences can be simultaneously envisaged [a winner-takes-all mechanism (see Feldman (1988) or Grossberg (1988)) selects the most satisfactory one]. The selected item will be transferred in STM for further treatments.

- larger implicitness: inferences are not programmed at a separated syntactic level, they are "materialized" in the network connections.

- in the well-known procedural/declarative controversy (Winograd, 1975), connectionism allows to modify the balance in favour of the declarative side. The procedural control of LTM retrieval is integrated in the hardware.

- regarding learning and since the connectionist network is "semantically transparent", nothing is, up to now, revolutionary in the localist approach.

2.2 The distributed approach

The distributed approach shows radical differences with GOFAI. So radical that various researchers regard the shift to distributed connectionism as a "paradigmatic revolution" (Smolensky, 1986) (For controversies risen by this view, see Boden (1988)). In the distributed methodology, the representation of any concept is indicated by the pattern of activation of a set of neurons whose activity does determine the representation of other concepts. Then a same neuron can be involved in the representation of various concepts (this is basically different from GOFAI and localist approaches).

Neurons are dedicated to the representation of concepts and others to the inferential mechanisms linking the concepts together. Since the "conceptual neurons" are totally distributed, the "inferential neurons", indispensable to the whole set of inferences, are difficult to "semantically interpret". Even if unavoidable, their precise function remains "un-explicitable". These intermediate neurons are generally called hidden units. In fig.3, NETtalk (Sejnowski & Rosenberg, 1986; the description of NETtalk is entirely extracted from (Schreter, 1988)) is shown as an application of distributed connectionism.

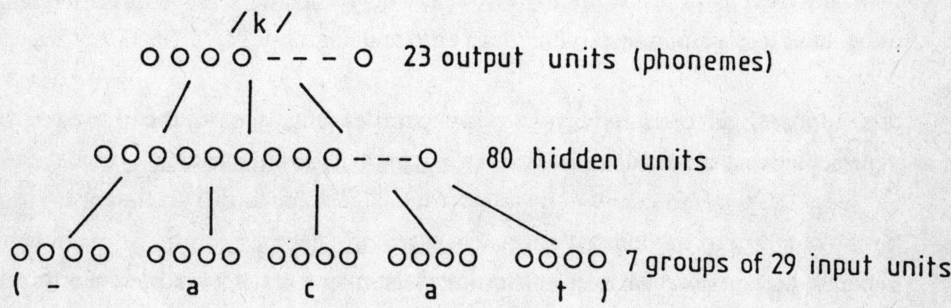

fig. 3 The NETtalk application

NETtalk can map patterns of letters into patterns of phonetic elements - that is, it can read loud. It has three neuron layers. The input layer contains seven group of nodes - one for each letter of the alphabet - plus 3 nodes representing punctuation. The second layer contains 80 nodes (the hidden units) which represent first-order combinations of the letter, and the output layer contains 23 nodes representing second-order combinations of the letter - at the same time phonetic elements.

The most distributed and implicit part of the network is the intermediate layer where the problem representation and control are completely diffused and nearly impossible to interpret. Each node in the input layer is connected to each node in the second layer, which are in turn connected to each node in the output layer - that is, there are connections only in one direction, from input towards output (also designated by: "one shot computing" (Fogelman Soulié, 1988)).

The distributed approach presents the same basic differences with GOFAI as the localist one: content-addressability, parallelism, implicitness and "declarativeness". However, its distributed nature is responsible for further differences which makes it more appealing to a large part of neuro and cognitive scientists:

- distributed network provides a greater degree of robustness or fault tolerance than GOFAI and localist methodologies. Damage to a few nodes or links thus need not impair overall performance significantly.

- a same robustness aspect compensates for minor variabilities (like noisy input) in characteristics of processing elements. The network can generalize up to a certain point for cases different from the original examples.

- these different aspects are equally characteristic of the holographic phenomenon, that, for long, has largely impressed numerous neuro and cognitive scientists (Dreyfus, 1972).

- this approach is obviously inherently parallel but, due to the presence of multi-representative and hidden units, is still more implicit than the localist one.

- on the contrary to the localist approach, learning becomes capital. It makes no sense to separate here performance from learning. Based on a set of examples, the learning takes place by synaptic modification in order to reproduce the examples. In the NETtalk application, each of the connections has a particular weight. The knowledge of the system about how to map letters into phonemes resides actually in these weights. These weights are set in a "training" phase. Training happens like this: an input pattern - seven letters - is presented to the system, and an output pattern is produced. In particular at the beginning of the learning process, the output will be almost certainly incorrect. However, there are fairly exact criteria about what the output should be. In NETtalk, for example, the pronunciation of the same input pattern by a child was used as criterion. This criterion is "told" the system, which then computes the difference between the actual activation pattern in the output and the criterion pattern. This learning technique will be described more precisely in the next chapter. In substance, learning is not conceptual. <u>The network adapts alone its morphology in order to reproduce the examples</u>.

- thus, <u>two processing phases take place sequentially</u> :

    1) <u>a learning phase</u> where the network adapts its connections in order to reproduce a set of behaviorist examples

    2) <u>a performing or retrieval phase</u> in which the network can obviously reproduce the examples but, above all, is capable to interpolate for cases not included in the basic set.

- since this adaptation process is kept out of the user's sight, <u>the distributed approach is a "black-box" one</u>. You see what the network can do but not the way it proceeds.

2.3 The two most famous distributed approaches

a) The back-propagation algorithm (McClelland & Rumelhart, 1986; Fogelman Soulié et al., 1988): this is the most prevalent learning algorithm for multi-layer feedforward networks. Since this algorithm has several variants, we will limit ourselves to the basic developments of McClelland and Rumelhart. NETtalk learning is based on this technique. The network is composed of an input layer (N units), several intermediate layers of hidden units and an output layer (M units). For a basic set of examples, it must learn to produce the desired output when an input vector is presented. The description of the back-propagation training algorithm is extracted form (Lippmann, 1987).

It is an iterative gradient technique designed to minimize the mean square error between the actual output of a multiplayer feed-forward perceptron and the desired output. It requires continuous differentiable non-linearities. In the following, a sigmoid logistic non-linearity is used with the function f(a) given by:

$$f(a) = \frac{1}{1 + e^{-a}}$$

Step 1. Initialize Weights and Offsets :
set all weights and node offsets to small random values.

Step 2. Present Inputs and Desired Outputs :
present a continuous valued input vector $X_0, X_1, ..., X_{N-1}$ and specify the desired output vector $D_0, D_1, ..., D_{M-1}$. If the net is used as a classifier then all desired outputs are typically set to zero except for that corresponding to the class the input is from. That desired output is 1.

Step 3. Calculate Actual Outputs :
use the sigmoid non-linearity from above to calculate outputs $Y_0, Y_1, ..., Y_{M-1}$ with $a_i = \sum_j w_{ij} a_j - \theta_i$ (θ_i is the threshold of unit i)

Step 4. Adapt Weights :
use a recursive algorithm starting at the output nodes and working back to the first hidden layer. Adjust weights by:

$$w_{ij}(t+1) = w_{ij}(t) + \eta \delta_j a_i$$

In this equation $w_{ij}(t)$ is the weight from node i to node j, a_i is either the output of node i or is an input, η is a gain term, and δ_j is an error term for node j. If node j is an output node, then

$$\delta_j = Y_j (1 - Y_j) (D_j - Y_j)$$

where D_j is the desired output of node j and Y_j is the actual output.
If node j is an internal hidden node, then

$$\delta_j = a_j (1-a_j) \sum_k \delta_k w_{jk}$$

where k is over all nodes in the layers above node j.

Step 5. Repeat by Going to Step 2

b) The Boltzmann Machine (BM) (this description is extracted from Sejnowski, 1988):
Hinton & Sejnowski introduced a stochastic network architecture, called the Boltzmann

machine, for solving optimization problems. The processing units are binary and are updated probabilistically using the function   :

$$P\,(a_i) = \frac{1}{1+e^{-\Sigma_j w_{ij} a_j - \theta_i /T}}$$

As a consequence, the internal state of the BM fluctuates even for a constant input pattern. The amount of fluctuation is controlled by a parameter - T - that is analogous to the temperature of a thermodynamic system. Fluctuations allow the system to escape from local traps into which it would get stuck if there were no noise in the system. All the units in the BM are symmetrically connected: this allows an "energy" ($E = -\Sigma_{i<j} a_i a_j w_{ij} + \Sigma_i a_i \theta_i$, with θ_i a threshold for unit i) to be defined for the network and insures that the network will relax to an equilibrium state which minimizes the energy.

BM has an interesting learning algorithm that allows "energy landscapes" to be created through training by example. The learning has two phases. In the training phase, a binary input pattern is imposed on the input group as well as the correct binary output pattern. The system is allowed to relax to equilibrium at a fixed "temperature" while the inputs and outputs are held fixed. In equilibrium, the average fraction of the time a pair of units is on together, the co-occurrence probability p_{ij}^+, is computed for each connection. In the test phase, the same procedure is followed with only the input units clamped, and the average co-occurrence probabilities, p_{ij}^-, are again computed. The weights are then updated according to:

$$\Delta w_{ij} = \eta\,(p_{ij}^+ - p_{ij}^-)\quad \text{where } \eta \text{ controls the rate of learning.}$$

2.4 The Competitive Learning algorithm

Competitive learning enters in the class of unsupervised learning techniques that can automatically categorize the environmental data by auto-organizing the network structure in order to reflect the experienced regularities. Quoting Kohonen (1988) - one of the fathers of competitive learning- :*"My suggestion is that in the development of artificial sensory systems, say, to feed speech and images to computers, one should not apply heuristically derived operations at all, but first to create an extremely adaptive network that then finds its own structures and parameters automatically, on the basis of presented observations"*. Indeed, Kohonen presents one such algorithm which produces a self-organizing feature maps (reproduced from (Lippmann, 1987)):

Step 1. Initialize Weights:

initialize weights from N inputs to the M output nodes to small random values. Define the initial radius of the neighbourhood.

Step 2. Present New Input

Step 3. Compute Distance to All Nodes:

Compute distances dj between the input and each output node j using:

$$dj = \sum_{i=0}^{N-1} (a_i(t) - w_{ij}(t))^2$$

where $a_i(t)$ is the input to node i at time t and $w_{ij}(t)$ is the weight from input node i to output node j at time t.

Step 4. Select Output Node with Minimum Distance:

Select node j* (the competition winner) as that output node with minimum distance dj.

Step 5. Update Weights to Node j* and Neighbours:

Weights are updated for node j* and all nodes in the neighbourhood. New weights are:

$$w_{ij}(t+1) = w_{ij}(t) + \eta(t)(a_i(t) - w_{ij}(t))$$

The term $\eta(t)$ is a gain term $(0<\eta(t)<1)$ that decreases in time.

Step 6. Repeat by going to Step 2

3. A brief critical analysis

This chapter will survey the most conclusive breakthroughs of connectionism in performing and learning, from a cognitive and an engineering point of view.

3.1 In Performing

3.1.1 From a cognitive point of view

From a cognitive perspective, it is today impossible to disregard the irreversible contributions of connectionism. For 25 years now, GOFAI and the greatest part of cognitive sciences have philosophically or methodologically respected the Newell's symbolic paradigm (Newell, 1982; Smolensky, 1986). Doing that, the cognitive domain privileged by their studies covers essentially explicit, propositional knowledge and reasoning. An important part of psychologists considers that this domain is not representative of the whole cognitive story. It does not account for the cognitive peripherals like perceptual processes or sensori-motor automatisms which cannot be separated from high-level processes. In spite of their intrinsic and "objective" complexity, (imagine how much information must be considered simultaneously in order to recognize a face), low-level processes are performed without any feeling of complexity.

Even when playing certain games (the historical favourite domain of AI), De Groot (1965) observed that important steps of the whole cognitive task are parallel and without explicit following of any rules. For chess masters, often a simple and brief snapshot of the game configuration is enough for selecting the next move. The pre-thinking perception of the game or, in general, the situation, is capital, including for hard problem-solving tasks. The direct interface of cognition with the world - that is perception and action - is supported by parallel and implicit processing fully integrated in the cognitive architecture. Now, the final step of perception consists of top-down or schematic processing, when the memory is explorated in parallel to find a schematic interpretation for the perceived stimuli. Without going too far, any memory exploration and extraction seem to be grounded on parallel and implicit processing. If I ask you to tell me the name of an assassinated American president, Kennedy will emerge into consciousness without having to think about Washington or Carter, even Lincoln and McKinley. That is an example of parallelism (every president was explorated simultaneously) and implicitness (only the most activated one appears consciously). The major part of cognitive scientists agrees today to distinguish two types of cognitive processing (see Reason, 1987):

- a controlled or conscious processing which is selective, resource-limited, slow, laborious and serial.

- an automatic or unconscious processing which is apparently unlimited, fast, effortless, parallel and based on two heuristics: "match like to like" and "select in favour of high-frequency items patterns .

Perception modalities, sensori-motor automatisms, memory exploration and extraction have to be classified in the second category of cognitive processing.

For argument's sake, a good old expert system can be roughly viewed as the combination of various mechanisms for LTM retrieval with a set of STM strategies for analyzing the task, organizing the facts (combining new informations with precedent ones), controlling (treatment of uncertainty and coherence) and chaining the reasoning. The current contribution of connectionism could be mainly as a more "cognitive-inspired" technique for LTM retrieval, allowing easily:

- to treat in one cycle multiple-premises rules, considerably reducing sequential "generate-and-test", search and backtracking.

- to perform approximate reasoning, like fuzzy matching or generalization, without the need for explicit mathematical tools.

Indeed, memory has much more importance than GOFAI suggests and search has much less importance than GOFAI suggests (Waltz, 1988). The procedural control of LTM retrieval is integrated in the hardware by means of simple insisting learning of examples of desired behaviours. However, although in a strongly reduced way, task analyzing, knowledge organizing and reasoning chaining, supported by GOFAI tools, will still be required. Therefore, because it reproduces in a more psychological and "architectural" way the subconscious sensori-motor and memory-based processing, a strong consensus emerges today among cognitive scientists. This consensus concerns the unavoidability of connectionist methodologies for any LTM extraction, either addressed to STM or straighforwardly leading to motor processes.

3.1.2 From an engineering point of view

Obviously, the privileged field of connectionist applications remains today the study of perception modalities and sensori-motor automatisms. Concerning the branch of engineering dealing with artificial perception, the success of connectionism will certainly depend on its adaptability to already well-proved stochastic and algorithmic instruments. In speech recognition (Lippmann, 1987; Kohonen, 1988) and vision studies (Waltz, 1988; Fukushima, 1988), <u>various researchers (Bridle, 1988) predict that the next generation of artificial perception systems will be based on a combination of connectionism with stochastic, algorithmic techniques and with mainstream perception knowledge used in a rather soft way to decide network structure</u>. Perceptual processing will take more and more importance in future AI.

However, this paper does not intend to tackle this large scientific domain but essentially to examine connectionist contributions in more classical AI fields dealing with knowledge, reasoning and natural language processing. Indeed, many of the problems only arise when the methodology is applied to these high-level cognitive functions. As we have already discussed, the main aspects making connectionism so appealing for cognitive scientists are: 1) memory is content-addressable, 2) parallelism and speed, 3) resistance to damages and noise, 4) implicitness (at different degree from localist to distributed approach), 5) generalization capabilities, 6) approximated performance.

Regarding the first aspect, nothing could prevent GOFAI disciples to treat memory in a more content-addressable way. The use of hash coding allows it. Secondly, when a great amount of information must be treated simultaneously in respecting a multitude of constraints, the advantages of parallelism are obvious. Indeed, more and more developments are achieved in order to implement various GOFAI applications in parallel computers architecture. It is the case for blackboard systems where a multitude of actors can act parallely on a common pool of knowledge (Nii, 1986). Various authors have explored the role of parallelism in the high-speed execution of production systems (Gupta, 1987). Finally, resistance to damages and noises is obviously a good thing but can be achieved in classical systems by means of redundant hardware and environmental filtering.

If the first three aspects can be transposed in a GOFAI spirit, the last three aspects remain more questionable when examining the potentialities of connectionism for the engineering of high-level functions. We will separate in two sub-chapters the problems that connectionism has to face. In a first part, a list of features that are easy for GOFAI models but pose, nowadays, difficult problems for connectionism (essentially the distributed approach), will be discussed. However, solutions are already proposed for these problems. Then, beyond these current but tractable difficulties, the second part will tackle a list of basic problems linking to the last three aspects. Since these problems are inherent in the methodological philosophy of connectionism, their resolution will be less a question of time than of faith in one methodology or another.

a) *Current difficulties faced by the distributed approach* (Pollack, 1989; Fodor and Pylyshyn, 1988; Hendler, 1988): we have seen that GOFAI tends to separate knowledge representation and knowledge treatments. The difficulties faced by connectionism for the high-level functions concern both these two points and their separation acceptance:

<u>For knowledge representation</u>: <u>cognitive knowledge is structured, hierarchical and schematized</u>. <u>None of these priorities appears in distributed approaches, due to a total lack of knowledge structuration</u>. The complete homogeneity of representation prevents networks from distinguishing classes from instance of classes and consequently from reasoning in a first order manner or from taking advantages of frame-based representation (inheritance, default reasoning, ...). This lack of structuration impedes the consideration of symmetry or transitivity in reasoning. Another disturbing aspect lies in <u>the cross-talk problem</u> when concepts blend their representative micro-features in an undesired way.

<u>For knowledge exploitation mechanisms:</u> <u>connectionism favours one-shot computing to long-chained reasoning</u>. Even if, by definition, the methodology "parallelize" a great amount of GOFAI sequentiality, reasoning remains an evolutive and linear process. Indeed, the first knowledge sample allows a second, more informed, sample to be made, and so on until the sampling process reaches quiescence. <u>Connectionism has no "generative capacity"</u> (the term is due to Chomsky who used it as a measure of the capacity of particular classes of formal grammars to generate natural language sentences) or "recurrent capability" that allows GOFAI to re-exploit infinitely a same syntactic rule for different knowledge items. It is largely problematic for simulating language production and prevents connectionism from being the only privileged road to the subconscious world.

<u>As regards separation between control and representation:</u> AI systems try to be the most portable possible and to allow a same inferential engine to apply to different situations by changing only the declarative representation of the domain. <u>Since control and memory are blended in a connectionist network, a same transportability seems out of reach</u>. A network configuration must be completely modified (re-learned from the beginning) even for a slight change in the task or just in the size of the task.

Facing these methodological difficulties, more and more authors advocate a <u>GOFAI/connectionism hybrid position</u> (Chandrasekaran, 1989; Pollack, 1988; Hendler, 1988; Clark, 1989). For instance, in expert system methodology, connectionism could be kept for its basic advantages - memory search and extraction - but problems of representations and controls should be left in the care of classical AI. As Clark underlines:

"The holy war between connectionism and classical AI may, I have suggested, have been constructed along an illusory frontier. That frontier is what I have called the Uniformity Assumption. According to such an assumption, there is a single computational architecture underlying the whole spread of genuinely cognitive achievements. Good psychological models, according to Uniformity, thus need avail themselves of just the basic resources and operations provided by the favoured option (either connectionist or GOFAI) (...) By contrast, the picture I am advocating is one of mind as comprising a richly interwoven stack of virtual machines, with accurate psychological models attending each level. The correct response to the holy war is, I believe, a healthy ecumenism"

In the hybrid perspective, the key problem will be to establish the precise frontier where connectionist methodology demonstrates useful capabilities beyond GOFAI style and refuses its neutrality in the implementation of the knowledge and reasoning level.

b) Basic conceptual problems: beyond these methodological problems that connectionism can expect to resolve by means of "hybridation" or further improvements, basic aspects, inherent in the methodological philosophy of connectionism, rise more delicate questions.

<u>Implicitness/Explicitness</u> : <u>The acceptance of some processing opacity is inherent in connectionism</u>. If this implicitness varies from localist to distributed approaches, each of them admits a certain architectural diffusion of either the concepts or the inferential rules. Like man for certain task (often, the task forming part of his expertise, that does not require figuring things out and using any kind of analysis - that is the reason why these tasks rise intractable problems for knowledge elicitation), connectionist network functions like a "<u>competent black box</u>". In an engineering perspective, to what extent can we allow a supporting tool to perform in a "subconscious" way ? For a lot of GOFAI disciples, the essence of AI lies in "transparency", explanation capabilities (see Kodratoff in "AI is a science of explanations not numbers", 1988). In the current state of the connectionist research, it appears extremely difficult to credit, discredit or debug parts of a distributed network, without avoiding to re-learn from the beginning. <u>Opacity is not very popular in AI</u>.

<u>Generalization and approximate performance:</u> Another problem concerns the connectionist approximate performances for problems in which explicit mathematical or algorithmic tools exist and can supply better solutions. For instance in the LTM extraction part

of expert system, connectionism allows interesting generalization and fuzzy reasoning without the need for explicit mathematical tools like fuzzy-set or uncertainty theories. In a sense, <u>connectionism favours speed and simplicity to optimality</u>. In some well-known NP-complete problems where optimality is intractable (like the traveling salesman problem), Boltzmann machine supplies rapidly a satisfactory solution. For complicated situations characterized by multi-data and multi-constraints, connectionism makes no delay in providing an acceptable solution. The key problem is to categorize the kind of situations where speed and simplicity is preferable to optimality. The replication of perceptual modalities could be eventually found in this category except for situations where even slight mistakes could have dangerous consequences like the control of an airplane or of a nuclear plant.

Then, if implicit approximation (easily implemented by the learning of a typical set of desired behaviours) can increase performance (like response delay), it should be in non-risky situations or for error-tolerant processes. For these processes, perhaps higher more "symbolic and explicit" systems should supervise and recover from faulty situations due to connectionist unadapted approximation or generalization. <u>There is a compromise to make between speed, simplicity, opacity, rigidity, and non-optimalization in a first hand and performance and reliability in a second hand</u>. In daily life, man is better for adapting to mistaken subconscious processes than for reasoning in order to predict and avoid those mistakes. From an engineering perspective, it is difficult to establish a priori to what extent should it be the case for an AI tool and to what extent part of a process could be the responsibility of a connectionist network.

3.2 In learning

The reason for making a superficial and rapid stop in the land of machine learning is the impossibility, when dealing with distributed connectionism, to separate learning from performance. <u>Connectionist performance is totally dependent on learning</u>. The most prevalent learning algorithms - Hebbian rule, Back-Propagation algorithm, Boltzmann Machine and Competitive Learning - have been briefly reviewed.

The tradition in GOFAI is to view learning as the accretion of symbolic structures like facts, rules or constraints (Partridge and Paap, 1988). On the other hand connectionist learning is a "biological-behaviorist" adjustment of link weights so that activity flows into the appropriate nodes at appropriate times in order to store experential (and not conceptual) knowledge. It is such a drastic shift that a trivial problem for GOFAI (as adding a new fact or a new relation) appears to be close to impossible (almost meaningless) in connectionism. However, cognition is able not only to improve performance through trial-and-error training but also, obviously, to increase easily (in one trial), when being told, declarative knowledge.

Therefore a discussion on this basic conceptual differences will be vain due to the different fields of application (<u>procedural learning in the first hand, declarative learning in the second</u>) and, there again, a hybrid approach will certainly be welcome. However, dealing more with the mechanisms than with the content of learning, one can throw a "philosophical" glance on the theoretical axis that distinguishes different learning approaches. This axis indicates the degree of external or environmental supervision and tutoring during the learning. At a first extremity, there is the <u>deeply-supervised learning</u> with an external tutor that clearly "hints" to the system the facts, rules or heuristics to add in response to the needs. That is typical of GOFAI construction, either achieved by a knowledge engineer or by the expert himself.

Laying somewhere in the middle, there is the <u>shallow-supervised type of learning</u> with an external tutor or some complementary process which just tells the system whether it is right or wrong. That is typical of cybernetic learning. It is also a GOFAI technique of learning launched by Holland (1986) and his views both on inductive learning and genetic algorithms. Back-propagation algorithm and Boltzmann machine emerge from this same inspiration. At the other extremity, there is the <u>unsupervised learning</u>, where the system just discovers regularities in its experiences and learn to cluster the data in order to evidentiate these regularities. That is typical of the Competitive learning algorithm and some clustering algorithmic techniques. What part of the axis is preferred by cognitive scientists and by engineers ?

3.2.1 From a cognitive point of view

<u>Cognitive scientists are certainly more attracted by the unsupervised style of learning</u>. In Piaget's views (Piaget, 1952; Drescher, 1986) on cognitive developments in infants, conceptual knowledge is acquired through innate sensori-motor experiences. This development of abstract representation is autonomous, without the need for any "conceptual tutoring". The "notion" of feedback in learning, although indispensable, poses a delicate problem. What could be the nature of this feedback ? In back-propagation algorithm and from a cognitive perspective, the feedback is totally unrealistic.

Unsupervision does not mean that the network must be initially tabula-rasa. Any non-trivial neural model requires a great deal of prior structure (Feldman, 1988). System must in some sense "intervene" in its learning, rejecting a completely passive and receptive attitude. An important problem lies in the quasi-ethical frontier between the part of the network being actually innate and the one being learned. However, a general feeling shared by cognitive scientists is to leave a great amount of autonomy and spontaneity in the learning mechanisms of their cognitive models. They certainly appreciate approaches like competitive learning (Grossberg, 1988; Kohonen, 1988) or selectionist learning (Holland, 1986; Edelman, 1987).

3.2.2 From an engineering point of view

This cognitive tendency for reduced supervision in learning is certainly regarded with much more circumspection by engineers. They certainly accept the challenging idea of systems capable of auto-organizing themselves in reaction to new events for which they are not adapted. However, they wouldn't understand the disregarding of contextual knowledge if this one is available. It is pure fiction to hope that, in any case, starting with a completely unstructured, tabula-rasa network, this one will, due to some really mystical emergent properties, discover alone the required contextual knowledge.

When knowledge is available, why not exploit and install it in some ways in the network? This healthy engineering attitude is more and more frequent in connectionist studies of perceptual modalities where networks are initially structurally constrained (Fogelman Soulié et al., 1988) (a constraint being a soft way for implementing contextual knowledge). An ideal solution should be an open system capable of flexibility and auto-adaptation to unforeseen situations but with all the engineering and prior knowledge necessary for adequately behaving in the majority of cases. Learning is an iterative and evolutive process. Learning is a function of what is already learned. Machine learning should complement and not re-learn any strong knowledge that has been acquired through years of humanity.

4. CONCLUSION

We have surveyed various connectionist methodologies both from an engineering and a cognitive perspective. Without any doubt, a lot of connectionist aspects have provoked essential conceptual shifts in cognitive sciences. This paper is more reserved on the adequacy of those same aspects in an engineering perspective. Optimality is wittingly beyond the scope of cognitive sciences. In harmony with Darwinist views, man is well fit to his environment but not necessarily the fittest. One of the few certainties is that this still immature connectionism re-emergence will not sweep away 25 years of GOFAI realizations. These two approaches will rather have to accept an opportunist cohabitation both in their apprehension of what learning and performing must be.

REFERENCES

Boden, M. (1988). *Computer Models of Mind.* Cambridge University Press

Bridle, J. (1988). Connectionist approaches to artificial perception: A speech pattern processing approach. In *Proceedings of Connectionism in Perspective.* University of Zurich, Switzerland - 10-13 October.

Chandrasekaran, B., Goel A., Allemang D. (1988). Connectionism and Information-Processing Abstractions. *AI magazine,* Winter 1988.

Churchland, P.S. (1986). *Neurophilosophy.* Cambridge, MA: MIT Press/Bradford Books.

Clark, A. (1989). Connectionism and the multiplicity of mind. *Artificial Intelligence Review.* Vol. 3, No 1

De Groot, A. (1965). *Thought and Choice in Chess.* Mouton.

Drescher, G.L. (1986). Genetic AI: Translating Piaget into LISP, *AI Memo No 890.* AI laboratory-MIT

Dreyfus, H. (1972). *What Computers can't do; A Critique of Artificial Reason.* New York: Harper & Row.

Dreyfus, H. and S. Dreyfus (1988). Making a Mind vs Modeling the Brain. in *The Artificial Intelligence Debate. False Starts, Real Foundations.* Stephen R. Graubard (Eds.) MIT Press

Edelman, G.M. (1987). *Neural Darwinism.* New York: Basic Books.

Fahlman S.E. (1979). *NETL: A System for Representing and Using Real-World Knowledge.* The MIT Press, Cambridge MA.

Feldman, J. and D. Waltz (eds.) (1988). *Connectionist Models and their implications.* Ablex Publishnig Corporations NJ.

Fodor, J.A. and Z.W. Pylyshyn (1988). Connectionism and Cognitive Architecture: A Critical Analysis. *Cognition,* 28.

Fogelman Soulié, F., Gallinari, P., Le Cun, Y., Thiria, S. (1988). Automata networks and Artificial Intelligence. In *Automata networks in computer science*; Fogelman Soulié, F., Robert, Y., Tchuente, M. (eds.), Manchester Univ. Press, Princeton Univ. Press.

Fukushima, K. (1988). A Neural Network for Visual Pattern Recognition. *IEEE Computer,* March.

Grossberg, S. (Eds) (1988). *Neural Networks and Natural Intelligence.* MIT Press, Cambridge Mass.

Gupta, A. (1987). *Parallelism in Production Systems.* Research Notes in AI. Pitman, London. Morgan Kaufmann Publishers.

Hebb, D.O. (1949). *The Organization of Behavior: A Neuropsychological Theory.* John Wiley & Sons, New York.

Hendler, J.A. (1988). Problem Solving and Reasoning: A Connectionist Perspective. in *Proceedings of Connectionism in Perspective.* University of Zurich, 10-13 oct. Switzerland.

Holland, J.H., Hollyak, K.J, Nisbett, R.E. and P.R. Thagard (1986). *Induction: Processes of inference, learning and discovery.* Cambridge: MIT Press.

Kodratoff, Y. (1988). Enlarging Symbols to more than Numbers or Artificial Intelligence is the Science of Explanations. in *Proceedings of Connectionism in Perspective.* University of Zurich, 10-13 oct. Switzerland

Kohonen, T. (1988). *Self organization and associative memory.* Springer Series in Information Sciences, vol. 8 Springer-Verlag, 2nd Edition

Lippmann, R.P. (1987). An Introduction to computing with neural nets. *IEEE ASSP Magazine.* April.

Newell, A. (1982). The knowledge level. *Artificial Intelligence,* 18(1).

Nii, H.P. (1986). Blackboard Systems. *AI Magazine,* 7(2) et 7(3).

Papert, S. (1987). One AI or many. in *The Artificial Intelligence Debate. False Starts, Real Foundations.* Stephen R. Graubard (Eds.) MIT Press

Partridge, D. , Paap, K. (1988). An Introduction to Learning. *Artificial Intelligence Review,* Vol. 2, No 2.

Piaget, J. (1952). *The Origins of Intelligence in the Child.* London: Montledge and Kegan Paul.

Pollack, J.B. (1989). Connectionism: past, present and future. *Artificial Intelligence Review,* Vol. 3, No 1.

Reason, J. (1987). Framework Models of Human Performance and Error: A Consumer Guide. in J. Rasmussen, K. Duncan and J. Leplat (Eds.) *New Technology and Human Errors* (J. Wiley and Sons).

Rumelhart, D.E. and J.L. McClelland (Eds) (1986). *Parallel Distributed Processing: Explorations in the Microstructure of Cognition.* Bradford Books. MIT Press.

Schreter, Z. (1988). Connectionist Models: Virtues and Problems. *SGAICO Newsletter*, No 20.

Sejnowski, T.J. and Rosenberg, C.R. (1986). NETtalk: A Parallel network that learns to read aloud. *JHU/EECS-86/01*: The Johns Hopkins University, Electrical Engineering and Computer Science Department

Sejnowski, T.J. and Rosenberg, C.R. (1988). Learning and Representation in Connectionist Models. *Perspective in Memory Research*, M.S. Gazzaniga (Eds.) MIT Press.

Shastri, L. (1988). *Semantic Networks: An Evidential Formalization and its Connectionist Realization*. Morgan Kaufmann Publishers, Inc., Los Altos, California.

Smolensky, P. (1986). Information Processing in dynamical Systems: foundations of harmony theory. in *Parallel Distributed Processing: Explorations in the Microstructure of Cognition*. (Rumelhart, D.E. and J.L. Mc Clelland, Eds). Bradford Books. MIT Press.

Waltz, D.L. (1988). The Prospects of Building Truly Intelligent Machines. in *The Artificial Intelligence Debate. False Starts, Real Foundations*. Stephen R. Graubard (Eds.) MIT Press

Winograd, T. (1975). Frames Representations and the Declarative/Procedural Controversy. In *Representation and Understanding: Studies in Cognitive Science*. D.G. Bobrow and A.M. Collins (Eds.) N.Y. Academic Press.

Lecture Notes in Engineering

Edited by C. A. Brebbia and S. A. Orszag

Lecture Notes in Engineering

Edited by C.A. Brebbia and S.A. Orszag